An Engineering Approach to ASYMPTOTICS and APPROXIMATIONS

WITH APPLICATIONS IN ELECTROMAGNETIC WAVE PROPAGATION AND IMAGING THROUGH TURBULENCE

An Engineering Approach to ASYMPTOTICS and APPROXIMATIONS

WITH APPLICATIONS IN ELECTROMAGNETIC WAVE PROPAGATION AND IMAGING THROUGH TURBULENCE

Larry C. Andrews • Melissa K. Beason

SPIE PRESS
Bellingham, Washington USA

Library of Congress Control Number: 2024058733

Published by
SPIE
P.O. Box 10
Bellingham, Washington 98227-0010 USA
Phone: +1 360.676.3290
Fax: +1 360.647.1445
Email: books@spie.org
Web: www.spie.org

The views expressed in this book are those of the authors and do not reflect the official guidance or position of the United States Government, the Department of Defense, the United States Air Force, or the United States Space Force.

The content of this book reflects the work and thoughts of the authors. Every effort has been made to publish reliable and accurate information herein, but the publisher is not responsible for the validity of the information or for any outcomes resulting from reliance thereon.

Cover image credits: satellite generated by Copilot AI; sky from Shutterstock.com - Milky way on a night sky - Hit1912; Earth from Shutterstock.com - Europe at night viewed from space with city lights - NicoElNino.

Printed in the United States of America.

First printing 2025.
For updates to this book, visit http://spie.org and type “PM387” in the search field.

SPIE.

Contents

Preface *xv*
Glossary of Symbols and Acronyms *xxi*

1 Review of Complex Variables **1**
1.0 Introduction 1
1.1 Complex Functions 2
1.1.1 Analytic functions 3
1.1.2 Elementary complex functions 3
1.2 Complex Integration 5
1.2.1 Deformation of contours 6
1.3 Taylor Series and Laurent Series 7
1.3.1 Analytic continuation 8
1.4 Singularities 9
1.4.1 Residues 10
1.5 Applications 10
1.6 Historical Remarks 16
Reference 17
Bibliography of Further Reading 17

Part I Special Functions, Asymptotic Analysis, and Approximations **19**

2 Basic Concepts **21**
2.0 Introduction 21
2.1 The Gamma Function 23
2.2 The Pochhammer Symbol 25
2.3 Asymptotic Series and the Cauchy Product 25
2.3.1 Large arguments of the function 26
2.3.2 Stirling's formula 26
2.3.3 The Cauchy product 27
2.4 The Mellin Transform 28
2.4.1 Residue theory 29
2.4.2 Convolution integral 33

2.5 Mellin–Barnes Integral 34
2.6 Zernike Polynomials 36
2.6.1 Application in optics 36
2.7 Applications 37
2.7.1 Asymptotic series 37
2.7.2 The Mellin transform 38
2.8 Historical Remarks 42
References 43

3 Generalized Hypergeometric Functions 45

3.0 Introduction 45
3.1 Generalized Hypergeometric Series 45
3.2 The Function ${}_1F_0$ 47
3.2.1 Large-argument asymptotic series 48
3.3 The Function ${}_1F_1$ 49
3.3.1 The Barnes integral 50
3.3.2 Large-argument asymptotic series 51
3.3.3 Elementary properties 52
3.4 The Function ${}_1F_k$ 53
3.4.1 Large-argument asymptotic series 53
3.5 The Function ${}_2F_1$ 54
3.5.1 The Barnes integral 55
3.5.2 Large-argument asymptotic series 55
3.5.3 Elementary properties 57
3.6 The Function ${}_2F_2$ 59
3.6.1 Large-argument asymptotic series 59
3.7 The Function ${}_pF_q$ 60
3.8 The Function $U(a; c; z)$ 61
3.8.1 The Barnes integral 61
3.8.2 Large-argument asymptotic series 63
3.9 Whittaker Functions 64
3.10 Asymptotic Behavior of Other Special Functions 64
3.10.1 Error functions 65
3.10.2 Fresnel integrals 67
3.10.3 Incomplete gamma functions 69
3.10.4 The exponential integral 70
3.10.5 Sine and cosine integrals 72
3.10.6 Complete elliptic integrals 74
3.11 Applications 76
3.11.1 Summing series 76
3.11.2 Probability and statistics 78
3.11.3 Elliptic integrals 79
3.12 Historical Remarks 80
References 82

4 Bessel Functions **83**
4.0 Introduction 83
4.1 Standard Bessel Functions 83
4.1.1 The function $J_p(z)$ 84
4.1.2 Large-argument asymptotic approximation 86
4.1.3 The function $Y_p(z)$ 88
4.1.4 Spherical Bessel functions 89
4.2 Modified Bessel Functions 91
4.2.1 The function $I_p(z)$ 91
4.2.2 The function $K_p(z)$ 94
4.2.3 Modified spherical Bessel functions 95
4.3 Other Bessel Functions 96
4.3.1 Hankel functions 96
4.3.2 Struve functions 98
4.3.3 Kelvin's functions 102
4.3.4 Airy functions 105
4.3.5 Bessel-integral function 108
4.3.6 Anger and Weber functions 109
4.3.7 Lommel functions 111
4.4 Applications 113
4.4.1 Differential equations related to Bessel's equation 113
4.4.2 Oscillations of a hanging chain 114
4.4.3 Other Bessel functions 115
4.5 Historical Remarks 116
References 117

5 Meijer *G*-Function **119**
5.0 Introduction 119
5.1 The Meijer *G*-Function Definitions 119
5.2 Elementary Properties 122
5.2.1 Reduction formulas 123
5.2.2 Integral formulas 125
5.3 Asymptotic Relations for Large Arguments 129
5.3.1 The function ${}_1F_1$ 130
5.3.2 The function ${}_1F_k$ 130
5.3.3 The function ${}_2F_1$ 131
5.3.4 The function ${}_3F_3$ 131
5.4 MacRobert *E*-Function 132
5.5 Comments on the *G*-Function 133
5.6 Historical Remarks 133
References 134

6 Approximations 135

6.0 Introduction 135
6.1 Approximations for the ${}_2F_1$ Function 135
6.2 Approximations for the ${}_1F_1$ Function 138
6.3 Other Functions 143
6.3.1 Error function 143
6.3.2 Bessel functions 144
6.3.3 Modified Bessel functions 146
6.3.4 Modified Struve function 148
6.3.5 Kelvin's functions 150
6.4 Discussion 151
Reference 152

Part II Applications 153

7 Integrals and Products 155

7.0 Introduction 155
7.1 Miscellaneous Integrals 156
7.1.1 Elementary functions 156
7.1.2 Bessel functions 157
7.1.3 Hypergeometric functions 161
7.2 Products of Special Functions 163
7.2.1 ${}_0F_1$ functions 163
7.2.2 ${}_1F_1$ and ${}_2F_1$ functions 165
7.2.3 Saalschütz's theorem 168
7.3 Products and Integrals of Bessel Functions 169
7.3.1 Product of two Bessel functions 169
7.3.2 Product of three Bessel functions 172
7.4 Laplace Transform 173
7.4.1 Error function 174
7.4.2 Bessel functions 176
7.4.3 Generalized hypergeometric functions 178
7.4.4 Miscellaneous functions 180
7.5 Mellin Transform 183
7.5.1 Elementary functions 183
7.5.2 Bessel functions 186
7.6 Hankel Transform 188
7.6.1 Elementary functions 189
7.6.2 Bessel functions 190
7.7 Applications of Integral Transforms 192
7.7.1 Heat conduction in a long rod 192
7.7.2 Probability and statistics 194
7.7.3 Summation of series 196
7.7.4 Optical wave propagation 198

7.7.5 Steady-state heat conduction in a cylinder 199
7.7.6 Fourier transform 201
7.7.7 Inverse transforms 205
7.8 Discussion 207
7.9 Historical Remarks 207
References 208

8 Electromagnetic Wave Propagation 211

8.0 Introduction 211
8.1 Atmospheric Turbulence 213
8.1.1 Random fields 214
8.1.2 Power spectrum models 215
8.1.3 Index-of-refraction structure function 216
8.2 Optical/IR Wave Propagation 220
8.2.1 Propagation in free space 220
8.2.2 Rytov approximation 222
8.2.3 Comments on the exponential spectrum model 224
8.3 Mutual Coherence Function 225
8.3.1 WSF for a plane wave 226
8.3.2 Outer-scale effect 230
8.3.3 WSF for a spherical wave 231
8.3.4 WSF for a Gaussian-beam wave 234
8.3.5 Long-term beam radius 239
8.4 Scintillation Index 241
8.4.1 Plane wave 241
8.4.2 Spherical wave 242
8.4.3 Gaussian-beam wave 244
8.5 Irradiance Covariance Function: Weak Fluctuations 249
8.5.1 Plane wave 249
8.5.2 Plane wave: aperture averaging 253
8.5.3 Spherical wave 254
8.5.4 Spherical wave: aperture averaging 257
8.5.5 Gaussian-beam wave 258
8.6 Irradiance Covariance Function: Strong Fluctuations 260
8.6.1 Plane wave 260
8.6.2 Spherical wave 262
8.7 Temporal Spectrum of Irradiance 264
8.7.1 Plane wave 265
8.7.2 Plane wave: aperture averaging 268
8.7.3 Spherical wave 271
8.8 Phase Statistics: Plane Wave 274
8.8.1 Phase structure function 274
8.8.2 Angle-of-arrival 276
8.8.3 Phase variance 277

8.8.4 Phase covariance function 278
8.8.5 Temporal spectrum of phase 279
8.9 Phase Statistics: Spherical Wave 280
8.9.1 Phase structure function 281
8.9.2 Phase covariance function 283
8.10 Discussion 284
8.11 Historical Remarks 285
References 286

9 Imaging Through Atmospheric Turbulence 289
9.0 Introduction 289
9.1 Imaging Performance Measures 290
9.1.1 Point spread function and modulation transfer function 291
9.1.2 Short-exposure imaging 296
9.1.3 Strehl ratio 298
9.1.4 Strehl ratio: outer-scale effect 300
9.1.5 Strehl ratio with tilt removed 301
9.1.6 Long-exposure resolution 303
9.1.7 Short-exposure resolution 303
9.2 Zernike Polynomials and Aperture Filter Functions 304
9.2.1 Zernike-tilt variance 308
9.2.2 Zernike-tilt variance: outer-scale effect 309
9.2.3 Gradient-tilt variance 311
9.2.4 Zernike-tilt power spectral density 311
9.3 Anisoplanatism 313
9.3.1 Laser guide star 313
9.4 Tilt Anisoplanatism 314
9.4.1 Multi-aperture case 314
9.4.2 Multi-aperture: outer-scale effect 317
9.4.3 Multi-source case 319
9.4.4 Multi-source: outer-scale effect 324
9.5 Angular and Focal Anisoplanatism 326
9.5.1 LGS anisoplanatic error: zero NGS offset 327
9.5.2 LGS anisoplanatic error: nonzero NGS offset 330
9.5.3 Focus anisoplanatism 333
9.6 Discussion 335
References 337

10 Non-Kolmogorov Turbulence 339
10.0 Introduction 339
10.1 Non-Kolmogorov Spectral Models 339
10.1.1 Isotropic models 340
10.1.2 Anisotropic models 340
10.1.3 Index-of-refraction structure function 341

10.2 Generalized Rytov Variance 343
10.3 Wave Structure Function: Isotropic Turbulence 344
10.3.1 Plane wave 344
10.3.2 Spherical wave 345
10.3.3 Gaussian-beam wave 347
10.4 Wave Structure Function: Anisotropic Turbulence 348
10.5 Scintillation Index: Weak Irradiance Fluctuations 349
10.5.1 Plane wave: isotropic turbulence 349
10.5.2 Spherical wave: isotropic turbulence 350
10.5.3 Gaussian-beam wave: anisotropic turbulence 351
10.6 Scintillation Index: Strong Irradiance Fluctuations 352
10.6.1 Plane wave: isotropic turbulence 353
10.6.2 Plane wave: anisotropic turbulence 356
10.6.3 Spherical wave: isotropic turbulence 358
10.6.4 Gaussian-beam wave: isotropic turbulence 360
10.7 Aperture Averaging: Weak Irradiance Fluctuations 361
10.7.1 Flux variance: plane wave 361
10.7.2 Flux variance: spherical wave 362
10.8 Irradiance Covariance Function 362
10.8.1 Plane wave 363
10.8.2 Spherical wave 364
10.9 Temporal Spectrum of Irradiance 366
10.9.1 Plane wave 366
10.9.2 Spherical wave 368
10.10 Phase Statistics: Plane Wave 370
10.10.1 Phase structure function 370
10.10.2 Phase variance 371
10.11 Imaging Performance Measures 371
10.11.1 Long-term beam radius: detector plane 371
10.11.2 Point spread function and modulation transfer function 372
10.11.3 Strehl ratio 372
10.12 Tilt Variances 373
10.12.1 Zernike-tilt variance 373
10.12.2 Gradient-tilt variance 374
10.12.3 Zernike-tilt power spectral density 375
10.13 Tilt Anisoplanatism 376
10.13.1 Multi-apertures 376
10.13.2 Multi-source 379
10.14 Focus Anisoplanatism 382
10.15 Discussion 384
References 384

11 Miscellaneous Applications **389**

11.0 Introduction 389
11.1 Statistical Communication Theory 389
11.1.1 Nonlinear devices 390
11.1.2 Ideal bandpass limiter 395
11.1.3 Joint moments of the envelopes 398
11.2 Cross Correlators 400
11.2.1 Cross correlator with limiter in one channel 404
11.2.2 Polarity coincidence correlator 405
11.3 Free-Space Optical Communication 407
11.3.1 Threshold detection 408
11.3.2 Performance measures 410
11.4 Fluid Mechanics 414
11.4.1 Unsteady hydrodynamic flow past an infinite plate 414
11.4.2 Irrotational flow of an ideal fluid 416
References 418

Appendix A: Special Function Identities **421**

(*G*) Gamma Function 421
(*PS*) Pochhammer Symbol 422
(*HY*) Hypergeometric Function 423
(*CH*) Confluent Hypergeometric Functions 425
(*BJ*) Bessel Function of the First Kind 427
(*BY*) Bessel Function of the Second Kind 428
(*BI*) Modified Bessel Function of the First Kind 429
(*BK*) Modified Bessel Function of the Second Kind 430
(*SBj*) Spherical Bessel Function of the First Kind 431
(*SBy*) Spherical Bessel Function of the Second Kind 432
(*MG*) Meijer *G*-Function 432

Appendix B: Formulas for the Generalized Hypergeometric Functions **435**

Appendix C: Formulas for the Meijer *G*-Function **439**

Appendix D: Integral Table **443**

Appendix E: Asymptotic Expressions **447**

Generalized Hypergeometric Functions 447
Error Functions 449
Fresnel Integrals 449
Incomplete Gamma Functions 449
Exponential Integral 449
Sine and Cosine Integrals 450
Complete Elliptic Integrals 450
Standard Bessel Functions 450

Spherical Bessel Functions 451
Modified Bessel Functions 451
Modified Spherical Bessel Functions 451
Struve Functions 452
Kelvin's Functions 452
Airy Functions 453
Bessel-Integral Function 454
Anger and Weber Functions 454
Lommel Functions 454

Index *455*

Preface

Prior to the 1960s, there was a strong research interest in applied mathematics at many universities. Much of this interest concerned subjects like partial differential equations, Fourier analysis, integral transforms, and special functions. However, during the 1960s and up to the present, the focus in the mathematics curriculums and research at American universities changed direction toward more abstract mathematics with applied mathematics playing a lesser role. Because of a reduced offering of applied mathematics courses at various universities, many engineering and physics students are not being exposed to much mathematics beyond calculus and basic differential equations. Yet, engineering and physics applications today demand a deeper understanding of advanced mathematical techniques than ever before. For example, advanced mathematical methods are required in application areas like fluid dynamics, heat conduction, advanced probability theory, statistical communication theory, and electromagnetic wave propagation.

Mathematical models developed in advanced engineering applications are often presented in the form of one or more *special functions* that are less familiar to engineers and applied scientists. These individuals may be acquainted with *gamma functions*, *error functions*, *Bessel functions*, and even some of the *orthogonal polynomials*—less well known are functions belonging to the class of *generalized hypergeometric functions*. Because members of the general hypergeometric-type family are occurring more frequently today in advanced science and engineering problems, some knowledge of them can be quite useful. Fortunately, several computer software packages are now available that incorporate such functions along with the standard elementary functions. This is very convenient for making numerical calculations, but these particular software packages are generally quite expensive and thus may not be available to researchers with limited resources. Moreover, numerical computation alone may not be adequate for interpreting some of the mathematical behavior of these functions when the parameters defining the function take on different values.

The objective of this text is twofold—to introduce advanced novel mathematical techniques and then to use them to solve a variety of problems that may arise in various applications. We first present an introduction to asymptotic analysis and approximations that are applied to many of the special functions. Although the concentration is largely on generalized hypergeometric functions, other special functions are also included. The mathematical techniques are later applied to problems in electromagnetic wave propagation, including imaging through atmospheric turbulence and non-Kolmogorov turbulence.

- Chapter 1: This optional chapter is included as a short review of complex variables to reintroduce the reader to terms like analytic function, analytic continuation, and residue theory that appear in subsequent chapters. Readers who have not previously studied complex variables might want to also consult one of the texts listed in the bibliography.

Chapters 2–6 comprise **Part I: Special Functions, Asymptotic Analysis, and Approximations**. These five chapters represent the mathematics portion of the text where we introduce the mathematical foundations used throughout the remainder of the text.

- Chapter 2: In this chapter we introduce the notions of *asymptotic analysis*, *Cauchy product*, and the *Mellin–Barnes integral* (often called simply the *Barnes integral*).
- Chapter 3: We use the Mellin–Barnes integral and residue theory to develop asymptotic series for both small and large arguments in our treatment of generalized hypergeometric functions. We then focus on a few more-familiar special functions, concentrating once again on the development of asymptotic results.
- Chapter 4: The general family of *Bessel functions* is introduced here. Bessel functions are so prominent in application areas that we have devoted an entire chapter to them, including many of the less-well known members of this large family. As in Chapter 3, our interest is mainly in the development of asymptotic relations.
- Chapter 5: The *Meijer G-function* is presented in this chapter along with some of its important properties. Knowledge of the operational properties of the *G*-function permits an easy development of large-argument asymptotic behavior without needing to resort back to the Barnes integral and residue theory utilized in earlier chapters. In addition, the Meijer *G*-function is used for solving many of the integrals that appear throughout the remaining chapters.
- Chapter 6: We next illustrate some simple *algebraic approximations* that bridge the asymptotic behavior between small and large arguments of hypergeometric-type functions. These kinds of approximation can often be developed with great accuracy to replace the generalized hypergeometric functions for numerical computation and interpretation.

Early studies of optical wave propagation were motivated mostly by astronomy applications, but with the introduction of the laser in 1960, those studies were expanded to include new areas of application like free-space optical communications, remote sensing, laser ranging, weaponry, target designation, and laser radar. The mathematical models associated with optical/IR wave propagation through random media have evolved from solving stochastic differential equations. Exact solutions of these equations have not been found so researchers resort to various approximation methods. The formalism required in some of the approximation methods is presented in detail in the textbook *Laser Beam Propagation Through Random Media*, 2nd edition, by L. C. Andrews and R. L. Phillips (SPIE Press, 2005). Some of this lengthy presentation leads to double and triple integrals that arise by applying the *Rytov approximation theory*. Details of solving these integrals are often omitted, even in journal papers, because they are generally deemed too lengthy or too complicated. For many engineers and applied scientists, this is adequate because they only need dependable models to apply in their work. For others who are doing research in the field, the mathematical details used in the derivation of these models can be most insightful for their own development of new models.

The last five chapters of the text represent **Part II: Applications**. The novel methods we use to solve various problems involve properties of the generalized hypergeometric functions developed in previous chapters. In many cases this approach leads to shorter derivations because much of the analysis is simply looking up useful identities and relations provided in the Appendices. As a consequence, along with developing some known results with this solution technique, some new math models are also obtained.

- Chapter 7: We first use an approach involving the Meijer *G*-function to evaluate a number of nonelementary integrals, some of which involve products of special functions. These same mathematical techniques are then applied in the evaluation of a few *Laplace*, *Mellin*, and *Hankel transforms*.
- Chapter 8: Here we provide the details on developing mathematical models through exact evaluation of the multiple integrals that arise in the study of *electromagnetic wave propagation*. We also use the algebraic approximation methods presented in Chapter 6 to simplify some of the results. Among the new models developed in this chapter are some based on the *exponential spectrum model* that includes both inner-scale and outer-scale parameters. For example, derivations of the *wave structure function* are provided using this spectrum model. In this chapter we also derive the *covariance function of irradiance* with and without aperture averaging and the corresponding *temporal power spectrum of irradiance* with and without aperture averaging. We end the

chapter with the development of several statistical models for the phase associated with plane waves and spherical waves.

- Chapter 9: The analysis in this chapter centers around problems from *imaging through turbulence* with the purpose of illustrating how to use the *ABCD ray matrix* approach in developing simple models for performance measures like the *Strehl ratio* and *image resolution*. Some of the models included here are well known—however, new formulations are also presented, including new expressions for *short-exposure imaging*. We also illustrate how to develop models for *anisoplanatic effects*, which generally involve products of two and three Bessel functions under the integral.
- Chapter 10: In this chapter we reexamine some of the optical wave propagation models from Chapters 8 and 9 under the assumption of *non-Kolmogorov turbulence*, including some development of models based on *anisotropic turbulence.* The topic of non-Kolmogorov turbulence is currently one of great interest to many researchers in the field. Again, we present a number of new models here, including irradiance covariance functions, and develop scintillation models under moderate-to-strong irradiance fluctuations. In addition, several of the imaging models from Chapter 9 are extended here to the case of non-Kolmogorov turbulence.
- Chapter 11: This last chapter is devoted to some conventional applications that also lead to special functions in areas like *statistical communication theory* and *fluid mechanics*. We illustrate how some of the techniques introduced in earlier chapters can be useful in the analysis of problems in these other application areas.

The emphasis in this book is on the mathematical techniques that lead to various models of electromagnetic wave propagation through optical turbulence. Although not required, it is helpful if the reader has a working knowledge of some of the special functions introduced here and is also somewhat acquainted with the basics of electromagnetic wave propagation. In writing this text, however, the authors have attempted to present the material in sufficient detail to reach an audience with limited knowledge of the subjects covered. To aid the reader in this regard, the text is composed of a large number of worked examples, sometimes offered at the end of a chapter. In addition, Chapters 6 and 7 are presented as a compilation of examples. Because complex variables play a central role in much of the analysis, we have included an optional Chapter 1 that presents a short review of this important subject. For easy reference, we have also included five Appendices at the end of the text that contain an abundance of special function relations. Appendix A is a summary of the most useful identities involving special functions; Appendix B presents a list of relations between generalized hypergeometric functions and other functions, including elementary functions; Appendix C provides the same type of relations as Appendix B but for the Meijer G-function; Appendix D consists of a short integral table for easy reference; and Appendix E is a summary of asymptotic expressions developed for various special functions throughout the text.

It is the authors hope that the material covered in this textbook will serve as a supplementary self-study reference for other books on engineering and physics applications, including the recently published book *Laser Beam Propagation in Random Media: New and Advanced Topics* by L. C. Andrews and M. K. Beason (SPIE Press, 2023).

Larry C. Andrews
Melissa K. Beason
April 2025

Glossary of Symbols and Acronyms

A_0	amplitude of optical wave
AO	adaptive optics
AOA	angle-of-arrival
AOI	adaptive optics imaging
$\mathrm{Ai}(z)$, $\mathrm{Ai}(-z)$	Airy functions
$\arg(z)$	argument of a complex variable
$(a)_n$	Pochhammer symbol
B	electrical bandwidth
BER	bit error rate
$B_I(\rho, L)$	spatial covariance function of irradiance
$B_S(\rho, L)$	covariance function of phase
$B_n(R)$	covariance function of refractive index
$B(x, y)$	beta function
$\mathrm{Bi}(z)$, $\mathrm{Bi}(-z)$	Airy functions
$\mathrm{ber}(z)$, $\mathrm{bei}(z)$	Kelvin functions
$C(z)$	Fresnel integral
$\mathrm{Ci}(z)$	cosine integral
C_n^2, $C_n^2(z)$	refractive-index structure parameter
$\tilde{C}_n^2$	refractive-index structure parameter for non-Kolmogorov turbulence
$\tilde{C}_{n,eff}^2$	empirical effective refractive-index structure parameter for anisotropic non-Kolmogorov turbulence
C_T^2	temperature structure parameter
$C_{\mathbf{z}}(u)$	characteristic function
D	diameter of receiver aperture
DE	differential equation
DM	deformable mirror
DOC	modulus of the complex degree of coherence
DP	detector plane

$D(\mathbf{r}_1, \mathbf{r}_2, L)$, $D(\rho, L)$	wave structure function
$D(\rho, \alpha)$	wave structure function in non-Kolmogorov turbulence
$D_S(\rho, L)$	phase structure function
$D_n(R)$	index of refraction structure function
D	hard aperture diameter of receiver $(= 2\sqrt{2}\, W_{Rx})$
D_{Tx}	hard aperture diameter of transmitter $(= 2\sqrt{2}\, W_0)$
$D_\chi(\rho, L)$	log-amplitude structure function
d	nondimensional diameter of receiver collecting lens $(= \sqrt{kD^2/4L})$
EO	electro-optical
$E(m)$	complete elliptic integral of the second kind
$\mathbf{E}_p(z)$	Weber function
$E_i(z)$, $E_1(z)$	exponential integral
$E_n(\mathbf{r}_1, \mathbf{r}_2)$, $n = 1, 2, 3$	second-order moments of the complex phase perturbations
erf(z), erfc(z)	error function, complementary error function
F	phase front radius of curvature of beam at receiver
FAR	false alarm rate
FSOC	free-space optical communication
F_0	phase front radius of curvature of beam at transmitter
F_{Rx}	focal length of lens
$F_j(\kappa)$, $F_j(\kappa, \phi)$	filter function
${}_1F_1(a; c; z)$	confluent hypergeomettric function of the first kind
${}_2F_1(a, b; c; z)$	hypergeometric function
${}_pF_q$	generalized hypergeometric function
$G(\mathbf{s}, \mathbf{r}; L)$	Green's function
$G_j(\kappa)$, $G_j(\kappa, \phi)$	complex filter function
$G_{p,q}^{m,n}$	Meijer G-function
$\mathcal{H}_\nu\{\}$	Hankel transform operator
H	altitude of receiver/transmitter
HAP	Hufnagel–Andrews–Phillips $C_n{}^2$ profile model
HV	Hufnagel–Valley $C_n{}^2$ profile model
$\mathbf{H}_p(z)$	Struve function
$H_n(z)$	Hermite polynomial
$H_p^{(1)}(z)$, $H_p^{(2)}(z)$	Hankel functions
$h_n^{(1)}(z)$, $h_n^{(2)}(z)$	spherical Hankel functions
$h(0, \mathbf{r})$	impulse response function
h	altitude
h_0	altitude of transmitter/receiver
Im	imaginary part of
IR	infrared

I_T	irradiance threshold in a receiver
$I_p(z)$	modified Bessel function of the first kind
$i_n(z)$	modified spherical Bessel function of the first kind
i, j	imaginary number $(=\sqrt{-1})$
$\mathbf{J}_p(z)$	Anger function
$J_p(z)$	standard Bessel function of the first kind
$j_n(z)$	spherical Bessel function of the first kind
Ji_p	Bessel-integral function
$\mathbf{K}$	vector spatial wave number
$K(m)$	complete elliptic integral of the first kind
$K_p(z)$	modified Bessel function of the second kind
$k_n(z)$	modified spherical Bessel function of the second kind
$\mathrm{ker}(z)$, $\mathrm{kei}(z)$	Kelvin functions
k	optical wave number of beam wave $(= 2\pi/\lambda)$
$\mathscr{L}\{\}$	Laplace transform operator
L	propagation path length
LED	light emitting diode
LGS	laser guide star
LSI	linear shift-invariant
$\mathbf{L}_p(z)$	modified Struve function
L_f	distance from receiver to photodetector
L_0	outer scale of turbulence
$L_n^{(m)}(z)$	associated Laguerre polynomial
$\mathrm{li}(z)$	logarithmic integral
l_0	inner scale of turbulence
$\mathcal{M}\{\}$	Mellin transform operator
MCF	mutual coherence function
MTF	modulation transfer function
$M_{k,m}(z)$	Whittaker function of the first kind
NGS	natural guide star
$n(\mathbf{R})$	index of refraction
N_0	output noise power
OOK	on-off keying
OTF	optical transfer function
PDF	probability density function
PIB	power in the bucket
PIF	power in the fiber
PSF	point spread function
$p_I(u)$	probability density function of irradiance
$P_n(z)$	Legendre polynomial
Pr_d	probability of detection
Pr_{fade}	probability of fade

Pr_{fa}	probability of false alarm
P_{Tx}	power of optical wave at transmitter
P_S, $P_{S,0}$	power of optical wave at receiver
$\mathbf{p}$	transverse vector between two observation points in beam
Q_l, Q_m	nondimensional inner-scale parameter ($= L\kappa_l^2/k$, $= L\kappa_m^2/L$)
Q_0	nondimensional outer-scale parameter ($= L\kappa_0^2/k$)
$\mathbf{R}(x, y, z)$	position vector in three dimensions
R	resolution
RF	radiofrequency
RP	receiver pupil plane
Re	real part of
Res	residue
Rx	receiver
$\mathbf{r}$	transverse position of observation point
r_0	atmospheric coherence width or Fried parameter
rms	root-mean-square
SNR	signal-to-noise ratio
SR	Strehl ratio
$S(z)$	Fresnel integral
$\mathrm{Si}(z)$	sine integral
$S_I(\omega)$	irradiance power spectral density
$S_S(\omega)$	phase power spectral density
$S_{T,Z}(\omega)$	Zernike-tilt power spectral density
S_0	output signal power
$S_{\mu,\nu}(z)$	Lommel function of the second kind
$s_{\mu,\nu}(z)$	Lommel function of the first kind
TEM	transverse electromagnetic mode
TNR	threshold-to-noise ratio
TR	tilt removed
Tx	transmitter
$T_n(z)$	Chebyshev polynomial of the first kind
T_G^2	two-axis gradient-tilt variance
T_Z^2	two-axis Zernike-tilt variance
UV	ultraviolet
$U(x)$	unit step function
$U_0(\mathbf{r}, L)$	complex amplitude of the field in free space
$U(\mathbf{r}, L)$	complex amplitude of the field in random medium
$U(a; c; z)$	confluent hypergeometric function of the second kind
$U_n(z)$	Chebyshev polynomial of the second kind
$\mathbf{V}_\perp$, $V_\perp$	transverse wind velocity, magnitude of wind velocity
WFS	wavefront sensor

WSF	wave structure function
W_0	beam radius at transmitter
W	beam radius in free space at receiver
$W_{k,m}(z)$	Whittaker function of the second kind
W_{Rx}	radius of Gaussian receiver lens
W_{LT}	long-term beam radius in random medium at receiver
W_{ST}	short-term beam radius in random medium at receiver
$Y_p(z)$	standard Bessel function of second kind
$y_n(z)$	spherical Bessel function of the second kind
$Z_n^m(r, \theta)$	Zernike polynomial
α	spectral index for non-Kolmogorov turbulence
α_0	complex parameter at transmitter $(= 2/kW_0^2 + i/F_0)$
$\sqrt{\langle\beta_a^2\rangle}$	rms angle-of-arrival
β_0^2	Rytov variance for a spherical wave
$\Gamma(z)$	gamma function
$\Gamma(a, z)$	incomplete gamma function of the second kind
$\Gamma_2(\mathbf{r}_1, \mathbf{r}_2, L)$	mutual coherence function
γ	Euler constant $(= 0.577216\ldots)$
$\gamma(a, z)$	incomplete gamma function of the first kind
θ_0	isoplanatic angle
Θ	beam curvature parameter at receiver
Θ_0	beam curvature parameter at transmitter
κ	scalar spatial wave number
κ_l	inner-scale wave number parameter $(= 3.3/l_0)$
κ_m	inner-scale wave number parameter $(= 5.92/l_0)$
κ_0	outer-scale wave number parameter $(= 2\pi/L_0)$
κ_X, κ_Y	large-scale and small-scale cutoff spatial frequencies
λ	wavelength
μ_x, μ_y	anisotropy parameters
$\boldsymbol{\nu}, \nu$	spatial frequency, magnitude of spatial frequency
ρ	scalar separation distance between observation points, correlation coefficient of noise
ρ_c	correlation width of irradiance fluctuations
$\rho_0, \rho_{pl}, \rho_{sph}$	transverse spatial coherence radius
$\sigma_I^2(r, L)$	scintillation index (normalized irradiance variance)
$\sigma_I^2(\alpha, D)$	irradiance flux variance
$\sigma_I^2(0, L)$	longitudinal scintillation index
$\sigma_{I,l}^2(\alpha; \mu_x, \mu_y)$	longitudinal scintillation index in anisotropic turbulence
$\sigma_{I,r}^2(r, L)$	radial scintillation index

$\sigma^2_{I,pl}$	plane wave scintillation index
$\sigma^2_{I,sph}$	spherical wave scintillation index
σ^2_{EFF}	effective LGS anisoplanatic error
$\sigma^2_{LED}(\alpha)$	G-tilt anisoplanatism for LED array
σ^2_G	variance of G-tilt anisoplanatism
σ^2_N	total noise power in detector current
σ^2_R	Rytov variance for a plane wave
$\tilde{\sigma}^2_R$	generalized Rytov variance for a plane wave
σ^2_S	phase variance
σ^2_Z	variance of Z-tilt anisoplanatism
σ^2_χ	log-amplitude variance (Rytov number)
$\sigma^2_r(r, L)$	mean irradiance profile in the transverse direction
$\Phi_n(\kappa)$	three-dimensional spatial power spectrum of refractive index
$\phi(x, y)$	velocity potential function
$\psi_1(\mathbf{r}, L)$	complex phase perturbations of Rytov approximation
$\psi_2(\mathbf{r}, L)$	
Ω_{Rx}	Fresnel ratio characterizing radius of Gaussian lens
ζ	zenith angle
$\zeta(s)$	Riemann zeta function
$\binom{a}{n}$	binomial coefficient $[=\Gamma(a+1)/n!\Gamma(a-n+1)]$
$\oint_C$	integral around a closed path C
Π	product symbol
$\vert\vert$	absolute value or magnitude
$\sim$	"…is asymptotic to"
$\langle\rangle$	ensemble average

Chapter 1
Review of Complex Variables

1.0 Introduction

The subject of complex variables is important in a variety of engineering and science applications such as electromagnetic theory, signal processing, control theory, harmonic functions, two-dimensional fluid flow, and optics. Also, in some applications there often occur real integrals that can't be evaluated by conventional integration methods but can be evaluated by complex integration methods. For example, to evaluate Fourier and Mellin transforms beyond the use of look-up tables, it may be necessary to perform integration in the complex plane using residue theory (e.g., see Section 2.4.1). This also applies to calculating the inverse Laplace transform and inverse Mellin transform, both of which are formulated in the complex plane. However, the use of complex variables goes far beyond these applications. In this optional first chapter we present a brief review of some of the most basic concepts and theorems from complex variables that have direct bearing on the material in this text. There is no attempt for completeness, and proofs of the theorems will generally not be provided.

A *complex number* is one that can be represented by

$$z = x + iy, \tag{1.1}$$

where x and y are real numbers and $i = \sqrt{-1}$. The variable x is called the *real part* of z, denoted by $x = \mathrm{Re}(z)$, and y is called the *imaginary part* of z, written as $y = \mathrm{Im}(z)$. Instead of writing $z = x + iy$, some authors use the *ordered pair* concept in which a complex number z can be defined by an ordered pair of real numbers x and y, i.e., $z = (x, y)$. For many purposes it is useful to represent the complex number $z = (x, y)$ as a point in the xy-plane. Each complex number corresponds to just one point in the plane, and each point in the plane corresponds to just one complex number. In this case it is customary to refer to this plane as the *complex plane*, which is illustrated in Fig. 1.1. The x-axis is called the *real axis*, whereas the y-axis is called the *imaginary axis*.

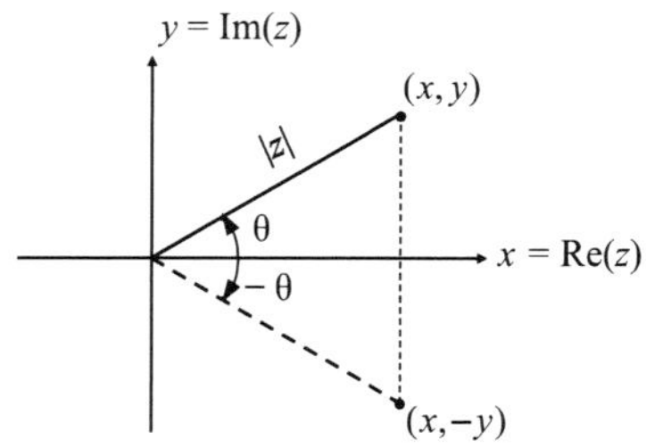

Figure 1.1 Complex plane.

The *complex conjugate* of $z = x + iy$ is denoted by $z^* = x - iy$. The modulus (also called the absolute value) geometrically represents the distance of the complex number $z = (x, y)$ from the origin, i.e., $|z| = \sqrt{x^2 + y^2}$.

In certain calculations it may be more convenient to use *polar coordinates* in place of rectangular coordinates to describe complex numbers. By introducing polar coordinates $z = (r, \theta)$ according to the standard equations of transformation, we have the alternate definition of a complex number:

$$z = r\cos\theta + i\,r\sin\theta = re^{i\theta}. \tag{1.2}$$

In this representation,

$$\begin{aligned} r &= \operatorname{mod}(z) = \sqrt{x^2 + y^2}, \\ \theta &= \arg(z) = \tan^{-1}\frac{y}{x}, \end{aligned} \tag{1.3}$$

where r is the *modulus* of z and the angle θ is called the *argument*. Clearly, $\arg(z)$ is not uniquely determined but is a multiple-valued quantity that we can express as

$$\arg(z) = \theta + 2k\pi, \quad k = 0, \pm 1, \pm 2, \ldots . \tag{1.4}$$

To avoid the multiple values associated with arg(z), some authors use the concept of a *principal angle* for θ, e.g., $-\pi < \theta_p \leq \pi$, written as $\theta_p = \operatorname{Arg}(z)$.

In cases where z is raised to a rational power, we define its roots by

$$z^{m/n} = \left(\sqrt[n]{|z|}\right)^m \exp\left[\frac{im}{n}(\theta_p + 2k\pi)\right], \quad k = 0, 1, 2, \ldots, (n-1). \tag{1.5}$$

1.1 Complex Functions

If f is a function of the complex variable $z = x + iy$, we say that $f(z)$ is a *complex function*. Such functions can always be expressed in the form

$$f(z) = u(x, y) + iv(x, y), \tag{1.6}$$

where u and v are real functions of two real variables. For example, consider the function $f(z) = z^2 = x^2 - y^2 + 2ixy$, where $u(x, y) = x^2 - y^2$ and $v(x, y) = 2xy$. Thus, the notion of a complex function is closely related to and parallels that of a real function, including the concepts of limit, continuity, and derivative.

1.1.1 Analytic functions

In discussing properties of a complex function, it is customary to talk about domains and regions. A *domain* D refers to an area in the complex plane inside a closed boundary curve such as $|z| < 1$. A *region* is a domain plus some or all of the boundary. If a region consists of a domain plus all of the boundary, e.g., $|z| \leq 1$, it is called a *closed region.*

The derivative of a complex function $f(z)$ is based on a limit, as it is for real functions, but the conditions under which a function has a derivative are stricter than for real functions. And unlike real functions, complex functions may be differentiable only at a single point or along some curve in the complex plane.

A complex function $f(z)$ is said to be *analytic* at a point z_0 if it is defined at z_0 and the derivative $f'(z)$ exists at all points in a neighborhood of z_0. A function cannot be analytic at just a single point or just along some curve. In fact, a necessary and sufficient condition for $f(z)$ to be an analytic function in a domain D is that its real and imaginary parts throughout D satisfy the *Cauchy–Riemann equations,*

$$\begin{aligned} \frac{\partial u}{\partial x} &= \frac{\partial v}{\partial y}, \\ \frac{\partial u}{\partial y} &= -\frac{\partial v}{\partial x}, \end{aligned} \tag{1.7}$$

where the partial derivatives are assumed to be continuous. An analytic function always has a derivative—in fact, it has infinitely many derivatives. A function that is analytic throughout the finite z-plane is called an *entire function.* A point where $f(z)$ is not analytic is called a *singular point.*

1.1.2 Elementary complex functions

When introducing complex functions, they should be defined in such a way that the definition is consistent with the real function when z is replaced by the real variable x.

Complex Exponential Function:

$$\begin{aligned}
e^z &= e^{x+iy} = e^x(\cos y + i\sin y),\\
|e^z| &= e^x,\\
|e^{iy}| &= 1, \quad y \text{ real},\\
\frac{d}{dz}e^z &= e^z.
\end{aligned} \tag{1.8}$$

Trigonometric and Hyperbolic Functions:

$$\begin{aligned}
\cos z &= \frac{1}{2}(e^{iz} + e^{-iz}), \qquad \sin z = \frac{1}{2i}(e^{iz} - e^{-iz}),\\
\cosh z &= \frac{1}{2}(e^{z} + e^{-z}), \qquad \sinh z = \frac{1}{2}(e^{z} - e^{-z}),\\
\cos z &= \cos x \cosh y - i \sin x \sinh y,\\
\sin z &= \sin x \cosh y + i \cos x \sinh y,\\
\cos iz &= \cosh z, \qquad \sin iz = i \sinh z,\\
\frac{d}{dz}\cos z &= -\sin z, \qquad \frac{d}{dz}\sin z = \cos z,\\
\frac{d}{dz}\cosh z &= \sinh z, \qquad \frac{d}{dz}\sinh z = \cosh z,\\
e^z &= \cosh z + \sinh z, \qquad e^{-z} = \cosh z - \sinh z.
\end{aligned} \tag{1.9}$$

The above examples are all single-valued functions. Below we introduce some multiple-valued functions. These multiple values are called *branches* of the function.

Complex Logarithm:

$$\begin{aligned}
\log z &= \ln r + i(\theta_p + 2k\pi), \quad k = 0, \pm 1, \pm 2, \ldots ,\\
e^{\log z} &= z,\\
\log(e^z) &= z + 2ik\pi, \quad k = 0, \pm 1, \pm 2, \ldots ,\\
\frac{d}{dz}\log z &= \frac{1}{z}, \quad z \neq 0, \quad -\pi < \arg(z) \leq \pi.
\end{aligned} \tag{1.10}$$

Complex Powers:

$$z^\alpha = \exp(\alpha \log z) = \exp\{\alpha[\ln r + i(\theta_p + 2k\pi)]\}, \quad k = 0, \pm 1, \pm 2, \ldots , \tag{1.11}$$

where α can be any complex number.

Inverse Trigonometric Functions:

$$\cos^{-1}z = -i\log\left[z + i(1-z^2)^{1/2}\right],\quad \sin^{-1}z = -i\log\left[iz + (1-z^2)^{1/2}\right],$$
$$\tan^{-1}z = \frac{i}{2}\log\left(\frac{i+z}{i-z}\right). \tag{1.12}$$

1.2 Complex Integration

The two-dimensional aspect of a complex variable also influences the theory of integration in the complex plane, requiring us to consider integrals along general curves in the plane rather than simply along segments of the x-axis. Because of this, we find that complex integrals behave more like *line integrals* from vector analysis than like standard Riemann integrals. In particular, if we let the real and imaginary parts of the complex variable z be dependent on some real parameter t, then we write $z(t) = x(t) + iy(t)$. In this case we can represent a curve C in the complex plane by

$$C\colon z(t) = x(t) + iy(t), \qquad a \le t \le b. \tag{1.13}$$

A *simple curve* or *arc* is one that doesn't intersect itself. If $z'(t)$ exists (nonzero) and is continuous on the interval $a \le t \le b$, we say the arc is *smooth*. A *contour* is a curve consisting of a finite number of connected smooth arcs.

It is common to declare all integration paths in the complex plane as a contour unless stated otherwise. Also, the *positive direction* along a closed contour is always taken to be *counterclockwise*. We can relate an integral in the complex plane to the sum of two real integrals by writing $f(z) = u(x, y) + iv(x, y)$, i.e.,

$$\int_C f(z)\,dz = \int_C u(x, y)dx - v(x, y)dy + i\int_C v(x, y)dx + u(x, y)dy, \tag{1.14}$$

where $dz = dx + idy$. An important result in the theory of complex integration is the extension of the *fundamental theorem of calculus* to contour integrals. In particular, we find that if $f(z)$ is analytic in a *simply connected* domain D (i.e., one with no "holes" or "donut shape"), and C is any contour contained entirely in D, then the integral $\int_C f(z)dz$ is independent of the contour C. Thus, such integrals satisfy the fundamental theorem of calculus, viz.,

$$\int_A^B f(z)dz = F(B) - F(A),$$

where $F'(z) = f(z)$.

Although there are many integral theorems associated with complex functions, the most fundamental integral theorem concerning analytic functions bears the name of Cauchy. Most of the results concerning contour integrals around closed paths are a consequence of the following theorem.

Theorem 1 (*Cauchy's integral theorem*). If $f(z)$ is analytic on and inside a simple closed contour C, then

$$\oint_C f(z)dz = 0.$$

1.2.1 Deformation of contours

If $f(z)$ is analytic on simple closed contours C_1 and C, with C_1 inside C, and is analytic in the annular domain between them, then it can be shown that

$$\oint_C f(z)dz = \oint_{C_1} f(z)dz. \tag{1.15}$$

To understand this result, let A and A_1 represent points on contour C, and B and B_1 points on contour C_1, as shown in Fig. 1.2, and connect them with a line. Starting at point A, we consider the closed contour integral described by

$$\begin{aligned} I &= \int_A^B f(z)dz - \oint_{C_1} f(z)dz + \int_{B_1}^{A_1} f(z)dz + \oint_C f(z)dz \\ &= \oint_C f(z)dz - \oint_{C_1} f(z)dz, \end{aligned} \tag{1.16}$$

where the integrals along the straight-line sections in opposite directions cancel each other as the gap is closed. Also, the remaining integrals in Eq. (1.16) sum to zero because of Cauchy's integral theorem. Hence, this establishes the validity of Eq. (1.15).

The result of Eq. (1.15) can be generalized to the case shown in Fig. 1.3. That is, if $f(z)$ is analytic on the simple closed contour C and on the simple closed contours $C_1, C_2, \ldots, C_n$, all of which lie within C, and $f(z)$ is analytic in the annular (shaded) domain between the contours, then

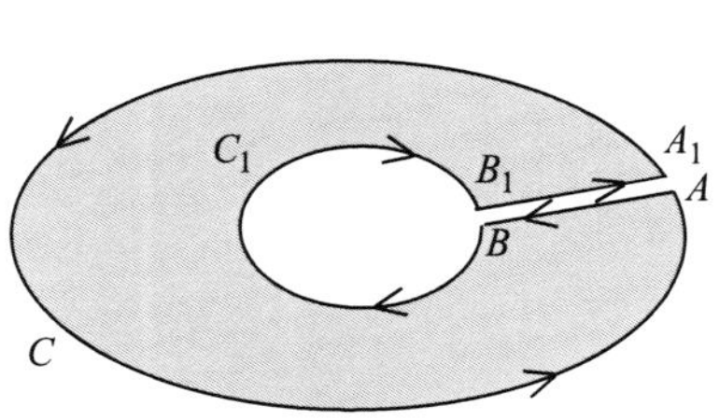

Figure 1.2 Annular domain.

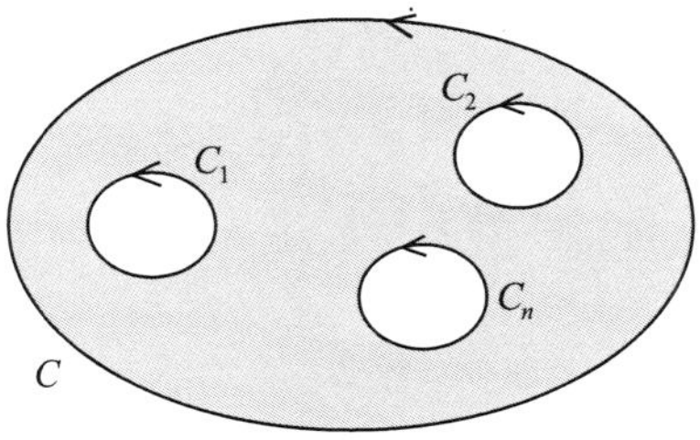

Figure 1.3 Multi-annular domain.

$$\oint_C f(z)dz = \oint_{C_1} f(z)dz + \oint_{C_2} f(z)dz + \cdots + \oint_{C_n} f(z)dz. \tag{1.17}$$

The proof of this result parallels that leading to Eq. (1.15).

1.3 Taylor Series and Laurent Series

Taylor series involving complex functions behave similarly to power series of functions of a real variable. That is, they look the same as series in real variables and satisfy the same properties such as convergence, termwise differentiation, and termwise integration under appropriate conditions. The important theorem concerning a Taylor series of the form

$$\begin{aligned} f(z) &= \sum_{n=0}^{\infty} A_n (z - z_0)^n, \quad |z - z_0| < \rho, \\ A_n &= \frac{f^{(n)}(z_0)}{n!}, \quad n = 0, 1, 2, \ldots, \end{aligned} \tag{1.18}$$

states that, if the series converges within some circle of convergence $|z - z_0| < \rho$, it converges to an analytic function within the circle. Moreover, it converges uniformly and absolutely in any domain within this circle of convergence. Real functions that have a convergent power series can also be written as a complex Taylor series such as Eq. (1.18).

Whereas Taylor series provide valid representations of functions that are analytic at some point $z = z_0$, the related *Laurent series* provide valid representations when the function is not analytic at $z = z_0$. These series are distinguished by the fact that they contain both positive and negative powers of $z - z_0$. Laurent series furnish us with a series representation for $f(z)$ in the annular domain between two concentric circles (see Fig. 1.4) provided the function is analytic everywhere on and between the two circles. Such series take the form

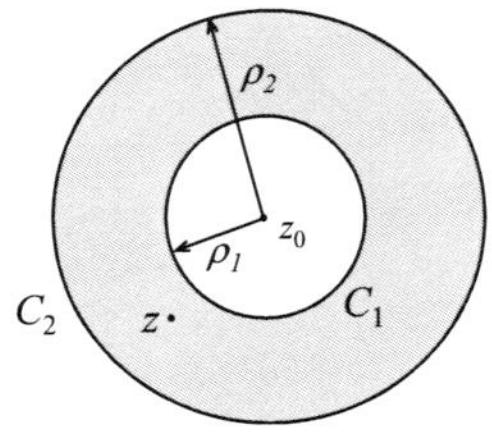

Figure 1.4 Annular domain for Laurent series about z_0.

$$f(z) = \sum_{n=-\infty}^{\infty} A_n(z - z_0)^n, \quad \rho_1 < |z - z_0| < \rho_2, \tag{1.19}$$

where the coefficients are defined by

$$A_n = \frac{1}{2\pi i}\oint_C \frac{f(\zeta)}{(\zeta - z_0)^{n+1}}d\zeta, \quad n = 0, \pm 1, \pm 2, \ldots . \tag{1.20}$$

In Eq. (1.20), the curve C is any closed contour lying within the annular domain of analyticity that contains a singularity z_0 in its interior.

1.3.1 Analytic continuation

There are power series that converge everywhere in the complex plane. These represent entire functions. Examples include the exponential function, sine function, and cosine function. Functions such as

$$f(z) = \frac{1}{1 - z} = \sum_{n=0}^{\infty} z^n, \quad |z| < 1,$$

have power series that converge only within a particular circle, but the range of analyticity of the function can be much wider than that.

To extend the domain of validity of a Taylor series representation of a given function $f(z)$, we can use a technique that extends the domain of the function beyond the circle of convergence of its Taylor series. Such a technique is commonly called *analytic continuation*. Suppose the Taylor series of a function $f(z)$ converges within the circle $|z - z_0| < \rho$. Starting with another point $z = z_1$ within the circle, we can obtain another Taylor series about that point that represents the same function (see Fig. 1.5). Over the common domain of the circles of convergence, both series will converge to the same value. The argument can be repeated over and over, producing more and more circles of convergence. Analytic continuation is always possible if the original function is analytic at least along an arc of the original circle of convergence. A common way in which analytic continuation is used is to extend the domain of a real

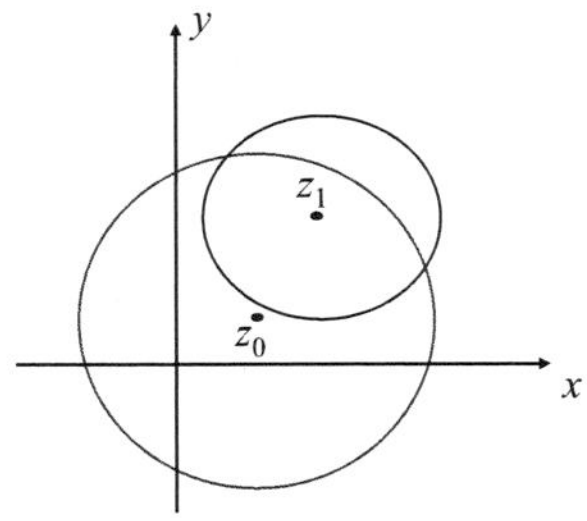

Figure 1.5 Analytic continuation.

function off the real axis into the complex plane. We will find the concept of analytic continuation useful in some of the later chapters.

1.4 Singularities

Laurent series are useful in the classification of various singularities of an analytic function. For example, if there is some neighborhood of a singular point $z = z_0$ of a function $f(z)$ throughout which $f(z)$ is analytic except at the point itself, then $z = z_0$ is called an *isolated singularity*. In this case, $f(z)$ has a Laurent series about $z = z_0$ that converges in the deleted neighborhood $0 < |z - z_0| < \rho$. We further classify an isolated singularity at $z = z_0$ for a function with a Laurent series such as Eq. (1.19) in the following way:

- $z = z_0$ is a *removable singularity* if $A_n = 0$ for all $n < 0$.
- $z = z_0$ is a *pole of order m* if $A_m \neq 0$, but $A_n = 0$ for all $n < m$.
- $z = z_0$ is an *essential singularity* if $A_n \neq 0$ for an infinite number of negative values of n.

The notion of a removable singularity is easily understood by considering the function $f(z) = \frac{\sin z}{z}$. This function has a singularity at $z = 0$ because it is not defined there. However, $\lim_{z\to 0} f(z) = 1$, so the singularity can be removed by defining $f(0) = 1$. Isolated singularities other than poles are called essential singularities. A classic example is $e^{1/z}$, which has an essential singularity at $z = 0$.

If a Laurent series is not available, there is another test for classifying singularities. That is, if $z = z_0$ is an isolated singularity, then using the limit concept, we can identify the three types of singularities by the following:

- $z = z_0$ is a *removable singularity* if $\lim_{z\to z_0} f(z)$ exists (finitely).
- $z = z_0$ is a *pole of order m* if $\lim_{z\to z_0} f(z) = \infty$, but $(z - z_0)^m f(z)$ has a removable singularity at $z = z_0$, provided m is the smallest integer for which this is true.
- $z = z_0$ is an *essential singularity* if $\lim_{z\to z_0} f(z)$ does not exist.

1.4.1 Residues

Every function $f(z)$ has a *residue* at each of its isolated singular points, the value of which may be zero. In general, the value of the residue is the value of the integral (taken in the counterclockwise direction)

$$A_{-1} = \text{Res}\{f(z); z_0\} = \frac{1}{2\pi i}\oint_C f(z)dz \tag{1.21}$$

around any closed contour C containing the isolated singularity. The coefficient A_{-1} refers to the Laurent series coefficient of the $1/(z - z_0)$ term in Eq. (1.19).

The most common type of isolated singularity is a *pole.* A pole of order one is called a *simple pole.* If $f(z) = p(z)/q(z)$ has a simple pole at $z = z_0$, then also

$$\text{Res}\{f(z); z_0\} = \frac{p(z_0)}{q'(z_0)}. \tag{1.22}$$

The residue for a function having a pole of order m at $z = z_0$ is given by

$$\text{Res}\{f(z); z_0\} = \frac{1}{(m-1)!}\lim_{z\to z_0}\frac{d^{m-1}}{dz^{m-1}}[(z - z_0)^m f(z)]. \tag{1.23}$$

The evaluation of an integral around a closed contour containing a finite number of poles of a given function $f(z)$ relies on the following theorem.

Theorem 2 (*Residue theorem*). If $f(z)$ is analytic on and inside a simple closed contour C, except for finitely many isolated singular points $a_1, a_2, \ldots, a_N$ inside C, then

$$\oint_C f(z)dz = 2\pi i\sum_{k=1}^{N}\text{Res}\{f(z); a_k\}.$$

1.5 Applications

In this section we will look at a few examples involving complex integrals and Laurent series of complex functions.

Example 1: Evaluate the integral I around a circle of radius R that is centered at $z = z_0$:

$$I = \oint_C (z - z_0)^n dz, \quad n = 0, \pm 1, \pm 2, \ldots .$$

A standard parameterization for the circle $|z - z_0| = R$ is

$$C:\ z(t) = z_0 + Re^{it}, \quad 0 \leq t < 2\pi. \tag{1.24}$$

Thus, the integral I in terms of t takes the form

$$I = \int_0^{2\pi} (Re^{it})^n iRe^{it}\,dt = iR^{n+1}\int_0^{2\pi} e^{i(n+1)}dt, \tag{1.25}$$

where we used $dz = iRe^{it}dt$. If $n \neq -1$, then

$$\int_0^{2\pi} e^{i(n+1)}dt = \frac{e^{i(n+1)t}}{i(n+1)}\bigg|_0^{2\pi} = 0, \tag{1.26}$$

whereas for $n = -1$, we have

$$\int_0^{2\pi} dt = 2\pi. \tag{1.27}$$

Consequently, we have shown that

$$I = \oint_C (z - z_0)^n dz = \begin{cases} 0, & n \neq -1 \\ 2\pi i, & n = -1. \end{cases} \tag{1.28}$$

When a branch point exists in the function to be integrated, we often need to impose a *branch cut* that cannot be crossed by the integration contour. This will generally eliminate part of one of the axes. Let us illustrate with the next example.

Example 2: Evaluate the integral $\int_{-1}^{1} z^{1/2}dz$ along a contour connecting the points $z = -1$ and $z = 1$.

The complex function

$$f(z) = z^{1/2} = \sqrt{r}e^{i\theta/2}$$

is multiple-valued (i.e., two values). Thus, it is not analytic in the entire z-plane. In fact, it has a *branch point* at the origin, i.e., a singular point common to all branches. If we restrict the contour to one of the branches illustrated in Fig. 1.6, it becomes an analytic function within the domain D_1 or domain D_2.

First, we form a branch cut along the negative imaginary axis and impose the restriction, $-\pi/2 < \theta < 3\pi/2$, so the function $f(z)$ is analytic in the domain D_1 shown in the figure. Then, along any contour C_1 lying inside D_1 that connects the points $(-1, 0)$ and $(1, 0)$, the integral leads to

$$\int_{-1}^{1} z^{1/2}dz = \frac{2}{3}z^{3/2}\bigg|_{-1}^{1} = \frac{2}{3}r\sqrt{r}e^{3i\theta/2}\bigg|_{(1,\pi)}^{(1,0)}. \tag{1.29}$$

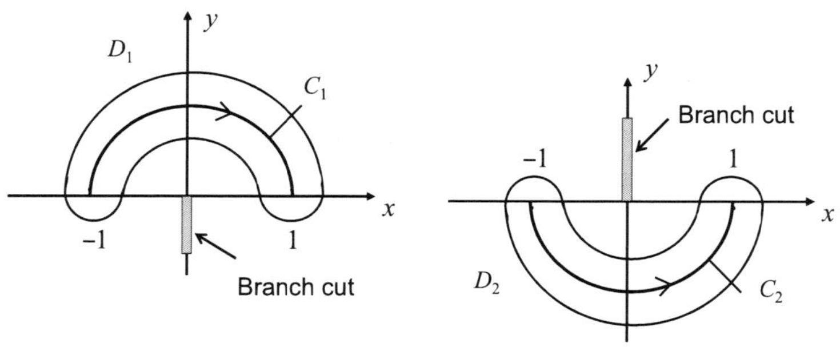

Figure 1.6 Contours of integration for different branches (adapted from Ref. 1).

Completing the process, we find that

$$\int_{-1}^{1} z^{1/2} dz = \frac{2}{3}(1 - e^{3\pi i/2}) = \frac{2}{3}(1 + i). \tag{1.30}$$

If we repeat the process with the restriction $-3\pi/2 < \theta < \pi/2$, this forms a different branch cut along the positive imaginary axis. The integral along C_2 within the domain D_2 now gives us

$$\int_{-1}^{1} z^{1/2} dz = \frac{2}{3} r\sqrt{r} e^{3i\theta/2}\Big|_{(1,-\pi)}^{(1,0)} = \frac{2}{3}(1 - i). \tag{1.31}$$

Note that we had to introduce polar coordinates to properly identify the points -1 and 1 associated with each branch cut.

Example 3: Find Laurent series expansions about the singular points $z = 0$ and $z = 1$ of the function

$$f(z) = \frac{1}{z^2(z-1)}.$$

Here we consider the singular point $z_0 = 0$ first. There are two possible Laurent series—one for $0 < |z| < 1$ and one for $|z| > 1$, as depicted in Fig. 1.7. For the domain $0 < |z| < 1$ we start with the Taylor series representation for $1/(z-1)$ from calculus given by

$$\frac{1}{z^2(z-1)} = -\frac{1}{z^2}\sum_{n=0}^{\infty} z^n = -\sum_{n=0}^{\infty} z^{n-2}.$$

By making the change of index $n \to n + 2$, we get the Laurent series

$$f(z) = \frac{1}{z^2(z-1)} = -\sum_{n=-2}^{\infty} z^n, \quad 0 < |z| < 1. \tag{1.32}$$

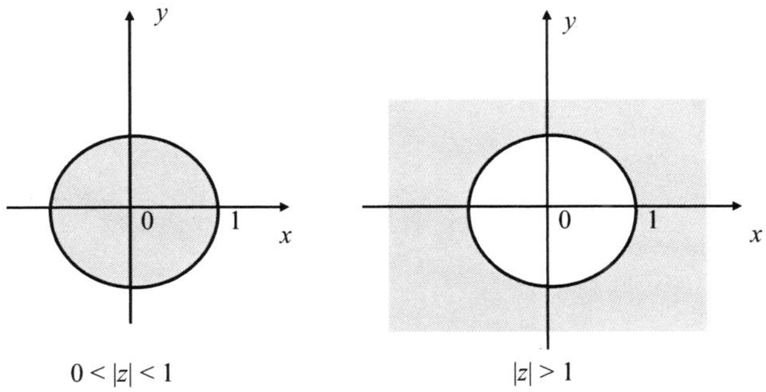

Figure 1.7 Domain for the Laurent series about $z_0 = 0$ (adapted from Ref. 1).

For the second domain $|z| > 1$, we have

$$f(z) = \frac{1}{z^2(z-1)} = \frac{1}{z^3(1-1/z)} = \frac{1}{z^3}\sum_{n=0}^{\infty}\left(\frac{1}{z}\right)^n, \tag{1.33}$$

and following the index change $n \to -(n-3)$, we arrive at

$$f(z) = \frac{1}{z^2(z-1)} = \sum_{n=-\infty}^{-3} z^n, \quad |z| > 1. \tag{1.34}$$

At the singular point $z = 1$ there are again two Laurent series corresponding to the domains $0 < |z-1| < 1$ and $|z-1| > 1$, as shown in Fig. 1.8. Here it is useful to first write $u = z - 1$, from which we obtain

$$f(z) = \frac{1}{z^2(z-1)} = \frac{1}{u(1+u)^2} = \frac{1}{u}\sum_{n=0}^{\infty}(-1)^n(n+1)u^n\bigg|_{u=z-1}. \tag{1.35}$$

Upon simplifying, this last expression gives us

$$f(z) = \frac{1}{z^2(z-1)} = \sum_{n=-1}^{\infty}(-1)^{n+1}(n+2)(z-1)^n, \quad 0 < |z-1| < 1, \tag{1.36}$$

where we have made the index change $n \to n+1$. Last, for $|z-1| > 1$, we write

$$f(z) = \frac{1}{z^2(z-1)} = \frac{1}{u^3(1+1/u)^2} = \frac{1}{u^3}\sum_{n=0}^{\infty}(-1)^n(n+1)\left(\frac{1}{u}\right)^n\bigg|_{u=z-1}, \tag{1.37}$$

from which we deduce that $[n \to -(n-3)]$,

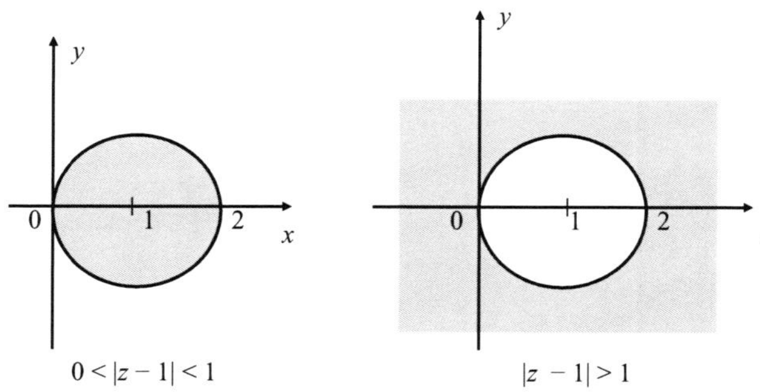

Figure 1.8 Domain for Laurent series about $z_0 = 1$ (adapted from Ref. 1).

$$f(z) = \frac{1}{z^2(z-1)} = \sum_{n=-\infty}^{-3} (-1)^n (n+2)(z-1)^n, \quad |z-1| > 1. \tag{1.38}$$

In later chapters we will have many occasions where we need to apply the residue theory in Theorem 2. We present two examples below.

Example 4: Use the residue theorem to evaluate the integral

$$I = \oint_C \frac{dz}{z(z-1)^2(z+3)}, \qquad C: \ |z| = 2.$$

We first note that $f(z)$ has a simple pole at $z = 0$ and a pole of order 2 at $z = 1$ inside the contour C. The pole at $z = -3$ lies outside C. From the residue theorem we see that

$$I = 2\pi i[\mathrm{Res}(0) + \mathrm{Res}(1)]. \tag{1.39}$$

Thus, we calculate the residues

$$\begin{aligned} \mathrm{Res}(0) &= \left.\frac{1}{(z-1)^2(z+3)}\right|_{z=0} = \frac{1}{3}, \\ \mathrm{Res}(1) &= \left.\frac{d}{dz}[(z-1)^2 f(z)]\right|_{z=1} = -\left.\frac{2z+3}{z^2(z+3)^2}\right|_{z=1} = -\frac{5}{16}, \end{aligned} \tag{1.40}$$

and consequently, obtain the result

$$\oint_C \frac{dz}{z(z-1)^2(z+3)} = 2\pi i\left(\frac{1}{3} - \frac{5}{16}\right) = \frac{\pi i}{24}. \tag{1.41}$$

Example 5: Use the residue theorem to evaluate the real integral

$$I = \int_0^\infty \frac{x^{m-1}}{1+x}\, dx, \quad 0 < m < 1.$$

With the prescribed restriction on m, it follows that the related complex function

$$f(z) = \frac{z^{m-1}}{1+z} \tag{1.42}$$

has a branch point at $z = 0$ in addition to a simple pole at $z = -1$.

The closed contour shown in Fig. 1.9 takes into account a branch cut along the positive real axis. To evaluate the integral I from 0 to ∞, we integrate the complex function $f(z)$ around a closed contour C that includes the positive real axis along both sides of the branch cut. Based on the residue theorem, we have that

$$\oint_C \frac{z^{m-1}}{1+z} dz = 2\pi i \operatorname{Res}(-1). \tag{1.43}$$

The simple pole at $z = -1$ is the only isolated singularity inside C, and we have indented around the branch point at the origin. Breaking the contour C into its various smooth arcs, we get

$$\begin{aligned} \oint_C f(z)dz = &-\int_{C_\rho} f(z)dz + \int_\rho^R f(x)dx \\ &+ \int_{C_R} f(z)dz + \int_R^\rho f(xe^{2\pi i})dx. \end{aligned} \tag{1.44}$$

It can be shown that both integrals over the circular paths C_R and C_ρ tend to zero as $R \to \infty$ and $\rho \to 0$, and therefore what remains is

$$\oint_C \frac{z^{m-1}}{1+z} dz = \int_0^\infty \frac{x^{m-1}}{1+x} dx + \int_\infty^0 \frac{(xe^{2\pi i})^{m-1}}{1+x} dx = 2\pi i \operatorname{Res}(-1). \tag{1.45}$$

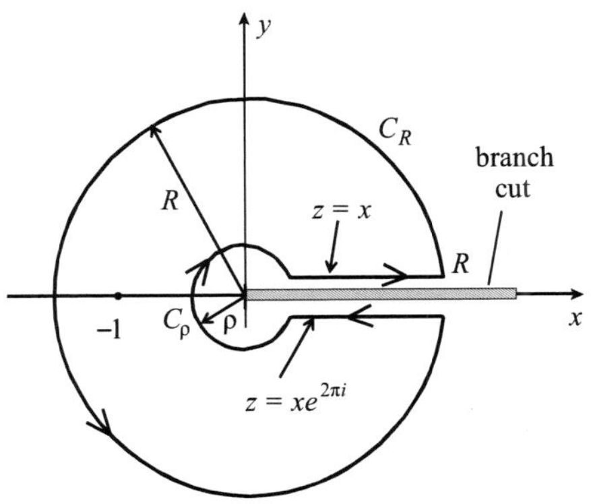

Figure 1.9 Contour for evaluating the integral in Eq. (1.37) (adapted from Ref. 1).

By combining the integrals along the real axis, we find that

$$(1 - e^{2im\pi})\int_0^\infty \frac{x^{m-1}}{1+x}dx = 2\pi i \operatorname{Res}(-1). \tag{1.46}$$

However, $1 - e^{2im\pi} = e^{im\pi}(e^{-im\pi} - e^{im\pi})$ and $\operatorname{Res}(-1) = (-1)^{m-1} = -e^{im\pi}$, which finally leads to

$$\int_0^\infty \frac{x^{m-1}}{1+x}dx = -\frac{2\pi i e^{im\pi}}{e^{im\pi}(e^{-im\pi} - e^{im\pi})} = \frac{\pi}{\sin m\pi}, \quad 0 < m < 1. \tag{1.47}$$

1.6 Historical Remarks

Perhaps the name most associated with the study of complex variables is that of the French mathematician Augustin-Louis Cauchy (1789–1857). In fact, several of the important integral theorems contain his name, as do the Cauchy–Riemann equations. Among other subjects studied by Cauchy were determinants, the theory of permutation groups in abstract algebra, and ordinary and partial differential equations; additionally, he developed the concept of convergence of an infinite series. Other prominent names associated with complex variables are those of the German mathematicians Carl Friedrich Gauss (1777–1855), Geor Friedrich Bernhard Riemann (1826–1866), who studied under Gauss, and Karl Theodor Wilhelm Weierstrass (1815–1897), the latter being known as the Father of modern analysis. Among many contributions to mathematics, Weierstrass is known for devising tests of convergence and for formalizing the definition of uniform convergence. Another famous name in complex variables is that of the French mathematician Pierre Alphonse Laurent (1813–1854), who is known for introducing a generalization of the Taylor series, now known as the Laurent series.

Although Cauchy is the most prominent name in the study of complex analysis, it was Gauss who gave real meaning to imaginary numbers by suggesting that the number *line* for real numbers be replaced by the concept of a *complex plane* for numbers such as $a + b\sqrt{-1}$, where a and b are real numbers. More generally, if a and b are real numbers and $i = \sqrt{-1}$, then the complex number $a + ib$ can be defined by the number-couple (a, b) that represents a point in the complex plane. Furthermore, such numbers obey the postulates of algebra. Gauss also introduced the concept of complex variable $z = x + iy$, which eventually led to the notion of a complex function $f(z)$. However, it was actually Leonhard Euler (1707–1783) who introduced the notation $i = \sqrt{-1}$. Last, it is interesting that although the symbol $\sqrt{-1}$ for imaginary number had been known by that name since the middle ages, such numbers were still called "imaginary" up to around 1800. However, the term is still commonly used today to identify numbers along the vertical axis in the complex plane.

Reference

1. L. C. Andrews and R. L. Phillips, *Mathematical Techniques for Engineers and Scientists*, SPIE Press, Bellingham, Washington, Chapters 6 and 7 (2003) [doi: 10.1117/3.467443].

Bibliography of Further Reading

Churchill, R. V., J. W. Brown, and R. F. Verhey, *Complex Variables and Applications*, McGraw Hill, New York (1974).

Kreyszig, E., *Advanced Engineering Mathematics*, John Wiley & Sons, New York, Chapters 12–17 (1993).

Saff, E. B. and A. D. Snider, *Fundamentals of Complex Analysis*, Prentice-Hall, Upper Saddle River, New Jersey (1976).

Wunsch, A. D., *Complex Variables with Applications*, 2nd Edition, Addison-Wesley, Massachusetts (1994).

Part I
Special Functions, Asymptotic Analysis, and Approximations

Chapter 2
Basic Concepts

2.0 Introduction

Functions beyond the elementary functions of calculus are called *special functions* or *higher transcendental functions.*[1–5] They commonly occur in areas of application such as heat conduction, communication systems, electro-optics, wave propagation, electromagnetic theory, quantum mechanics, fluid mechanics, probability theory, and electric circuit theory, among others. Still, many engineers and scientists are not well acquainted with them. A few of the special functions, like *Bessel functions* and *error functions*, may be introduced in certain education courses in engineering and physics, but *generalized hypergeometric functions* are seldom discussed in these courses. Numerical computation presents a further difficulty. Until recent times, the numerical evaluation of all special functions was performed by hand, using look-up tables, or with self-written computer programs. Today, however, there are several commercial software programs available to the user that include many or most of the special functions for easy evaluation.

In beginning calculus courses, the properties of elementary functions are studied through direct means or through examination of their power series. For example, some typical power series representations from calculus are

$$\frac{1}{1-x} = \sum_{n=0}^{\infty} x^n, \quad -1 < x < 1, \tag{2.1}$$

$$e^x = \sum_{n=0}^{\infty} \frac{x^n}{n!}, \quad -\infty < x < \infty, \tag{2.2}$$

$$\cos x = \sum_{n=0}^{\infty} \frac{(-1)^n x^{2n}}{(2n)!}, \quad -\infty < x < \infty, \tag{2.3}$$

$$\sin^{-1} x = \sum_{n=0}^{\infty} \frac{(2n)!x^{2n+1}}{2^{2n}(n!)^2(2n+1)}, \quad -1 < x < 1, \tag{2.4}$$

$$\ln x = \sum_{n=1}^{\infty} \frac{(-1)^{n-1}}{n}(x-1)^n, \quad 0 < x \le 2. \tag{2.5}$$

Power series are also used to study the properties of many of the special functions. Among others, these include the *error function*,

$$\text{erf}(x) = \frac{2}{\sqrt{\pi}} \sum_{n=0}^{\infty} \frac{(-1)^n x^{2n+1}}{n!(2n+1)}, \quad -\infty < x < \infty, \tag{2.6}$$

the *complete elliptic integral of the first kind*,

$$K(x) = \frac{\pi}{2} \sum_{n=0}^{\infty} \binom{-\frac{1}{2}}{n}^2 x^{2n}, \quad -1 < x < 1, \tag{2.7}$$

where $\binom{-\frac{1}{2}}{n}$ is the binomial coefficient, and the *Bessel function of the first kind*,

$$J_0(x) = \sum_{n=0}^{\infty} \frac{(-1)^n}{(n!)^2} \left(\frac{x}{2}\right)^{2n}, \quad -\infty < x < \infty. \tag{2.8}$$

Alternatively, we might study some of these elementary and special functions through their integral representations, such as

$$\ln x = \int_1^x \frac{dt}{t}, \tag{2.9}$$

$$\text{erf}(x) = \frac{2}{\sqrt{\pi}} \int_0^x e^{-t^2} dt, \tag{2.10}$$

$$K(x) = \int_0^{\pi/2} \frac{d\phi}{\sqrt{1 - x^2 \sin^2\phi}}, \tag{2.11}$$

$$J_0(x) = \frac{1}{\pi} \int_0^{\pi} \cos(x \sin\phi) d\phi. \tag{2.12}$$

In some cases, the function of interest may be defined beyond the interval of convergence of its power series by a procedure called *analytic continuation*. For example, the function on the left in Eq. (2.1) is defined for all $x \neq 1$, including complex values, but the power series representation is limited to $|x| < 1$. Also, by restricting the study of elementary and special functions to either a series or integral representation, we may not discover that they have certain characteristics in common with other functions. For instance, it is not clear from their power series representation alone that the geometric series (2.1), inverse sine function (2.4), and the complete elliptic integral (2.7) all belong to the same class of functions, as do the cosine function (2.3) and Bessel function (2.8). By identifying a function as belonging to a particular type of generalized hypergeometric function, we can study the basic properties of all of such functions simultaneously through the general class rather than just on an individual basis.

Topics introduced in this chapter include *asymptotic series*, the *Cauchy product*, the *Mellin transform*, and the related *Mellin–Barnes integral*. The Mellin transform is similar to the Laplace transform and is used in mathematical statistics and in the theory of asymptotic expansions. In physics and engineering, the most famous application of the Mellin transform is finding the solution of a potential problem in a wedge-shaped region where the unknown function satisfies Laplace's equation with given boundary conditions on the edges. Lastly, it is also used in the study of electromagnetic wave propagation through turbulence. The Mellin–Barnes integral is closely linked to generalized hypergeometric functions, asymptotic series, and high-energy physics. The gamma function and the Pochhammer symbol used in describing such generalized hypergeometric functions are powerful tools that lend themselves to helping identify a certain series as a known hypergeometric-type function that we might otherwise not recognize. Thus, before taking up other topics mentioned above, we first recognize the central role played by the gamma function and the Pochhammer symbol by starting with a short review of these important functions.

2.1 The Gamma Function

The usual way the *gamma function* is introduced is by the integral

$$\Gamma(z) = \int_0^\infty e^{-t} t^{z-1} dt, \qquad \mathrm{Re}(z) > 0, \tag{2.13}$$

where z may be real or complex. Under the restriction $\mathrm{Re}(z) > 0$, it can be shown that the integral converges uniformly and that the function $\Gamma(z)$ is analytic throughout this domain.

Although the definition in Eq. (2.13) is restricted to $\mathrm{Re}(z) > 0$, there are more complete definitions that extend this function to many other values of z. For example, there is the limit definition

$$\Gamma(z) = \lim_{n\to\infty} \frac{n!n^z}{z(z+1)(z+2)\cdots(z+n)}, \qquad z \neq 0, -1, -2, \ldots. \tag{2.14}$$

This limit defines the gamma function for all z except for simple poles at zero and the negative integers.

Using either of the above definitions for $\Gamma(z)$, it is easily shown that this function satisfies the following properties:

$$\Gamma(z+1) = z\Gamma(z), \tag{2.15}$$

$$\Gamma(n+1) = n!, \quad n = 0, 1, 2, \ldots. \tag{2.16}$$

From Eq. (2.16) it can be seen that the gamma function is a generalization of the factorial function $n!$ to all complex numbers with the exclusions already noted in Eq. (2.14). It also follows from Eq. (2.14) that the function $1/\Gamma(z)$ is an *entire function*. In addition to these properties, there are numerous other identities associated with the gamma function, some of which are tabulated in Appendix A. The graph of the gamma function for real x, $-4 < x \leq 4$, is shown in Fig. 2.1.

An important property of the gamma function involves the product

$$\Gamma(x)\Gamma(y) = 2\int_0^\infty e^{-u^2}u^{2x-1}du \cdot 2\int_0^\infty e^{-v^2}v^{2y-1}dv, \quad x > 0,\ y > 0, \tag{2.17}$$

where we have set $t = u^2$ and $t = v^2$ in Eq. (2.13). Converting to polar coordinates

$$u = r\cos\theta, \quad v = r\sin\theta, \quad r \geq 0,\ 0 \leq \theta \leq \frac{\pi}{2},$$

and doing some rearranging, we obtain

$$\begin{aligned} \Gamma(x)\Gamma(y) &= 4\int_0^\infty e^{-r^2}r^{2(x+y)-1}dr \cdot \int_0^{\pi/2} \cos^{2x-1}\theta\ \sin^{2y-1}\theta\, d\theta \\ &= 2\Gamma(x+y)\int_0^{\pi/2} \cos^{2x-1}\theta\ \sin^{2y-1}\theta\, d\theta, \end{aligned} \tag{2.18}$$

from which we deduce the integral relation

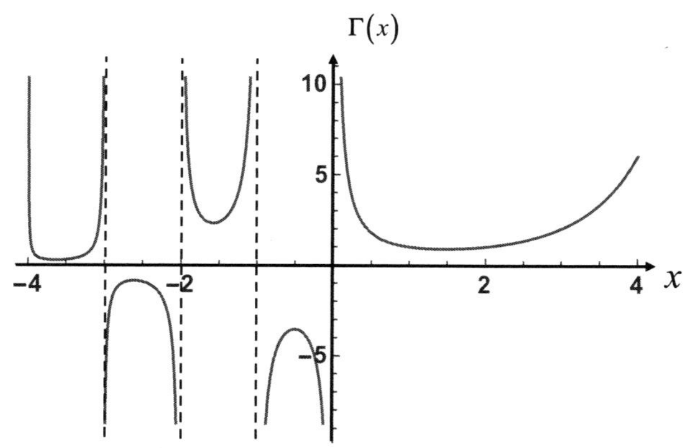

Figure 2.1 The gamma function.

$$\int_0^{\pi/2} \cos^{2x-1}\theta \sin^{2y-1}\theta \, d\theta = \frac{\Gamma(x)\Gamma(y)}{2\Gamma(x+y)}, \quad x > 0, \ y > 0. \tag{2.19}$$

A function of two variables that is directly related to the gamma function is the *beta function*:

$$B(x, y) = \begin{cases} \int_0^1 t^{x-1}(1-t)^{y-1}\,dt, & x > 0, \ y > 0, \\ \int_0^\infty \frac{t^{x-1}}{(1+t)^{x+y}}\,dt, & x > 0, \ y > 0. \end{cases} \tag{2.20}$$

By making the change of variable $t = \cos^2\theta$ in the upper expression of Eq. (2.20), it readily follows from Eq. (2.19) that

$$B(x, y) = 2\int_0^{\pi/2} \cos^{2x-1}\theta \sin^{2y-1}\theta \, d\theta = \frac{\Gamma(x)\Gamma(y)}{\Gamma(x+y)}, \quad x > 0, \ y > 0. \tag{2.21}$$

2.2 The Pochhammer Symbol

In dealing with certain product forms, factorials, and gamma functions, it can be useful to introduce the notation (Ref. 3, pp. 358–360)

$$(a)_0 = 1; \quad (a)_n = a(a+1)\cdots(a+n-1), \quad n = 1, 2, 3, \ldots, \tag{2.22}$$

called the *Pochhammer symbol*. The parameter a can be real or complex, but most often we assume it is real. In terms of the gamma function, this symbol can also be defined by

$$(a)_n = \frac{\Gamma(a+n)}{\Gamma(a)}, \quad n = 0, 1, 2, \ldots. \tag{2.23}$$

Appendix A contains a list of useful properties for this symbol.

2.3 Asymptotic Series and the Cauchy Product

For computational purposes, we often seek to represent a given function $f(z)$ by some simpler function, say $g(z)$, that accurately describes the numerical values of $f(z)$ in the vicinity of a particular point. In many cases that point is $z = 0$. If a function has a convergent Maclaurin series representation like those in Eqs. (2.1) through (2.4), then the first few terms of the series will provide the required asymptotic approximation $g(z)$. By taking a sufficient number of terms in the series, the approximation to $f(z)$ can be made as precise as desired. However, if the power series representation about $z = 0$ is not a Maclaurin series, then it may not converge. In such cases, for fixed $\arg(z)$, we call a series such as

$$f(z) \sim \sum_{n=0}^{\infty} c_n z^n, \quad z \to 0 \tag{2.24}$$

an *asymptotic series* for $f(z)$ as $z \to 0$ if and only if, for each n,

$$\lim_{z \to 0} \frac{|f(z) - S_n(z)|}{|z|^n} = 0; \quad S_n(z) = \sum_{k=0}^{n} c_k z^k. \tag{2.25}$$

The symbol "$\sim$" in Eq. (2.24) means "is asymptotic to." By the condition given in Eq. (2.25), we are requiring the magnitude of the difference between $f(z)$ and the partial sum $S_n(z)$ to approach zero more rapidly than $|z|^n$, as z itself tends to zero.

2.3.1 Large arguments of the function

Given a function $f(z)$, we might also seek a useful approximation for large values of z. This often leads to an expression of the form

$$f(z) \sim h(z) \sum_{n=0}^{\infty} \frac{c_n}{z^n}, \quad z \to \infty, \tag{2.26}$$

where $h(z)$ is a function that may be unity. We say that Eq. (2.26) is an *asymptotic series* for large arguments if and only if, for fixed arg(z) and for every n,

$$\lim_{z \to \infty} |z|^n \left| \frac{f(z)}{h(z)} - S_n(z) \right| = 0; \quad S_n(z) = \sum_{k=0}^{n} \frac{c_k}{z^k}. \tag{2.27}$$

If a given asymptotic series diverges, the accuracy of computation for fixed z is directly tied to the number n. That is, after a certain number of terms, the accuracy of computation may actually get *worse* instead of better with increasing n. That makes series such as Eq. (2.26) intriguing because they often *diverge* for all values of z. Nonetheless, they may still be useful for computation if one does not take too many terms. That is, if we truncate the series, it will converge. As a "rule of thumb," *one should stop adding terms just before the terms begin to increase*. The error incurred in most cases turns out to be less than the first term omitted in the approximation.

We should point out that not all functions have an asymptotic series as $z \to \infty$, such as the exponential function, sine function, and cosine function, among others. Nonetheless, there are many functions that do.

2.3.2 Stirling's formula

To illustrate the derivation of a large-argument asymptotic approximation from an integral definition, we consider the asymptotic result for the gamma function called Stirling's formula. Let us start with the gamma function integral written as

$$\begin{aligned}\Gamma(x+1) &= \int_0^\infty t^x e^{-t}\,dt \\ &= x^x e^{-x}\int_{-x}^\infty e^{x\ln(1+s/x)}e^{-s}\,ds,\end{aligned} \tag{2.28}$$

where the second step follows from the substitution $t = x + s$. For $x \gg s$, we have

$$\ln\left(1+\frac{s}{x}\right) \sim \frac{s}{x} - \frac{s^2}{2x^2} + \cdots, \qquad x \gg s,$$

which leads to

$$\begin{aligned}\Gamma(x+1) &\sim x^x e^{-x}\int_{-\infty}^\infty e^{-s^2/2x}\,ds \\ &\sim \sqrt{2x}\,x^x e^{-x}\int_{-\infty}^\infty e^{-u^2}\,du,\end{aligned} \tag{2.29}$$

where we have now made the substitution $u = s/\sqrt{2x}$. However, it readily follows from properties of the gamma function that [cf. ($G2$) in Appendix A]

$$\int_{-\infty}^\infty e^{-u^2}du = 2\int_0^\infty e^{-u^2}du = \Gamma\left(\frac{1}{2}\right) = \sqrt{\pi}. \tag{2.30}$$

Hence, by combining the results of Eqs. (2.29) and (2.30), we arrive at *Stirling's formula*:

$$\Gamma(x+1) \sim \sqrt{2\pi x}\,x^x e^{-x}, \qquad x \to \infty. \tag{2.31}$$

Stirling's formula is remarkably accurate for even small values of x. For example, when $x = 3$, the error is less than 3%; for $x = 6$, the error is 1.4%.

2.3.3 The Cauchy product

In some applications the need arises to combine various series by the operations of addition and multiplication. Two absolutely convergent series may be summed termwise, and the resulting series will converge within the common interval (or circle) of convergence. The same is true for the product of two absolutely convergent series. In some cases, we can extend this concept to other series, including asymptotic series as well.

In forming the product of two Maclaurin power series representing complex functions $f(z)$ and $g(z)$, we are led to

$$f(z)g(z) = \sum_{m=0}^{\infty} a_m z^m \cdot \sum_{k=0}^{\infty} b_k z^k. \tag{2.32}$$

If we make the change of index $m = n - k$, $k = 0, 1, \ldots, n$, then the product in Eq. (2.32) becomes

$$f(z)g(z) = \sum_{n=0}^{\infty} \sum_{k=0}^{n} a_{n-k} b_k z^n, \tag{2.33}$$

which involves a finite series within each term of a single infinite series. This property, known as the *Cauchy product*, is particularly important in making numerical calculations. It is known that the Cauchy product may still be valid when only one of the series converges absolutely and the other converges conditionally.

2.4 The Mellin Transform

The Mellin transform and its inversion formula can be formally derived from the *Fourier integral theorem*, which, for piecewise continuous and absolutely integrable functions $g(t)$, can be expressed in the form (Ref. 6, p. 245)

$$g(u) = \frac{1}{2\pi} \int_{-\infty}^{\infty} e^{-i\xi u} \int_{-\infty}^{\infty} g(t) e^{i\xi t} dt\, d\xi, \tag{2.34}$$

where $i = \sqrt{-1}$. By making the change of variables $x = e^t$, $y = e^u$, and $s = c + i\xi$, where c is a real constant, Eq. (2.34) becomes

$$g(\ln y) y^{-c} = \frac{1}{2\pi i} \int_{c-i\infty}^{c+i\infty} y^{-s} \int_0^{\infty} g(\ln x) x^{-c} x^{s-1} dx\, ds. \tag{2.35}$$

If we now introduce the function $f(x) = g(\ln x) x^{-c}$, then Eq. (2.35) leads to the pair of transform integrals

$$\mathcal{M}\{f(x)\} = F(s) = \int_0^{\infty} x^{s-1} f(x) dx, \tag{2.36}$$

$$\mathcal{M}^{-1}\{F(s)\} = f(x) = \frac{1}{2\pi i} \int_{c-i\infty}^{c+i\infty} x^{-s} F(s) ds, \quad x > 0. \tag{2.37}$$

We call Eq. (2.36) the *Mellin transform* of $f(x)$, whereas Eq. (2.37) is the inversion formula or *inverse Mellin transform* of $F(s)$, where integration is in the complex s-plane.

If the integrals

$$\int_0^\infty x^{a-1} f(x)dx, \qquad \int_0^\infty x^{b-1} f(x)dx,$$

both converge for real a and b $(a < b)$, then the Mellin transform of $f(x)$ converges uniformly to $F(s)$ in any finite domain within the infinite vertical strip $a < \mathrm{Re}(s) < b$. In such cases, the transform function $F(s)$ is analytic in this vertical strip. Also, if $x^{k-1} f(x)$ is absolutely integrable on the positive real axis for some $k > 0$, then the inversion formula (2.37) is valid for any $c > k$.

As a simple example, let us consider the Mellin transform of $f(x) = e^{-x}$,

$$\mathcal{M}\{e^{-x}\} = \int_0^\infty x^{s-1} e^{-x} dx = \Gamma(s), \quad \mathrm{Re}(s) > 0, \tag{2.38}$$

where we are recalling Eq. (2.13). The inverse Mellin transform is therefore

$$\mathcal{M}^{-1}\{\Gamma(s)\} = \frac{1}{2\pi i}\int_{c-i\infty}^{c+i\infty} x^{-s}\Gamma(s)ds = e^{-x}, \quad x > 0. \tag{2.39}$$

As another example, consider the function

$$f(x) = \begin{cases} (1-x)^{t-1}, & 0 < x < 1 \\ 0, & x \ge 1. \end{cases} \tag{2.40}$$

This time the Mellin transform leads to the beta function ($G17$) in Appendix A,

$$\mathcal{M}\{f(x)\} = \int_0^1 x^{s-1}(1-x)^{t-1} dx = B(s, t). \tag{2.41}$$

Thus, the corresponding inversion formula is given by

$$\mathcal{M}^{-1}\{B(s, t)\} = \frac{1}{2\pi i}\int_{c-i\infty}^{c+i\infty} x^{-s} B(s, t)\, ds = \frac{\Gamma(t)}{2\pi i}\int_{c-i\infty}^{c+i\infty} x^{-s}\frac{\Gamma(s)}{\Gamma(s+t)}\, ds, \tag{2.42}$$

which can be shown by the residue theory (see Section 2.4.1) to reduce to $f(x)$ as defined by Eq. (2.40).

2.4.1 Residue theory

When Cauchy's integral theorem is extended to functions having isolated singularities such as poles, the value of the integral is determined by the sum of the residues at each singularity. For example, if $f(z)$ has a *simple pole* at $z = z_0$, its residue is defined by (see Chapter 1 for a review of complex variables)

$$\text{Res}\{f(z); z_0\} = \lim_{z \to z_0} (z - z_0) f(z). \tag{2.43}$$

However, if $f(z)$ has the quotient form $f(z) = p(z)/q(z)$, where $p(z)$ and $q(z)$ are analytic at $z = z_0$ but $f(z)$ still has a simple pole at $z = z_0$, then

$$\text{Res}\{f(z); z_0\} = \lim_{z \to z_0} (z - z_0) \frac{p(z)}{q(z)} = \frac{p(z_0)}{q'(z_0)}. \tag{2.44}$$

If $f(z)$ has a pole at $z = z_0$ of order m, then

$$\text{Res}\{f(z); z_0\} = \lim_{z \to z_0} \frac{1}{(m-1)!} \frac{d^{m-1}}{dz^{m-1}} [(z - z_0)^m f(z)]. \tag{2.45}$$

Direct evaluation of the Mellin transform and/or its inverse can often be accomplished more readily by the use of the residue theory from complex variables. To illustrate, let us assume that $f(z)$ is a rational function having no poles on the positive real axis but having N poles elsewhere at $z = a_1, a_2, \ldots, a_N$. We further assume there exists real constants σ_1 and σ_2 such that

$$\begin{aligned} &\lim_{z \to 0} z^{s-1} f(z) = 0, \quad \sigma_1 < \text{Re}(s) < \sigma_2, \\ &\lim_{|z| \to \infty} z^{s-1} f(z) = 0, \quad \sigma_1 < \text{Re}(s) < \sigma_2. \end{aligned} \tag{2.46}$$

By integrating the function $z^{s-1} f(z)$ around the closed contour C shown in Fig. 2.2, which encloses all poles of $f(z)$ but doesn't include the branch point at the origin or the positive x-axis, we find by application of the residue theorem that

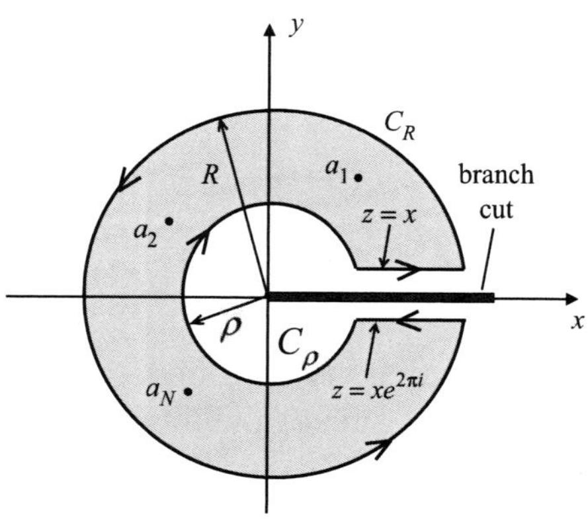

Figure 2.2 Contour of integration with branch cut (adapted from Ref. 7).

$$\oint_C z^{s-1}f(z)dz = 2\pi i \sum_{k=1}^{N} \text{Res}\{z^{s-1}f(z);\, a_k\}, \tag{2.47}$$

or, more explicitly,

$$\int_\rho^R x^{s-1}f(x)dx + \int_{C_R} z^{s-1}f(z)dz + \int_R^\rho (xe^{2\pi i})^{s-1}f(x)dx - \int_{C_\rho} z^{s-1}f(z)dz$$
$$= 2\pi i \sum_{k=1}^{N} \text{Res}\{z^{s-1}f(z);\, a_k\}. \tag{2.48}$$

Because of the limit relations in Eq. (2.46), it can be shown that the integral around the large circle C_R of radius R in Fig. 2.2 vanishes in the limit as $R \to \infty$. Similarly, the integral around the small circle C_ρ of radius ρ also vanishes as $\rho \to 0$. Hence, Eq. (2.48) reduces to

$$(1 - e^{2\pi is}) \int_0^\infty x^{s-1}f(x)dx = 2\pi i \sum_{k=1}^{N} \text{Res}\{z^{s-1}f(z);\, a_k\}. \tag{2.49}$$

However, some algebraic rearranging leads to

$$1 - e^{2\pi is} = e^{i\pi s}(e^{-i\pi s} - e^{i\pi s}) = -2ie^{i\pi s}\sin \pi s,$$

and thus, Eq. (2.49) becomes

$$\int_0^\infty x^{s-1}f(x)dx = -\frac{\pi e^{-i\pi s}}{\sin \pi s} \sum_{k=1}^{N} \text{Res}\{z^{s-1}f(z);\, a_k\}. \tag{2.50}$$

As an example of evaluating the Mellin transform based on residue theory, we consider the transform integral

$$\mathcal{M}\left\{\frac{1}{1+x^2}\right\} = \int_0^\infty \frac{x^{s-1}}{1+x^2}dx, \quad 0 < \text{Re}(s) < 2. \tag{2.51}$$

Note that the related complex function $f(z) = 1/(1+z^2)$ has simple poles at $z = \pm i$, leading to the residues

$$\begin{aligned} \text{Res}\left\{\frac{z^{s-1}}{1+z^2};\, i\right\} &= \frac{i^{s-1}}{2i} = -\frac{1}{2}e^{i\pi s/2}, \\ \text{Res}\left\{\frac{z^{s-1}}{1+z^2};\, -i\right\} &= \frac{(-i)^{s-1}}{-2i} = -\frac{1}{2}e^{3i\pi s/2}, \end{aligned} \tag{2.52}$$

where we have written $i = e^{i\pi/2}$ and $-i = e^{3i\pi/2}$. Combining the residues in Eq. (2.50), which are defined by Eq. (2.52), followed by some algebraic simplification, we obtain

$$\int_0^\infty \frac{x^{s-1}}{1+x^2}dx = \frac{\pi\cos(\pi s/2)}{\sin\pi s} = \frac{\pi}{2\sin(\pi s/2)}, \quad 0 < \mathrm{Re}(s) < 2. \tag{2.53}$$

The restriction on Re(s) is required to satisfy the limits in Eq. (2.46).

The use of residue calculus in evaluating Eq. (2.51) was based on the contour shown in Fig. 2.2. In the case of the inverse Mellin transform, however, we may need to consider more than one contour in the complex s-plane, as shown in Fig. 2.3. That is, if $F(s)$ has poles in the left-half plane, we use the contour shown in Fig. 2.3(a), which yields the inverse transform for the restricted interval $0 < x < x_0$, where x_0 is determined such that the integral along the circular arc in Fig. 2.3(a) vanishes as the radius tends to infinity. If the number of poles is infinite, this will lead to an infinite *ascending series* of the inverse transform. In some cases, the function $F(s)$ may also have poles in the right-half plane, which we then enclose by a contour, as shown in Fig. 2.3(b). This time the result will be valid for $x > x_0$, and will be a *descending series*.

To further illustrate the use of residues, let us consider evaluating the inverse Mellin transform of $\Gamma(s)$ around the contour in Fig. 2.3(a) using residue theory,

$$\mathcal{M}^{-1}\{\Gamma(s)\} = \frac{1}{2\pi i}\int_{c-i\infty}^{c+i\infty} x^{-s}\Gamma(s)ds, \quad x > 0. \tag{2.54}$$

Note that the poles of $\Gamma(s)$ all occur in the left-half plane at $s = 0, -1, -2, \ldots$. Hence, we deduce that

$$\begin{aligned}\mathcal{M}^{-1}\{\Gamma(s)\} &= \sum_{n=0}^{\infty}\mathrm{Res}\{x^{-s}\Gamma(s);\,-n\}\\ &= \sum_{n=0}^{\infty}\mathrm{Res}\left\{x^{-s}\frac{\pi}{\Gamma(1-s)\sin\pi s};\,-n\right\},\end{aligned} \tag{2.55}$$

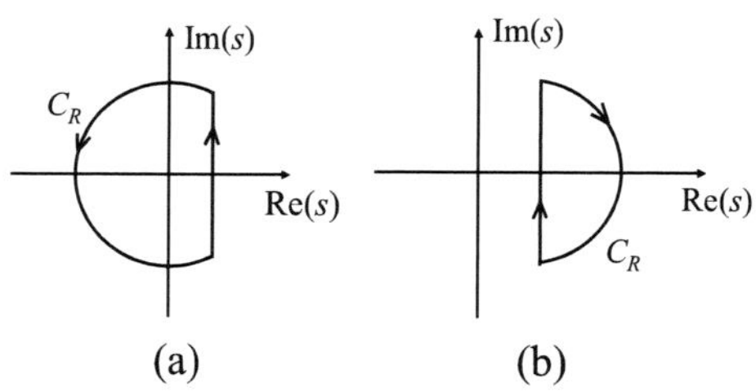

Figure 2.3 Contours of integration for inverse Mellin transforms.

where we have used the identity $\Gamma(s)\Gamma(1-s) = \pi/\sin\pi s$ from ($G7$) in Appendix A. The poles of $\Gamma(s)$ in Eq. (2.55) are simple poles in the left-half plane that correspond to the zeros of $\sin\pi s$. Hence, it follows that

$$\begin{aligned}\mathcal{M}^{-1}\{\Gamma(s)\} &= \sum_{n=0}^{\infty} \frac{x^{-s}}{\Gamma(1-s)\cos\pi s}\bigg|_{s=-n} \\ &= \sum_{n=0}^{\infty} \frac{x^n}{\Gamma(1+n)\cos(-n\pi)},\end{aligned} \tag{2.56}$$

from which we obtain

$$\mathcal{M}^{-1}\{\Gamma(s)\} = \sum_{n=0}^{\infty} \frac{(-1)^n x^n}{n!} = e^{-x}, \quad 0 \le x < \infty. \tag{2.57}$$

Other examples of the Mellin transform and its inversion formula calculated by the residue theory are provided in Refs. 6–8. Also, we illustrate another technique in Chapter 7 for evaluating the Mellin transform along with other integral transforms.

2.4.2 Convolution integral

One of the more useful properties of both the Fourier and Laplace integral transforms is that of convolution. This also applies to the Mellin transform. Convolution is often introduced in terms of finding the inverse transform of a product of transform functions. For the Mellin transform, the *convolution integral* takes the form

$$\mathcal{M}^{-1}\{F(s)G(s)\} = \int_0^{\infty} f\left(\frac{x}{u}\right) g(u) \frac{du}{u}. \tag{2.58}$$

To derive this property, we start with the Mellin transform of the right-hand side,

$$\mathcal{M}\left\{\int_0^{\infty} f\left(\frac{x}{u}\right) g(u) \frac{du}{u}\right\} = \int_0^{\infty} x^{s-1} \int_0^{\infty} f\left(\frac{x}{u}\right) g(u) \frac{du}{u}\, dx. \tag{2.59}$$

Next, we make the change of variable $x = tu$, which leads to

$$\begin{aligned}\mathcal{M}\left\{\int_0^{\infty} f\left(\frac{x}{u}\right) g(u) \frac{du}{u}\right\} &= \int_0^{\infty} t^{s-1} f(t)\, dt \cdot \int_0^{\infty} u^{s-1}\, g(u)\, du \\ &= F(s)G(s).\end{aligned} \tag{2.60}$$

By applying the inverse Mellin transform to this last result, we get Eq. (2.58).

2.5 Mellin–Barnes Integral

A *Mellin–Barnes integral*, most often called simply a *Barnes integral*, is a contour integral in the complex plane that involves a finite product of gamma functions in the numerator.[9,10] These integrals are generalizations of the inverse Mellin transform and are closely related to generalized hypergeometric functions. Among other roles, they are useful in deriving asymptotic series for both small and large arguments of special functions. In our analysis we will replace s with $-s$ and the real variable x with the complex variable z. This requires the use of a contour in the right-half plane such as that in Fig. 2.3(b) for ascending series and such as that in Fig. 2.3(a) for descending series.

An example of a typical Barnes integral is provided by

$$\begin{aligned} f(a_1, \ldots, a_p; c_1, \ldots, c_q; z) &= \lim_{n\to\infty} \frac{1}{2\pi i} \int_{B_n+C_n} \frac{\prod_{j=1}^{p} \Gamma(a_j+s)}{\prod_{j=1}^{q} \Gamma(c_j+s)} \Gamma(-s) z^s \, ds \\ &= \frac{1}{2\pi i} \int_{-i\infty}^{i\infty} \frac{\prod_{j=1}^{p} \Gamma(a_j+s)}{\prod_{j=1}^{q} \Gamma(c_j+s)} \Gamma(-s) z^s \, ds, \end{aligned} \tag{2.61}$$

where p and q are non-negative integers. In this notation, semicolons are used to separate the first set of parameters $a_1, \ldots, a_p$ from the second set of parameters $c_1, \ldots, c_q$, and the argument z in the function $f(a_1, \ldots, a_p; c_1, \ldots, c_q; z)$. Also, the symbol Π denotes a product. One particular path of integration for Eq. (2.61) is shown in Fig. 2.4, where B_n represents the vertical line along the imaginary axis, looping around the poles of $\Gamma(a_1+s), \ldots, \Gamma(a_p+s)$ to the left of the path and the poles of $\Gamma(-s)$ to

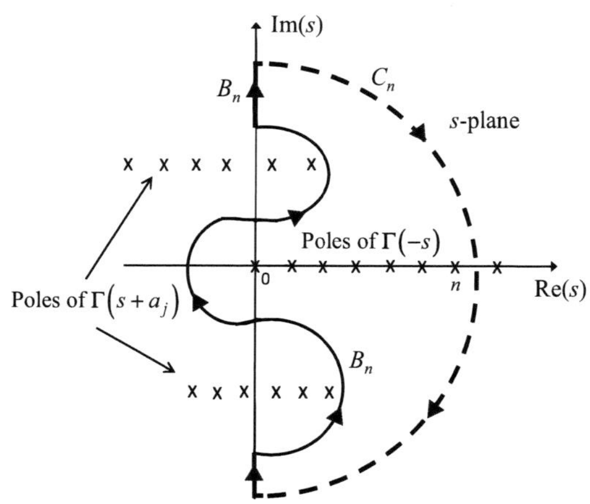

Figure 2.4 Contour of integration $B_n + C_n$ for the Barnes integral defined by Eq. (2.61). The enclosed poles of $\Gamma(-s)$ are shown on the positive real axis only up to $s = n$.

the right of the path. The semi-circle C_n in Fig. 2.4 is defined by $s = (n + \frac{1}{2})e^{i\theta}$, $-\pi/2 \le \theta \le \pi/2$. Enclosing only the poles of $\Gamma(-s)$ as the radius of C_n becomes unbounded $(n \to \infty)$ will lead to an ascending series for the function $f(a_1, \ldots, a_p; c_1, \ldots, c_q; z)$.

In our analysis we assume that the poles of $\Gamma(a_1 + s), \ldots, \Gamma(a_p + s)$ are distinct from those of $\Gamma(-s)$ and that $a_m - c_j \neq 0, 1, 2, \ldots$ for all m and j. Based on Stirling's asymptotic formula [Eq. (2.31)], it can be shown that the integral around the semi-circle C_n will vanish as the radius becomes unbounded $(n \to \infty)$. Thus, we can replace the path $B_n + C_n$ with that in Fig. 2.5, showing only B_n along the vertical axis, starting at $-i\infty$ and running to $+i\infty$, looping around the poles as necessary. If $p = q + 1$, the function defined by a Barnes integral is an analytic function in the cut plane $|\arg(z)| < \pi$. If $p = q$, the function is analytic in the half-plane $|\arg(z)| < \pi/2$. In general, the integral converges absolutely to an analytic function if

$$|\arg(z)| < (1 + p - q)\frac{\pi}{2}. \tag{2.62}$$

In Chapter 3 we will show that functions like $f(a_1, \ldots, a_p; c_1, \ldots, c_q; z)$ defined by Eq. (2.61) are related to the class of generalized hypergeometric functions denoted by ${}_pF_q$ [e.g., see Eq. (3.3)].

One of the simplest examples of a function like those defined in Eq. (2.61) is given by the exponential function [recall Eq. (2.39)]:

$$e^{-z} = \frac{1}{2\pi i}\int_{-i\infty}^{i\infty} \Gamma(-s)z^s ds. \tag{2.63}$$

Here we see that no gamma functions appear other than $\Gamma(-s)$, which we denote by $p = q = 0$ in Eq. (2.61). Hence, it can be shown that

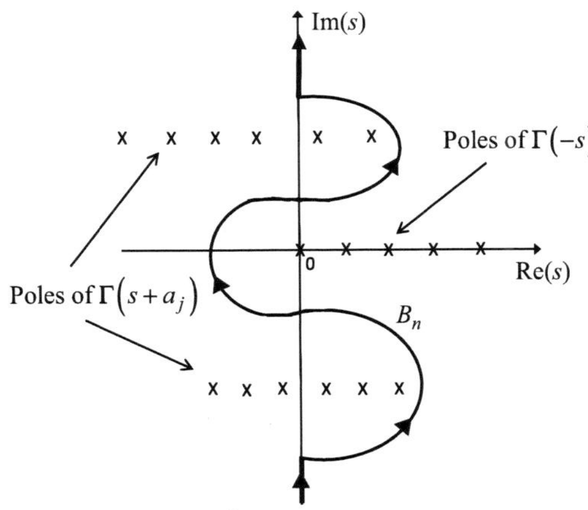

Figure 2.5 Integration path B_n from $-i\infty$ to $+i\infty$ for the Barnes integral (2.61).

$$f(-;\,-;\,z) = {}_0F_0(-;\,-;\,-z) = e^{-z}. \tag{2.64}$$

It is customary to use a dash in the absence of a parameter, as we have illustrated in Eq. (2.64).

The function ${}_0F_0$ appearing in Eq. (2.64) is a simple example of a generalized hypergeometric function. In the next chapter we will explore various properties of other generalized hypergeometric functions ($p \geq 0,\ q \geq 0$) by using the Barnes integral representation as introduced here. In particular, we will concentrate on developing asymptotic behavior for both small and large arguments of these functions.

2.6 Zernike Polynomials

Zernike polynomials, first introduced by F. Zernike in 1934, have been extensively used to describe fixed aberrations in optical systems (Ref. 11, pp. 523–527). In particular, they provide a mathematical way to describe the phase of an optical wave. These polynomials represent a set of functions of two variables that are orthogonal over a circle with unit radius. It is customary to define the polynomials as a product of two functions—one depending only on a radial variable r and the other depending only on an azimuthal (angular) variable θ. The complex Zernike polynomials are defined by

$$Z_n^m(r,\,\theta) = R_n^m(r)e^{im\theta}, \tag{2.65}$$

where both m and n are integers, $n \geq 0$, $-n \leq m \leq n$, $n \pm |m|$ is even, and

$$R_n^m(r) = \sum_{k=0}^{(n-|m|)/2} \frac{(-1)^k(n-k)!}{\left(\dfrac{n+m}{2}-k\right)!\left(\dfrac{n-m}{2}-k\right)!}\frac{r^{n-2k}}{k!}. \tag{2.66}$$

Clearly, the polynomial $R_n^m(r)$ has degree n and contains no power of r less than $|m|$, where m is the angular dependence. Also, the polynomial is even if m is even and odd if m is odd. The orthogonality property of the Zernike polynomials is

$$\int_0^1\int_0^{2\pi} Z_n^m(r,\,\theta)Z_k^{m\,*}(r,\,\theta)r\,d\theta\,dr = \begin{cases} \dfrac{\pi}{n+1}, & k = n \\ 0, & k \neq n, \end{cases} \tag{2.67}$$

where * denotes complex conjugate.

2.6.1 Application in optics

One of the principal uses of the Zernike polynomials is to represent fixed aberrations in optical systems in the form of a generalized Fourier series.

In doing so, researchers have been able to study how aberrations affect various optical systems. They are also useful in adaptive optics systems designed for atmospheric turbulence decomposition (see Section 9.2). The lower-order polynomials are called piston, tilt, focus, astigmatism, coma, and so forth.

Virtually any realistic wavefront $\Phi(r, \theta)$ can be represented in a two-dimensional series of Zernike polynomials by the following expression,

$$\Phi(r, \theta) = \sum_{n=0}^{\infty} \sum_{m=-n}^{n} C_{mn} R_n^m(r) e^{im\theta}, \tag{2.68}$$

where C_{mn} are the Fourier coefficients. They are obtained by multiplying both sides of Eq. (2.68) by $rR_n^{|m|}(r)e^{-im\theta}$ and then integrating over the unit disk. For example, when $m = n = 0$, we obtain the coefficient for piston, viz.,

$$C_{00} = \frac{1}{\pi} \int_0^1 \int_0^{2\pi} r\,\Phi(r, \theta)\, d\theta\, dr = \langle \Phi(r, \theta) \rangle. \tag{2.69}$$

The brackets $\langle\,\rangle$ indicates an average value of the wave (phase) front.

A useful property of the Zernike series in Eq. (2.68) is that the root-mean-square (rms) wavefront error due to primary (lower-order) aberrations can be readily calculated. That is, if all of the Fourier coefficients are known, the geometric sum of the non-piston terms yields the wavefront variance

$$(\Delta\Phi)^2 = \frac{1}{\pi} \int_0^1 \int_0^{2\pi} r[\Phi(r, \theta) - \langle \Phi(r, \theta) \rangle]^2 \, d\theta\, dr = \sum_{n=1}^{\infty} \sum_{m=-n}^{n} \frac{|C_{mn}|^2}{n+1}, \tag{2.70}$$

the square root of which yields the *rms wavefront error.*

2.7 Applications

In this last section we wish to illustrate the application of topics introduced in this chapter. Several examples of the Cauchy product are presented in Chapter 7, so we refer the interested reader to Section 7.2. Also, applications of the Zernike polynomials and related Zernike filter functions in atmospheric turbulence are presented in Chapter 9.

2.7.1 Asymptotic series

In Section 2.3.2 we developed Stirling's formula for the gamma function. Let us now develop a large-argument $(x \to \infty)$ asymptotic series for another function defined by an integral.

Example 1: Find a large-argument asymptotic series for

$$f(x)=\int_0^\infty \frac{e^{-xt}}{1+t^2}dt, \quad x>0.$$

We start by making the change of variable $s=xt$, which leads to the expression

$$f(x)=\frac{1}{x}\int_0^\infty e^{-s}\left(1+\frac{s^2}{x^2}\right)^{-1}ds. \tag{2.71}$$

For $x>s$, we can use the geometric series in Eq. (2.1) to obtain

$$\left(1+\frac{s^2}{x^2}\right)^{-1}=\sum_{n=0}^\infty (-1)^n\frac{s^{2n}}{x^{2n}}. \tag{2.72}$$

After inserting this series into the integral and using termwise integration, we arrive at

$$\begin{aligned} f(x) &\sim \sum_{n=0}^\infty (-1)^n\frac{1}{x^{2n+1}}\int_0^\infty e^{-s}s^{2n}ds \\ &\sim \sum_{n=0}^\infty (-1)^n\frac{(2n)!}{x^{2n+1}}, \quad x\to\infty, \end{aligned} \tag{2.73}$$

where the last step follows from the integral defining the gamma function. Although a finite number of terms from Eq. (2.73) may provide a good approximation for large values of x in the above integral, the asymptotic series in Eq. (2.73) *diverges* for all finite values of x. The reason for this divergence is that the series (2.72) requires $s<x$, but we are allowing s to be arbitrarily large in Eq. (2.73). This is a common occurrence in asymptotic expansions of this kind.

2.7.2 The Mellin transform

It is known that the Mellin transform is an efficient tool for obtaining the distribution function of products of independent random variables.[12] For example, consider the product $\mathbf{z}=\mathbf{xy}$, where $\mathbf{x}$ and $\mathbf{y}$ are real and positive independent random variables (bold letters denote random variables). The probability density function (PDF) of $\mathbf{z}$ is related to the PDFs of $\mathbf{x}$ and $\mathbf{y}$ by (Ref. 7, p. 677)

$$p_{\mathbf{z}}(z)=\frac{d}{dz}\int_0^\infty\int_0^{z/x} p_{\mathbf{x}}(x)p_{\mathbf{y}}(y)\,dy\,dx=\int_0^\infty p_{\mathbf{x}}(x)p_{\mathbf{y}}\left(\frac{z}{x}\right)\frac{dx}{x}. \tag{2.74}$$

We recognize the last integral in Eq. (2.74) as the convolution integral (2.58) of the Mellin transform. Hence, we can also write Eq. (2.74) as

$$p_{\mathbf{z}}(z) = \mathcal{M}^{-1}\{F(s)G(s)\}, \tag{2.75}$$

where $F(s)$ and $G(s)$ are the Mellin transforms of the PDFs $p_{\mathbf{x}}(x)$ and $p_{\mathbf{y}}(y)$, respectively.

Example 2: Find the PDF for the product $\mathbf{z} = \mathbf{xy}$, where $\mathbf{x}$ and $\mathbf{y}$ have PDFs defined by

$$p_{\mathbf{x}}(x) = \frac{2}{\pi\sqrt{1-x^2}}, \quad 0 < x < 1; \quad p_{\mathbf{y}}(y) = \frac{y}{\sigma^2}e^{-y^2/2\sigma^2}, \quad y > 0.$$

The Mellin transform of $p_{\mathbf{x}}(x)$ is

$$\mathcal{M}\{p_{\mathbf{x}}(x)\} = \frac{2}{\pi}\int_0^1 x^{s-1}(1-x^2)^{-1/2}dx = \frac{1}{\sqrt{\pi}}\frac{\Gamma\left(\frac{s}{2}\right)}{\Gamma\left(\frac{s+1}{2}\right)}, \tag{2.76}$$

whereas

$$\mathcal{M}\{p_{\mathbf{y}}(y)\} = \frac{1}{\sigma^2}\int_0^\infty y^s e^{-y^2/2\sigma^2}dy = \frac{1}{\sqrt{2}\,\sigma}(2\sigma^2)^{s/2}\Gamma\left(\frac{s+1}{2}\right). \tag{2.77}$$

The integral in Eq. (2.76) was evaluated by setting $x = \sqrt{t}$ and recalling the beta function ($G17$) in Appendix A. For the integral in Eq. (2.77), we used ($D2$) in Appendix D. By calculating the inverse Mellin transform of the product of transforms using the residue theory, we have

$$\begin{aligned} p_{\mathbf{z}}(z) &= \mathcal{M}^{-1}\left\{\frac{\Gamma\left(\frac{s}{2}\right)}{\sqrt{2\pi}\,\sigma}(2\sigma^2)^{s/2}\right\} \\ &= \frac{1}{\sqrt{2\pi}\,\sigma}\frac{1}{2\pi i}\int_{c-i\infty}^{c+i\infty} z^{-s}(2\sigma^2)^{s/2}\Gamma\left(\frac{s}{2}\right)ds \\ &= \frac{1}{\sqrt{2\pi}\,\sigma}\sum_{n=0}^{\infty}\mathrm{Res}\left\{z^{-s}(2\sigma^2)^{s/2}\Gamma\left(\frac{s}{2}\right);\ -2n\right\} \\ &= \frac{1}{\sqrt{2\pi}\,\sigma}\sum_{n=0}^{\infty}\mathrm{Res}\left\{\frac{\pi z^{-s}(2\sigma^2)^{s/2}}{\Gamma(1-s/2)\sin(\pi s/2)};\ -2n\right\}, \end{aligned} \tag{2.78}$$

where we have identified the poles at $s = -2n$, $n = 0, 1, 2, \ldots$ and used ($G7$) in Appendix A. Last, by calculating the residues in Eq. (2.78), we arrive at

$$p_{\mathbf{z}}(z) = \frac{2}{\sqrt{2\pi}\,\sigma}\sum_{n=0}^{\infty}\frac{(-1)^n}{n!}\left(\frac{z^2}{2\sigma^2}\right)^n$$
$$= \frac{1}{\sigma}\sqrt{\frac{2}{\pi}}\exp\left(-\frac{z^2}{2\sigma^2}\right), \quad z > 0. \tag{2.79}$$

Equation (2.79) is recognized as a *one-sided Gaussian distribution.* It will become the standard Gaussian distribution over $-\infty < z < \infty$ if we extend the definition of $p_{\mathbf{x}}(x)$ to the interval $-1 < x < 1$ and thus divide Eq. (2.79) by 2.

Another form of the Mellin convolution integral is provided by

$$\mathcal{M}^{-1}\{F(s)G(2-s)\} = \int_0^{\infty} f(xu)g(u)u\,du. \tag{2.80}$$

The proof of this result is very similar to that of the convolution integral defined by Eq. (2.58). The identity (2.80) is useful in calculating the PDF of quotients like $\mathbf{z} = \mathbf{x}/\mathbf{y}$, where $\mathbf{x}$ and $\mathbf{y}$ are independent positive random variables.

Example 3: Determine the PDF for the quotient $\mathbf{z} = \mathbf{x}/\mathbf{y}$ given that $\mathbf{x}$ and $\mathbf{y}$ are independent random variables with marginal PDFs

$$p_{\mathbf{x}}(x) = ae^{-ax},\ x > 0, \quad p_{\mathbf{y}}(y) = be^{-by},\ y > 0, \quad a, b > 0.$$

We start with the relation from probability theory:

$$p_{\mathbf{z}}(z) = \frac{d}{dz}\int_0^{\infty}\int_0^{yz} p_{\mathbf{x}}(x)p_{\mathbf{y}}(y)\,dx\,dy = \int_0^{\infty} p_{\mathbf{x}}(yz)p_{\mathbf{y}}(y)y\,dy. \tag{2.81}$$

Hence, it follows from Eqs. (2.80) and (2.81) that

$$p_{\mathbf{z}}(z) = \mathcal{M}^{-1}\{F(s)G(2-s)\}, \tag{2.82}$$

where $F(s)$ and $G(s)$ are the Mellin transforms of the PDFs $p_{\mathbf{x}}(x)$ and $p_{\mathbf{y}}(y)$, respectively. In this case we use ($G19$) in Appendix A to find that

$$\mathcal{M}\{ae^{-ax}\} = F(s) = \frac{\Gamma(s)}{a^{s-1}};$$
$$G(2-s) = b^{s-1}\Gamma(2-s). \tag{2.83}$$

From Eq. (2.82), we now obtain

$$p_{\mathbf{z}}(z) = \mathcal{M}^{-1}\{F(s)G(2-s)\}$$
$$= \mathcal{M}^{-1}\left\{\left(\frac{b}{a}\right)^{s-1}\Gamma(s)\Gamma(2-s)\right\}$$
$$= \sum_{n=0}^{\infty}\text{Res}\left\{\left(\frac{b}{a}\right)^{s-1}\frac{\pi\Gamma(2-s)z^{-s}}{\Gamma(1-s)\sin\pi s}; -n\right\}, \tag{2.84}$$

where we have used ($G7$) in Appendix A. After identifying simple poles of $\Gamma(s)$ at $s = -n$ ($n = 0, 1, 2, \ldots$), this last expression leads to

$$p_{\mathbf{z}}(z) = \frac{a}{b}\sum_{n=0}^{\infty}\frac{(-1)^n(2)_n}{n!}\left(\frac{az}{b}\right)^n = \frac{a}{b}\sum_{n=0}^{\infty}\binom{-2}{n}\left(\frac{az}{b}\right)^n. \tag{2.85}$$

To reach this result, we used ($PS9$) in Appendix A to recognize the binomial series from calculus. Our final result is therefore

$$\begin{aligned} p_{\mathbf{z}}(z) &= \frac{a}{b}(1 + az/b)^{-2} \\ &= \frac{ab}{(b+az)^2}, \quad z > 0. \end{aligned} \tag{2.86}$$

Example 4: Use the residue theory to calculate $\mathcal{M}^{-1}\left\{\frac{\pi\cos(s\phi)}{\sin(s\pi)}\right\}, \ -\frac{\pi}{2} < \phi < \frac{\pi}{2}$.

We start by noting that simple poles exist at $s = \pm n$, $n = 0, 1, 2, \ldots$. However, because $\mathrm{Re}(s) > 0$, we only need to consider non-negative n values. Hence, it follows that

$$\begin{aligned} \mathcal{M}^{-1}\left\{\frac{\pi\cos(s\phi)}{\sin(s\pi)}\right\} &= \frac{\pi}{2\pi i}\int_{c-i\infty}^{c+i\infty} x^{-s}\frac{\cos(s\phi)}{\sin(s\pi)}\,ds \\ &= \pi\sum_{n=0}^{\infty}\mathrm{Res}\left\{x^{-s}\frac{\cos(s\phi)}{\sin(s\pi)};\ n\right\} \\ &= \sum_{n=0}^{\infty}x^{-n}\frac{\cos(n\phi)}{\cos(n\pi)}. \end{aligned} \tag{2.87}$$

To sum the series, we first write $\cos(n\phi) = \mathrm{Re}(e^{-in\phi})$, from which it follows that

$$\begin{aligned} \mathcal{M}^{-1}\left\{\frac{\pi\cos(s\phi)}{\sin(s\pi)}\right\} &= \mathrm{Re}\sum_{n=0}^{\infty}(-1)^n\left(\frac{e^{-i\phi}}{x}\right)^n \\ &= \mathrm{Re}\left(\frac{1}{1+xe^{i\phi}}\right) \\ &= \mathrm{Re}\left[\frac{1}{(1+x\cos\phi)+ix\sin\phi}\right], \end{aligned} \tag{2.88}$$

where we recognize the series as the geometric series in Eq. (2.1). If we now multiply the numerator and denominator by the complex conjugate of the denominator and simplify, we are led to

$$\mathcal{M}^{-1}\left\{\frac{\pi\cos(s\phi)}{\sin(s\pi)}\right\} = \frac{1+x\cos\phi}{1+2x\cos\phi+x^2}. \tag{2.89}$$

2.8 Historical Remarks

The Scottish mathematician Colin Maclaurin (1698–1746) is known for his contributions to geometry and algebra. He also contributed to the study of elliptic integrals, although he is best known for the power series now bearing his name. Maclaurin series are a specialization of Taylor series, named after the British mathematician Brook Taylor (1685–1731). Taylor is also known for inventing an area of mathematics called the "calculus of finite differences" and for introducing integration by parts.

In the eighteenth century, the Swiss mathematician, physicist, and astronomer Leonhard Euler (1707–1783) concerned himself with the problem of interpolating between the factorials $n!, n = 0, 1, 2, \ldots$, which led to the gamma function. His result gave meaning to $x!$ when x is not a positive integer. The notation "Γ" that is now widely accepted for the gamma function is not due to Euler, however, but was introduced by the French mathematician Adrien Marie Legendre (1752–1833), who was also responsible for the duplication formula of the gamma function. Another useful function founded by Euler is the beta function, often called the Eulerian integral of the first kind. Forty-three years later, Euler found that the beta function was actually a particular combination of gamma functions. Euler also developed the theory of trigonometric functions, made significant contributions to analytic geometry and number theory, introduced the symbol "e" for the base of the natural logarithm, and presented a theory for logarithms of complex numbers. Euler's work in mathematics is so vast that his collected works fill more than 70 volumes—he is considered by many to be the most prolific writer of mathematics of all time. Nearly 150 years after Euler's discovery of the gamma function, the theory concerning the gamma function was greatly expanded by means of the theory of entire functions developed by the German mathematician Karl Theodor Wilhelm Weierstrass (1815–1897). The Scottish mathematician Sir James Frazer Stirling (1692–1770) is perhaps most noted for introducing in 1730 the asymptotic formula for $n!$ now known as Stirling's formula. The Pochhammer symbol is named after the Prussian mathematician Leo August Pochhammer (1841–1920), who used this notation for the rising factorial. The French mathematician Jules Henri Poincaré (1854–1912) made many contributions to pure and applied mathematics. Among other contributions, he was one of the founders of topology and chaos theory and is credited with providing a precise definition of asymptotic series in 1886.

The Mellin transform was introduced by the Finnish mathematician Hjalmar Mellin (1854–1933), whose research was primarily in the area of the theory of functions. His integral transform was utilized in his study of the gamma function and hypergeometric function. He also applied his transform to solving partial differential equations and the inverse transform for developing asymptotic series expansions. Ernest William Barnes (1874–1953) was a British mathematician whose early mathematical work was concerned with the gamma function. Later he turned his attention to integral functions that led to his fascination with the integrals of Mellin and eventually to the Mellin–Barnes integral for generalized hypergeometric functions. Like many other early mathematicians, Barnes also became heavily involved in ministry throughout his career.

References

1. A. Erdélyi, W. Magnus, F. Oberhettinger, and F. G. Tricomi, *Higher Transcendental Functions*, Vols. I and II, Bateman Manuscript Project, McGraw Hill, New York (1953).
2. E. D. Rainville, *Special Functions*, Academic Press, New York (1960).
3. L. C. Andrews, *Special Functions of Mathematics for Engineers*, 2nd Edition, SPIE Press, Bellingham, Washington (1998) [doi: 10.1117/3.270709].
4. L. C. Andrews, *Field Guide to Special Functions for Engineers*, SPIE Press, Bellingham, Washington (2011) [doi: 10.1117/3.885484].
5. F. W. J. Olver, *Asymptotics and Special Functions*, 2nd Edition, A K Peters/CRC Press, New York (1997).
6. L. C. Andrews and B. K. Shivamoggi, *Integral Transforms for Engineers*, SPIE Press, Bellingham, Washington (1999) [doi: 10.1117/3.339204].
7. L. C. Andrews and R. L. Phillips, *Mathematical Techniques for Engineers and Scientists*, SPIE Press, Bellingham, Washington (2003) [doi: 10.1117/3/467443].
8. I. N. Sneddon, *The Use of Integral Transforms*, McGraw Hill, New York (1972).
9. E. W. Barnes, "A new development of the theory of the hypergeometric function," *Proc. London Math. Soc.* Volume **s2-6**(1), 141–177 (1908).
10. E. W. Barnes, "A transformation of generalized hypergeometric series," *Quarterly Journal of Math.* **41**, 136–140 (1910).
11. M. Born and E. Wolf, *Principles of Optics*, 7th Edition, Cambridge University Press, New York (1999).
12. M. D. Springer and W. E. Thompson, "The distribution of products of beta, gamma, and Gaussian random variables," *SIAM J. Appl. Math.* **18**, 721–737 (1970).

Chapter 3
Generalized Hypergeometric Functions

3.0 Introduction

The first systematic study of generalized hypergeometric functions was carried out by C. F. Gauss[1] on the hypergeometric series and published in his memoir of 1812. Years later, E. E. Kummer[2] developed the confluent hypergeometric function of the first kind and published his work in 1836. Still today, these latter functions are often referred to as *Kummer functions* by some authors. Other generalized hypergeometric functions were studied throughout the late nineteenth century and first half of the twentieth century. The importance of studying generalized hypergeometric functions is the fact that various other special functions and elementary functions are adaptations of them (cf. Appendices B and C), and thus each property that is developed becomes a master formula for many of these other functions. Also, the use of generalized hypergeometric functions often facilitates the analysis by permitting complex expressions such as products of functions to be represented more simply in terms of a single generalized function.

3.1 Generalized Hypergeometric Series

Although hypergeometric-type functions often arise as solutions of a differential equation,[3,4] they also arise in other settings—in particular, as solutions of certain integrals. The most common of these functions is the *hypergeometric function*, which is normally introduced by its power series:

$$\begin{aligned}{}_2F_1(a, b; c; z) &= 1 + \frac{ab}{c}\frac{z}{1!} + \frac{a(a+1)b(b+1)}{c(c+1)}\frac{z^2}{2!} + \cdots \\ &= \sum_{n=0}^{\infty} \frac{(a)_n (b)_n}{(c)_n}\frac{z^n}{n!}, \quad |z| < 1.\end{aligned} \tag{3.1}$$

The parameters a, b, and c may be real or complex but are generally considered real numbers with the restriction $c \neq 0, -1, -2, \ldots$. Also, the argument z can be real or complex. The infinite series in Eq. (3.1) converges for $|z| < 1$ and will also converge on the boundary $|z| = 1$ if $\mathrm{Re}(c - a - b) > 0$. The hypergeometric function is related to several other special functions and some elementary functions such as $1/(1 - z)$, $\sin^{-1}z$, and $\ln(1 - z)$, among others. The hypergeometric series (3.1) gets its name because, for $a = 1$ and $b = c$, it reduces to the elementary *geometric series* defined by Eq. (2.1).

Another function that occurs in many applications and is defined by a similar power series is the *confluent hypergeometric function of the first kind*:

$$ {}_1F_1(a; c; z) = \sum_{n=0}^{\infty} \frac{(a)_n}{(c)_n} \frac{z^n}{n!}, \quad |z| < \infty. \tag{3.2}$$

Again, it is assumed that $c \neq 0, -1, -2, \ldots$ and, as before, the argument z can be real or complex. Because the series converges for all finite z values in the complex plane, the confluent hypergeometric function is an *entire function.*

The functions defined by Eqs. (3.1) and (3.2) are special cases of the *generalized hypergeometric function*, which has the power series representation

$$ {}_pF_q(a_1, \ldots, a_p; c_1, \ldots, c_q; z) = \sum_{n=0}^{\infty} \frac{(a_1)_n \cdots (a_p)_n}{(c_1)_n \cdots (c_q)_n} \frac{z^n}{n!}, \tag{3.3}$$

where p and q are non-negative integers and no c_k $(k = 1, 2, \ldots, q)$ value is zero or a negative integer. If a numerator parameter $a_j (j = 1, \ldots, p)$ is zero or a negative integer, the series terminates at that point. Provided the series does not terminate, it can be established by the ratio test of calculus that

- If $p \leq q$, the series *converges* for all $|z| < \infty$.
- If $p = q + 1$, the series *converges* for $|z| < 1$ and *diverges* for $|z| > 1$.
- If $p > q + 1$, the series *diverges* for all z except $z = 0$.

Based on the third bullet above, the series for ${}_pF_q$ is therefore meaningful in the case when $p > q + 1$ only if it terminates. Attempts over many years to give meaning to the function ${}_pF_q$ when $p > q + 1$ finally led to further generalizations known as the *MacRobert E-function* and the *Meijer G-function* (see Chapter 5).

The Barnes integral introduced by Eq. (2.61) in Chapter 2 turns out to be a representation of hypergeometric-type functions that for $p = q + 1$ can extend the definition of these functions beyond their power series representation. In this chapter we develop the Barnes integral for generalized hypergeometric functions that will be used to derive asymptotic results for both small and large arguments of several of these generalized hypergeometric functions and related functions.

3.2 The Function ${}_1F_0$

Let us begin by considering the Barnes integral definition (2.61) for the special case when $p = 1$ and $q = 0$. Thus, we have

$$f(a;\,-;\,z) = \frac{1}{2\pi i}\int_{-i\infty}^{i\infty}\Gamma(a+s)\Gamma(-s)z^s ds, \qquad |\arg(z)| < \pi, \tag{3.4}$$

where it is assumed that the poles of $\Gamma(a+s)$ at $s = -(a+n)$, $n = 0, 1, 2, \ldots$ are all distinct from those of $\Gamma(-s)$, which occur at $s = 0, 1, 2, \ldots$.

By use of the residue calculus, we first evaluate Eq. (3.4) for a contour like that in Fig. 2.4—around only (and all of) the poles of $\Gamma(-s)$,—from which we get in the limit as the radius of C_n becomes infinite:

$$\begin{aligned} f(a;\,-;\,z) &= -\sum_{n=0}^{\infty}\text{Res}\{\Gamma(a+s)\Gamma(-s)z^s;\, n\} \\ &= \sum_{n=0}^{\infty}\text{Res}\left\{\frac{\pi\Gamma(a+s)z^s}{\Gamma(s+1)\sin\pi s};\, n\right\}. \end{aligned} \tag{3.5}$$

The negative sign in front of the first summation is due to integrating in the negative direction along the closed contour in Fig. 2.4. Also, we have used the identity $\Gamma(-s)\Gamma(1+s) = -\pi/\sin\pi s$ from (*G7*) in Appendix A to obtain the second line in Eq. (3.5). Computing the residues, we obtain the *ascending series*

$$\begin{aligned} f(a;\,-;\,z) &= \sum_{n=0}^{\infty}\frac{\Gamma(a+n)z^n}{\Gamma(n+1)\cos n\pi} \\ &= \Gamma(a)\sum_{n=0}^{\infty}(-1)^n(a)_n\frac{z^n}{n!}, \qquad |z| < 1, \end{aligned} \tag{3.6}$$

which, based on Eq. (3.3), can be written equivalently as

$$f(a;\,-;\,z) = \Gamma(a)\,{}_1F_0(a;\,-;\,-z). \tag{3.7}$$

In arriving at Eqs. (3.6) and (3.7), we used the Pochhammer symbol definition given by Eq. (2.23) and recognized the series from Eq. (3.3) as that of the generalized hypergeometric function ${}_1F_0$.

Now, by using the identity (*PS9*) in Appendix A, written as

$$(a)_n = (-1)^n\binom{-a}{n}n!,$$

the series in Eq. (3.6) can be expressed in a more familiar form,

$$\sum_{n=0}^{\infty}(-1)^n(a)_n\frac{z^n}{n!}=\sum_{n=0}^{\infty}\binom{-a}{n}z^n=\frac{1}{(1+z)^a}, \tag{3.8}$$

which is the *binomial series*. Thus, it follows that

$${}_1F_0(a;\,-;\,-z)=\frac{1}{(1+z)^a}, \qquad z\neq -1. \tag{3.9}$$

For small arguments, we can simply use the first few terms of the series in Eq. (3.8) to obtain the asymptotic result (where N is finite)

$${}_1F_0(a;\,-;\,-z)\sim 1-az+\cdots+\frac{(-1)^N}{N!}(a)_N z^N, \qquad z\to 0. \tag{3.10}$$

Clearly, the function in Eq. (3.9) is well defined for all $z\neq -1$ even though its series is restricted to $|z|<1$. And although the function in Eq. (3.9) is elementary and its properties are well known from calculus, we find it instructive to further pursue the analysis of this function from the Barnes integral point of view. In particular, we want to develop its *descending asymptotic series* for large arguments.

3.2.1 Large-argument asymptotic series

The generalized hypergeometric function identified in Eq. (3.7) can now be expressed in terms of the Barnes integral from Eq. (3.4), i.e.,

$${}_1F_0(a;\,-;\,-z)=\frac{1}{\Gamma(a)}\frac{1}{2\pi i}\int_{-i\infty}^{i\infty}\Gamma(a+s)\,\Gamma(-s)z^s ds. \tag{3.11}$$

This formulation represents an analytic function of z in the cut plane $|\arg(z)|<\pi$. As previously pointed out, the series in Eq. (3.6) is not valid for $|z|>1$ so we use the Barnes integral in Eq. (3.11) to investigate the behavior of the function ${}_1F_0$ for large arguments. We start with the change of variable $s\to -(s+a)$,

$$\begin{aligned}{}_1F_0(a;\,-;\,-z)&=\frac{1}{\Gamma(a)}\frac{1}{2\pi i}\int_{-i\infty}^{i\infty}\Gamma(-s)\Gamma(a+s)\,z^{-(s+a)}ds\\ &=\frac{z^{-a}}{\Gamma(a)}\frac{1}{2\pi i}\int_{-i\infty}^{i\infty}\Gamma(a+s)\,\Gamma(-s)\left(\frac{1}{z}\right)^s ds,\end{aligned} \tag{3.12}$$

where the contour path again encloses the poles of $\Gamma(-s)$ transferred from the change of variable, i.e., $\Gamma(a+s)|_{s\to-(s+a)}=\Gamma(-s)$. Based on Eqs. (3.11) and (3.12), we have just derived the identity

$$ {}_1F_0(a;\,-;\,-z) = z^{-a}\sum_{n=0}^{\infty}(-1)^n(a)_n\frac{z^{-n}}{n!} = z^{-a}{}_1F_0\left(a;\,-;\,-\frac{1}{z}\right). \tag{3.13} $$

Notice that the change of variable $s \to -(s+a)$ causes a change in z^n to $1/z^n$. Also, the series representation (3.6) for the function on the left is restricted to $|z| < 1$, whereas the power series representation for the expression on the right converges for $|z| > 1$. Moreover, Eq. (3.13) has led to the obvious identity

$$ (1+z)^{-a} = z^{-a}\left(1+\frac{1}{z}\right)^{-a}, \quad z \neq -1. \tag{3.14} $$

Taking the first few terms of the power series in Eq. (3.13) produces the finite asymptotic series

$$ {}_1F_0(a;\,-;\,-z) \sim z^{-a}\left(1-\frac{a}{z}+\cdots+\frac{(-1)^N(a)_N}{N!\,z^N}\right), \quad z \to \infty. \tag{3.15} $$

The above Barnes integral exercise has led to the asymptotic series in Eq. (3.15) for ${}_1F_0(a;\,-;\,-z)$ as $z \to \infty$. Although the results derived here are trivial because of the simple nature of the elementary function in question, this analysis provides the path for converting other generalized hypergeometric functions whose argument is $\pm z$ into one with argument $\pm 1/z$.

3.3 The Function ${}_1F_1$

In the literature on special functions, it is common practice to introduce them by their power series. For example, we introduced the confluent hypergeometric function by the power series (3.2). From this power series, it can be shown that the function $w(z) = {}_1F_1(a;\,c;\,z)$ is a solution of the differential equation (DE)

$$ z\frac{d^2w}{dz^2} + (c-z)\frac{dw}{dz} - aw = 0. \tag{3.16} $$

However, such functions may also arise as solutions of certain integrals. In the case of the ${}_1F_1$ function, we can derive such an integral from the series itself. To see this, let us recall the beta function identity ($G17$) in Appendix A:

$$ \frac{\Gamma(x)\Gamma(y)}{\Gamma(x+y)} = \int_0^1 t^{x-1}(1-t)^{y-1}dt, \quad x > 0,\, y > 0. \tag{3.17} $$

By setting $x = n + a$ and $y = c - a$, it follows that

$$\frac{(a)_n}{(c)_n} = \frac{\Gamma(c)}{\Gamma(a)\Gamma(c-a)} \int_0^1 t^{n+a-1}(1-t)^{c-a-1} dt, \quad c > a > 0. \tag{3.18}$$

Thus, using Eq. (3.18) to replace $(a)_n/(c)_n$ in Eq. (3.2), we have

$$\begin{aligned} {}_1F_1(a;\, c;\, z) &= \frac{\Gamma(c)}{\Gamma(a)\Gamma(c-a)} \sum_{n=0}^{\infty} \frac{z^n}{n!} \int_0^1 t^{n+a-1}(1-t)^{c-a-1} dt \\ &= \frac{\Gamma(c)}{\Gamma(a)\Gamma(c-a)} \int_0^1 \left[\sum_{n=0}^{\infty} \frac{(zt)^n}{n!} \right] t^{a-1}(1-t)^{c-a-1} dt, \end{aligned} \tag{3.19}$$

where we interchanged the order of summation and integration in the second step. The infinite sum in Eq. (3.19) is an exponential function [recall Eq. (2.2)] that leads to the integral representation

$${}_1F_1(a;\, c;\, z) = \frac{\Gamma(c)}{\Gamma(a)\Gamma(c-a)} \int_0^1 e^{zt} t^{a-1}(1-t)^{c-a-1} dt, \quad c > a > 0. \tag{3.20}$$

Replacing t with $1-t$ in Eq. (3.20) yields the additional integral representation

$${}_1F_1(a;\, c;\, z) = \frac{\Gamma(c)e^z}{\Gamma(a)\Gamma(c-a)} \int_0^1 e^{-zt} t^{c-a-1}(1-t)^{a-1} dt, \quad c > a > 0. \tag{3.21}$$

3.3.1 The Barnes integral

Let us now consider the Barnes integral representation from Eq. (2.61),

$$f(a;\, c;\, z) = \frac{1}{2\pi i} \int_{-i\infty}^{i\infty} \frac{\Gamma(a+s)}{\Gamma(c+s)} \Gamma(-s) z^s ds, \quad |\arg(z)| < \frac{\pi}{2}, \tag{3.22}$$

which features one gamma function in the numerator [besides $\Gamma(-s)$] and one gamma function in the denominator (i.e., $p = q = 1$). Based on the residue calculus for simple poles at $s = n,\ n = 0,\ 1,\ 2,\ \ldots,$ we first find that

$$\begin{aligned} f(a;\, c;\, z) &= -\sum_{n=0}^{\infty} \operatorname{Res}\left\{ \frac{\Gamma(a+s)}{\Gamma(c+s)} \Gamma(-s) z^s;\, n \right\} \\ &= \sum_{n=0}^{\infty} \operatorname{Res}\left\{ \frac{\pi\Gamma(a+s) z^s}{\Gamma(c+s)\Gamma(s+1)\sin \pi s};\, n \right\}. \end{aligned} \tag{3.23}$$

Upon calculating the residues in Eq. (3.23), we obtain

$$
\begin{aligned}
f(a;\,c;\,z) &= \sum_{n=0}^{\infty} \frac{\Gamma(a+n)z^n}{\Gamma(c+n)\Gamma(n+1)\cos n\pi} \\
&= \frac{\Gamma(a)}{\Gamma(c)} \sum_{n=0}^{\infty} \frac{(-1)^n (a)_n}{(c)_n} \frac{z^n}{n!} \\
&= \frac{\Gamma(a)}{\Gamma(c)}\, {}_1F_1(a;\,c;\,-z), \qquad |z| < \infty.
\end{aligned} \tag{3.24}
$$

The infinite series in Eq. (3.24) has been identified as the ${}_1F_1$ function in Eq. (3.2), but now with a negative argument. From this series representation, we readily produce the small-argument finite asymptotic series

$$
{}_1F_1(a;\,c;\,-z) \sim 1 + \sum_{n=1}^{N} \frac{(-1)^n (a)_n}{(c)_n} \frac{z^n}{n!}, \qquad z \to 0. \tag{3.25}
$$

3.3.2 Large-argument asymptotic series

To derive an asymptotic series for large arguments, we see from Eqs. (3.22) and (3.24) that we can write ${}_1F_1$ as a Barnes integral,

$$
{}_1F_1(a;\,c;\,-z) = \frac{\Gamma(c)}{\Gamma(a)} \frac{1}{2\pi i} \int_{-i\infty}^{i\infty} \frac{\Gamma(a+s)}{\Gamma(c+s)} \Gamma(-s) z^s ds, \tag{3.26}
$$

which has simple poles at $s = -(a+n)$, $n = 0,\,1,\,2...$, which are assumed to be distinct from those of $\Gamma(-s)$. As before, we start with the change of variable $s \to -(s+a)$, from which we get

$$
{}_1F_1(a;\,c;\,-z) = \frac{\Gamma(c)}{\Gamma(a)} \frac{z^{-a}}{2\pi i} \int_{-i\infty}^{i\infty} \frac{\Gamma(a+s)}{\Gamma(c-a-s)} \Gamma(-s) z^{-s} ds. \tag{3.27}
$$

The evaluation of Eq. (3.27) for the poles of $\Gamma(-s)$ through use of the residue theory leads to

$$
\begin{aligned}
{}_1F_1(a;\,c;\,-z) &= -\frac{\Gamma(c) z^{-a}}{\Gamma(a)} \sum_{n=0}^{\infty} \operatorname{Res}\left\{ \frac{\Gamma(a+s)\Gamma(-s)z^{-s}}{\Gamma(c-a-s)};\, n \right\} \\
&= \frac{\Gamma(c)}{\Gamma(a)} z^{-a} \sum_{n=0}^{\infty} \operatorname{Res}\left\{ \frac{\pi\Gamma(a+s)z^{-s}}{\Gamma(c-a-s)\Gamma(1+s)\sin \pi s};\, n \right\},
\end{aligned} \tag{3.28}
$$

which takes the form of a descending asymptotic series,

$$
\begin{aligned}
{}_1F_1(a;\, c;\, -z) &\sim \frac{\Gamma(c)}{\Gamma(c-a)} z^{-a} \sum_{n=0}^{\infty} (a)_n (1+a-c)_n \frac{z^{-n}}{n!} \\
&\sim \frac{\Gamma(c)}{\Gamma(c-a)} z^{-a} {}_2F_0\left(a,\, 1+a-c;\, -;\, \frac{1}{z}\right).
\end{aligned}
\tag{3.29}
$$

In arriving at this last result, we made use of $(PS4)$ in Appendix A,

$$
\Gamma(a-n) = \Gamma(a)(a)_{-n} = \frac{(-1)^n \Gamma(a)}{(1-a)_n}.
$$

Last, by writing out the first few terms of the series for ${}_2F_0$, it follows that

$$
{}_1F_1(a;\, c;\, -z) \sim \frac{\Gamma(c)}{\Gamma(c-a)} z^{-a} \left[1 + \sum_{n=1}^{N} \frac{(a)_n (1+a-c)_n}{n!} \left(\frac{1}{z}\right)^n\right], \qquad z \to \infty,
\tag{3.30}
$$

which, for sufficiently large z, can be shortened to

$$
{}_1F_1(a;\, c;\, -z) \sim \frac{\Gamma(c)}{\Gamma(c-a)} z^{-a}, \qquad z \to \infty.
\tag{3.31}
$$

3.3.3 Elementary properties

The confluent hypergeometric function has many properties, a few of which are presented below. Some of these and others in Appendices A and B will be useful in our further development throughout the text:

$$
\frac{d}{dz}\, {}_1F_1(a;\, c;\, z) = \frac{a}{c}\, {}_1F_1(a+1;\, c+1;\, z),
\tag{3.32}
$$

$$
{}_1F_1(a;\, c;\, z) = e^z {}_1F_1(c-a;\, c;\, -z),
\tag{3.33}
$$

$$
{}_1F_1(a;\, a;\, z) = e^z.
\tag{3.34}
$$

To prove Eq. (3.32), we simply recognize that

$$
\frac{d}{dz}\, {}_1F_1(a;\, c;\, z) = \underbrace{\sum_{n=1}^{\infty}}_{n \to n+1} \frac{(a)_n}{(c)_n} \frac{z^{n-1}}{(n-1)!} = \sum_{n=0}^{\infty} \frac{(a)_{n+1}}{(c)_{n+1}} \frac{z^n}{n!}.
\tag{3.35}
$$

From $(PS2)$ in Appendix A, it follows that $(a)_{n+1} = a(a+1)_n$ and likewise that $(c)_{n+1} = c(c+1)_n$, from which we obtain Eq. (3.32). Equation (3.33) is a simple consequence of Eq. (3.21), and Eq. (3.34) is self-evident from Eq. (3.33).

3.4 The Function ${}_1F_k$

Next, let us consider the Barnes integral where there exists a product of k gamma functions in the denominator, i.e.,

$$f(a;\, c_1,\, \dots,\, c_k;\, z) = \frac{1}{2\pi i}\int_{-i\infty}^{i\infty} \frac{\Gamma(a+s)}{\prod_{j=1}^{k}\Gamma(c_j+s)}\Gamma(-s)z^s ds. \tag{3.36}$$

Here we have $p = 1$ and $q = k$, $k = 2, 3, 4, \dots$. Following the above analysis for ${}_1F_1$, the residues from $\Gamma(-s)$ in Eq. (3.36) can be written as

$$f(a;\, c_1,\, \dots,\, c_k;\, z) = \sum_{n=0}^{\infty} \mathrm{Res}\left\{\frac{\pi\,\Gamma(a+s)z^s}{\prod_{j=1}^{k}\Gamma(c_j+s)\Gamma(s+1)\sin\pi s};\, n\right\}. \tag{3.37}$$

Upon calculating the residues in Eq. (3.37), this expression reduces to

$$\begin{aligned} f(a;\, c_1,\, \dots,\, c_k;\, z) &= \frac{\Gamma(a)}{\prod_{j=1}^{k}\Gamma(c_j)}\sum_{n=0}^{\infty}\frac{(-1)^n(a)_n}{\prod_{j=1}^{k}(c_j)_n}\frac{z^n}{n!} \\ &= \frac{\Gamma(a)}{\prod_{j=1}^{k}\Gamma(c_j)}\,{}_1F_k(a;\, c_1,\, \dots,\, c_k;\, -z), \qquad |z| < \infty. \end{aligned} \tag{3.38}$$

Clearly, ${}_1F_k$ is an entire function because the series converges for all finite z values whenever $p \le q$. Based on Eq. (3.38), the subsequent small-argument finite asymptotic series becomes

$${}_1F_k(a;\, c_1,\, \dots,\, c_k;\, -z) \sim 1 + \sum_{n=1}^{N}\frac{(-1)^n(a)_n}{\prod_{j=1}^{k}(c_j)_n}\frac{z^n}{n!}, \qquad z \to 0. \tag{3.39}$$

3.4.1 Large-argument asymptotic series

To derive an asymptotic series for large arguments, we first write

$$\begin{aligned} {}_1F_k(a;\, c_1,\, \dots,\, c_k;\, -z) &= \frac{\prod_{j=1}^{k}\Gamma(c_j)}{\Gamma(a)} f(a;\, c_1,\, \dots,\, c_k;\, z) \\ &= \frac{\prod_{j=1}^{k}\Gamma(c_j)}{\Gamma(a)}\frac{1}{2\pi i}\int_{-i\infty}^{i\infty}\frac{\Gamma(a+s)}{\prod_{j=1}^{k}\Gamma(c_j+s)}\Gamma(-s)z^s ds, \end{aligned} \tag{3.40}$$

followed by the change of variable $s \to -(s+a)$. Doing this, we find that

$${}_1F_k(a; c_1, \dots, c_k; -z) = \frac{\prod_{j=1}^{k}\Gamma(c_j)}{\Gamma(a)}\frac{z^{-a}}{2\pi i}\int_{-i\infty}^{i\infty}\frac{\Gamma(a+s)}{\prod_{j=1}^{k}\Gamma(c_j-a-s)}\Gamma(-s)z^{-s} ds, \tag{3.41}$$

and from residue theory, it follows that

$$
\begin{aligned}
{}_1F_k(a; c_1, \ldots, c_k; -z) &\sim \frac{\prod_{j=1}^{k}\Gamma(c_j)}{\Gamma(a)} z^{-a} \sum_{n=0}^{\infty} \frac{\Gamma(a+n)z^{-n}}{\Gamma(1+n)\cos n\pi \prod_{j=1}^{k}\Gamma(c_j-a-n)} \\
&\sim \frac{\prod_{j=1}^{k}\Gamma(c_j)}{\prod_{j=1}^{k}\Gamma(c_j-a)} z^{-a} \sum_{n=0}^{\infty} (a)_n \prod_{j=1}^{k}(1-c_j+a)_n \frac{z^{-n}}{n!}.
\end{aligned}
\tag{3.42}
$$

Hence, we can identify the infinite series in Eq. (3.42) as

$$
\begin{aligned}
&{}_1F_k(a; c_1, \ldots, c_k; -z) \\
&\quad \sim \frac{\prod_{j=1}^{k}\Gamma(c_j)}{\prod_{j=1}^{k}\Gamma(c_j-a)} z^{-a}\, {}_{k+1}F_0\left(a,\, 1-c_1+a,\, \ldots,\, 1-c_k+a;\, -;\, \frac{1}{z}\right),
\end{aligned}
\tag{3.43}
$$

from which we deduce the large-argument finite asymptotic series

$$
\begin{aligned}
&{}_1F_k(a; c_1, \ldots, c_k; -z) \\
&\quad \sim \frac{\prod_{j=1}^{k}\Gamma(c_j)}{\prod_{j=1}^{k}\Gamma(c_j-a)} z^{-a} \left[1 + \sum_{n=1}^{N} (a)_n \prod_{j=1}^{k}(1-c_j+a)_n \frac{z^{-n}}{n!}\right], \quad z \to \infty.
\end{aligned}
\tag{3.44}
$$

3.5 The Function ${}_2F_1$

Like the confluent hypergeometric function, the hypergeometric function defined by Eq. (3.1) also satisfies a second-order DE:

$$
z(1-z)\frac{d^2w}{dz^2} + [c-(a+b+1)z]\frac{dw}{dz} - abw = 0. \tag{3.45}
$$

In many publications, this is considered one of the basic properties of the function $w(z) = {}_2F_1(a, b; c; z)$ and is proved by substituting Eq. (3.1) into Eq. (3.45). In addition to this relation, we can derive an integral representation following a scheme analogous to that leading to Eq. (3.20). Here we start with Eq. (3.18) and the series in Eq. (3.1) to find that

$$
\begin{aligned}
{}_2F_1(a, b; c; z) &= \frac{\Gamma(c)}{\Gamma(a)\Gamma(c-a)} \sum_{n=0}^{\infty} \frac{(b)_n}{n!} z^n \int_0^1 t^{n+a-1}(1-t)^{c-a-1} dt \\
&= \frac{\Gamma(c)}{\Gamma(a)\Gamma(c-a)} \int_0^1 t^{a-1}(1-t)^{c-a-1} \left[\sum_{n=0}^{\infty} \frac{(b)_n}{n!}(zt)^n\right] dt
\end{aligned}
\tag{3.46}
$$

or, recalling the binomial power series in Eq. (3.8),

$$
{}_2F_1(a, b; c; z) = \frac{\Gamma(c)}{\Gamma(a)\Gamma(c-a)} \int_0^1 t^{a-1}(1-t)^{c-a-1}(1-zt)^{-b}dt, \quad c > a > 0. \tag{3.47}
$$

3.5.1 The Barnes integral

The Barnes integral with a product of two gamma functions in the numerator and one in the denominator leads to the hypergeometric function ${}_2F_1$. To find the connection, we start with

$$
f(a, b; c; z) = \frac{1}{2\pi i} \int_{-i\infty}^{i\infty} \frac{\Gamma(a+s)\Gamma(b+s)}{\Gamma(c+s)} \Gamma(-s) z^s ds, \quad |\arg(z)| < \pi. \tag{3.48}
$$

Applying the residue calculus, we obtain

$$
\begin{aligned}
f(a, b; c; z) &= -\sum_{n=0}^{\infty} \text{Res}\left\{\frac{\Gamma(a+s)\Gamma(b+s)}{\Gamma(c+s)}\Gamma(-s)z^s; n\right\} \\
&= \sum_{n=0}^{\infty} \text{Res}\left\{\frac{\pi\Gamma(a+s)\Gamma(b+s)z^s}{\Gamma(c+s)\Gamma(s+1)\sin\pi s}; n\right\},
\end{aligned} \tag{3.49}
$$

which reduces to

$$
\begin{aligned}
f(a, b; c; z) &= \frac{\Gamma(a)\Gamma(b)}{\Gamma(c)} \sum_{n=0}^{\infty} \frac{(-1)^n (a)_n (b)_n}{(c)_n} \frac{z^n}{n!} \\
&= \frac{\Gamma(a)\Gamma(b)}{\Gamma(c)} {}_2F_1(a, b; c; -z), \quad |z| < 1.
\end{aligned} \tag{3.50}
$$

In the last step of Eq. (3.50) we have recognized the infinite series with a negative argument as the ${}_2F_1$ hypergeometric function in Eq. (3.1). Thus, the first few terms of the series lead to the small-argument asymptotic result

$$
{}_2F_1(a, b; c; -z) \sim 1 + \sum_{n=1}^{N} \frac{(-1)^n (a)_n (b)_n}{(c)_n} \frac{z^n}{n!}, \quad z \to 0. \tag{3.51}
$$

3.5.2 Large-argument asymptotic series

Based on Eqs. (3.48) and (3.50), it becomes clear that the Barnes integral representation of the hypergeometric function is

$$
{}_2F_1(a, b; c; -z) = \frac{\Gamma(c)}{\Gamma(a)\Gamma(b)} \frac{1}{2\pi i} \int_{-i\infty}^{i\infty} \frac{\Gamma(a+s)\Gamma(b+s)}{\Gamma(c+s)} \Gamma(-s) z^s ds. \tag{3.52}
$$

Because there are two gamma functions in the numerator of Eq. (3.52), there are two sets of poles in the left-half plane corresponding to $\Gamma(a+s)$ and

$\Gamma(b+s)$. For the large-argument asymptotic series, therefore, we must consider both sets of poles, i.e.,

$$ {}_2F_1(a, b; c; -z) = \frac{\Gamma(c)}{\Gamma(a)\Gamma(b)}\Big\{\sum \mathrm{Res}[-(a+n)] + \sum \mathrm{Res}[-(b+n)]\Big\}. \tag{3.53} $$

Let us consider the set of poles corresponding to $\Gamma(a+s)$, denoted below by the notation ${}_2F_1(a, b; c; -z)|_a$. To begin, we make the change of variable $s \to -(s+a)$, which produces

$$ {}_2F_1(a, b; c; -z)|_a = \frac{\Gamma(c)}{\Gamma(a)\Gamma(b)}\frac{z^{-a}}{2\pi i}\int_{-i\infty}^{i\infty} \frac{\Gamma(a+s)\Gamma(b-a-s)}{\Gamma(c-a-s)}\Gamma(-s)z^{-s}ds. \tag{3.54} $$

Under the same assumptions as in previous sections, we now find that

$$ \begin{aligned} {}_2F_1(a, b; c; -z)|_a &= -\frac{\Gamma(c)z^{-a}}{\Gamma(a)\Gamma(b)}\sum_{n=0}^{\infty}\mathrm{Res}\left\{\frac{\Gamma(a+s)\Gamma(b-a-s)}{\Gamma(c-a-s)}\Gamma(-s)z^{-s};\, n\right\} \\ &= \frac{\Gamma(c)z^{-a}}{\Gamma(a)\Gamma(b)}\sum_{n=0}^{\infty}\mathrm{Res}\left\{\frac{\pi\Gamma(a+s)\Gamma(b-a-s)}{\Gamma(c-a-s)\Gamma(s+1)\sin\pi s}z^{-s};\, n\right\}, \end{aligned} \tag{3.55} $$

and finally,

$$ \begin{aligned} {}_2F_1(a, b; c; -z)|_a &= \frac{\Gamma(c)z^{-a}}{\Gamma(a)\Gamma(b)}\sum_{n=0}^{\infty}\frac{(-1)^n\Gamma(a+n)\Gamma(b-a-n)}{\Gamma(c-a-n)\Gamma(n+1)}z^{-n} \\ &= \frac{\Gamma(c)\Gamma(b-a)z^{-a}}{\Gamma(b)\Gamma(c-a)}\sum_{n=0}^{\infty}\frac{(a)_n(1-c+a)_n}{(1-b+a)_n n!}(-z)^{-n} \\ &= \frac{\Gamma(c)\Gamma(b-a)z^{-a}}{\Gamma(b)\Gamma(c-a)}\,{}_2F_1\left(a,\, 1-c+a;\, 1-b+a;\, -\frac{1}{z}\right). \end{aligned} \tag{3.56} $$

Now we need to repeat the above analysis interchanging the roles of a and b. By doing so and combining the result with Eq. (3.56), we get

$$\begin{aligned}
{}_2F_1(a,\, b;\, c;\, -z) &= {}_2F_1(a,\, b;\, c;\, -z)|_a + {}_2F_1(a,\, b;\, c;\, -z)|_b \\
&= \frac{\Gamma(c)\Gamma(b-a)z^{-a}}{\Gamma(b)\Gamma(c-a)}\, {}_2F_1\left(a,\, 1-c+a;\, 1-b+a;\, -\frac{1}{z}\right) \\
&\quad + \frac{\Gamma(c)\Gamma(a-b)z^{-b}}{\Gamma(a)\Gamma(c-b)}\, {}_2F_1\left(b,\, 1-c+b;\, 1-a+b;\, -\frac{1}{z}\right) \\
&\sim \frac{\Gamma(c)\Gamma(b-a)z^{-a}}{\Gamma(b)\Gamma(c-a)} + \frac{\Gamma(c)\Gamma(a-b)z^{-b}}{\Gamma(a)\Gamma(c-b)}, \qquad z \to \infty.
\end{aligned} \tag{3.57}$$

Unlike some of the asymptotic series, both infinite series for the ${}_2F_1$ functions on the right-hand side of Eq. (3.57) converge for $|z| > 1$.

3.5.3 Elementary properties

Like the confluent hypergeometric function, the hypergeometric function enjoys many properties, a few of which are listed below (cf. also Appendices A and B):

$$ {}_2F_1(a,\, b;\, c;\, 1) = \frac{\Gamma(c)\Gamma(c-a-b)}{\Gamma(c-a)\Gamma(c-b)}, \tag{3.58}$$

$$ {}_2F_1(a,\, b;\, c;\, -z) = (1+z)^{-a}\, {}_2F_1\left(a,\, c-b;\, c;\, \frac{z}{1+z}\right), \tag{3.59}$$

$$ {}_2F_1(c-a,\, c-b;\, c;\, z) = (1-z)^{a+b-c}\, {}_2F_1(a,\, b;\, c;\, z). \tag{3.60}$$

Because it can be instructive to do so, we will prove all three of these identities. For Eq. (3.58), we use the integral representation (3.47), which leads to

$$\begin{aligned}
{}_2F_1(a,\, b;\, c;\, 1) &= \frac{\Gamma(c)}{\Gamma(a)\Gamma(c-a)} \int_0^1 t^{a-1}(1-t)^{c-a-1}(1-t)^{-b}dt \\
&= \frac{\Gamma(c)}{\Gamma(a)\Gamma(c-a)} \int_0^1 t^{a-1}(1-t)^{c-a-b-1}dt \\
&= \frac{\Gamma(c)\Gamma(c-a-b)}{\Gamma(c-a)\Gamma(c-b)},
\end{aligned} \tag{3.61}$$

the last step being a consequence of the beta function ($G17$) in Appendix A. For the proof of Eq. (3.59), we first form the product

$$\begin{aligned}(1+z)^{-a}{}_2F_1\left(a,\, c-b;\, c;\, \frac{z}{1+z}\right) &= \sum_{k=0}^{\infty} \frac{(a)_k(c-b)_k}{(c)_k k!} \frac{z^k}{(1+z)^{a+k}} \\ &= \sum_{k=0}^{\infty}\sum_{m=0}^{\infty} \frac{(a)_k(c-b)_k}{(c)_k k!} \frac{(a+k)_m}{m!}(-1)^m z^{m+k},\end{aligned} \tag{3.62}$$

where we have used the series in Eq. (3.8) to get the second step. Now, recalling the Pochhammer symbol identity (*PS*2) from Appendix A,

$$(a)_k(a+k)_m = (a)_{m+k},$$

we then switch the order of summation, make a change of index $m = n - k$ for the Cauchy product [Eq. (2.33)], and finally apply (*PS*7) from Appendix A to get

$$\begin{aligned}(1+z)^{-a}{}_2F_1\left(a,\, c-b;\, c;\, \frac{z}{1+z}\right) &= \sum_{n=0}^{\infty}\sum_{k=0}^{n} \frac{(c-b)_k}{(c)_k k!(n-k)!}(a)_n(-1)^{n-k}z^n \\ &= \sum_{n=0}^{\infty}(-1)^n(a)_n\frac{z^n}{n!}\sum_{k}\frac{(c-b)_k(-n)_k}{(c)_k k!} \\ &= \sum_{n=0}^{\infty}(-1)^n(a)_n\frac{z^n}{n!}\,{}_2F_1(-n,\, c-b;\, c;\, 1).\end{aligned} \tag{3.63}$$

However, from Eq. (3.58), it follows that

$${}_2F_1(-n,\, c-b;\, c;\, 1) = \frac{\Gamma(c)\Gamma(b+n)}{\Gamma(c+n)\Gamma(b)} = \frac{(b)_n}{(c)_n}, \tag{3.64}$$

and thus, we are led to the intended result:

$$\begin{aligned}(1+z)^{-a}{}_2F_1\left(a,\, c-b;\, c;\, \frac{z}{1+z}\right) &= \sum_{n=0}^{\infty}\frac{(-1)^n(a)_n(b)_n}{(c)_n}\frac{z^n}{n!} \\ &= {}_2F_1(a,\, b;\, c;\, -z).\end{aligned} \tag{3.65}$$

To prove Eq. (3.60), we start with the integral relation of Eq. (3.47) and make the substitution $t = (1-s)/(1-sz)$. By doing this and performing some algebraic simplification, we arrive at

$$
\begin{aligned}
{}_2F_1(a, b; c; z) &= \frac{\Gamma(c)(1-z)^{c-a-b}}{\Gamma(a)\Gamma(c-a)} \int_0^1 s^{c-a-1}(1-s)^{a-1}(1-zs)^{b-c} ds \\
&= (1-z)^{c-a-b}\, {}_2F_1(c-a, c-b; c; z).
\end{aligned}
\tag{3.66}
$$

3.6 The Function ${}_2F_2$

Next in the analysis of generalized hypergeometric functions we consider a function with two gamma functions in the numerator and two in the denominator ($p = q = 2$), i.e.,

$$
f(a, b; c, d; z) = \frac{1}{2\pi i} \int_{-i\infty}^{i\infty} \frac{\Gamma(a+s)\Gamma(b+s)}{\Gamma(c+s)\Gamma(d+s)} \Gamma(-s) z^s ds, \qquad |\arg(z)| < \frac{\pi}{2}.
\tag{3.67}
$$

Based on the residue calculus for the poles of $\Gamma(-s)$, we obtain

$$
\begin{aligned}
f(a, b; c, d; z) &= -\sum_{n=0}^{\infty} \operatorname{Res}\left\{ \frac{\Gamma(a+s)\Gamma(b+s)}{\Gamma(c+s)\Gamma(d+s)} \Gamma(-s) z^s; n \right\} \\
&= \sum_{n=0}^{\infty} \operatorname{Res}\left\{ \frac{\pi\Gamma(a+s)\Gamma(b+s) z^s}{\Gamma(c+s)\Gamma(d+s)\Gamma(s+1)\sin \pi s}; n \right\},
\end{aligned}
\tag{3.68}
$$

which reduces to

$$
\begin{aligned}
f(a, b; c, d; z) &= \frac{\Gamma(a)\Gamma(b)}{\Gamma(c)\Gamma(d)} \sum_{n=0}^{\infty} \frac{(-1)^n (a)_n (b)_n}{(c)_n (d)_n} \frac{z^n}{n!} \\
&= \frac{\Gamma(a)\Gamma(b)}{\Gamma(c)\Gamma(d)}\, {}_2F_2(a, b; c, d; -z), \qquad |z| < \infty.
\end{aligned}
\tag{3.69}
$$

The series in Eq. (3.69) is recognized as the generalized hypergeometric function ${}_2F_2$ from which we obtain the small argument finite asymptotic series,

$$
{}_2F_2(a, b; c, d; -z) \sim 1 + \sum_{n=1}^{N} \frac{(-1)^n (a)_n (b)_n}{(c)_n (d)_n} \frac{z^n}{n!}, \qquad z \to 0.
\tag{3.70}
$$

3.6.1 Large-argument asymptotic series

Analogous to the treatment in Section 3.5.2, we again have two gamma functions in the numerator of Eq. (3.67). Consequently, we have two sets of poles in the left-half plane corresponding to $\Gamma(a+s)$ and $\Gamma(b+s)$. Following the same procedure used in Section 3.5.2, it is straightforward to show that

$$
\begin{aligned}
{}_2F_2(a, b; c, d; -z) \sim & \frac{\Gamma(c)\Gamma(d)\Gamma(b-a)}{\Gamma(b)\Gamma(c-a)\Gamma(d-a)} z^{-a} \\
& \times {}_3F_1\left(a, 1-c+a, 1-d+a; 1-b+a; -\frac{1}{z}\right) \\
& + \frac{\Gamma(c)\Gamma(d)\Gamma(a-b)}{\Gamma(a)\Gamma(c-b)\Gamma(d-b)} z^{-b} \\
& \times {}_3F_1\left(b, 1-c+b, 1-d+b; 1-a+b; -\frac{1}{z}\right) \\
\sim & \frac{\Gamma(c)\Gamma(d)\Gamma(b-a)}{\Gamma(b)\Gamma(c-a)\Gamma(d-a)} z^{-a} \\
& + \frac{\Gamma(c)\Gamma(d)\Gamma(a-b)}{\Gamma(a)\Gamma(c-b)\Gamma(d-b)} z^{-b}, \quad z \to \infty.
\end{aligned}
\tag{3.71}
$$

We have introduced the symbol "∼" in the first line of Eq. (3.71) because the relation is not exact as it was for the hypergeometric function relation of Eq. (3.56). That is, the series for ${}_3F_1$ diverges $(p > q + 1)$.

3.7 The Function ${}_pF_q$

From the pattern taken in previous sections for other generalized hypergeometric functions, we expect the Barnes integral for the general ${}_pF_q$ function to be proportional to

$$
f(a_1, \ldots, a_p; c_1, \ldots, c_q; z) = \frac{1}{2\pi i} \int_{-i\infty}^{i\infty} \frac{\prod_{j=1}^{p} \Gamma(a_j + s)}{\prod_{j=1}^{q} \Gamma(c_j + s)} \Gamma(-s) z^s ds. \tag{3.72}
$$

By summing the residues for the poles of $\Gamma(-s)$, we obtain

$$
\begin{aligned}
f(a_1, \ldots, a_p; c_1, \ldots, c_q; z) &= -\sum_{n=0}^{\infty} \operatorname{Res}\left\{\frac{\prod_{j=1}^{p} \Gamma(a_j + s)}{\prod_{j=1}^{q} \Gamma(c_j + s)} \Gamma(-s) z^s; n\right\} \\
&= \frac{\prod_{j=1}^{p} \Gamma(a_j)}{\prod_{j=1}^{q} \Gamma(c_j)} \sum_{n=0}^{\infty} \frac{(a_1)_n \cdots (a_p)_n}{(c_1)_n \cdots (c_q)_n} \frac{(-z)^n}{n!} \\
&= \frac{\prod_{j=1}^{p} \Gamma(a_j)}{\prod_{j=1}^{q} \Gamma(c_j)} {}_pF_q(a_1, \ldots, a_p; c_1, \ldots, c_q; -z),
\end{aligned}
\tag{3.73}
$$

where we have recognized the series as that of Eq. (3.3). Based on this last result, it can be seen that the Barnes integral for ${}_pF_q$ is given by

$$
{}_pF_q(a_1, ..., a_p; c_1, ..., c_q; -z) = \frac{\prod_{j=1}^q \Gamma(c_j)}{\prod_{j=1}^p \Gamma(a_j)} \frac{1}{2\pi i} \int_{-i\infty}^{i\infty} \frac{\prod_{j=1}^p \Gamma(a_j + s)}{\prod_{j=1}^q \Gamma(c_j + s)} \Gamma(-s) z^s ds. \tag{3.74}
$$

We will find this Barnes integral representation useful in Chapter 5, where we introduce the Meijer G-function.

3.8 The Function *U*(*a*; *c*; *z*)

The functions $w_1(z) = {}_1F_1(a; c; z)$ and $w_2(z) = z^{1-c}{}_1F_1(1 + a - c; 2 - c; z)$ are known linearly independent solutions of the DE (3.16), provided that $c \neq 0, \pm 1, \pm 2, \ldots$. Thus, it follows that $w = C_1 w_1(z) + C_2 w_2(z)$ is a general solution, where C_1 and C_2 are arbitrary constants. The function $U(a; c; z)$, introduced in 1947 by F. Tricomi,[5] is related to $w_1(z)$ and $w_2(z)$ through the expression

$$
U(a; c; z) = \frac{\Gamma(1 - c)}{\Gamma(1 + a - c)} {}_1F_1(a; c; z) + \frac{\Gamma(c - 1)}{\Gamma(a)} z^{1-c} {}_1F_1(1 + a - c; 2 - c; z). \tag{3.75}
$$

Hence, the function $U(a; c; z)$ represents another solution of Eq. (3.16) called the *confluent hypergeometric function of the second kind.*

For small arguments, we can use the first term of each ${}_1F_1$ function in Eq. (3.75) to arrive at the asymptotic relation

$$
U(a; c; z) \sim \frac{\Gamma(1 - c)}{\Gamma(1 + a - c)} + \frac{\Gamma(c - 1)}{\Gamma(a)} z^{1-c}, \quad z \to 0. \tag{3.76}
$$

Also, if we replace a with $1 + a - c$ and c with $2 - c$ in Eq. (3.75), we find that

$$
\begin{aligned}
U(1 + a - c; 2 - c; z) = &\frac{\Gamma(c - 1)}{\Gamma(a)} {}_1F_1(1 + a - c; 2 - c; z) \\
&+ \frac{\Gamma(1 - c)}{\Gamma(1 + a - c)} z^{c-1} {}_1F_1(a; c; z),
\end{aligned} \tag{3.77}
$$

and by comparison with Eq. (3.75), we deduce the identity

$$
U(a; c; z) = z^{1-c} U(1 + a - c; 2 - c; z). \tag{3.78}
$$

3.8.1 The Barnes integral

In some applications, the confluent hypergeometric function $U(a; c; z)$ may arise through the integral representation (Ref. 6, pp. 267–268):

$$U(a;\, c;\, z) = \frac{1}{\Gamma(a)} \int_0^{\infty} e^{-zt} t^{a-1} (1+t)^{c-a-1} dt, \quad \mathrm{Re}(a) > 0, \quad \mathrm{Re}(z) > 0. \tag{3.79}$$

However, it is the Barnes integral representation for $U(a;\, c;\, x)$ that we wish to derive here. To start, we first examine a Barnes integral suggested by the two ${}_1F_1$ functions in Eq. (3.75):

$$f(a;\, c;\, z) = \frac{1}{2\pi i} \int_{-i\infty}^{+i\infty} \Gamma(a+s)\Gamma(1+a-c+s)\Gamma(-s) z^{-s} ds, \quad |\arg(z)| < \frac{3\pi}{2}. \tag{3.80}$$

In Eq. (3.80) we are writing z^{-s} rather than z^s in anticipation of developing the asymptotic results in Section 3.8.2. Also, we see that there are two sets of simple poles in the left-half plane, similar to that of the hypergeometric function. Our objective is to relate Eq. (3.80) to the right-hand side of Eq. (3.75).

For the first set of simple poles $s = -(a+n)$, $n = 0, 1, 2, \ldots$ in Eq. (3.80), let us make the change of variable $s \to -(s+a)$, which leads to

$$f(a;\, c;\, z) = \frac{z^a}{2\pi i} \int_{-i\infty}^{+i\infty} \Gamma(a+s)\Gamma(1-c-s)\Gamma(-s) z^s ds. \tag{3.81}$$

By use of the residue theory followed by the identity (*PS*4) in Appendix A, we find that

$$\begin{aligned} f(a;\, c;\, z)|_a &= z^a \sum_{n=0}^{\infty} \mathrm{Res}\left\{ \frac{\pi\Gamma(a+s)\Gamma(1-c-s)z^s}{\Gamma(1+s)\sin \pi s};\, n \right\} \\ &= z^a \Gamma(a)\Gamma(1-c) \sum_{n=0}^{\infty} \frac{(a)_n}{(c)_n} \frac{z^n}{n!}. \end{aligned} \tag{3.82}$$

For the second set of poles at $s = -(1+a-c+n)$, $n = 0, 1, 2, \ldots$, we introduce the change of variable $s \to -(s+1+a-c)$ in Eq. (3.80). Then, in an analogous manner to that above, we arrive at

$$\begin{aligned} f(a;\, c;\, z)|_{1+a-c} &= \frac{z^{1+a-c}}{2\pi i} \int_{-i\infty}^{+i\infty} \Gamma(1+a-c+s)\Gamma(c-1-s)\Gamma(-s) z^s ds \\ &= z^{1+a-c} \sum_{n=0}^{\infty} \mathrm{Res}\left\{ \frac{\pi\Gamma(1+a-c+s)\Gamma(c-1-s)z^s}{\Gamma(s+1)\sin \pi s};\, n \right\} \\ &= z^{1+a-c}\Gamma(1+a-c)\Gamma(c-1) \sum_{n=0}^{\infty} \frac{(1+a-c)_n}{(2-c)_n} \frac{z^n}{n!}. \end{aligned} \tag{3.83}$$

By writing the series in Eqs. (3.82) and (3.83) as confluent hypergeometric functions of the first kind and summing the resulting expressions divided by the product $\Gamma(a)\Gamma(1+a-c)z^a$, we obtain

$$\frac{f(a;\ c;\ z)|_a + f(a;\ c;\ z)|_{1+a-c}}{\Gamma(a)\Gamma(1+a-c)z^a} = \frac{\Gamma(1-c)}{\Gamma(1+a-c)}\,{}_1F_1(a;\ c;\ z) + \frac{\Gamma(c-1)}{\Gamma(a)} z^{1-c}\,{}_1F_1(1+a-c;\ 2-c;\ z). \tag{3.84}$$

The right-hand side of Eq. (3.84) is the same as that in Eq. (3.75). Hence, based on Eqs. (3.80) and (3.84), we have derived the Barnes integral representation,

$$U(a;\ c;\ z) = \frac{z^{-a}}{\Gamma(a)\Gamma(1+a-c)} \frac{1}{2\pi i} \int_{-i\infty}^{+i\infty} \Gamma(a+s)\Gamma(1+a-c+s)\Gamma(-s)z^{-s} ds, \tag{3.85}$$

where it is assumed that $a \neq 0, -1, -2, \ldots$ and $1 + a - c \neq 0, -1, -2, \ldots$.

3.8.2 Large-argument asymptotic series

By applying the residue theory to Eq. (3.85) based on $\Gamma(-s)$, we have

$$\begin{aligned} U(a;\ c;\ z) &\sim -\frac{z^{-a}}{\Gamma(a)\Gamma(1+a-c)} \sum_{n=0}^{\infty} \mathrm{Res}\{\Gamma(a+s)\Gamma(1+a-c+s)\Gamma(-s)z^{-s};\ n\} \\ &\sim \frac{z^{-a}}{\Gamma(a)\Gamma(1+a-c)} \sum_{n=0}^{\infty} \frac{\Gamma(a+n)\Gamma(1+a-c+n)z^{-n}}{\Gamma(1+n)\cos n\pi} \\ &\sim z^{-a} \sum_{n=0}^{\infty} (a)_n (1+a-c)_n \frac{(-1)^n}{n!} \left(\frac{1}{z}\right)^n \\ &\sim z^{-a}\,{}_2F_0\left(a,\ 1+a-c;\ -;\ -\frac{1}{z}\right), \qquad z \to \infty. \end{aligned} \tag{3.86}$$

Then, retaining only the first term of the ${}_2F_0$ series, we are left with the simple asymptotic result,

$$U(a;\ c;\ z) \sim z^{-a}, \qquad z \to \infty. \tag{3.87}$$

Additional terms in the asymptotic series can be obtained from Eq. (3.86).

3.9 Whittaker Functions

Some of the development concerning confluent hypergeometric functions during the early 1900s concentrated on the related Whittaker functions (Ref. 7, pp. 337–354). These functions often arise as solutions of

$$\frac{d^2w}{dz^2} + \left(-\frac{1}{4} + \frac{k}{z} + \frac{\frac{1}{4} - m^2}{z^2}\right) w = 0. \tag{3.88}$$

Equation (3.88) is considered to be the *Liouville standard form* of the DE (3.16) (cf. Ref. 8, pp. 403–405).

Based on the relation between Eqs. (3.16) and (3.88), it follows that the *Whittaker functions of the first* and *second kind* are defined, respectively, by

$$M_{k,m}(z) = e^{-z/2} z^{m+1/2} \, {}_1F_1\left(\frac{1}{2} + m - k;\, 2m+1;\, z\right), \tag{3.89}$$

$$W_{k,m}(z) = e^{-z/2} z^{m+1/2} U\left(\frac{1}{2} + m - k;\, 2m+1;\, z\right). \tag{3.90}$$

Many of the properties associated with these functions can be derived directly from properties already established for the confluent hypergeometric functions. For example, the small- and large-argument approximations for $M_{k,m}(z)$ lead to

$$M_{k,m}(z) \sim \begin{cases} z^{m+1/2}, & z \to 0 \\ \dfrac{\Gamma(2m+1)}{\Gamma\left(\frac{1}{2} + m - k\right)} z^{-k} e^{z/2}, & z \to \infty, \end{cases} \tag{3.91}$$

and similarly, for $W_{k,m}(z)$,

$$W_{k,m}(z) \sim \begin{cases} \dfrac{\Gamma(2m)}{\Gamma\left(\frac{1}{2} + m - k\right)} z^{1/2-m} + \dfrac{\Gamma(-2m)}{\Gamma\left(\frac{1}{2} - m - k\right)} z^{1/2+m}, & z \to 0 \\ z^k e^{-z/2}, & z \to \infty. \end{cases} \tag{3.92}$$

Because of a direct connection with the confluent hypergeometric functions, it follows that the Whittaker functions are related to a variety of elementary and other special functions.

3.10 Asymptotic Behavior of Other Special Functions

By establishing the connecting relations between other special functions and generalized hypergeometric functions (cf. Appendix B), we can easily develop asymptotic approximations for some of these other special functions. Most of these relations can be proven by comparing series or integral representations.

3.10.1 Error functions

The *error function* and *complementary error function* are defined respectively, by

$$\operatorname{erf}(z) = \frac{2}{\sqrt{\pi}} \int_0^z e^{-t^2} dt, \quad |z| < \infty, \tag{3.93}$$

$$\operatorname{erfc}(z) = \frac{2}{\sqrt{\pi}} \int_z^\infty e^{-t^2} dt, \quad |z| < \infty. \tag{3.94}$$

Error functions arise in calculating detection probabilities and bit error rates in communication systems (see Chapter 11), and they are closely connected with other special functions like the Fresnel integrals and incomplete gamma function. These functions are also important in the theory of errors, heat conduction, and other areas of mathematical physics. From Eq. (3.93), it can readily be shown that $\operatorname{erf}(-z) = -\operatorname{erf}(z)$, which illustrates that $\operatorname{erf}(z)$ satisfies the condition of an *odd function* in the real variable $z = x$. It also follows from properties of the gamma function that the sum of Eqs. (3.93) and (3.94) leads to $\operatorname{erf}(z) + \operatorname{erfc}(z) = 1$, or

$$\operatorname{erfc}(z) = 1 - \operatorname{erf}(z). \tag{3.95}$$

The plots of both error functions for real x values are shown in Fig. 3.1.

By expressing e^{-t^2} in Eq. (3.93) in a power series and interchanging the order of integration and summation followed by termwise integration, we find the power series representation:

$$\operatorname{erf}(z) = \frac{2}{\sqrt{\pi}} \sum_{n=0}^{\infty} \frac{(-1)^n z^{2n+1}}{(2n+1)n!}, \quad |z| < \infty. \tag{3.96}$$

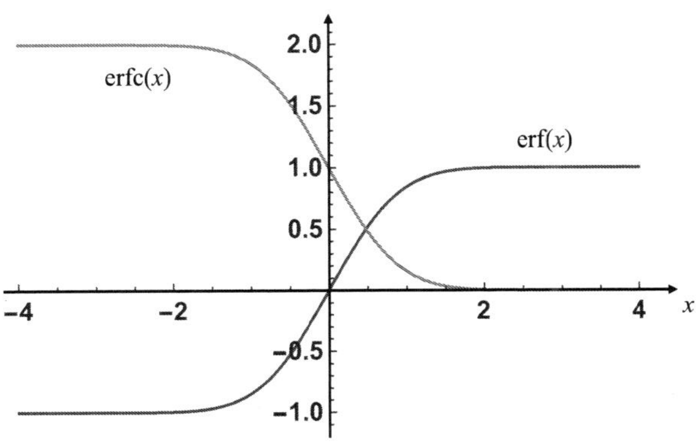

Figure 3.1 Plots of the error function and complementary error function.

From the Pochhammer symbol identities in Appendix A, it can be shown that

$$\frac{1}{2n+1} = \frac{(2n)!}{(2n+1)!} = \frac{\left(\frac{1}{2}\right)_n}{\left(\frac{3}{2}\right)_n},$$

and thus, Eq. (3.96) becomes

$$\begin{aligned} \operatorname{erf}(z) &= \frac{2z}{\sqrt{\pi}} \sum_{n=0}^{\infty} \frac{(-1)^n \left(\frac{1}{2}\right)_n}{\left(\frac{3}{2}\right)_n} \frac{z^{2n}}{n!} \\ &= \frac{2z}{\sqrt{\pi}} \, {}_1F_1\left(\frac{1}{2}; \frac{3}{2}; -z^2\right), \quad |z| < \infty. \end{aligned} \tag{3.97}$$

Based on Eq. (3.97) and asymptotic properties of the confluent hypergeometric function found in Appendix E, it can be shown that the error function $\operatorname{erf}(z)$ has the following asymptotic relations:

$$\operatorname{erf}(z) \sim \begin{cases} \dfrac{2z}{\sqrt{\pi}}\left(1 - \dfrac{z^2}{3} + \cdots\right), & z \to 0 \\ 1, \quad z \to \infty. \end{cases} \tag{3.98}$$

For the complementary error function, we start by making the change of variable $t = z\sqrt{1+\tau}$ in Eq. (3.94), which leads to

$$\operatorname{erfc}(z) = \frac{z}{\sqrt{\pi}} e^{-z^2} \int_0^\infty e^{-z^2\tau}(1+\tau)^{-1/2} d\tau. \tag{3.99}$$

If we compare Eq. (3.99) with Eq. (3.79), we see that

$$\begin{aligned} \operatorname{erfc}(z) &= \frac{z}{\sqrt{\pi}} e^{-z^2} U\left(1; \frac{3}{2}; z^2\right) \\ &= \frac{1}{\sqrt{\pi}} e^{-z^2} U\left(\frac{1}{2}; \frac{1}{2}; z^2\right), \end{aligned} \tag{3.100}$$

where we used ($CH11$) in Appendix A to get the second result. Now from Eq. (3.86), we obtain the large-argument asymptotic relation

$$\operatorname{erfc}(z) \sim \frac{1}{z\sqrt{\pi}} e^{-z^2} \left[1 + \sum_{n=1}^{N} \frac{\left(\frac{1}{2}\right)_n (1)_n}{n!} \left(-\frac{1}{z^2}\right)^n\right], \quad z \to \infty. \tag{3.101}$$

Note that the asymptotic expression (3.101) can be obtained from either of the confluent hypergeometric functions of the second kind in Eq. (3.100). Moreover, by using the lead term from Eq. (3.101) in the identity (3.95), we obtain a more revealing asymptotic result for $\mathrm{erf}(z)$ with a large argument than that provided in Eq. (3.98), viz.,

$$\mathrm{erf}(z) \sim 1 - \frac{1}{z\sqrt{\pi}} e^{-z^2}, \quad z \to \infty. \tag{3.102}$$

3.10.2 Fresnel integrals

Closely associated with the error function are the *Fresnel integrals*:

$$C(z) = \int_0^z \cos\left(\frac{1}{2}\pi t^2\right) dt, \tag{3.103}$$

$$S(z) = \int_0^z \sin\left(\frac{1}{2}\pi t^2\right) dt. \tag{3.104}$$

These functions arise in the near-field Fresnel diffraction patterns and the theory of vibrations, among other areas. Both Fresnel integrals have an oscillatory behavior for real arguments, as shown in the plots of these functions in Fig. 3.2.

From the power series for the cosine function in Eq. (3.103), termwise integration leads to

$$C(z) = \sum_{n=0}^{\infty} \frac{(-1)^n (\pi/2)^{2n}}{(2n)!(4n+1)} z^{4n+1}. \tag{3.105}$$

Applying Pochhammer symbol identities from Appendix A, we have

$$(2n)! = 2^{2n} \left(\frac{1}{2}\right)_n n!,$$

$$\frac{1}{4n+1} = \frac{\frac{1}{4}}{\left(n+\frac{1}{4}\right)} = \frac{\frac{1}{4}\Gamma\left(n+\frac{1}{4}\right)}{\Gamma\left(n+\frac{5}{4}\right)} = \frac{\frac{1}{4}\Gamma\left(\frac{1}{4}\right)\left(\frac{1}{4}\right)_n}{\frac{1}{4}\Gamma\left(\frac{1}{4}\right)\left(\frac{5}{4}\right)_n} = \frac{\left(\frac{1}{4}\right)_n}{\left(\frac{5}{4}\right)_n},$$

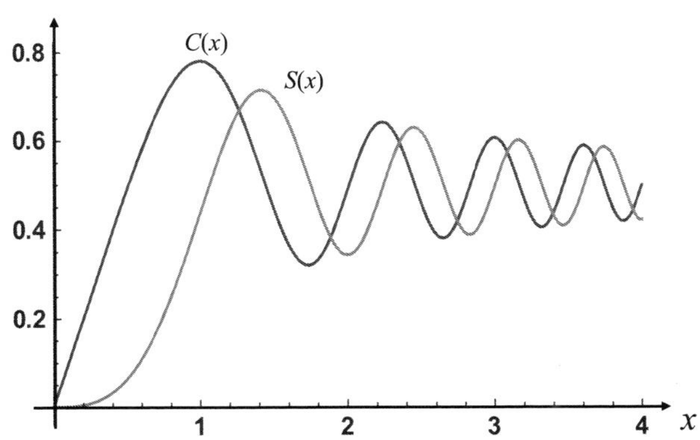

Figure 3.2 The Fresnel integrals.

and therefore

$$
\begin{aligned}
C(z) &= z\sum_{n=0}^{\infty}\frac{(-1)^n\left(\dfrac{1}{4}\right)_n}{\left(\dfrac{1}{2}\right)_n\left(\dfrac{5}{4}\right)_n}\left(\frac{\pi}{4}\right)^{2n}\frac{z^{4n}}{n!} \\
&= z\,{}_1F_2\left(\frac{1}{4};\,\frac{1}{2},\,\frac{5}{4};\,-\frac{\pi^2 z^4}{16}\right).
\end{aligned}
\tag{3.106}
$$

By using Eqs. (3.40) and (3.45), we are led to the asymptotic results

$$
C(z) \sim \begin{cases} z\left(1-\dfrac{\pi^2 z^4}{40}+\cdots\right), & z\to 0 \\ \dfrac{1}{2}, & z\to\infty. \end{cases}
\tag{3.107}
$$

In a manner similar to that used for $C(z)$, it can be shown that

$$
\begin{aligned}
S(z) &= \frac{\pi z^3}{2}\sum_{n=0}^{\infty}\frac{(-1)^n(\pi/2)^{2n}}{(2n+1)!(4n+3)}z^{4n} \\
&= \frac{\pi z^3}{6}\,{}_1F_2\left(\frac{3}{4};\,\frac{3}{2},\,\frac{7}{4};\,-\frac{\pi^2 z^4}{16}\right).
\end{aligned}
\tag{3.108}
$$

Consequently, we can derive the similar asymptotic relations

$$
S(z) \sim \begin{cases} \dfrac{\pi z^3}{6}\left(1-\dfrac{\pi^2 z^4}{56}+\cdots\right), & z\to 0 \\ \dfrac{1}{2}, & z\to\infty. \end{cases}
\tag{3.109}
$$

Last, another interesting relation involving the Fresnel integrals arises by considering the combination

$$
\begin{aligned}
C(z) - iS(z) &= \int_0^z \exp\left(-\frac{i\pi t^2}{2}\right)dt \\
&= \frac{1}{\sqrt{2i}}\operatorname{erf}\left(z\sqrt{\frac{i\pi}{2}}\right) \\
&= z\,{}_1F_1\left(\frac{1}{2};\,\frac{3}{2};\,-\frac{i\pi z^2}{2}\right),
\end{aligned}
\tag{3.110}
$$

which shows a direct relationship between the Fresnel integrals and the error function. Also, by the use of Eq. (3.97), we have derived a direct relationship with the confluent hypergeometric function in the last line of Eq. (3.110).

3.10.3 Incomplete gamma functions

The *incomplete gamma function* and *complementary incomplete gamma function* are common in probability theory. They are defined, respectively, by

$$\gamma(a, z) = \int_0^z e^{-t} t^{a-1} dt, \quad \mathrm{Re}(a) > 0, \tag{3.111}$$

$$\Gamma(a, z) = \int_z^\infty e^{-t} t^{a-1} dt, \quad \mathrm{Re}(a) > 0. \tag{3.112}$$

Note that by combining Eqs. (3.111) and (3.112), we obtain the identity

$$\gamma(a, z) + \Gamma(a, z) = \Gamma(a). \tag{3.113}$$

If we write the exponential function e^{-t} in Eq. (3.111) as a power series and employ termwise integration, we get the power series representation

$$\gamma(a, z) = z^a \sum_{n=0}^\infty \frac{(-1)^n z^n}{(n+a)n!}. \tag{3.114}$$

Then, from the identity ($PS13$) in Appendix A,

$$\frac{1}{n+a} = \frac{\Gamma(n+a)}{\Gamma(n+a+1)} = \frac{\Gamma(a)(a)_n}{\Gamma(a+1)(a+1)_n} = \frac{(a)_n}{a(a+1)_n},$$

we are led to the confluent hypergeometric function representation,

$$\begin{aligned}\gamma(a, z) &= \frac{z^a}{a} \sum_{n=0}^\infty \frac{(-1)^n (a)_n z^n}{(a+1)_n n!} \\ &= \frac{z^a}{a} \, {}_1F_1(a; a+1; -z).\end{aligned} \tag{3.115}$$

Hence, based on properties of the confluent hypergeometric function, we deduce the asymptotic relations

$$\gamma(a, z) \sim \begin{cases} \dfrac{z^a}{a}\left(1 - \dfrac{az}{a+1} + \cdots\right), & z \to 0 \\ \Gamma(a), \quad z \to \infty. \end{cases} \tag{3.116}$$

For the complementary incomplete gamma function, the substitution $t = z(1+\tau)$ into Eq. (3.112) yields [recall ($CH16$) in Appendix A]

$$\begin{aligned}\Gamma(a, z) &= z^a e^{-z} \int_0^\infty e^{-z\tau}(1+\tau)^{a-1} d\tau \\ &= z^a e^{-z} U(1;\, a+1;\, z),\end{aligned} \tag{3.117}$$

and finally, by applying Eq. (3.86), we obtain the finite asymptotic series

$$\Gamma(a, z) \sim z^{a-1} e^{-z}\left(1 + \frac{a-1}{z} + \cdots\right), \qquad z \to \infty. \tag{3.118}$$

Combining Eq. (3.113) with the first term of Eq. (3.118), we obtain the more detailed large-argument asymptotic result

$$\gamma(a, z) \sim \Gamma(a) - z^{a-1} e^{-z}, \qquad z \to \infty. \tag{3.119}$$

The complementary incomplete gamma function and the complementary error function are related. To see this, observe that by setting $a = \frac{1}{2}$ and changing $z \to z^2$, Eq. (3.117) reduces to

$$\begin{aligned}\Gamma\left(\frac{1}{2}, z^2\right) &= z e^{-z^2} U\left(1;\, \frac{3}{2};\, z^2\right) \\ &= \sqrt{\pi}\,\mathrm{erfc}(z).\end{aligned} \tag{3.120}$$

This last result is based on Eq. (3.100).

3.10.4 The exponential integral

The *exponential integral,* which arises in antenna theory and some astrophysical problems, is defined by

$$\mathrm{Ei}(z) = \int_{-\infty}^{z} \frac{e^t}{t}\, dt. \tag{3.121}$$

If we replace z with $-z$ and t with $-t$, Eq. (3.121) becomes

$$-\mathrm{Ei}(-z) \equiv E_1(z) = \int_z^\infty \frac{e^{-t}}{t}\, dt. \tag{3.122}$$

The function $E_1(x)$ is plotted in Fig. 3.3 for real values of x.

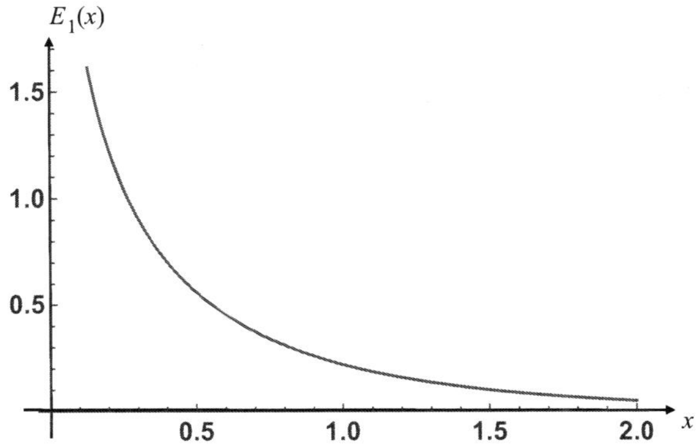

Figure 3.3 The exponential integral.

By comparing Eq. (3.122) with Eq. (3.112), we are led to

$$\begin{aligned} E_1(z) = \Gamma(0, z) &= \lim_{a\to 0}[\Gamma(a) - \gamma(a, z)] \\ &= \lim_{a\to 0}\left[\Gamma(a) - z^a\sum_{n=0}^{\infty}\frac{(-1)^n z^n}{(n+a)n!}\right] \\ &= \lim_{a\to 0}\left[\frac{a\Gamma(a) - z^a}{a}\right] - \sum_{n=1}^{\infty}\frac{(-1)^n z^n}{n\,n!}, \end{aligned} \tag{3.123}$$

and by application of L'Hôpital's rule from calculus, Eq. (3.123) reduces to (Ref. 8, p. 127)

$$E_1(z) = -\gamma - \ln z - \sum_{n=1}^{\infty}\frac{(-1)^n z^n}{nn!}. \tag{3.124}$$

In Eq. (3.124) the symbol γ is *Euler's constant* defined by

$$\gamma = -\Gamma'(1) = -\int_0^\infty e^{-t}\ln t\,dt = 0.577216\ldots\,. \tag{3.125}$$

The substitution $t = z(1+\tau)$ in Eq. (3.122) yields

$$\begin{aligned} E_1(z) &= e^{-z}\int_0^\infty e^{-z\tau}(1+\tau)^{-1}\,d\tau \\ &= e^{-z}U(1; 1; z). \end{aligned} \tag{3.126}$$

Thus, from Eq. (3.124) and the asymptotic relation of Eq. (3.87), we see that

$$E_1(z) \sim \begin{cases} -\ln z, & z \to 0 \\ \dfrac{e^{-z}}{z}, & z \to \infty. \end{cases} \tag{3.127}$$

3.10.5 Sine and cosine integrals

Related to the exponential integral in Section 3.10.4 are the *sine* and *cosine integrals*,

$$\mathrm{Si}(z) = \int_0^z \frac{\sin t}{t}\, dt, \tag{3.128}$$

$$\mathrm{Ci}(z) = -\int_z^\infty \frac{\cos t}{t}\, dt. \tag{3.129}$$

In Fig. 3.4 we plot the sine and cosine integrals for real values of x. Observe the small oscillatory behavior of each function.

Here again, to derive the asymptotic relations for these functions, we start by writing the sine function in Eq. (3.128) in a power series followed by termwise integration, which leads to

$$\mathrm{Si}(z) = \sum_{n=0}^{\infty} \frac{(-1)^n z^{2n+1}}{(2n+1)(2n+1)!}. \tag{3.130}$$

Then, by following the analysis used in previous sections, we can express Eq. (3.130) in terms of Pochhammer symbols and obtain

$$\begin{aligned} \mathrm{Si}(z) &= z \sum_{n=0}^{\infty} \frac{(-1)^n \left(\frac{1}{2}\right)_n (z/2)^{2n}}{\left(\frac{3}{2}\right)_n \left(\frac{3}{2}\right)_n n!} \\ &= z\,{}_1F_2\left(\frac{1}{2}; \frac{3}{2}, \frac{3}{2}; -\frac{z^2}{4}\right). \end{aligned} \tag{3.131}$$

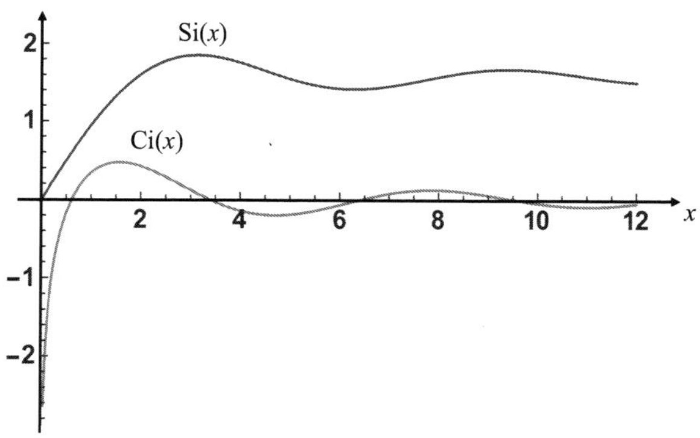

Figure 3.4 The sine and cosine integrals.

From results presented in Section 3.4, we generate the asymptotic formulas,

$$\mathrm{Si}(z) \sim \begin{cases} z\left(1 - \dfrac{z^2}{18} + \cdots\right), & z \to 0 \\ \dfrac{\pi}{2}, & z \to \infty. \end{cases} \tag{3.132}$$

For the cosine integral, we first observe that (Ref. 6, p. 36)

$$\int_0^z \frac{\cos t - 1}{t}\, dt = -\gamma - \ln z + \mathrm{Ci}(z). \tag{3.133}$$

Rearranging the terms in Eq. (3.133) and expanding the integrand of the integral in a power series, we are led to

$$\mathrm{Ci}(z) = \gamma + \ln z + \sum_{n=1}^{\infty} \frac{(-1)^n z^{2n}}{2n(2n)!}. \tag{3.134}$$

However, by changing the index $n \to n + 1$, we find, after some algebraic manipulation, that

$$\sum_{n=1}^{\infty} \frac{(-1)^n z^{2n}}{2n(2n)!} = -\frac{z^2}{4} \sum_{n=0}^{\infty} \frac{(-1)^n (1)_n (1)_n (z/2)^{2n}}{\left(\frac{3}{2}\right)_n (2)_n (2)_n n!}, \tag{3.135}$$

such that we conclude that

$$\mathrm{Ci}(z) = \gamma + \ln z - \frac{1}{4} z^2 {}_2F_3\left(1, 1; \frac{3}{2}, 2, 2; -\frac{z^2}{4}\right). \tag{3.136}$$

Hence, we see that Ci(z) is a multi-valued function with a branch point at $z = 0$.

Theoretically, we can deduce the behavior of Ci(z) for $z \to \infty$ from Eq. (3.136), although we can obtain a simpler expression by using integration by parts on Eq. (3.129) to find the large-argument behavior

$$\begin{aligned} \mathrm{Ci}(z) &= -\frac{\sin t}{t}\bigg|_z^{\infty} - \int_z^{\infty} \frac{\sin t}{t^2} dt \\ &= \frac{\sin z}{z} - \int_z^{\infty} \frac{\sin t}{t^2} dt. \end{aligned} \tag{3.137}$$

Consequently, we obtain the following asymptotic relations from Eqs. (3.136) and (3.137):

$$\mathrm{Ci}(z) \sim \begin{cases} \ln z, & z \to 0 \\ \dfrac{\sin z}{z}, & z \to \infty. \end{cases} \tag{3.138}$$

3.10.6 Complete elliptic integrals

The circular (trigonometric) functions sine and cosine are well known examples of periodic functions having one real period of 2π. Also, the complex exponential function has one imaginary period of $2\pi i$. However, there exist functions, called *Jacobi elliptic functions*, that are doubly periodic with one real period and one imaginary period (Ref. 4, Vol. II, p. 94). Among other areas of application, elliptic functions are important in solving the pendulum problem (Ref. 8, pp. 135–138). Although we won't pursue them here, elliptic functions are actually inverses of the elliptic integrals discussed below.

The parametric equations for an elliptic arc are given by $(b > a)$

$$\begin{aligned} x &= a\cos\theta, \\ y &= b\sin\theta, \quad 0 \le \theta \le \phi. \end{aligned} \tag{3.139}$$

Based on the formula for arc length from calculus, we find that the length of the elliptic arc defined by Eq. (3.139) leads to the integral

$$\begin{aligned} L &= \int_0^{\phi} \sqrt{a^2\sin^2\theta + b^2\cos^2\theta}\, d\theta \\ &= b\int_0^{\phi} \sqrt{1 - e^2\sin^2\theta}\, d\theta, \end{aligned} \tag{3.140}$$

where e is the eccentricity of the ellipse, defined by

$$e = \frac{1}{b}\sqrt{b^2 - a^2}. \tag{3.141}$$

The integral in Eq. (3.140) is clearly nonelementary—it is called an *elliptic integral* because of its origin.

Complete elliptic integrals of the first and *second kind* are defined, respectively, by

$$K(m) = \int_0^{\pi/2} \frac{d\theta}{\sqrt{1 - m^2\sin^2\theta}}, \quad 0 \le m < 1, \tag{3.142}$$

$$E(m) = \int_0^{\pi/2} \sqrt{1 - m^2\sin^2\theta}\, d\theta, \quad 0 \le m < 1. \tag{3.143}$$

If the upper limit of integration is an arbitrary angle ϕ like that in Eq. (3.140), both Eqs. (3.142) and (3.143) are called elliptic integrals of the first and second kind. Only when $\phi = \pi/2$ are they called "complete."

By expanding the integrand of Eq. (3.142) in a power series,

$$(1 - m^2\sin^2\theta)^{-1/2} = \sum_{n=0}^{\infty} \binom{-\frac{1}{2}}{n} (-1)^n m^{2n} \sin^{2n}\theta, \tag{3.144}$$

and using the integration formula (*G*16) in Appendix A along with (*PS*9), we get

$$\int_0^{\pi/2} \sin^{2n}\theta \, d\theta = \frac{\pi}{2}(-1)^n \binom{-\frac{1}{2}}{n}. \tag{3.145}$$

Thus, we deduce the series representation:

$$\begin{aligned} K(m) &= \frac{\pi}{2}\sum_{n=0}^{\infty} \binom{-\frac{1}{2}}{n}^2 m^{2n} \\ &= \frac{\pi}{2}\, {}_2F_1\left(\frac{1}{2}, \frac{1}{2}; 1; m^2\right). \end{aligned} \tag{3.146}$$

In a similar manner, we find that

$$\begin{aligned} E(m) &= \frac{\pi}{2}\sum_{n=0}^{\infty} \binom{-\frac{1}{2}}{n}\binom{\frac{1}{2}}{n} m^{2n} \\ &= \frac{\pi}{2}\, {}_2F_1\left(-\frac{1}{2}, \frac{1}{2}; 1; m^2\right). \end{aligned} \tag{3.147}$$

The two complete elliptic integrals (3.142) and (3.143) are plotted over the interval $0 \le m < 1$ in Fig. 3.5.

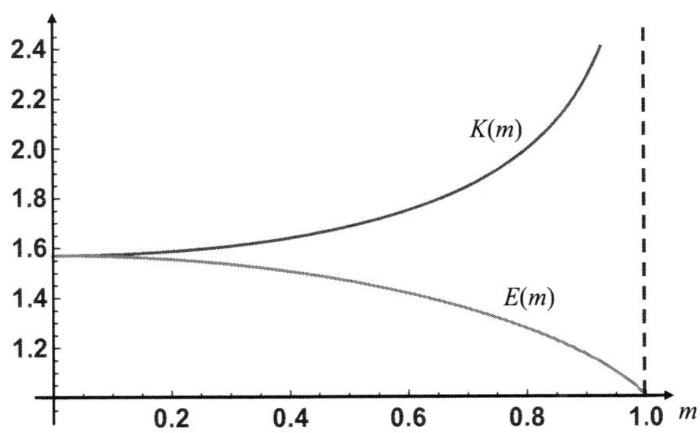

Figure 3.5 The complete elliptic integrals.

Based on the representations of Eqs. (3.146) and (3.147) in terms of the hypergeometric function, we readily obtain the asymptotic relations

$$K(m) \sim \begin{cases} \frac{\pi}{2}\left(1 + \frac{m^2}{4} + \cdots\right), & m \to 0 \\ \infty, & m \to 1 \end{cases} \tag{3.148}$$

$$E(m) \sim \begin{cases} \frac{\pi}{2}\left(1 - \frac{m^2}{4} + \cdots\right), & m \to 0 \\ 1, & m \to 1. \end{cases} \tag{3.149}$$

Remark: We caution the reader that some computer software programs are based on $k = m^2$ for calculation of the complete elliptic integrals: $K(k) = \frac{\pi}{2}\,{}_2F_1\left(\frac{1}{2}, \frac{1}{2}; k\right)$ and $E(k) = \frac{\pi}{2}\,{}_2F_1\left(-\frac{1}{2}, \frac{1}{2}; k\right)$.

3.11 Applications

The generalized hypergeometric functions and related special functions occur in a wide variety of applications. In this short section, we will look at some examples that involve these special functions. Several of the generalized hypergeometric functions will also appear in later chapters (Part II) of the text, where we consider additional applications. For a deeper understanding of special functions and their applications, the reader should consult Refs. 4, and 6–12.

3.11.1 Summing series

Among other applications, the generalized hypergeometric functions are useful for summing certain series, both finite and infinite. In the first example we consider summing up a finite series.

Example 1: Derive the combinatorial relation

$$\sum_{k=0}^{n} (-1)^k \binom{n}{k} \binom{a+k}{m} = (-1)^n \binom{a}{m-n}, \quad m = 0, 1, 2, \ldots.$$

From the Pochhammer symbol identities (*PS*9) and (*PS*14) in Appendix A, we have

$$\binom{n}{k} = \frac{(-1)^k}{k!}(-n)_k,$$
$$\binom{a+k}{m} = \frac{(-1)^m(-a)_m(a+1)_k}{m!(a+1-m)_k},$$

and therefore,

$$\sum_{k=0}^{n}(-1)^k\binom{n}{k}\binom{a+k}{m} = \frac{(-1)^m(-a)_m}{m!}\sum_{k=0}^{n}\frac{(-n)_k(a+1)_k}{(a+1-m)_k k!}$$
$$= \frac{(-1)^m(-a)_m}{m!}\,{}_2F_1(-n,\, a+1;\, a+1-m;\, 1). \tag{3.150}$$

By recalling Eq. (3.58), we see that

$$\begin{aligned} {}_2F_1(-n,\, a+1;\, a+1-m;\, 1) &= \frac{\Gamma(a+1-m)\Gamma(n-m)}{\Gamma(a+1+n-m)\Gamma(-m)} \\ &= \frac{(-1)^n m!}{(m-n)!(a+1-m)_n}. \end{aligned} \tag{3.151}$$

In arriving at the last step in Eq. (3.151), we used Pochhammer symbol identities from Appendix A to find that

$$\frac{\Gamma(a+1-m)}{\Gamma(a+1+n-m)} = \frac{1}{(a+1-m)_n},$$
$$\frac{\Gamma(n-m)}{\Gamma(-m)} = \frac{(-1)^n m!}{(m-n)!}.$$

Also, from (*PS*3) in Appendix A, it follows that

$$(-a)_m = (-1)^n(-a)_{m-n}(a+1-m)_n,$$

so by combining results, we arrive at

$$\begin{aligned} \sum_{k=0}^{n}(-1)^k\binom{n}{k}\binom{a+k}{m} &= \frac{(-1)^m}{(m-n)!}(-a)_{m-n} \\ &= (-1)^n\binom{a}{m-n}, \qquad m = 0, 1, 2, \ldots. \end{aligned} \tag{3.152}$$

In the next example we illustrate again the use of Pochhammer symbols to recognize a given series as a generalized hypergeometric function.

Example 2: Sum the series $\sum_{n=0}^{\infty}\frac{n!}{(2n+1)!}x^{2n+1}$.

From the Pochhammer relations (*PS*1) and (*PS*11) in Appendix A, we obtain

$$\begin{aligned} \sum_{n=0}^{\infty}\frac{n!}{(2n+1)!}x^{2n+1} &= x\sum_{n=0}^{\infty}\frac{(1)_n}{\left(\frac{3}{2}\right)_n}\frac{(x/2)^{2n}}{n!} \\ &= x\,{}_1F_1\left(1;\, \frac{3}{2};\, \frac{x^2}{4}\right) \\ &= xe^{x^2/4}\,{}_1F_1\left(\frac{1}{2};\, \frac{3}{2};\, -\frac{x^2}{4}\right). \end{aligned} \tag{3.153}$$

We have applied property (*CH*4) in Appendix A in the last step of Eq. (3.153). However, from Eq. (3.97), we can replace this last expression with

$$\sum_{n=0}^{\infty} \frac{n!}{(2n+1)!} x^{2n+1} = \sqrt{\pi} e^{x^2/4} \operatorname{erf}(x/2). \tag{3.154}$$

3.11.2 Probability and statistics

Some of the functions studied in this chapter are closely associated with problems in probability and statistics. Two of the classic examples concern the cumulative distribution function (CDF).

Example 3: Find the probability that the random variable $\mathbf{x} \leq x$, given that the PDF is the Gaussian distribution

$$p_{\mathbf{x}}(x) = \frac{1}{\sqrt{2\pi}\,\sigma} \exp\left[-\frac{(x-m)^2}{2\sigma^2}\right], \quad -\infty < x < \infty,$$

where m is the mean and σ^2 is the variance.

The CDF is defined by

$$\begin{aligned} \Pr(\mathbf{x} \leq x) &= \frac{1}{\sqrt{2\pi}\sigma} \int_{-\infty}^{x} \exp\left[-\frac{(u-m)^2}{2\sigma^2}\right] du \\ &= \frac{1}{\sqrt{2\pi}\sigma} \int_{-\infty}^{\infty} \exp\left[-\frac{(u-m)^2}{2\sigma^2}\right] du \\ &\quad - \frac{1}{\sqrt{2\pi}\sigma} \int_{x}^{\infty} \exp\left[-\frac{(u-m)^2}{2\sigma^2}\right] du \\ &= 1 - \frac{1}{2} \operatorname{erfc}\left(\frac{x-m}{\sqrt{2}\sigma}\right), \end{aligned} \tag{3.155}$$

where we made the change of variable $t = (u-m)/(\sqrt{2}\,\sigma)$ to evaluate the last two integrals. By using the identity $\operatorname{erfc}(z) = 1 - \operatorname{erf}(z)$, we can rewrite the last line in Eq. (3.155) as

$$\Pr(\mathbf{x} \leq x) = \frac{1}{2}\left[1 + \operatorname{erf}\left(\frac{x-m}{\sqrt{2}\,\sigma}\right)\right]. \tag{3.156}$$

Example 4: Find the CDF and the moments $\langle x^n \rangle$ for the gamma distribution,

$$p_{\mathbf{x}}(x) = \frac{(\alpha/m)^{\alpha}}{\Gamma(\alpha)} x^{\alpha-1} \exp\left(-\frac{\alpha x}{m}\right), \quad x > 0,\ \alpha > 0,$$

for which m is the mean and α is a parameter inversely related to the variance.

From the definition, we have

$$\begin{aligned}\Pr(\mathbf{x} \le x) &= \frac{(\alpha/m)^\alpha}{\Gamma(\alpha)} \int_0^x u^{\alpha-1} \exp\left(-\frac{\alpha u}{m}\right) du \\ &= \frac{1}{\Gamma(\alpha)} \int_0^{\alpha x/m} t^{\alpha-1} e^{-t}\, dt,\end{aligned} \tag{3.157}$$

where we made the substitution $t = \alpha u/m$. The last integral is recognized as the incomplete gamma function such that

$$\Pr(\mathbf{x} \le x) = \frac{\gamma(\alpha, \alpha x/m)}{\Gamma(\alpha)}. \tag{3.158}$$

The statistical moments are given by

$$\begin{aligned}\langle x^n \rangle &= \frac{(\alpha/m)^\alpha}{\Gamma(\alpha)} \int_0^\infty x^{n+\alpha-1} \exp\left(-\frac{\alpha x}{m}\right) dx \\ &= \frac{(m/\alpha)^n}{\Gamma(\alpha)} \int_0^\infty t^{n+\alpha-1} e^{-t}\, dt \\ &= \frac{(m/\alpha)^n}{\Gamma(\alpha)} \Gamma(n+\alpha), \qquad n = 1, 2, 3, \ldots,\end{aligned} \tag{3.159}$$

where $\langle\rangle$ denotes an ensemble average. For $n = 1$ and $n = 2$, we find that $\langle x \rangle = m$ and $\langle x^2 \rangle = m^2(1 + 1/\alpha)$, and thus, the variance is

$$\sigma^2 = \langle x^2 \rangle - \langle x \rangle^2 = \frac{m^2}{\alpha}. \tag{3.160}$$

3.11.3 Elliptic integrals

Elliptic integrals arise in a variety of settings besides finding the arc length of an ellipse or some other curve. Consider the next example.

Example 5: Determine the function $f(y)$ defined by the integral

$$f(y) = \int_0^\pi \frac{\cos(t)dt}{\sqrt{a^2 + b^2 + y^2 - 2ab\cos(t)}}.$$

In the following analysis, we will introduce the parameter

$$m^2 = \frac{4ab}{(a+b)^2 + y^2},$$

and make use of the trigonometric identity $\cos(2x) = 2\cos^2 x - 1$. We start with the change of variable $t = 2x$ and some algebraic rearranging to obtain

$$
\begin{aligned}
f(y) &= 2\int_0^{\pi/2} \frac{\cos(2x)dx}{\sqrt{a^2+2ab+b^2+y^2-2ab[1+\cos(2x)]}} \\
&= \frac{2}{\sqrt{(a+b)^2+y^2}}\int_0^{\pi/2} \frac{2\cos^2 x-1}{\sqrt{1-m^2\cos^2 x}}\,dx.
\end{aligned}
\tag{3.161}
$$

Now let $x=\frac{\pi}{2}-\theta$, which leads to

$$
\begin{aligned}
f(y) &= \frac{2}{m\sqrt{ab}}\int_0^{\pi/2} \frac{m^2\sin^2\theta}{\sqrt{1-m^2\sin^2\theta}}\,d\theta - \frac{m}{\sqrt{ab}}\int_0^{\pi/2} \frac{1}{\sqrt{1-m^2\sin^2\theta}}\,d\theta \\
&= -\frac{2}{m\sqrt{ab}}\int_0^{\pi/2} \frac{1-m^2\sin^2\theta}{\sqrt{1-m^2\sin^2\theta}}\,d\theta + \left(\frac{2}{m\sqrt{ab}} - \frac{m}{\sqrt{ab}}\right)K(m),
\end{aligned}
\tag{3.162}
$$

or

$$
f(y) = \frac{2}{m\sqrt{ab}}\left[\left(1-\frac{m^2}{2}\right)K(m) - E(m)\right], \tag{3.163}
$$

where

$$
m = 2\sqrt{\frac{ab}{(a+b)^2+y^2}}. \tag{3.164}
$$

3.12 Historical Remarks

The term "hypergeometric series" appears to have first been used in 1655 by the English mathematician John Wallis (1616–1703) in his book *Arithmetica Infinitorum*. Wallis was also the first to introduce the symbol "∞" for infinity. Nonetheless, the hypergeometric series as we know it today was actually introduced by Leonhard Euler (1707–1783), who also discovered the governing differential equation. Years later, Carl Friedrich Gauss (1777–1855) performed the first systematic analysis of the hypergeometric series and is the name we most associate today with the hypergeometric function. The mathematical achievements of Gauss, however, go far beyond the hypergeometric series and, in fact, he is generally regarded as one of the greatest mathematicians of all time for his many contributions to the fields of mathematics, astronomy, and physics that include number theory, differential geometry, probability theory, geodesy, planetary astronomy, the theory of functions, and potential theory.

Further analysis on the generalized hypergeometric functions that was done in the nineteenth and early twentieth centuries is associated with a host of names like Ernst Eduard Kummer (1810–1893), the German mathematician who introduced the confluent hypergeometric function of the first kind and showed that there exist 24 solutions of the hypergeometric differential equation. Kummer was also the doctoral advisor of Leo August Pochhammer (1841–1920), who introduced the Pochhammer symbol. Other prominent names include Geor Friedrich Bernhard Riemann (1826–1866), who was a student of Gauss and, like Gauss, is considered by many to be one of the greatest mathematicians of all time for his numerous mathematical achievements, including the use of complex analytical methods to present solutions of the differential equation satisfied by the hypergeometric function in the complex plane. The French mathematician Paul Émile Appell (1855–1930) introduced a set of four hypergeometric series of two variables, and the Italian mathematician Giuseppe Lauricella (1867–1913) introduced hypergeometric functions of three or more variables.

Some of the famous more recent names associated with special functions include Ernest William Barnes (1874–1953), who made extensive use of the integrals studied by Hjalmar Mellin (1854–1933), in which the integral involves gamma functions of the variable of integration. In addition, Sir Edmund Taylor Whittaker (1873–1956) is known for the functions named after him and for his treatise with George Neville Watson (1886–1965) called *A Course of Modern Analysis.*[7] Whittaker also made contributions in mathematical physics and numerical analysis. Thomas Murray MacRobert (1884–1962) was a Scottish mathematician who introduced the MacRobert E-function in an effort to address generalized hypergeometric functions for the case $p > q + 1$. For the same reason, the Dutch mathematician Cornelis Simon Meijer (1904–1974) introduced the Meijer G-function; he also introduced generalizations of the Laplace transform, which are now referred to as Meijer transforms. Arthur Erdélyi (1908–1977), after the death of Harry Bateman in 1946, was given the enormous task of finishing and publishing Bateman's manuscripts—the *Bateman Manuscript Project.*[4] In 1947, the Italian mathematician Francesco G. Tricomi (1897–1978) introduced the confluent hypergeometric function of the second kind and was one of the research associates on the Bateman Manuscript Project along with Wilhelm Magnus (1907–1990) and Fritz Oberhettinger (1911–1993).

Other special functions introduced here were investigated by a number of scientists. Adrien Marie Legendre (1752–1833), who studied mathematics, mechanics, and astronomy, introduced the symbol "Γ" for the gamma function and was also responsible for the duplication formula of the gamma function. His name is probably more well known, however, for his work in celestial mechanics, which led to the polynomials now bearing his name. The first investigation of the incomplete gamma function is also accredited to Legendre. The incomplete gamma functions can be directly related to the exponential and logarithmic integrals, sine and cosine integrals, error functions, and the Fresnel integrals. A form of the complementary error function was originally introduced in 1808 by Christian Kramp (1760–1826), who called it the "Faddeeva function." Among other applications, the Faddeeva function was used in the analysis of electromagnetic waves of the type used in AM radio. However, Kramp is better known for his work on factorials, where he was the first to use the notation $n!$ for a factorial. The Fresnel integrals used in optics are named after the civil engineer and physicist Augustin-Jean Fresnel (1788–1827). Elliptic integrals were encountered over the years by several mathematicians such as John Wallis and Leonhard Euler, but much of the theory of elliptic integrals was developed by Legendre over a period spanning 40 years.

References

1. C. F. Gauss, "Disquisitiones Generales circa Seriem Infinitam...," *Comment. Soc. Reg. Sci. Gottingensis Recent.* **2**, 1–42 (1812).
2. E. E. Kummer, "Über die Hypergeometrische Reihe $F(a;b;x)$," *J. Reine Angew. Math.* **15** (39–83), 127–172 (1836).
3. F. W. J. Olver, *Asymptotics and Special Functions*, A K Peters/CRC Press, New York (1997).
4. A. Erdélyi, W. Magnus, F. Oberhettinger, and F. G. Tricomi, *Higher Transcendental Functions*, Vols. I and II, Bateman Manuscript Project, McGraw Hill, New York (1953).
5. F. Tricomi, "Sulle funzioni ipergeometriche confluent," *Annali di Matematica Pura ed Applicata*, Series 4 (in Italian) **26**, 141–175 (1947).
6. N. N. Lebedev, *Special Functions and Their Applications*, translated and edited by R. A. Silverman, Prentice-Hall, Upper Saddle River, New Jersey (1965).
7. E. T. Whittaker and G. N. Watson, *A Course of Modern Analysis*, 4th Edition, Cambridge University Press, London (1965).
8. L. C. Andrews, *Special Functions of Mathematics for Engineers*, 2nd Edition, SPIE Press, Bellingham, Washington (1998) [doi: 10.1117/3.270709].
9. R. A. Askey, Editor, *Theory and Application of Special Functions*, Academic Press, New York (1975).
10. L. J. Slater, *Generalized Hypergeometric Functions*, Cambridge University Press, Cambridge, UK (1966).
11. Y. L. Luke, *The Special Functions and Their Applications*, 2 vols., Academic Press, New York (1969).
12. V. Lakshminarayanan and L. S. Varadharajan, *Special Functions for Optical Science and Engineering*, SPIE Press, Bellingham, Washington (2015) [doi: 10.1117/3.2207310].

Chapter 4
Bessel Functions

4.0 Introduction

Thus far we have used the notion of a Barnes integral to derive asymptotic results for small and large arguments of several generalized hypergeometric functions and a few related special functions. Here we wish to continue that analysis in connection with various Bessel functions, based on derived asymptotic results in Chapter 3. In particular, we will first derive asymptotic formulas for the standard and modified Bessel functions and then move on to less well-known functions from the Bessel function family. All of the Bessel functions satisfy a large number of identities, some of which are listed here and in Appendix A. For a more complete treatment of Bessel functions, the reader should consult Refs. 1–6.

4.1 Standard Bessel Functions

The *standard Bessel functions of the first* and *second kind* are often associated with problems having circular or cylindrical symmetry. Because of this association they are sometimes called *cylinder functions*. However, beyond their connection to cylindrical domains, they also occur in a wide variety of physics and engineering applications, including electromagnetic theory, heat conduction, vibrations of a thin plate, fiber optics, and optical wave propagation. In fact, they are considered the most important functions beyond the elementary functions of calculus.

The Bessel functions $J_p(z)$, $J_{-p}(z)$, and $Y_p(z)$ are all solutions of the second-order differential equation (DE)

$$z^2\frac{d^2w}{dz^2} + z\frac{dw}{dz} + (z^2 - p^2)w = 0. \tag{4.1}$$

To include all values of p, the general solution is usually written in terms of $J_p(z)$ and $Y_p(z)$, i.e.,

$$w(z) = C_1 J_p(z) + C_2 Y_p(z), \tag{4.2}$$

where C_1 and C_2 are arbitrary constants. In cases where p is not an integer, the general solution can also be written as $w(z) = C_1 J_p(z) + C_2 J_{-p}(z)$.

4.1.1 The function $J_p(z)$

A series solution of Eq. (4.1) for the standard Bessel function of the first kind and order p is given by

$$\begin{aligned} J_p(z) &= \left(\frac{z}{2}\right)^p \sum_{k=0}^{\infty} \frac{(-1)^k (z/2)^{2k}}{\Gamma(1+p+k)k!} \\ &= \frac{(z/2)^p}{\Gamma(1+p)}\, {}_0F_1\left(-;\, 1+p;\, -\frac{z^2}{4}\right), \qquad |z| < \infty, \end{aligned} \tag{4.3}$$

where in the second line the series is expressed in terms of a generalized hypergeometric function. From Eq. (4.3) we obtain the small-argument asymptotic approximation

$$J_p(z) \sim \frac{(z/2)^p}{\Gamma(1+p)}\left[1 - \frac{z^2}{4(1+p)} + \cdots\right],\; p \neq -1,\, -2,\, -3, \ldots,\; z \to 0. \tag{4.4}$$

The series representation of Eq. (4.3) for non-integer negative order is obtained by replacing p with $-p$. For negative integer order, we replace p with $-n$, $n = 1, 2, 3, \ldots$, which leads to

$$J_{-n}(z) = \sum_{k=0}^{\infty} \frac{(-1)^k (z/2)^{2k-n}}{\Gamma(1-n+k)k!} = \sum_{k=n}^{\infty} \frac{(-1)^k (z/2)^{2k-n}}{\Gamma(1-n+k)k!}, \tag{4.5}$$

where we have used the result $1/\Gamma(1-n+k) = 0$, $k = 0, 1, \ldots, n-1$ by virtue of ($G6$) in Appendix A. Now we make the change of index $k = m + n$ to get

$$J_{-n}(z) = \sum_{m=0}^{\infty} \frac{(-1)^{m+n} (z/2)^{2m+n}}{\Gamma(1+n+m)m!} = (-1)^n J_n(z). \tag{4.6}$$

When p is not an integer, the Bessel function (4.3) exhibits a branch point at $z = 0$ because of the factor $(z/2)^p$.

The graphs of the first few Bessel functions of the first kind for integer order $p = 0, 1, 2$ and real $z = x$ are shown in Fig. 4.1. These functions have an oscillatory behavior similar to that of the trigonometric functions. However, the zeros of the Bessel functions are not evenly spaced like those of the trigonometric functions, and the maximum amplitude decreases with increasing values of x.

The Bessel function $J_p(z)$ has several integral representations. One of the standard integrals for integer order n is Bessel's original definition of $J_n(z)$, viz.,

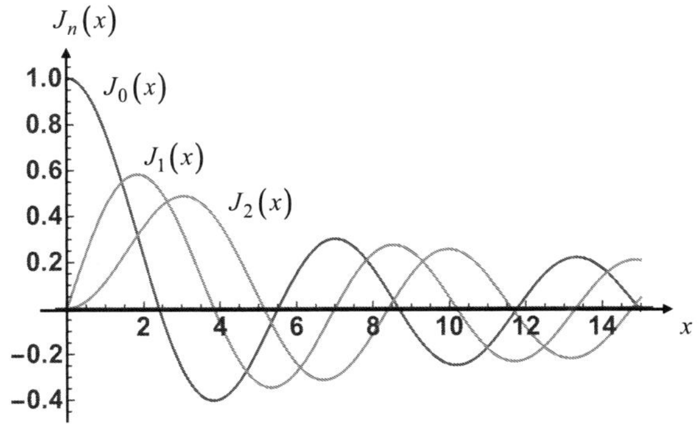

Figure 4.1 Standard Bessel functions of the first kind: $n = 0, 1, 2$.

$$J_n(z) = \frac{1}{\pi}\int_0^{\pi} \cos(n\theta - z\sin\theta)\,d\theta, \quad n = 0, 1, 2, \ldots \tag{4.7}$$

Although we won't derive Eq. (4.7), we will derive a more general form provided by the integral

$$J_p(z) = \frac{(z/2)^p}{\sqrt{\pi}\,\Gamma\left(p+\frac{1}{2}\right)}\int_{-1}^{1} e^{izt}(1-t^2)^{p-\frac{1}{2}}dt, \quad p > -\frac{1}{2}. \tag{4.8}$$

To develop this integral representation, we start with the beta function relation ($G17$) in Appendix A, a change of variable $t \to t^2$, and some rearrangement to get

$$\frac{1}{\Gamma(k+p+1)} = \frac{1}{\Gamma\left(k+\frac{1}{2}\right)\Gamma\left(p+\frac{1}{2}\right)}\int_{-1}^{1} t^{2k}(1-t^2)^{p-\frac{1}{2}}dt. \tag{4.9}$$

If we substitute the right-hand side of this result for $1/\Gamma(k+p+1)$ in the series shown in Eq. (4.3), and then interchange the order of summation and integration, we are led to

$$\begin{aligned} J_p(z) &= \frac{(z/2)^p}{\Gamma\left(p+\frac{1}{2}\right)}\int_{-1}^{1}(1-t^2)^{p-\frac{1}{2}}\sum_{k=0}^{\infty}\frac{(-1)^k(tz/2)^{2k}}{\Gamma\left(k+\frac{1}{2}\right)k!}dt \\ &= \frac{(z/2)^p}{\sqrt{\pi}\,\Gamma\left(p+\frac{1}{2}\right)}\int_{-1}^{1}(1-t^2)^{p-\frac{1}{2}}\sum_{k=0}^{\infty}\frac{(-1)^k(tz)^{2k}}{(2k)!}dt \\ &= \frac{(z/2)^p}{\sqrt{\pi}\,\Gamma\left(p+\frac{1}{2}\right)}\int_{-1}^{1}(1-t^2)^{p-\frac{1}{2}}\cos(zt)dt. \end{aligned} \tag{4.10}$$

Property ($G5$) of the gamma function in Appendix A was used to obtain the middle expression in Eq. (4.10). Also, the cosine function can be replaced by e^{izt} with no change in value of the integral.

If we make the substitution $t \to 2t - 1$ in Eq. (4.8), we arrive at a form reminiscent of the confluent hypergeometric function,

$$J_p(z) = \frac{(2z)^p}{\sqrt{\pi}\Gamma\left(p + \frac{1}{2}\right)} e^{-iz} \int_0^1 e^{2izt} t^{p-\frac{1}{2}} (1 - t)^{p-\frac{1}{2}} dt. \tag{4.11}$$

That is, we recognize the above integral as a special case of the confluent hypergeometric function relation ($CH2$) in Appendix A,

$${}_1F_1(a; c; z) = \frac{\Gamma(c)}{\Gamma(a)\Gamma(c-a)} \int_0^1 e^{zt} t^{a-1} (1 - t)^{c-a-1} dt. \tag{4.12}$$

By comparing Eq. (4.11) with Eq. (4.12), we find equivalence in the two integrals with $a = p + \frac{1}{2}$ and $c = 2p + 1$. Thus, Eq. (4.11) can be written as

$$J_p(z) = \frac{(2z)^p \Gamma\left(p + \frac{1}{2}\right)}{\sqrt{\pi}\,\Gamma(2p+1)} e^{-iz} {}_1F_1\left(p + \frac{1}{2}; 2p + 1; 2iz\right). \tag{4.13}$$

From the formulas ($G10$) and ($CH4$) in Appendix A, we can further rewrite Eq. (4.13) as

$$J_p(z) = \frac{(z/2)^p}{\Gamma(p+1)} e^{iz} {}_1F_1\left(p + \frac{1}{2}; 2p + 1; -2iz\right). \tag{4.14}$$

We end this section by displaying two special cases of the Bessel function that make a simple connection to the trigonometric functions (cf. also Section 4.1.4):

$$J_{1/2}(z) = \sqrt{\frac{2}{\pi z}} \sin z, \quad J_{-1/2}(z) = \sqrt{\frac{2}{\pi z}} \cos z. \tag{4.15}$$

4.1.2 Large-argument asymptotic approximation

In the following we will use a process called *analytic continuation*, which is a technique used to extend the domain of definition of a given analytic function. For the Bessel function this can take the form $e^{imp\pi} J_p(z) = J_p(ze^{im\pi})$, which we rewrite as

$$J_p(z) = e^{-imp\pi} J_p(ze^{im\pi}), \quad m = 0, \pm 1, \pm 2, \ldots. \tag{4.16}$$

Note that the argument $ze^{im\pi}$ is the same for all even values of m or all odd values of m. Thus, we can find the large-argument approximation by summing the asymptotic expressions for $m = 0$ and $m = 1$. To start, we set $m = 0$ in Eq. (4.16) and recall Eq. (4.13). We then use identity ($CH4$) in Appendix A to get ($p \geq 0$)

$$\begin{aligned} J_p(z)|_{m=0} &= \frac{(2z)^p \Gamma\left(p+\frac{1}{2}\right)}{\sqrt{\pi}\,\Gamma(2p+1)} e^{iz}\,{}_1F_1\left(p+\frac{1}{2};\, 2p+1;\, -2iz\right) \\ &\sim \frac{(2z)^p}{\sqrt{\pi}} e^{iz}(2iz)^{-\left(p+\frac{1}{2}\right)}, \quad z \to \infty, \end{aligned} \tag{4.17}$$

the last step of which comes from ($CH6$) in Appendix A. By also setting $i^{-\left(p+\frac{1}{2}\right)} = \exp\left[-i\left(p+\frac{1}{2}\right)\frac{\pi}{2}\right]$, we arrive at

$$J_p(z)|_{m=0} \sim \frac{1}{\sqrt{2\pi z}} \exp\left\{i\left[z - \frac{(2p+1)\pi}{4}\right]\right\}, \quad z \to \infty. \tag{4.18}$$

Equation (4.18) is part of the large-argument expression corresponding to $m = 0$ in Eq. (4.16). By returning to Eq. (4.16) and setting $m = 1$, we find that

$$\begin{aligned} J_p(z)|_{m=1} &= e^{-ip\pi} J_p(-z) \\ &= \frac{(-2z)^p \Gamma\left(p+\frac{1}{2}\right)}{\sqrt{\pi}\,\Gamma(2p+1)} e^{-ip\pi} e^{-iz}\,{}_1F_1\left(p+\frac{1}{2};\, 2p+1;\, 2iz\right) \\ &\sim \frac{(-2z)^p}{\sqrt{\pi}} e^{-i(z+p\pi)}(-2iz)^{-\left(p+\frac{1}{2}\right)}, \quad z \to \infty, \end{aligned} \tag{4.19}$$

where we have again used ($CH6$) in Appendix A in the last step. Now, from complex variables we see that

$$(-1)^p = e^{ip\pi}, \quad (-i)^{-\left(p+\frac{1}{2}\right)} = \exp\left[i\left(p+\frac{1}{2}\right)\frac{\pi}{2}\right],$$

and thus, it follows that Eq. (4.19) becomes

$$J_p(z)|_{m=1} \sim \frac{1}{\sqrt{2\pi z}} \exp\left\{-i\left[z - \frac{(2p+1)\pi}{4}\right]\right\}, \quad z \to \infty. \tag{4.20}$$

Hence, by summing Eqs. (4.18) and (4.20), we obtain the asymptotic result,

$$
\begin{aligned}
J_p(z) &\sim \frac{1}{\sqrt{2\pi z}}\exp\left\{i\left[z-\frac{(2p+1)\pi}{4}\right]\right\}+\frac{1}{\sqrt{2\pi z}}\exp\left\{-i\left[z-\frac{(2p+1)\pi}{4}\right]\right\}\\
&\sim \sqrt{\frac{2}{\pi z}}\cos\left[z-\frac{(2p+1)\pi}{4}\right], \quad p\geq 0, \quad z\to\infty.
\end{aligned}
\tag{4.21}
$$

4.1.3 The function $Y_p(z)$

The standard Bessel function of the second kind is defined by

$$
Y_p(z)=\frac{(\cos p\pi)J_p(z)-J_{-p}(z)}{\sin p\pi}, \quad p>0. \tag{4.22}
$$

When $p = n$, $n = 0, 1, 2, \ldots$, the expression (4.22) requires further investigation. For example, when $p = 0$ we employ L'Hôpital's rule to get (p. 274–276 of Ref. 4)

$$
\begin{aligned}
Y_0(z) &= \lim_{p\to 0}\frac{1}{\pi}\left[\frac{\partial}{\partial p}J_p(z)-\frac{\partial}{\partial p}J_{-p}(z)\right]\\
&= \frac{2}{\pi}J_0(z)\left(\gamma+\ln\frac{z}{2}\right)-\frac{2}{\pi}\sum_{k=1}^{\infty}\frac{(-1)^k(z/2)^{2k}}{(k!)^2}\left(1+\frac{1}{2}+\cdots+\frac{1}{k}\right).
\end{aligned}
\tag{4.23}
$$

The constant γ is the Euler constant defined by ($G20$) in Appendix A. For general integer n, the expression for $Y_n(z)$ is similar to that of Eq. (4.23). Also, it can be shown that $Y_{-n}(z) = (-1)^n Y_n(z)$, $n = 1, 2, 3, \ldots$. The functions $Y_p(x)$ have oscillatory characteristics like those of $J_p(x)$, but the asymptotic behavior near zero is more like that of a logarithmic function. Graphs of the first few Bessel functions of the second kind for integer $p = 0, 1, 2$ and real $z = x$ are shown in Fig. 4.2.

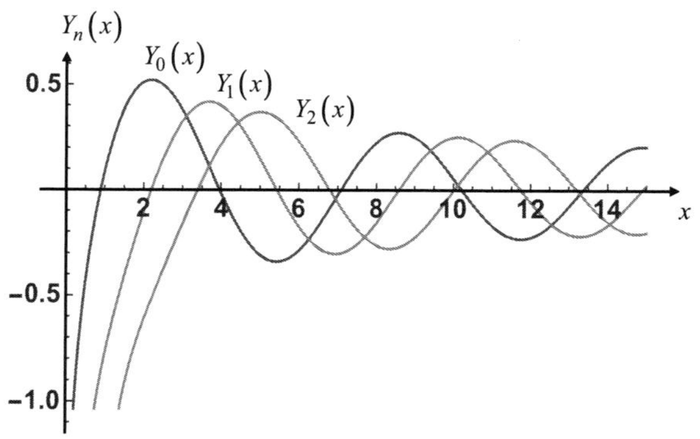

Figure 4.2 Standard Bessel functions of the second kind: $n = 0, 1, 2$.

To derive the asymptotic behavior for small arguments of $Y_p(z)$, we note that $J_p(z) \sim 0$ for $p > 0$ and $z \to 0$; thus, Eq. (4.22) becomes

$$\begin{aligned} Y_p(z) &\sim -\frac{J_{-p}(z)}{\sin p\pi} \\ &\sim -\frac{(z/2)^{-p}}{\Gamma(1-p)\sin p\pi}, \quad p > 0, \quad z \to 0. \end{aligned} \tag{4.24}$$

By use of the identity ($G7$) in Appendix A, we then arrive at the simpler asymptotic relation

$$Y_p(z) \sim -\frac{\Gamma(p)}{\pi}\left(\frac{2}{z}\right)^p, \quad p > 0, \quad z \to 0. \tag{4.25}$$

Also, for small arguments with $p = 0$, it follows from Eq. (4.23) that

$$Y_0(z) \sim \frac{2}{\pi}\ln z, \quad z \to 0. \tag{4.26}$$

The large-argument expression for $Y_p(z)$ is similar to that of $J_p(z)$, although we won't derive it here. A derivation based on the Hankel function relationships is provided in Section 4.3.1, which leads to

$$Y_p(z) \sim \sqrt{\frac{2}{\pi z}}\sin\left[z - \frac{(2p+1)\pi}{4}\right], \quad p \geq 0, \quad z \to \infty. \tag{4.27}$$

Both of the standard Bessel functions $J_p(z)$ and $Y_p(z)$ satisfy a number of similar identities (cf. Appendix A).

4.1.4 Spherical Bessel functions

The *spherical Bessel functions* are associated with the standard Bessel functions of half-integer order. They often arise as solutions of the Helmholtz equation in spherical coordinates. Through separation of variables, such problems may lead to the second-order DE,

$$z^2\frac{d^2w}{dz^2} + 2z\frac{dw}{dz} + [k^2z^2 - n(n+1)]w = 0, \quad n = 0, 1, 2, \ldots, \tag{4.28}$$

which has the general solution (cf. Section 4.4.1)

$$w(z) = \frac{1}{\sqrt{z}}[C_1J_{n+1/2}(kz) + C_2Y_{n+1/2}(kz)]. \tag{4.29}$$

The spherical Bessel functions of the first kind are defined by

$$j_n(z) = \sqrt{\frac{\pi}{2z}} J_{n+1/2}(z), \quad n = 0, 1, 2, ..., \tag{4.30}$$

and those of the second kind by

$$y_n(z) = \sqrt{\frac{\pi}{2z}} Y_{n+1/2}(z), \quad n = 0, 1, 2, ... \tag{4.31}$$

Thus, the general solution of Eq. (4.28) is also $w(z) = C_3 j_n(kz) + C_4 y_n(kz)$. The graphs of Eqs. (4.30) and (4.31) for $n = 0, 1, 2$ are provided in Fig. 4.3. Note the similarity to the standard Bessel functions plotted in Figs. 4.1 and 4.2.

Last, from the series definition for $J_{n+1/2}(z)$ and $Y_{n+1/2}(z)$, it is easily shown that the first two spherical Bessel functions of each kind are

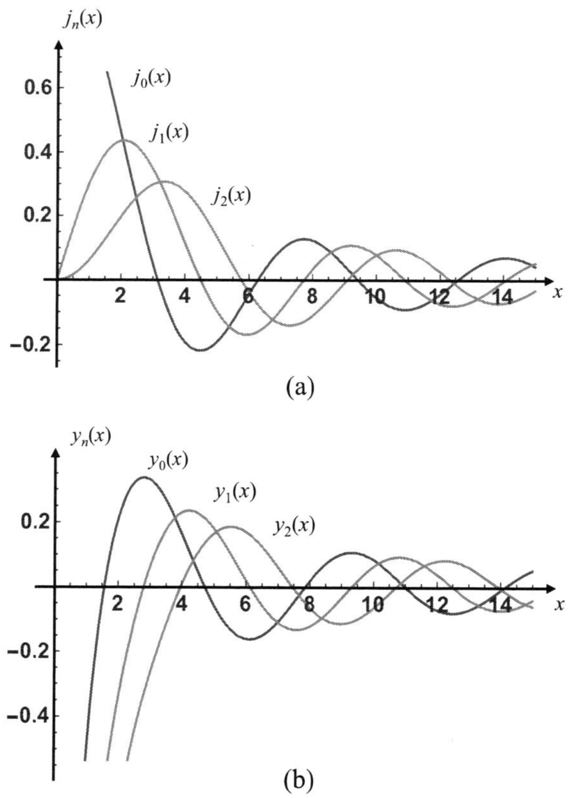

Figure 4.3 Spherical Bessel functions (a) of the first kind defined by Eq. (4.30) and (b) of the second kind defined by Eq. (4.31): $n = 0, 1, 2$.

$$j_0(z) = \frac{\sin z}{z}, \quad j_1(z) = \frac{\sin z}{z^2} - \frac{\cos z}{z},$$
$$y_0(z) = -\frac{\cos z}{z}, \quad y_1(z) = -\frac{\cos z}{z^2} - \frac{\sin z}{z}.$$

The small- and large-argument behavior of these functions can be deduced directly from that of the standard Bessel functions. That is,

$$j_n(z) \sim \begin{cases} \dfrac{n!(2z)^n}{(2n+1)!}, & n = 0, 1, 2, ..., \quad z \to 0 \\ \dfrac{1}{z}\sin\left(z - \dfrac{n\pi}{2}\right), & n = 0, 1, 2, ..., \quad z \to \infty, \end{cases} \tag{4.32}$$

$$y_n(z) \sim \begin{cases} -\dfrac{2(2n)!}{n!(2z)^{n+1}}, & n = 0, 1, 2, ..., \quad z \to 0 \\ -\dfrac{1}{z}\cos\left(z - \dfrac{n\pi}{2}\right), & n = 0, 1, 2, ..., \quad z \to \infty. \end{cases} \tag{4.33}$$

4.2 Modified Bessel Functions

Modified Bessel functions of the first and *second kind* display a behavior somewhat similar to that of the hyperbolic functions rather than that of the trigonometric functions. For this reason, they are sometimes called the *hyperbolic Bessel functions.* These functions are commonly associated with probability theory (e.g., see Sections 7.7.2 and 11.1) but also occur in many other applications. The Bessel functions $I_p(z)$ and $K_p(z)$ are linearly independent solutions of the second-order DE,

$$z^2 \frac{d^2w}{dz^2} + z\frac{dw}{dz} - (z^2 + p^2)w = 0, \tag{4.34}$$

from which we acquire the general solution $w(z) = C_1 I_p(z) + C_2 K_p(z)$. Below we present series representations and integral representations of these functions, similar to our treatment of the standard Bessel functions.

4.2.1 The function $I_p(z)$

The series representation for the modified Bessel function of the first kind and order p is

$$\begin{aligned} I_p(z) = i^{-p} J_p(iz) &= \left(\frac{z}{2}\right)^p \sum_{k=0}^{\infty} \frac{(z/2)^{2k}}{\Gamma(1+p+k)k!} \\ &= \frac{(z/2)^p}{\Gamma(1+p)}\, {}_0F_1\left(-; 1+p; \frac{z^2}{4}\right), \quad |z| < \infty, \end{aligned} \tag{4.35}$$

where we have also represented the series in terms of a generalized hypergeometric function. From the series in Eq. (4.35), we readily obtain the small argument approximation

$$I_p(z) \sim \frac{(z/2)^p}{\Gamma(1+p)}\left[1+\frac{z^2}{4(1+p)}+\cdots\right], p \neq -1, -2, -3, ..., z \to 0. \quad (4.36)$$

When $p = -n$, $n = 1, 2, 3, ...$, we find that $I_{-n}(z) = I_n(z)$. Like the standard Bessel functions, the modified Bessel functions $I_p(z)$ satisfy a large number of identities, some of which are listed in Appendix A.

Because each term of the power series (4.35) is positive for real $z = x$, the modified Bessel functions cannot have an oscillatory behavior. Graphs of the first few functions with $p = 0, 1, 2$ and real $z = x$ are shown in Fig. 4.4.

One of several integral representations of the modified Bessel function of the first kind, similar to that of the standard Bessel function, is

$$I_p(z) = \frac{(z/2)^p}{\sqrt{\pi}\Gamma\left(p+\frac{1}{2}\right)}\int_{-1}^{1} e^{-zt}(1-t^2)^{p-\frac{1}{2}}dt, \quad p > -\frac{1}{2}. \quad (4.37)$$

To derive this result, we can follow the derivation given for Eq. (4.8), which leads to

$$I_p(z) = \frac{(z/2)^p}{\sqrt{\pi}\Gamma\left(p+\frac{1}{2}\right)}\int_{-1}^{1} (1-t^2)^{p-\frac{1}{2}}\cosh(zt)dt. \quad (4.38)$$

However, the hyperbolic cosine function can be replaced by e^{-zt} with no change in the numerical results.

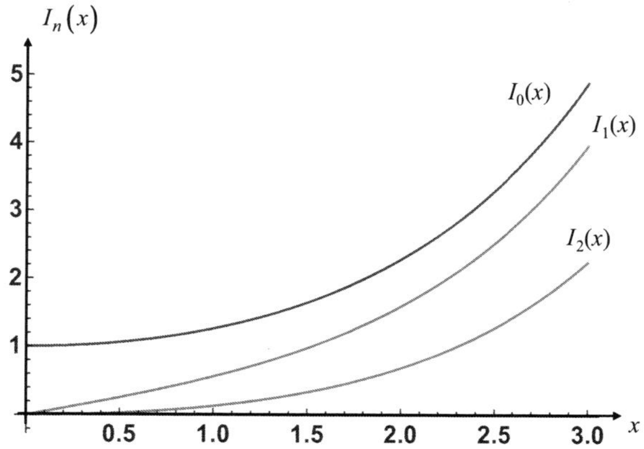

Figure 4.4 Modified Bessel functions of the first kind: $n = 0, 1, 2$.

If we make the substitution $t \to 1 - 2t$ in Eq. (4.37), the equation becomes

$$I_p(z) = \frac{(2z)^p}{\sqrt{\pi}\Gamma\left(p + \frac{1}{2}\right)} e^{-z} \int_0^1 e^{2zt} t^{p-\frac{1}{2}} (1-t)^{p-\frac{1}{2}} dt. \tag{4.39}$$

To compare the integral in Eq. (4.39) with that in Eq. (4.12), we see that $a = p + \frac{1}{2}$ and $c = 2p + 1$. Thus, Eq. (4.39) can be written as

$$I_p(z) = \frac{(2z)^p \Gamma\left(p + \frac{1}{2}\right)}{\sqrt{\pi}\Gamma(2p+1)} e^{-z} {}_1F_1\left(p + \frac{1}{2};\ 2p+1;\ 2z\right). \tag{4.40}$$

Then, with identity ($CH4$) in Appendix A, we arrive at

$$I_p(z) = \frac{(2z)^p \Gamma\left(p + \frac{1}{2}\right)}{\sqrt{\pi}\Gamma(2p+1)} e^{z} {}_1F_1\left(p + \frac{1}{2};\ 2p+1;\ -2z\right), \tag{4.41}$$

and by applying Eq. (3.29), it follows that Eq. (4.41) further reduces to the asymptotic form for large argument, given by

$$\begin{aligned} I_p(z) &\sim \frac{e^z}{\sqrt{2\pi z}}\, {}_2F_0\left(p + \frac{1}{2}, \frac{1}{2} - p;\ -;\ \frac{1}{2z}\right) \\ &\sim \frac{e^z}{\sqrt{2\pi z}}\left(1 - \frac{4p^2 - 1}{8z} + \cdots\right), \qquad z \to \infty. \end{aligned} \tag{4.42}$$

Here we have retained only the first couple of terms in the series for ${}_2F_0$ in Eq. (4.42), but by taking just the first term of the series, we have the simple relation

$$I_p(z) \sim \frac{e^z}{\sqrt{2\pi z}}, \qquad z \to \infty. \tag{4.43}$$

We end this section by displaying two special cases of the modified Bessel function that make a direct connection with the hyperbolic functions,

$$I_{1/2}(z) = \sqrt{\frac{2}{\pi z}} \sinh z, \quad I_{-1/2}(z) = \sqrt{\frac{2}{\pi z}} \cosh z. \tag{4.44}$$

In fact, all of the half-integer-order modified Bessel functions $I_{n \pm 1/2}(z)$ are related to various combinations of the hyperbolic functions (Section 4.2.3).

4.2.2 The function $K_p(z)$

The modified Bessel function of the second kind and order p is defined by

$$K_p(z) = \frac{\pi}{2}\frac{I_{-p}(z) - I_p(z)}{\sin p\pi}, \quad p \neq 0, \quad |\arg(z)| < \pi. \tag{4.45}$$

From definition, it follows that $K_{-p}(z) = K_p(z)$. The special case when $p = 0$ is obtained from the limit, $K_0(z) = \lim_{p\to 0} K_p(z)$, which leads to

$$K_0(z) = -I_0(z)\left(\gamma + \ln\frac{z}{2}\right) + \sum_{k=1}^{\infty}\frac{(z/2)^{2k}}{(k!)^2}\left(1 + \frac{1}{2} + \cdots + \frac{1}{k}\right), \tag{4.46}$$

where the constant γ is defined by ($G20$) in Appendix A. Graphs of the first few modified Bessel functions of the second kind with integer $p = 0, 1, 2$ and real $z = x$ are plotted in Fig. 4.5. Observe the exponential decay for increasing x, which is in contrast with the modified Bessel functions of the first kind, which exhibit exponential growth.

A power series representation for $K_p(z)$ can be deduced directly from the series representations for $I_p(z)$ and $I_{-p}(z)$. Also, for small arguments we can use $I_{-p}(z)$ in Eq. (4.45) and ($G7$) in Appendix A to obtain

$$K_p(z) \sim \frac{\Gamma(p)}{2}\left(\frac{2}{z}\right)^p, \quad p > 0, \quad z \to 0. \tag{4.47}$$

The small-argument special case $p = 0$ deduced from Eq. (4.46) is

$$K_0(z) \sim -\ln z, \quad z \to 0. \tag{4.48}$$

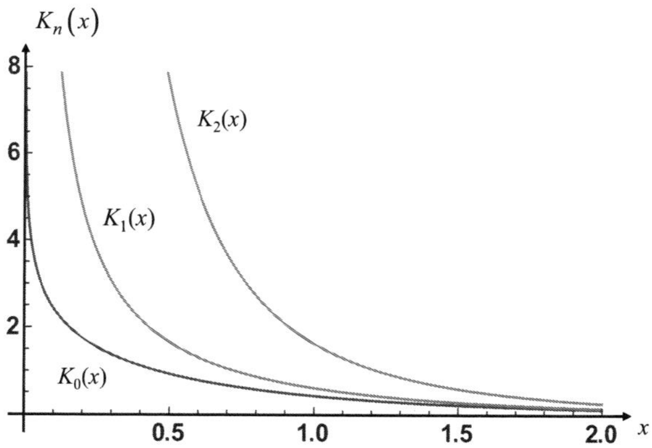

Figure 4.5 Modified Bessel functions of the second kind: $n = 0, 1, 2$.

The Bessel function $K_p(z)$ has an integral representation similar to that in Eq. (4.37), viz.,

$$K_p(z) = \frac{\sqrt{\pi}(z/2)^p}{\Gamma\left(p+\frac{1}{2}\right)} \int_1^\infty e^{-zt}(t^2-1)^{p-\frac{1}{2}}dt, \quad p > -\frac{1}{2}. \tag{4.49}$$

If we make the substitution $t \to 1 + 2t$, then Eq. (4.49) becomes

$$K_p(z) = \frac{\sqrt{\pi}\,(2z)^p}{\Gamma(p+\frac{1}{2})} e^{-z} \int_0^\infty e^{-2zt} t^{p-\frac{1}{2}}(1+t)^{p-\frac{1}{2}}dt. \tag{4.50}$$

This time we compare the above integral with the integral representation of the confluent hypergeometric function of the second kind ($CH10$) in Appendix A,

$$U(a;\, c;\, z) = \frac{1}{\Gamma(a)} \int_0^\infty e^{-zt} t^{a-1}(1+t)^{c-a-1}dt. \tag{4.51}$$

Clearly, if we choose $a = p + \frac{1}{2}$ and $c = 2p + 1$ in Eq. (4.51), it follows that

$$\begin{aligned} K_p(z) &= \sqrt{\pi}\,(2z)^p e^{-z} U\left(p+\frac{1}{2};\, 2p+1;\, 2z\right) \\ &\sim \sqrt{\frac{\pi}{2z}} e^{-z}\left(1 - \frac{4p^2-1}{8z} + \cdots\right), \quad p > -\frac{1}{2}, \quad z \to \infty. \end{aligned} \tag{4.52}$$

The last step in Eq. (4.52) is a consequence of Eq. (3.86). Taking only the first term of Eq. (4.52) yields

$$K_p(z) \sim \sqrt{\frac{\pi}{2z}} e^{-z}, \quad z \to \infty. \tag{4.53}$$

The final result here is the special case

$$K_{1/2}(z) = K_{-1/2}(z) = \sqrt{\frac{\pi}{2z}} e^{-z}. \tag{4.54}$$

4.2.3 Modified spherical Bessel functions

Spherical Bessel functions are directly related to both standard Bessel functions and modified Bessel functions. The *modified spherical Bessel functions of the first kind* and *second kind* are defined, respectively, by

$$i_n(z) = \sqrt{\frac{\pi}{2z}}\, I_{n+1/2}(z), \quad n = 0, 1, 2, \ldots, \tag{4.55}$$

$$k_n(z) = \sqrt{\frac{2}{\pi z}}\, K_{n+1/2}(z), \qquad n = 0, 1, 2, \ldots . \tag{4.56}$$

The asymptotic behavior of these functions for small arguments and for large arguments is readily deduced from the behavior of $I_p(z)$ and $K_p(z)$, i.e.,

$$i_n(z) \sim \begin{cases} \dfrac{n!(2z)^n}{(2n+1)!}, & n = 0, 1, 2, \ldots, \quad z \to 0 \\ \dfrac{e^z}{2z}, & n = 0, 1, 2, \ldots, \quad z \to \infty, \end{cases} \tag{4.57}$$

$$k_n(z) \sim \begin{cases} \dfrac{2(2n)!}{n!(2z)^{n+1}}, & n = 0, 1, 2, \ldots, \quad z \to 0 \\ \dfrac{e^{-z}}{z}, & n = 0, 1, 2, \ldots, \quad z \to \infty. \end{cases} \tag{4.58}$$

4.3 Other Bessel Functions

Up to this point we have concentrated on the most common members of the Bessel function family. Here we wish to consider some of the additional Bessel functions that are generally not as well-known.

4.3.1 Hankel functions

Analogous to the Euler formulas

$$e^{\pm iz} = \cos z \pm i \sin z,$$

we introduce the *Hankel functions*

$$H_p^{(1)}(z) = J_p(z) + iY_p(z), \quad p \ge 0, \tag{4.59}$$

$$H_p^{(2)}(z) = J_p(z) - iY_p(z), \quad p \ge 0. \tag{4.60}$$

These functions are also called *Bessel functions of the third kind.* Properties of these functions clearly follow directly from properties developed for the standard Bessel functions of the first and second kind.

The Hankel functions can also be related to the modified Bessel function of the second kind. In particular, we have $(z \to iz)$

$$\begin{aligned}
H_p^{(1)}(iz) &= J_p(iz) + iY_p(iz)\\
&= J_p(iz) + \frac{i}{\sin p\pi}[(\cos p\pi)J_p(iz) - J_{-p}(iz)]\\
&= \frac{J_{-p}(iz) - e^{-ip\pi}J_p(iz)}{i\sin p\pi}\\
&= e^{-ip\pi/2}\frac{I_{-p}(z) - I_p(z)}{i\sin p\pi},
\end{aligned} \tag{4.61}$$

which reduces to

$$H_p^{(1)}(iz) = \frac{2}{i\pi}e^{-ip\pi/2}K_p(z), \quad p \geq 0. \tag{4.62}$$

Now, by recalling the asymptotic relation of Eq. (4.53), it follows that $(z \to -iz)$

$$\begin{aligned}
H_p^{(1)}(z) &\sim \frac{2}{i\pi}e^{-ip\pi/2}\sqrt{\frac{\pi}{-2iz}}e^{iz}\\
&\sim \sqrt{\frac{2}{\pi z}}\exp\left\{i\left[z - \frac{(2p+1)\pi}{4}\right]\right\}\\
&\sim \sqrt{\frac{2}{\pi z}}\left\{\cos\left[z - \frac{(2p+1)\pi}{4}\right] + i\sin\left[z - \frac{(2p+1)\pi}{4}\right]\right\}, \quad z \to \infty.
\end{aligned} \tag{4.63}$$

Note that the real and imaginary parts of this last expression are the asymptotic results of Eqs. (4.21) and (4.27) for the standard Bessel functions.

The Hankel functions with negative order satisfy the relations

$$\begin{aligned}
H_{-p}^{(1)}(z) &= e^{ip\pi}H_p^{(1)}(z),\\
H_{-p}^{(2)}(z) &= e^{-ip\pi}H_p^{(2)}(z).
\end{aligned} \tag{4.64}$$

Last, because of the relations given in Eqs. (4.59) and (4.60), it is common to likewise introduce the *spherical Hankel functions*,

$$h_n^{(1)}(z) = j_n(z) + iy_n(z), \quad n = 0, 1, 2, \ldots, \tag{4.65}$$

$$h_n^{(2)}(z) = j_n(z) - iy_n(z), \quad n = 0, 1, 2, \ldots. \tag{4.66}$$

Naturally, the properties of these functions are determined from properties of the standard spherical Bessel functions (see Section 4.1.4).

4.3.2 Struve functions

Struve functions are important in certain problems in electrodynamics, potential theory, optics, and in loudspeaker design. They appear as solutions of the nonhomogeneous standard Bessel equation[7]

$$z^2\frac{d^2w}{dz^2}+z\frac{dw}{dz}+(z^2-p^2)w=\frac{4(z/2)^{p+1}}{\sqrt{\pi}\,\Gamma\left(p+\frac{1}{2}\right)}. \tag{4.67}$$

The general solution, including homogeneous and particular solutions, is given by

$$w(z)=C_1J_p(z)+C_2Y_p(z)+\mathbf{H}_p(z), \tag{4.68}$$

where C_1 and C_2 are arbitrary constants and the particular solution $\mathbf{H}_p(z)$ is the *Struve function* of order p.

The Struve function $\mathbf{H}_p(z)$ has the series representation

$$\begin{aligned}\mathbf{H}_p(z)&=\sum_{k=0}^{\infty}\frac{(-1)^k(z/2)^{2k+p+1}}{\Gamma\left(k+\frac{3}{2}\right)\Gamma\left(k+p+\frac{3}{2}\right)}\\&=\frac{2(z/2)^{p+1}}{\sqrt{\pi}\,\Gamma\left(p+\frac{3}{2}\right)}\sum_{k=0}^{\infty}\frac{(-1)^k(1)_k}{\left(\frac{3}{2}\right)_k\left(p+\frac{3}{2}\right)_k}\frac{(z/2)^{2k}}{k!},\end{aligned} \tag{4.69}$$

which can be recognized as that of a generalized hypergeometric function

$$\mathbf{H}_p(z)=\frac{2(z/2)^{p+1}}{\sqrt{\pi}\,\Gamma\left(p+\frac{3}{2}\right)}\,{}_1F_2\left(1;\,\frac{3}{2},\,p+\frac{3}{2};\,-\frac{z^2}{4}\right). \tag{4.70}$$

Thus, the Struve function has the restriction that $p+\frac{3}{2}\neq 0,\,-1,\,-2,\ldots.$

We plot Struve functions $\mathbf{H}_0(x)$, $\mathbf{H}_1(x)$, $\mathbf{H}_2(x)$, and $\mathbf{H}_4(x)$ in Fig. 4.6 for real x over the interval $0\leq x\leq 15$. Observe the oscillatory behavior in the lower-order functions. However, as the order of the function increases, the curves tend to grow more exponentially, and the oscillations tend to be less prominent.

To derive an integral representation for the Struve function, we start with the beta function relation from ($G17$) in Appendix A as well as ($G10$),

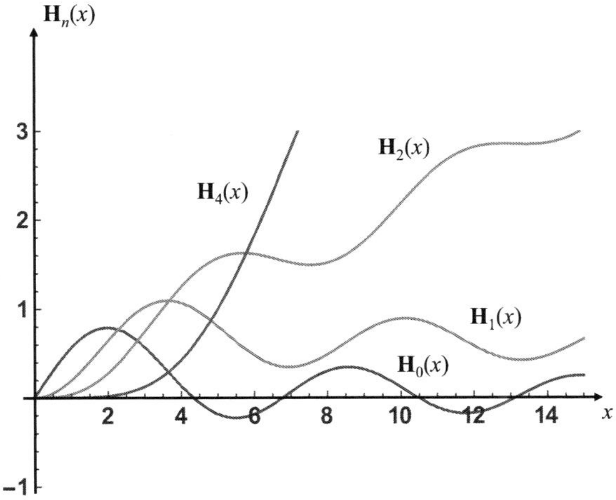

Figure 4.6 Struve functions: $n = 0,1,2,4$.

$$\frac{1}{\Gamma\left(k+\frac{3}{2}\right)\Gamma\left(k+p+\frac{3}{2}\right)} = \frac{2^{2k+1}}{\sqrt{\pi}\,(2k+1)!\Gamma\left(p+\frac{1}{2}\right)}\int_0^1 \tau^k(1-\tau)^{p-\frac{1}{2}}\,d\tau, \tag{4.71}$$

where $p > -\frac{1}{2}$. If we substitute this relation into the upper expression in Eq. (4.69) and let $\tau = t^2$, we arrive at the integral representation

$$\mathbf{H}_p(z) = \frac{2(z/2)^p}{\sqrt{\pi}\Gamma\left(p+\frac{1}{2}\right)}\int_0^1 \sin(zt)(1-t^2)^{p-\frac{1}{2}}dt, \quad p > -\frac{1}{2}. \tag{4.72}$$

For small arguments, we obtain from Eq. (4.70) the asymptotic result

$$\mathbf{H}_p(z) \sim \frac{z^{p+1}}{2^p\sqrt{\pi}\,\Gamma\left(p+\frac{3}{2}\right)}, \quad z \to 0. \tag{4.73}$$

For large arguments, we use Eqs. (4.70) and (3.45) to get

$$\mathbf{H}_p(z) \sim \frac{(z/2)^{p-1}}{\sqrt{\pi}\Gamma\left(p+\frac{1}{2}\right)}, \quad p > -\frac{1}{2}, \quad z \to \infty. \tag{4.74}$$

However, another asymptotic relation can be found involving the Bessel function of the second kind $Y_p(z)$. That is, we start with the integral representation for $Y_p(z)$ given by (p. 18 of Ref. 2)

$$Y_p(z) = \frac{2(z/2)^p}{\sqrt{\pi}\Gamma\left(p+\frac{1}{2}\right)}\left[\int_0^1 \sin(z\tau)(1-\tau^2)^{p-\frac{1}{2}}d\tau - \int_0^\infty e^{-z\tau}(1+\tau^2)^{p-\frac{1}{2}}d\tau\right]. \tag{4.75}$$

If we replace the first integral above with $\mathbf{H}_p(z)$ and rearrange terms, we obtain

$$\mathbf{H}_p(z) - Y_p(z) = \frac{2(z/2)^p}{\sqrt{\pi}\Gamma\left(p+\frac{1}{2}\right)}\int_0^\infty e^{-z\tau}\left(1+\tau^2\right)^{p-\frac{1}{2}}d\tau. \tag{4.76}$$

Then, under the substitution $s = z\tau$, the integral in Eq. (4.76) for large z can be written as

$$\begin{aligned}\int_0^\infty e^{-z\tau}(1+\tau^2)^{p-\frac{1}{2}}d\tau &= \frac{1}{z}\int_0^\infty e^{-s}\left(1+\frac{s^2}{z^2}\right)^{p-\frac{1}{2}}ds \\ &\sim \frac{1}{z}\int_0^\infty e^{-s}ds, \quad \mathrm{Re}(z) \gg s,\end{aligned} \tag{4.77}$$

which simplifies to $1/z$, and we have the asymptotic result

$$\mathbf{H}_p(z) - Y_p(z) \sim \frac{(z/2)^{p-1}}{\sqrt{\pi}\Gamma\left(p+\frac{1}{2}\right)}, \quad p > -\frac{1}{2}, \quad z \to \infty. \tag{4.78}$$

Note that the right-hand side of Eq. (4.78) is the same as Eq. (4.74). This follows from the fact that the Bessel function of the second kind $Y_p(z) \to 0$ as $z \to \infty$.

The related function

$$\mathbf{L}_p(z) = i^{-(p+1)}\mathbf{H}_p(iz) = \sum_{k=0}^\infty \frac{(z/2)^{2k+p+1}}{\Gamma\left(k+\frac{3}{2}\right)\Gamma\left(k+p+\frac{3}{2}\right)}, \quad p > -\frac{1}{2}, \tag{4.79}$$

is called the *modified Struve function* of order p. In terms of generalized hypergeometric functions, the series in Eq. (4.79) can be written as

$$\mathbf{L}_p(z) = \frac{2(z/2)^{p+1}}{\sqrt{\pi}\Gamma\left(p+\frac{3}{2}\right)}\, {}_1F_2\left(1; \frac{3}{2}, p+\frac{3}{2}; \frac{z^2}{4}\right). \tag{4.80}$$

In Fig. 4.7 we plot modified Struve functions $\mathbf{L}_0(x)$, $\mathbf{L}_1(x)$, and $\mathbf{L}_2(x)$ for real x over the interval $0 \le x \le 4$. Observe that none of these functions exhibit

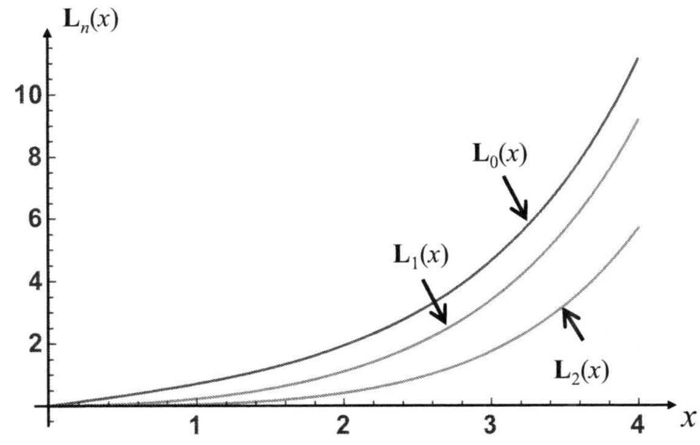

Figure 4.7 Modified Struve functions: $n = 0, 1, 2$.

an oscillatory behavior. In this regard the modified Struve functions are similar to the modified Bessel functions of the first kind.

For small arguments, it follows directly from Eq. (4.80) that

$$\mathbf{L}_p(z) \sim \frac{z^{p+1}}{2^p\sqrt{\pi}\,\Gamma\left(p+\frac{3}{2}\right)}, \quad z \to 0, \tag{4.81}$$

which is the same as Eq. (4.73). To derive an asymptotic approximation for large arguments, we first recall from (*BI7*) in Appendix A that

$$\begin{aligned} I_p(z) &= \frac{(z/2)^p}{\sqrt{\pi}\Gamma\left(p+\frac{1}{2}\right)}\int_{-1}^{1} e^{-zt}(1-t^2)^{p-\frac{1}{2}}dt \\ &= \frac{2(z/2)^p}{\sqrt{\pi}\Gamma\left(p+\frac{1}{2}\right)}\int_0^1 \cosh(zt)(1-t^2)^{p-\frac{1}{2}}dt. \end{aligned} \tag{4.82}$$

Also, from Eqs. (4.72) and (4.79), we obtain

$$\begin{aligned} \mathbf{L}_p(z) &= i^{-(p+1)}\mathbf{H}_p(iz) \\ &= \frac{2(z/2)^p}{\sqrt{\pi}\Gamma\left(p+\frac{1}{2}\right)}\int_0^1 \sinh(zt)(1-t^2)^{p-\frac{1}{2}}dt. \end{aligned} \tag{4.83}$$

Subtracting Eq. (4.83) from Eq. (4.82) and recalling that $\cosh(zt) - \sinh(zt) = e^{-zt}$, we arrive at

$$\begin{aligned} I_p(z) - \mathbf{L}_p(z) &= \frac{z^p}{2^{p-1}\sqrt{\pi}\,\Gamma\left(p+\dfrac{1}{2}\right)} \int_0^1 e^{-zt}(1-t^2)^{p-\frac{1}{2}} dt \\ &= \frac{z^p}{2^{p-1}\sqrt{\pi}\,\Gamma\left(p+\dfrac{1}{2}\right)} \frac{1}{z} \int_0^z e^{-s}\left(1-\frac{s^2}{z^2}\right)^{p-\frac{1}{2}} ds. \end{aligned} \tag{4.84}$$

For $z \to \infty$, the last integral simplifies to unity in a fashion similar to Eq. (4.77), and we are left with the asymptotic relationship

$$I_p(z) - \mathbf{L}_p(z) \sim \frac{(z/2)^{p-1}}{\sqrt{\pi}\,\Gamma\left(p+\dfrac{1}{2}\right)}, \quad p > -\frac{1}{2}, \quad z \to \infty. \tag{4.85}$$

Last, Struve functions of negative half-integer order are related to the standard and modified Bessel functions by

$$\mathbf{H}_{-(n+1/2)}(z) = (-1)^n J_{n+1/2}(z), \quad n = 0, 1, 2, \ldots, \tag{4.86}$$

$$\mathbf{L}_{-(n+1/2)}(z) = I_{n+1/2}(z), \quad n = 0, 1, 2, \ldots. \tag{4.87}$$

4.3.3 Kelvin's functions

The DE (p. 379 of Ref. 6)

$$z\frac{d^2w}{dz^2} + \frac{dw}{dz} - i\kappa^2 zw = 0, \quad \kappa \text{ real}, \tag{4.88}$$

arises in the problem of obtaining the current density in a wire carrying an alternating current. The solutions are also used in the design of heat exchangers, acoustic wave propagation, and modeling electromagnetic fields. The general solution can be written in terms of the modified Bessel functions, viz.,

$$w(z) = C_1 I_0(\kappa z\sqrt{i}) + C_2 K_0(\kappa z\sqrt{i}). \tag{4.89}$$

This particular form of the modified Bessel functions with complex argument is used to define four additional functions known as *Kelvin's functions*. For example, by setting $\kappa = 1$,

$$I_0(z\sqrt{i}) = \sum_{m=0}^{\infty} \frac{i^m (z/2)^{2m}}{(m!)^2} = \mathrm{ber}(z) + i\,\mathrm{bei}(z), \tag{4.90}$$

where

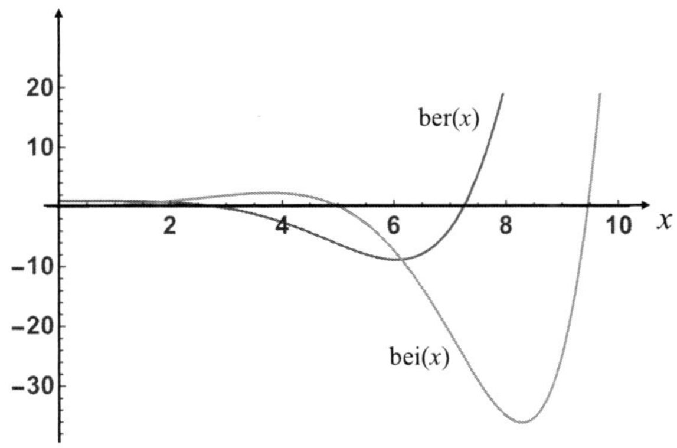

Figure 4.8 Kelvin's functions: ber(x) and bei(x).

$$\text{ber}(z) = \sum_{n=0}^{\infty} \frac{(-1)^n (z/2)^{4n}}{[(2n)!]^2} = {}_0F_3\left(-; \frac{1}{2}, \frac{1}{2}, 1; -\frac{z^4}{4^4}\right), \tag{4.91}$$

$$\text{bei}(z) = \left(\frac{z}{2}\right)^2 \sum_{n=0}^{\infty} \frac{(-1)^n (z/2)^{4n}}{[(2n+1)!]^2} = \left(\frac{z}{2}\right)^2 {}_0F_3\left(-; \frac{3}{2}, \frac{3}{2}, 1; -\frac{z^4}{4^4}\right). \tag{4.92}$$

The notation for these functions is based on Lord Kelvin's notation; i.e., ber is "Bessel real" and bei is "Bessel imaginary." We plot these functions in Fig. 4.8 for real x over the interval $0 \le x < 10$.

The small-argument behavior of these functions is obtained from the generalized hypergeometric functions—this leads to

$$\text{ber}(z) \sim 1 - \frac{z^4}{64} + \cdots, \quad z \to 0, \tag{4.93}$$

$$\text{bei}(z) \sim \frac{z^2}{4}\left(1 - \frac{z^4}{576} + \cdots\right), \quad z \to 0. \tag{4.94}$$

For large arguments, we can use the asymptotic results from the modified Bessel function. That is,

$$\begin{aligned} I_0(z\sqrt{i}) &\sim \frac{e^{z\sqrt{i}}}{\sqrt{2\pi z\sqrt{i}}} \\ &\sim \frac{1}{\sqrt{2\pi z}} \exp\left[\frac{z}{\sqrt{2}}(1+i) - \frac{i\pi}{8}\right], \quad z \to \infty, \end{aligned} \tag{4.95}$$

and by separating this expression into its real and imaginary parts, we have, respectively,

$$\begin{aligned}\operatorname{ber}(z) &\sim \operatorname{Re}\left(\frac{e^{z\sqrt{i}}}{\sqrt{2\pi z\sqrt{i}}}\right)\\ &\sim \frac{e^{z/\sqrt{2}}}{\sqrt{2\pi z}}\cos\left(\frac{z}{\sqrt{2}}-\frac{\pi}{8}\right), \qquad z\to\infty,\end{aligned} \tag{4.96}$$

$$\begin{aligned}\operatorname{bei}(z) &\sim \operatorname{Im}\left(\frac{e^{z\sqrt{i}}}{\sqrt{2\pi z\sqrt{i}}}\right)\\ &\sim \frac{e^{z/\sqrt{2}}}{\sqrt{2\pi z}}\sin\left(\frac{z}{\sqrt{2}}-\frac{\pi}{8}\right), \qquad z\to\infty.\end{aligned} \tag{4.97}$$

A second set of Kelvin functions arises from the relation

$$K_0(z\sqrt{i}) = \operatorname{ker}(z) + i\operatorname{kei}(z), \tag{4.98}$$

where $K_0(z)$ is defined by Eq. (4.46). Replacing z by $z\sqrt{i}$ in Eq. (4.46) and the separation of real and imaginary parts lead to the definitions

$$\operatorname{ker}(z) = -\ln\left(\frac{z}{2}\right)\operatorname{ber}(z) + \frac{\pi}{4}\operatorname{bei}(z) + \sum_{k=0}^{\infty}\frac{(-1)^k\psi(2k+1)}{[(2k)!]^2}\left(\frac{z^2}{4}\right)^{2k}, \tag{4.99}$$

$$\operatorname{kei}(z) = -\ln\left(\frac{z}{2}\right)\operatorname{bei}(z) - \frac{\pi}{4}\operatorname{ber}(z) + \sum_{k=0}^{\infty}\frac{(-1)^k\psi(2k+2)}{[(2k+1)!]^2}\left(\frac{z^2}{4}\right)^{2k+1}, \tag{4.100}$$

where

$$\psi(1) = -\gamma, \quad \psi(n) = -\gamma + \sum_{k=1}^{n-1}\frac{1}{k}, \qquad n\geq 2. \tag{4.101}$$

The functions (4.99) and (4.100) are plotted in Fig. 4.9 for real x over the interval $0 < x \leq 5$.

For small values of the argument, Eqs. (4.99) and (4.100) lead to

$$\operatorname{ker}(z) = -\gamma - \ln\left(\frac{z}{2}\right)\left(1-\frac{z^4}{64}+\cdots\right) + \frac{\pi z^2}{16}\left(1-\frac{z^4}{576}+\cdots\right), \qquad z\to 0, \tag{4.102}$$

$$\operatorname{kei}(z) = -\ln\left(\frac{z}{2}\right)\left(\frac{z^2}{4}\right)\left(1-\frac{z^4}{576}+\cdots\right) - \frac{\pi}{4}\left(1-\frac{z^4}{64}+\cdots\right), \qquad z\to 0. \tag{4.103}$$

For large arguments we use the asymptotic relation (4.53)

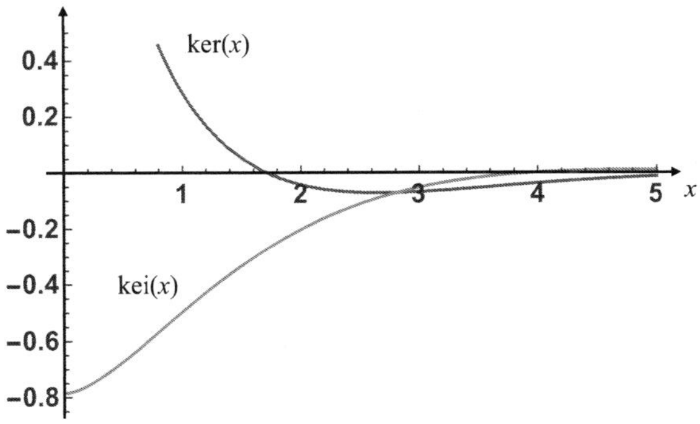

Figure 4.9 Kelvin's functions: ker(x) and kei(x).

$$K_0(z\sqrt{i}) \sim \sqrt{\frac{\pi}{2z\sqrt{i}}}e^{-z\sqrt{i}}, \qquad z \to \infty. \tag{4.104}$$

The asymptotic relations for $\mathrm{ker}(z)$ and $\mathrm{kei}(z)$ are found by taking the real and imaginary parts of Eq. (4.104). However, both functions rapidly approach zero for $x > 5$, as shown in Fig. 4.9.

4.3.4 Airy functions

The general solution of the DE

$$\frac{d^2w}{dz^2} - zw = 0 \tag{4.105}$$

is given by

$$w = C_1\mathrm{Ai}(z) + C_2\mathrm{Bi}(z), \tag{4.106}$$

where $\mathrm{Ai}(z)$ and $\mathrm{Bi}(z)$ are called *Airy functions of the first* and *second kind.* These functions are important in several application areas, including the diffraction of radio waves around the Earth's surface and the construction of asymptotic results near a two-dimensional caustic in the study of optical fields.[8] They also arise in classical and quantum physics. The Airy functions are defined by the modified Bessel functions through the relations

$$\begin{aligned}\mathrm{Ai}(z) &= \frac{1}{\pi}\sqrt{\frac{z}{3}}K_{1/3}\left(\frac{2}{3}z^{3/2}\right)\\ &= \frac{\sqrt{z}}{3}\left[I_{-1/3}\left(\frac{2}{3}z^{3/2}\right) - I_{1/3}\left(\frac{2}{3}z^{3/2}\right)\right],\end{aligned} \tag{4.107}$$

$$\mathrm{Bi}(z) = \sqrt{\frac{z}{3}}\left[I_{-1/3}\left(\frac{2}{3}z^{3/2}\right) + I_{1/3}\left(\frac{2}{3}z^{3/2}\right)\right]. \tag{4.108}$$

When the argument is negative, the Airy functions are defined by

$$\mathrm{Ai}(-z) = \frac{\sqrt{z}}{3}\left[J_{-1/3}\left(\frac{2}{3}z^{3/2}\right) + J_{1/3}\left(\frac{2}{3}z^{3/2}\right)\right], \tag{4.109}$$

$$\mathrm{Bi}(-z) = \sqrt{\frac{z}{3}}\left[J_{-1/3}\left(\frac{2}{3}z^{3/2}\right) - J_{1/3}\left(\frac{2}{3}z^{3/2}\right)\right]. \tag{4.110}$$

The additional Airy functions $\mathrm{Ai}(-z)$ and $\mathrm{Bi}(-z)$, defined by Eqs. (4.109) and (4.110), represent solutions of Eq. (4.105) when $-zw$ is replaced by zw (cf. Section 4.4.1). Graphs of all four Airy functions are shown for real x in Fig. 4.10 over various ranges. For real x, the results for $\mathrm{Ai}(-x)$, $\mathrm{Bi}(-x)$ are mirror images across the vertical axis of the results for $\mathrm{Ai}(x)$, $\mathrm{Bi}(x)$.

From properties of the modified Bessel function, we readily obtain the relation between the Airy functions and the generalized hypergeometric functions,

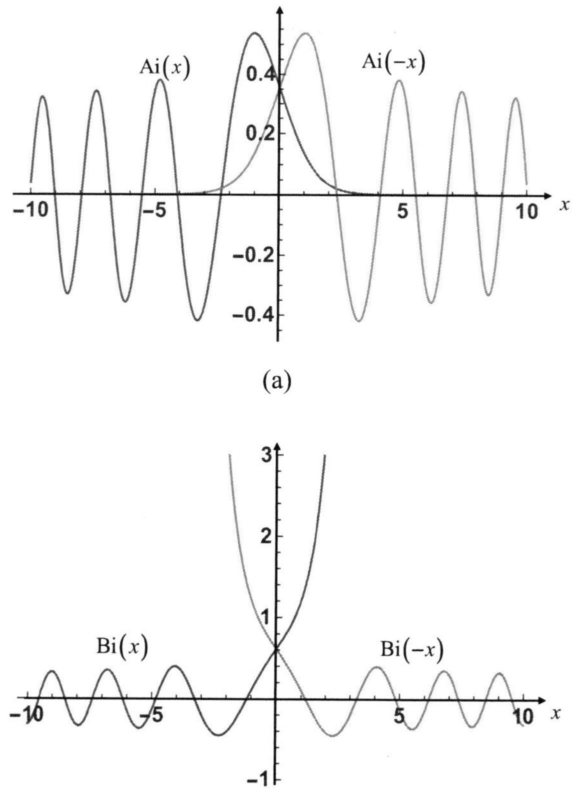

Figure 4.10 Graphs of the Airy functions: (a) Ai(x) and Ai(x); (b) Bi(x) and Bi(x).

$$\mathrm{Ai}(z) = \frac{1}{3^{2/3}\Gamma\left(\frac{2}{3}\right)}\, {}_0F_1\left(-;\frac{2}{3};\frac{z^3}{9}\right) - \frac{z}{3^{1/3}\Gamma\left(\frac{1}{3}\right)}\, {}_0F_1\left(-;\frac{4}{3};\frac{z^3}{9}\right), \tag{4.111}$$

$$\mathrm{Bi}(z) = \frac{1}{3^{1/6}\Gamma\left(\frac{2}{3}\right)}\, {}_0F_1\left(-;\frac{2}{3};\frac{z^3}{9}\right) + \frac{3^{1/6}z}{\Gamma\left(\frac{1}{3}\right)}\, {}_0F_1\left(-;\frac{4}{3};\frac{z^3}{9}\right). \tag{4.112}$$

From these last expressions, we can derive asymptotic results for the case of small arguments and large arguments. For example, for small arguments it readily follows from Eqs. (4.111) and (4.112) that

$$\mathrm{Ai}(z) = \frac{1}{3^{2/3}\Gamma\left(\frac{2}{3}\right)}\left(1+\frac{z^3}{6}+\cdots\right) - \frac{z}{3^{1/3}\Gamma\left(\frac{1}{3}\right)}\left(1+\frac{z^3}{12}+\cdots\right), \quad z \to 0, \tag{4.113}$$

$$\mathrm{Bi}(z) = \frac{1}{3^{1/6}\Gamma\left(\frac{2}{3}\right)}\left(1+\frac{z^3}{6}+\cdots\right) + \frac{3^{1/6}z}{\Gamma\left(\frac{1}{3}\right)}\left(1+\frac{z^3}{12}+\cdots\right), \quad z \to 0. \tag{4.114}$$

For large arguments we refer to the modified Bessel functions of the first and second kind to obtain

$$\mathrm{Ai}(z) \sim \frac{1}{2\sqrt{\pi}\,z^{1/4}} \exp\left(-\frac{2}{3}z^{2/3}\right), \quad z \to \infty, \tag{4.115}$$

$$\mathrm{Bi}(z) \sim \frac{1}{\sqrt{\pi}\,z^{1/4}} \exp\left(\frac{2}{3}z^{2/3}\right), \quad z \to \infty. \tag{4.116}$$

Also, from the behavior of $J_{\pm 1/3}(z)$ it can be shown that

$$\mathrm{Ai}(-z) \sim \frac{1}{\sqrt{\pi}\,z^{1/4}}\left[\sin\left(\frac{2}{3}z^{3/2}+\frac{\pi}{4}\right) - \frac{15}{144\,z^{3/2}}\cos\left(\frac{2}{3}z^{3/2}+\frac{\pi}{4}\right)\right], \quad z \to \infty, \tag{4.117}$$

$$\mathrm{Bi}(-z) \sim \frac{1}{\sqrt{\pi}\,z^{1/4}}\left[\cos\left(\frac{2}{3}z^{3/2}+\frac{\pi}{4}\right) + \frac{15}{144\,z^{3/2}}\sin\left(\frac{2}{3}z^{3/2}+\frac{\pi}{4}\right)\right], \quad z \to \infty. \tag{4.118}$$

In addition, the Airy functions have integral representations that are sometimes used to introduce these functions. In particular, for real variable x, we have

$$\mathrm{Ai}(x) = \frac{1}{\pi}\int_0^\infty \cos\left(\frac{1}{3}t^3 + xt\right) dt, \tag{4.119}$$

$$\mathrm{Ai}(-x) = \frac{1}{\pi}\int_0^\infty \cos\left(\frac{1}{3}t^3 - xt\right) dt. \tag{4.120}$$

A particular form of the integral in Eq. (4.120) was introduced by G. Airy[8] in the middle 1800s, and the integral in Eq. (4.119) was introduced many years later by H. Jeffreys[9] in 1928.

4.3.5 Bessel-integral function

The *Bessel-integral function* of order p is defined by

$$\mathrm{Ji}_p(z) = -\int_z^\infty \frac{J_p(t)}{t}\, dt, \tag{4.121}$$

which can also be arranged as

$$\mathrm{Ji}_p(z) = \int_0^z \frac{J_p(t)}{t}\, dt - \frac{1}{p}, \quad p > 0. \tag{4.122}$$

Note the similarity between the Bessel-integral function and the sine and cosine integrals [Eqs. (3.128) and (3.129)]. By writing the Bessel function in Eq. (4.122) in a power series and integrating termwise, we obtain

$$\begin{aligned}\mathrm{Ji}_p(z) &= -\frac{1}{p} + \sum_{k=0}^\infty \frac{(-1)^k (z/2)^{2k+p}}{(2k+p)\Gamma(k+p+1)k!}\\ &= -\frac{1}{p} + \frac{(z/2)^p}{p\Gamma(1+p)}\, {}_1F_2\left(\frac{p}{2};\, 1+\frac{p}{2},\, 1+p;\, -\frac{z^2}{4}\right), \quad p > 0.\end{aligned} \tag{4.123}$$

From this last expression we deduce the asymptotic results

$$\mathrm{Ji}_p(z) \sim \begin{cases} -\dfrac{1}{p} + \dfrac{(z/2)^p}{p\Gamma(1+p)}, & p > 0, \quad z \to 0\\ 0, \quad \mathrm{Re}(z) > 0, & z \to \infty. \end{cases} \tag{4.124}$$

For the special case when $p = 0$, it can be shown that

$$\mathrm{Ji}_0(z) = \gamma + \ln\frac{z}{2} - \frac{1}{8}z^2\, {}_2F_3\left(1,\, 1;\, 2,\, 2,\, 2;\, -\frac{z^2}{4}\right). \tag{4.125}$$

In this instance the asymptotic results are

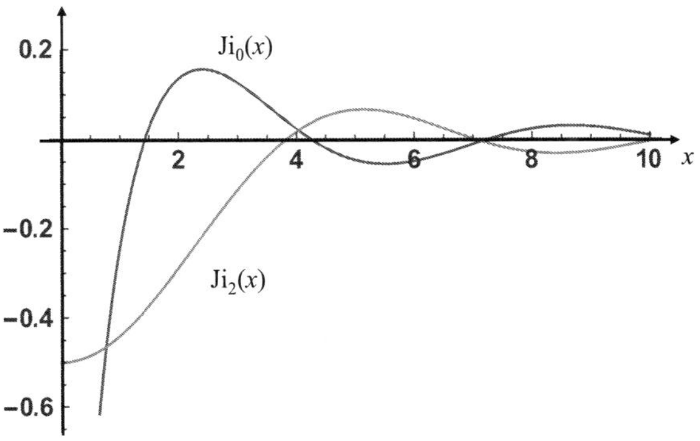

Figure 4.11 Bessel-integral function with $p = 0$ and $p = 2$.

$$\text{Ji}_0(z) \sim \begin{cases} \ln z, & z \to 0 \\ 0, & z \to \infty. \end{cases} \tag{4.126}$$

In Fig. 4.11 we plot the special cases $p = 0$ and $p = 2$ of the Bessel-integral function (4.121) for real x over the interval $0 < x \leq 10$.

4.3.6 Anger and Weber functions

The *Anger function* $\mathbf{J}_p(z)$ introduced by C. T. Anger in 1855 is a solution of the nonhomogeneous Bessel DE (p. 498 of Ref. 6)

$$z^2 \frac{d^2w}{dz^2} + z\frac{dw}{dz} + (z^2 - p^2)w = \frac{(z-p)}{\pi} \sin p\pi, \tag{4.127}$$

whereas the *Weber function* $\mathbf{E}_p(z)$ introduced in 1879 by H. F. Weber satisfies

$$z^2 \frac{d^2w}{dz^2} + z\frac{dw}{dz} + (z^2 - p^2)w = \frac{1}{\pi}[z + p + (z-p)\cos p\pi]. \tag{4.128}$$

These functions are defined, respectively, by the integrals

$$\mathbf{J}_p(z) = \frac{1}{\pi}\int_0^\pi \cos(p\theta - z\sin\theta)\, d\theta, \quad \text{Re}(z) \geq 0, \tag{4.129}$$

$$\mathbf{E}_p(z) = \frac{1}{\pi}\int_0^\pi \sin(p\theta - z\sin\theta)\, d\theta, \quad \text{Re}(z) \geq 0. \tag{4.130}$$

Comparing the integral in Eq. (4.129) for $p = n$, $n = 0, 1, 2, \ldots$ with Eq. (4.7), it is clear that $\mathbf{J}_n(z) = J_n(z)$. In this special case the asymptotic behavior of $\mathbf{J}_n(z)$ is exactly the same as that for $J_n(z)$. On the other hand, Weber's function for $p = n$ is directly related to the Struve function $\mathbf{H}_n(z)$. For example, it can be shown that

$$\mathbf{E}_0(z) = -\mathbf{H}_0(z), \quad \mathbf{E}_1(z) = \frac{2}{\pi} - \mathbf{H}_1(z), \quad \mathbf{E}_2(z) = \frac{2z}{3\pi} - \mathbf{H}_2(z), \tag{4.131}$$

whereas, in general,

$$\mathbf{E}_n(z) = -\mathbf{H}_n(z) + \frac{1}{\pi} \sum_{k=0}^{[(n-1)/2]} \frac{\Gamma\left(k+\frac{1}{2}\right)(z/2)^{n-2k-1}}{\Gamma\left(n-k+\frac{1}{2}\right)}, \quad n = 0, 1, 2, \ldots. \tag{4.132}$$

The bracket notation in Eq. (4.132) is defined by $[n/2] = n/2$ if n is even and by $(n-1)/2$ if n is odd. Some particular values of importance as $z \to 0$ are

$$\mathbf{J}_p(0) = \frac{\sin p\pi}{p\pi}; \quad \mathbf{E}_p(0) = \frac{1-\cos p\pi}{p\pi}. \tag{4.133}$$

Power series representations of these functions are given by

$$\begin{aligned} \mathbf{J}_p(z) &= \cos\left(\frac{p\pi}{2}\right) S_1(p, z) + \sin\left(\frac{p\pi}{2}\right) S_2(p, z), \\ \mathbf{E}_p(z) &= \sin\left(\frac{p\pi}{2}\right) S_1(p, z) - \cos\left(\frac{p\pi}{2}\right) S_2(p, z), \end{aligned} \tag{4.134}$$

where

$$\begin{aligned} S_1(p, z) &= \frac{1}{\Gamma\left(1+\frac{p}{2}\right)\Gamma\left(1-\frac{p}{2}\right)} \sum_{k=0}^{\infty} \frac{(-1)^k (1)_k (z/2)^{2k}}{\left(1+\frac{p}{2}\right)_k \left(1-\frac{p}{2}\right)_k k!} \\ &= \frac{1}{\Gamma\left(1+\frac{p}{2}\right)\Gamma\left(1-\frac{p}{2}\right)} \, {}_1F_2\left(1; 1+\frac{p}{2}, 1-\frac{p}{2}; -\frac{z^2}{4}\right), \end{aligned} \tag{4.135}$$

$$\begin{aligned} S_2(p, z) &= \frac{z}{2\Gamma\left(\frac{3+p}{2}\right)\Gamma\left(\frac{3-p}{2}\right)} \sum_{k=0}^{\infty} \frac{(-1)^k (1)_k (z/2)^{2k}}{\left(\frac{3+p}{2}\right)_k \left(\frac{3-p}{2}\right)_k k!} \\ &= \frac{z}{2\Gamma\left(\frac{3+p}{2}\right)\Gamma\left(\frac{3-p}{2}\right)} \, {}_1F_2\left(1; \frac{3+p}{2}, \frac{3-p}{2}; -\frac{z^2}{4}\right). \end{aligned} \tag{4.136}$$

Large-argument asymptotic results for Eqs. (4.135) and (4.136) are readily obtained from Eq. (3.44) and leads to

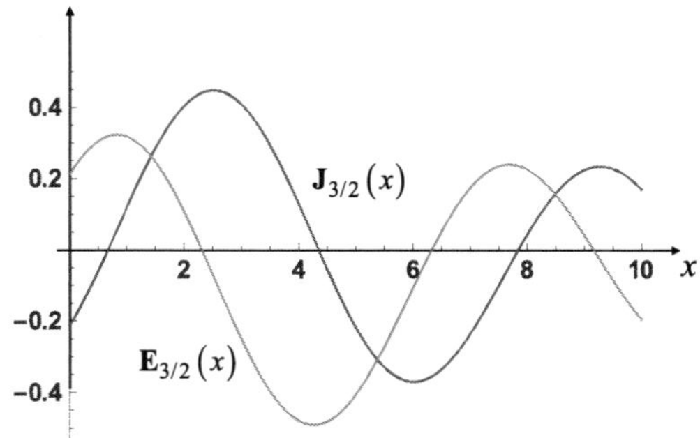

Figure 4.12 Anger and Weber functions with $p = 3/2$.

$$
\begin{aligned}
S_1(p, z) &\sim \frac{4}{\Gamma\left(\frac{p}{2}\right)\Gamma\left(-\frac{p}{2}\right)z^2}, \qquad z \to \infty, \\
S_2(p, z) &\sim \frac{2}{\Gamma\left(\frac{1+p}{2}\right)\Gamma\left(\frac{1-p}{2}\right)z}, \qquad z \to \infty.
\end{aligned} \tag{4.137}
$$

Plots of the Anger and Weber functions for real x are shown in Fig. 4.12 with $p = \frac{3}{2}$.

4.3.7 Lommel functions

Lommel functions[10] arise in physical problems such as in acoustic radiation and in describing the intensity distribution of light diffracted through a circular aperture, among other areas of physics. Solutions of the nonhomogeneous Bessel equation

$$
z^2\frac{d^2w}{dz^2} + z\frac{dw}{dz} + (z^2 - \nu^2)w = z^{\mu+1} \tag{4.138}
$$

are given by

$$
w(z) = C_1 J_\nu(z) + C_2 Y_\nu(z) + s_{\mu,\nu}(z), \tag{4.139}
$$

where $s_{\mu,\nu}(z)$ is the *Lommel function of the first kind.* It has the series representation

$$s_{\mu,\nu}(z) = \frac{1}{4}z^{\mu+1}\sum_{k=0}^{\infty}\frac{(-1)^k(z/2)^{2k}}{\left(\frac{\mu-\nu+1}{2}\right)_{k+1}\left(\frac{\mu+\nu+1}{2}\right)_{k+1}}$$
$$= \frac{z^{\mu+1}}{(\mu-\nu+1)(\mu+\nu+1)}\,{}_1F_2\left(1;\ \frac{\mu-\nu+3}{2},\ \frac{\mu+\nu+3}{2};\ -\frac{z^2}{4}\right), \tag{4.140}$$

where $\mu \pm \nu \neq -1, -3, -5, \ldots$. Clearly, $s_{\mu,\nu}(z) = s_{\mu,-\nu}(z)$.

A particular case of interest for these functions is that in which $\mu = \nu$. Then, unless ν is half of an odd negative integer, we have the relation

$$s_{\nu,\nu}(z) = 2^{\nu-1}\sqrt{\pi}\,\Gamma\left(\nu+\frac{1}{2}\right)\mathbf{H}_\nu(z), \tag{4.141}$$

where $\mathbf{H}_\nu(z)$ is the Struve function of order ν (Section 4.3.2).

From Eq. (4.140) and properties of the generalized hypergeometric function provided in Appendix E, we find the small- and large-argument approximations,

$$s_{\mu,\nu}(z) \sim \begin{cases} \dfrac{z^{\mu+1}}{(\mu-\nu+1)(\mu+\nu+1)}, & z \to 0 \\ z^{\mu-1}, & z \to \infty. \end{cases} \tag{4.142}$$

Two integrals that reduce to special cases of the Lommel function are given by

$$\int_0^\pi \sin(z\sin\theta)\sin\nu\theta\, d\theta = \sin(\nu\pi)s_{0,\nu}(z), \tag{4.143}$$

$$\int_0^\pi \sin(z\sin\theta)\cos\nu\theta\, d\theta = (1+\cos\nu\pi)s_{0,\nu}(z). \tag{4.144}$$

The *Lommel function of the second kind* is defined by

$$S_{\mu,\nu}(z) = s_{\mu,\nu}(z) + 2^{\mu-1}\Gamma\left(\frac{\mu-\nu+1}{2}\right)\Gamma\left(\frac{\mu+\nu+1}{2}\right)$$
$$\times\left\{\sin\left[\frac{(\mu-\nu)\pi}{2}\right]J_\nu(z) - \cos\left[\frac{(\mu-\nu)\pi}{2}\right]Y_\nu(z)\right\}. \tag{4.145}$$

It has the large argument asymptotic relation

$$S_{\mu,\nu}(z) \sim z^{\mu-1}, \quad \mu > \frac{1}{2}, \quad z \to \infty. \tag{4.146}$$

4.4 Applications

Bessel functions are prominent in numerous applications, many of which are formulated as partial DEs. Through separation of variables, it is often found that one of the resulting ordinary DEs is either Bessel's equation (4.1) or is related to Bessel's equation through certain transformations. In this section we first present some examples involving variations of Bessel's equation.

4.4.1 Differential equations related to Bessel's equation

Differential equations involving Bessel functions as solutions are quite common but are often disguised by appearing different from the standard Bessel equation. For example, some DEs in physics and engineering applications are specializations of

$$x^2y'' + (1-2a)xy' + \left[b^2c^2x^{2c} + \left(a^2 - c^2p^2\right)\right]y = 0, \quad p \geq 0, \quad b > 0. \quad (4.147)$$

To solve this DE, we reduce it to Eq. (4.1) through a series of transformations.

First, we set $y = x^a z$, from which we obtain

$$\begin{aligned} xy' &= x^{a+1}z' + ax^a z, \\ x^2y'' &= x^{a+2}z'' + 2ax^{a+1}z' + a(a-1)x^a z. \end{aligned} \quad (4.148)$$

By substituting these expressions into Eq. (4.147) and simplifying, we get

$$x^2z'' + xz' + (b^2c^2x^{2c} - c^2p^2)z = 0. \quad (4.149)$$

Next, we make the change of variable $t = x^c$, which through use of the chain rule of calculus, leads to

$$\begin{aligned} xz' &= cx^c\frac{dz}{dt}, \\ x^2z'' &= c(c-1)x^c\frac{dz}{dt} + c^2x^{2c}\frac{d^2z}{dt^2}. \end{aligned} \quad (4.150)$$

Hence, Eq. (4.149) reduces to

$$t^2\frac{d^2z}{dt^2} + t\frac{dz}{dt} + \left(b^2t^2 - p^2\right)z = 0. \quad (4.151)$$

We leave it to the reader to further reduce this DE to Eq. (4.1) by setting $s = bt$. Therefore, the general solution of Eq. (4.151) is

$$z(t) = C_1J_p(bt) + C_2Y_p(bt). \quad (4.152)$$

Last, by transforming back to the original variables x and y, we arrive at the solution of Eq. (4.147) given by

$$y = x^a\left[C_1 J_p(bx^c) + C_2 Y_p(bx^c)\right]. \tag{4.153}$$

Example 1: Find a general solution of the Airy equation

$$y'' + xy = 0.$$

To compare Airy's DE with Eq. (4.147), we first multiply through by x^2, $x \neq 0$, which gives us the form

$$x^2 y'' + x^3 y = 0. \tag{4.154}$$

By comparison with Eq. (4.147), we find that $a = \frac{1}{2}$, $b = \frac{2}{3}$, $c = \frac{3}{2}$, and $p = \frac{1}{3}$. Because p is not an integer, we can write the general solution as

$$y = \sqrt{x}\left[C_1 J_{1/3}\left(\frac{2}{3}x^{3/2}\right) + C_2 J_{-1/3}\left(\frac{2}{3}x^{3/2}\right)\right]. \tag{4.155}$$

Then, by recalling that

$$\begin{aligned} \mathrm{Ai}(-x) &= \frac{1}{3}\sqrt{x}\left[J_{-1/3}\left(\frac{2}{3}x^{3/2}\right) + J_{1/3}\left(\frac{2}{3}x^{3/2}\right)\right], \\ \mathrm{Bi}(-x) &= \sqrt{\frac{x}{3}}\left[J_{-1/3}\left(\frac{2}{3}x^{3/2}\right) - J_{1/3}\left(\frac{2}{3}x^{3/2}\right)\right], \end{aligned} \tag{4.156}$$

it follows that a general solution of Airy's equation above can also be expressed in terms of the Airy functions with negative argument, i.e.,

$$y = C_3 \mathrm{Ai}(-x) + C_4 \mathrm{Bi}(-x). \tag{4.157}$$

4.4.2 Oscillations of a hanging chain

The classical problem of a flexible hanging chain was first examined in 1732 by Daniel Bernoulli. Consider the oscillations of a heavy hanging chain fixed at the upper end at $x = L$ and free at the end of the chain at $x = 0$ (see Fig. 4.13). Let T be the tension in the chain due only to the weight of the chain below x; thus, $T = \rho g x$, where ρ is mass per unit length and $g = 32$ ft/s^2 is the gravitational constant. The governing DE for small oscillations eventually leads to a variation of Bessel's equation (4.1) as shown in Example 2.

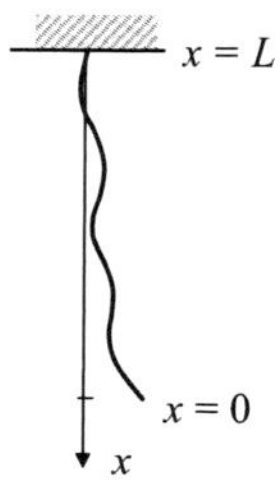

Figure 4.13 A heavy hanging chain.

Example 2: The small oscillations of the chain in Fig. 4.13 are governed by

$$\frac{\partial}{\partial x}\left(T\frac{\partial y}{\partial x}\right) = \rho\frac{\partial^2 y}{\partial t^2}, \quad |y(0, t)| < \infty, \; y(L, t) = 0.$$

Given that the chain is 10 ft long, find the lowest frequency of oscillation.

Let us assume that the oscillations are described by

$$y(x, t) = R(x)\cos(\omega t - \varphi), \tag{4.158}$$

where R is amplitude, ω is angular frequency, and φ is an unimportant phase term. By substituting $y(x, t)$ into the governing DE, we obtain after simplification

$$xR'' + R' + k^2R = 0, \quad |R(0)| < \infty, \; R(L) = 0, \tag{4.159}$$

where $k^2 = \omega^2/g$. By multiplying the DE in Eq. (4.159) by x, we see from Eqs. (4.147) and (4.153) that the general solution is

$$R(x) = C_1J_0(2k\sqrt{x}) + C_2Y_0(2k\sqrt{x}). \tag{4.160}$$

To satisfy the boundary conditions, we set $C_2 = 0$ and $R(L) = C_1J_0(2k\sqrt{L}) = 0$. The first zero of the Bessel function $J_0(u)$ occurs at $u = 2.405$. If $L = 10\,\text{ft}$, it follows that $2k\sqrt{10} = 2.405$, or $k = 0.380$. Thus, we can now calculate the lowest frequency of oscillation,

$$f = 2\pi\omega = 2\pi k\sqrt{g} = 13.5\,\text{Hz}. \tag{4.161}$$

4.4.3 Other Bessel functions

In this section we feature some Bessel functions other than just the standard Bessel functions $J_p(z)$ and $Y_p(z)$.

Example 3: Use a form of the Cauchy product to derive the *generating function* relation for the modified Bessel function of the first kind,

$$\exp\left[\frac{1}{2}x\left(t + \frac{1}{t}\right)\right] = \sum_{n=-\infty}^{\infty} I_n(x)t^n, \quad t \neq 0.$$

We begin by writing the exponential function as a product,

$$\begin{aligned}\exp\left[\frac{1}{2}x\left(t + \frac{1}{t}\right)\right] &= e^{xt/2} \cdot e^{x/2t}\\ &= \sum_{j=0}^{\infty}\frac{(xt/2)^j}{j!}\sum_{k=0}^{\infty}\frac{(x/2t)^k}{k!}\\ &= \sum_{j=0}^{\infty}\sum_{k=0}^{\infty}\frac{(x/2)^{j+k}}{j!k!}t^{j-k}.\end{aligned} \tag{4.162}$$

Next, we make the change of index $n = j - k$, and because both j and k have an infinite range of values, it follows that $-\infty < n < \infty$. Consequently, Eq. (4.162) becomes

$$\exp\left[\frac{1}{2}x\left(t+\frac{1}{t}\right)\right] = \sum_{n=-\infty}^{\infty}\sum_{k=0}^{\infty}\frac{(x/2)^{2k+n}}{k!(k+n)!}t^n = \sum_{n=-\infty}^{\infty} I_n(x)t^n. \tag{4.163}$$

Example 4: By directly performing the integration, show that

$$\frac{2}{\pi}\int_0^1 \sin(zt)(1-t^2)^{-\frac{1}{2}}dt = \mathbf{H}_0(z).$$

If we represent the sine function in its power series, we obtain the result

$$\begin{aligned}\frac{2}{\pi}\int_0^1 \sin(zt)(1-t^2)^{-\frac{1}{2}}dt &= \frac{2}{\pi}\sum_{k=0}^{\infty}\frac{(-1)^k z^{2k+1}}{(2k+1!)}\int_0^1 t^{2k+1}(1-t^2)^{-\frac{1}{2}}dt\\ &= \frac{1}{\pi}\sum_{k=0}^{\infty}\frac{(-1)^k z^{2k+1}}{(2k+1!)}\int_0^1 u^k(1-u)^{-\frac{1}{2}}du,\end{aligned} \tag{4.164}$$

where we interchanged the order of integration and summation and made the change of variable $u = t^2$. From the beta function relationship ($G17$) in Appendix A, we now obtain

$$\begin{aligned}\frac{2}{\pi}\int_0^1 \sin(zt)(1-t^2)^{-\frac{1}{2}}dt &= \frac{1}{\pi}\sum_{k=0}^{\infty}\frac{(-1)^k z^{2k+1}}{(2k+1!)}\frac{k!\sqrt{\pi}}{\Gamma\left(k+\frac{3}{2}\right)}\\ &= \sum_{k=0}^{\infty}\frac{(-1)^k (z/2)^{2k+1}}{\Gamma^2\left(k+\frac{3}{2}\right)}\\ &= \mathbf{H}_0(z).\end{aligned} \tag{4.165}$$

In obtaining the last result, we used the identity ($PS11$) in Appendix A,

$$(2k+1)! = 2^{2k}\left(\frac{3}{2}\right)_k k! = \frac{2^{2k+1}}{\sqrt{\pi}}\Gamma\left(k+\frac{3}{2}\right)k!.$$

4.5 Historical Remarks

As evidenced by the variety of Bessel functions introduced in this chapter, the family of Bessel functions is quite large. Historically, the family is so-named because of the close connection to the functions originally studied by Friedrich W. Bessel (1784–1846) and published in his memoir of 1824. Bessel was a German astronomer who first achieved fame by computing the orbit of Halley's comet. He is also credited with the derivation of the second-order differential equation that now bears his name. One of the early problems that intrigued Bessel was one proposed much earlier by Johann Kepler (1571–1630) concerning elliptic planetary motion. Bessel's solution to the problem involved a variation of the integral given by Eq. (4.7) that we now recognize as a Bessel function. His solution led to a systematic study of some of the basic properties of this function. Oddly enough, Bessel functions were observed earlier in 1732 by Daniel Bernoulli (1700–1782) in connection with his series solution for the displacements of a heavy oscillating chain (see Example 2 in Section 4.4). Leonhard Euler (1707–1783) in 1781 and Joseph Fourier (1768–1836) in 1822 also had some early connection with Bessel functions, but it was Bessel himself who first recognized their special properties.

In 1882, the German astronomer Hermann Struve (1854–1920) introduced the Struve function as a solution of a nonhomogeneous Bessel differential equation. His work in astronomy included establishing the movements of single and double stars. He also studied the planets. Kelvin's functions are named in honor of the Scottish-Irish physicist William Thompson (1824–1907), better known as Lord Kelvin. Among many other achievements, he formulated the first and second laws of thermodynamics. Early work of the British mathematician and astronomer George Biddell Airy (1801–1892) concerned the mass of Jupiter and the irregular motions of the Earth and Venus. Airy also worked on calculating the polar and equatorial lengths of the Earth. He is known in optics for the Airy disk, which describes the resolution of optical devices limited by diffraction. Perhaps the greatest achievement of Airy was the Airy Transit Circle at the Royal Greenwich Observatory. It is still the principal instrument of its class at the observatory. The Dutch physicist Balthasar van der Pol (1889–1959) introduced the Bessel-integral function and is also associated with the differential equation now bearing his name. Van der Pol made many contributions in mathematics, radio, and electrical engineering. Eugen von Lommel (1837–1899) was a German physicist, notable for introducing the Lommel polynomial, the Lommel functions, the Weber or Lommel–Weber function, and the Lommel differential equation.

Although there are many books written about Bessel functions, the most complete reference for all the functions in the Bessel function family is the treatise[1] by English mathematician George Neville Watson (1886–1965).

References

1. G. N. Watson, *A Treatise on the Theory of Bessel Functions*, 2nd Edition, Cambridge University Press, London (1952).

2. A. Erdelyi, W. Magnus, F. Oberhettinger, and F. G. Tricomi, *Higher Transcendental Functions* Vol. II, Bateman Manuscript Project, McGraw Hill, New York (1953).
3. F. W. J. Olver, *Asymptotics and Special Functions*, A K Peters, Massachusetts (1997).
4. L. C. Andrews, *Special Functions of Mathematics for Engineers*, 2nd Edition, SPIE Press, Bellingham, Washington (1998) [doi: 10.1117/3.270709].
5. L. C. Andrews, *Field Guide to Special Functions for Engineers*, SPIE Press, Bellingham, Washington (2011) [doi: 10.1117/3.885484].
6. M. Abramowitz and I. Stegun, Eds., *Handbook of Mathematical Functions*, National Bureau of Standards Applied Mathematics Series 55, U.S. Government Printing Office, Washington, D.C. (1964).
7. H. Struve, "Beitrag zur Theorie der Diffraction an Fernröhren," *Annalen der Physik und Chemie* **17**(13), 1008–1016 (1882).
8. G. B. Airy, "On the intensity of light in the neighbourhood of a caustic," *Trans. Cambridge Philos. Soc.* **6**, 379–407 (1838).
9. H. Jeffreys, "The effect on Love waves of heterogeneity in the lower layer," *Monthly Notices Roy. Astronom. Soc. Geophys. Suppl.* **2**, 101–111 (1928).
10. E. Lommel, "Ueber eine mit den Besseischen Functionen verwandte Function," *Math. Ann.* **9**, 425–444 (1876).

Chapter 5
Meijer *G*-Function

5.0 Introduction

Following the development of the hypergeometric function by Gauss and later the confluent hypergeometric function by Kummer, there was a considerable research effort during the late 1800s and first half of the 1900s to develop more general hypergeometric-type functions and their properties. In particular, much of the activity in the 1900s concerned the use of a Barnes integral to give meaning to the ${}_pF_q$ functions for the case of $p > q + 1$. Fortunately, this research effort eventually led to the *MacRobert E-function* introduced by T. M. MacRobert and the *Meijer G-function* introduced by C. S. Meijer. Both the MacRobert *E*-function and the Meijer *G*-function satisfied the criteria of defining the ${}_pF_q$ functions for the case of $p > q + 1$. This permitted a definition of these functions without the limitation imposed by a power series representation. However, because the MacRobert *E*-function turned out to be a special case of the Meijer *G*-function, we won't give separate treatment of it beyond its definition and relation to the Meijer *G*-function.

5.1 The Meijer *G*-Function Definitions

Meijer's first definition of the *G*-function in 1936 was in terms of a finite series of generalized hypergeometric functions given by[1]

$$G_{p,q}^{m,n}\left(z\left|\begin{matrix}a_1,\ldots a_p\\ c_1,\ldots,c_q\end{matrix}\right.\right)=\sum_{k=1}^{m}\frac{\prod'^{m}_{j=1}\Gamma(c_j-c_k)\prod_{j=1}^{n}\Gamma(1+c_k-a_j)}{\prod_{j=m+1}^{q}\Gamma(1+c_k-c_j)\prod_{j=n+1}^{p}\Gamma(a_j-c_k)}z^{c_k}$$
$$\times\ {}_pF_{q-1}[1+c_k-a_1,\ldots,1+c_k-a_p;$$
$$1+c_k-c_1,\ldots,*,\ldots,1+c_k-c_q;\ (-1)^{p-m-n}z], \tag{5.1}$$

where $z \neq 0$, $1 \leq m \leq q$, $0 \leq n \leq p \leq q - 1$, and no two of the c_k's $(k = 1, 2, \ldots, m)$ differ by zero or an integer (i.e., all poles are simple). Also, $a_j - c_k \neq 1, 2, 3, \ldots$ for $j = 1, 2, \ldots, n$ and $k = 1, 2, \ldots, m$. If $p = q$, we restrict $|z| < 1$. The prime in the product symbol $\prod'$ denotes omission of the term when $j = k$. An empty product of gamma functions is interpreted as unity, and the parameter corresponding to $1 + c_k - c_k$, $k = 1, 2, \ldots, m$, indicated by * in the ${}_pF_{q-1}$ function, is to be omitted.

For notational convenience when confusion is unlikely, it can sometimes be useful to introduce the shortened symbol notation

$$G_{p,q}^{m,n}\left(z\,\middle|\,\begin{matrix} a_p \\ c_q \end{matrix}\right) \equiv G_{p,q}^{m,n}\left(z\,\middle|\,\begin{matrix} a_1, \ldots, a_p \\ c_1, \ldots, c_q \end{matrix}\right). \tag{5.2}$$

Also, some authors make no distinction between the representations

$$G_{p,q}^{m,n}\left(z\,\middle|\,\begin{matrix} a_1, \ldots, a_p \\ c_1, \ldots, c_q \end{matrix}\right) \equiv G_{pq}^{mn}\left(z\,\middle|\,\begin{matrix} a_1, \ldots, a_p \\ c_1, \ldots, c_q \end{matrix}\right), \tag{5.3}$$

where the form on the right-hand side contains no commas separating the parameters m, n and p, q when they assume specific numerical values.

In 1941, Meijer proposed the Barnes integral definition[2]

$$G_{p,q}^{m,n}\left(z\,\middle|\,\begin{matrix} a_1, \ldots, a_p \\ c_1, \ldots, c_q \end{matrix}\right) = \frac{1}{2\pi i}\int_C \frac{\prod_{j=1}^{m}\Gamma(c_j - s)\prod_{j=1}^{n}\Gamma(1 - a_j + s)}{\prod_{j=m+1}^{q}\Gamma(1 - c_j + s)\prod_{j=n+1}^{p}\Gamma(a_j - s)} z^s ds, \tag{5.4}$$

where $0 \le n \le p$, $0 \le m \le q$ and the contour C is described by path (i) below. As before, an empty product is interpreted as unity, and it is assumed that no pole of $\Gamma(c_j - s)$, $j = 1, \ldots, m$, coincides with any pole of $\Gamma(1 - a_j + s)$, $j = 1, \ldots, n$. There are two additional contours C described by paths (ii) and (iii) below that were introduced by Erdélyi[3] in 1953 to complete Meijer's definition of Eq. (5.4). These additional paths allowed the G-function to be extended to all non-negative integer values of p and q, even for $p > q + 1$.

(i) The path C is a loop starting and ending at $+\infty$. It encircles the poles of $\Gamma(c_j - s)$ once in the negative direction while the poles of $\Gamma(1 - a_j + s)$ remain outside the loop. The integral converges if $q \ge 1$ and either $p < q$ or $p = q$ and $|z| < 1$.

(ii) The path C runs from $-i\infty$ to $i\infty$, keeping all poles of $\Gamma(c_j - s)$ to the right of C and all poles of $\Gamma(1 - a_j + s)$ to the left of C. The integral converges if $m + n - p/2 - q/2 > 0$ and $|\arg(z)| < (m + n - p/2 - q/2)\pi$.

(iii) The path is a loop starting and ending at $-\infty$. It encircles the poles of $\Gamma(1 - a_j + s)$ once in the positive direction while the poles of $\Gamma(c_j - s)$ remain outside the loop. The integral converges if $p \ge 1$ and either $p > q$ or $p = q$ and $|z| > 1$.

The integration contour C for path (i) is shown in Fig. 5.1(a), whereas the coutours for paths (ii) and (iii) are shown in Figs. 5.1(b) and (c), respectively. The open circles in (a) represent the poles of $\Gamma(1 - a_j + s)$ outside the contour C; in (c) the open circles represent the poles of $\Gamma(c_j - s)$ outside C. One of the above integration paths always makes sense. When more than one of the

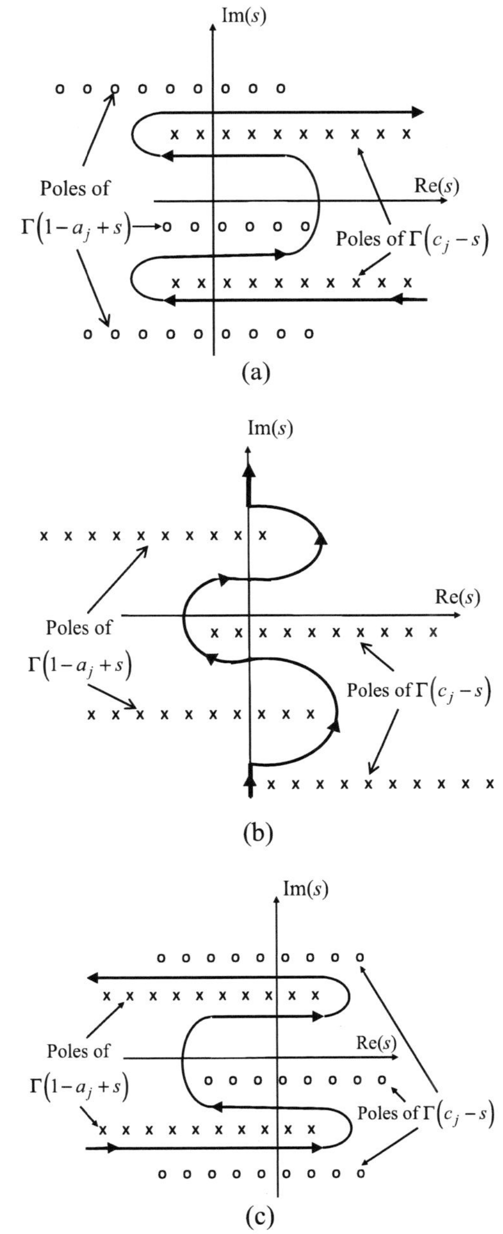

Figure 5.1 Integration contours C for the Meijer G-function based on (a) path (i), (b) path (ii), and (c) path (iii).

above paths make sense for a particular choice of parameters, they will lead to the same result—no ambiguity. In most cases it is assumed that all poles are simple.

Summing the residues associated with Eq. (5.4) over path (i) above leads to Meijer's original definition given by Eq. (5.1). With path (iii) the sum of residues assumes that no two a_j's ($j = 1, 2, \ldots, n$) differ by an integer. In that case,

$$G_{p,q}^{m,n}\left(z\left|\begin{matrix} a_1, \ldots a_p \\ c_1, \ldots, c_q \end{matrix}\right.\right) = \sum_{k=1}^{n} \frac{\prod'^{n}_{j=1}\Gamma(a_k - a_j)\prod_{j=1}^{m}\Gamma(1 + c_j - a_k)}{\prod_{j=n+1}^{p}\Gamma(1 + a_j - a_k)\prod_{j=m+1}^{q}\Gamma(a_k - c_j)} \frac{1}{z^{a_k}}$$
$$\times\ {}_qF_{p-1}\left[1 + c_1 - a_k, \ldots, 1 + c_q - a_k;\right.$$
$$\left. 1 + a_1 - a_k, \ldots, *, \ldots, 1 + a_p - a_k;\ (-1)^{q-m-n}\frac{1}{z}\right], \tag{5.5}$$

where $p \geq 1$ and either $p > q$ or $p = q$ and $|z| > 1$.

5.2 Elementary Properties

The Meijer G-function enjoys a number of useful properties, four of which are listed below. These properties and others are also tabulated in Appendix A.

$$z^r G_{p,q}^{m,n}\left(z\left|\begin{matrix} a_p \\ c_q \end{matrix}\right.\right) = G_{p,q}^{m,n}\left(z\left|\begin{matrix} a_p + r \\ c_q + r \end{matrix}\right.\right), \tag{5.6}$$

$$\frac{d}{dz}\left[z^{-c_1} G_{p,q}^{m,n}\left(z\left|\begin{matrix} a_p \\ c_q \end{matrix}\right.\right)\right] = -z^{-1-c_1} G_{p,q}^{m,n}\left(z\left|\begin{matrix} a_1, \ldots, a_p \\ c_1 + 1, c_2, \ldots, c_q \end{matrix}\right.\right), \tag{5.7}$$

$$z\frac{d}{dz}\left[G_{p,q}^{m,n}\left(z\left|\begin{matrix} a_p \\ c_q \end{matrix}\right.\right)\right] = G_{p,q}^{m,n}\left(z\left|\begin{matrix} a_1 - 1, a_2, \ldots, a_p \\ c_1, \ldots, c_q \end{matrix}\right.\right) + (a_1 - 1)G_{p,q}^{m,n}\left(z\left|\begin{matrix} a_p \\ c_q \end{matrix}\right.\right), \tag{5.8}$$

$$G_{p,q}^{m,n}\left(z\left|\begin{matrix} a_p \\ c_q \end{matrix}\right.\right) = G_{q,p}^{n,m}\left(\frac{1}{z}\left|\begin{matrix} 1 - c_q \\ 1 - a_p \end{matrix}\right.\right), \quad \arg\left(\frac{1}{z}\right) = -\arg(z). \tag{5.9}$$

For the proof of Eq. (5.6), we simply multiply both sides of Eq. (5.4) by z^r, which leads to

$$z^r G_{p,q}^{m,n}\left(z \middle| \begin{matrix} a_1, \ldots, a_p \\ c_1, \ldots, c_q \end{matrix}\right) = \frac{1}{2\pi i}\int_C \frac{\prod_{j=1}^m \Gamma(c_j - s)\prod_{j=1}^n \Gamma(1 - a_j + s)}{\prod_{j=m+1}^q \Gamma(1 - c_j + s)\prod_{j=n+1}^p \Gamma(a_j - s)} z^{r+s} ds$$
$$= \frac{1}{2\pi i}\int_C \frac{\prod_{j=1}^m \Gamma(c_j + r - s)\prod_{j=1}^n \Gamma(1 - a_j - r + s)}{\prod_{j=m+1}^q \Gamma(1 - c_j - r + s)\prod_{j=n+1}^p \Gamma(a_j + r - s)} z^s ds, \tag{5.10}$$

where we set $s \to s - r$ in the second step. Thus, we have arrived at Eq. (5.6). The implication in Eq. (5.6) is that each numerator and denominator parameter is increased by r. The proofs of Eqs. (5.7) and (5.8) are not difficult but will not be presented here. To prove the identity (5.9), we simply make the change of variable $s \to -s$, which leads to

$$G_{p,q}^{m,n}\left(z \middle| \begin{matrix} a_p \\ c_q \end{matrix}\right) = -\frac{1}{2\pi i}\int_C \frac{\prod_{j=1}^m \Gamma(c_j + s)\prod_{j=1}^n \Gamma(1 - a_j - s)}{\prod_{j=m+1}^q \Gamma(1 - c_j - s)\prod_{j=n+1}^p \Gamma(a_j + s)} z^{-s}(-ds)$$
$$= \frac{1}{2\pi i}\int_C \frac{\prod_{j=1}^n \Gamma(1 - a_j - s)\prod_{j=1}^m \Gamma(c_j + s)}{\prod_{j=n+1}^p \Gamma(a_j + s)\prod_{j=m+1}^q \Gamma(1 - c_j - s)} z^{-s} ds. \tag{5.11}$$

Upon comparison with Eq. (5.4), we now see that the parameters c_j and a_j have essentially traded places so that $c_j \to 1 - a_j$ and $a_j \to 1 - c_j$ to match Eq. (5.4). Also, the roles of m and n are reversed, as are the roles of p and q. Last, the argument z is now $1/z$, and this completes the proof. Equation (5.9) will be important in the development of large-argument asymptotic expressions as illustrated by several examples in Section 5.3.

5.2.1 Reduction formulas

If one of the parameters in the numerator set coincides with one of the parameters in the denominator set, the order of the G-function may decrease. For example, if $a_j = c_k$ for some $j = 1, \ldots, n$ and some $k = m + 1, \ldots, q$, then

$$G_{p,q}^{m,n}\left(z \middle| \begin{matrix} a_p \\ c_q \end{matrix}\right) = G_{p-1,q-1}^{m,n-1}\left(z \middle| \begin{matrix} a_1, \ldots, a_{j-1}, a_{j+1}, \ldots, a_p \\ c_1, \ldots, c_{k-1}, c_{k+1}, \ldots, c_q \end{matrix}\right). \tag{5.12}$$

The proof of this reduction formula falls out from the Barnes integral [Eq. (5.4)]. That is, assuming that $a_j = c_k$ for some $j = 1, \ldots, n$ and some $k = m + 1, \ldots, q$,

$$G_{p,q}^{m,n}\left(z\left|\begin{matrix}a_p\\c_q\end{matrix}\right.\right)=\frac{1}{2\pi i}\int_C\frac{\prod_{\ell=1}^m\Gamma(c_\ell-s)\prod_{\ell=1}^n\Gamma(1-a_\ell+s)}{\prod_{\ell=m+1}^q\Gamma(1-c_\ell+s)\prod_{\ell=n+1}^p\Gamma(a_\ell-s)}z^s\,ds$$

$$=\frac{1}{2\pi i}\int_C\frac{\Gamma(1-a_j+s)\prod_{\ell=1}^m\Gamma(c_\ell-s)\prod_{\substack{\ell=1\\ \ell\neq j}}^n\Gamma(1-a_\ell+s)}{\Gamma(1-c_k+s)\prod_{\substack{\ell=m+1\\ \ell\neq k}}^q\Gamma(1-c_\ell+s)\prod_{\ell=n+1}^p\Gamma(a_\ell-s)}z^s\,ds$$

$$=G_{p-1,q-1}^{m,n-1}\left(z\left|\begin{matrix}a_1,\ldots,a_{j-1},a_{j+1},\ldots,a_p\\c_1,\ldots,c_{k-1},c_{k+1},\ldots,c_q\end{matrix}\right.\right).\tag{5.13}$$

Also, if $a_j = c_k$ for some $j = n+1, \ldots, p$ and some $k = 1, \ldots, m,$ then

$$G_{p,q}^{m,n}\left(z\left|\begin{matrix}a_p\\c_q\end{matrix}\right.\right)=G_{p-1,q-1}^{m-1,n}\left(z\left|\begin{matrix}a_1,\ldots,a_{j-1},a_{j+1},\ldots,a_p\\c_1,\ldots,c_{k-1},c_{k+1},\ldots,c_q\end{matrix}\right.\right).\tag{5.14}$$

The proof of this second case is basically identical to that given for Eq. (5.12).

Last, if $a_j = c_k$ for some $j = n+1, \ldots, p$ and some $k = m+1, \ldots, q,$ then

$$G_{p,q}^{m,n}\left(z\left|\begin{matrix}a_p\\c_q\end{matrix}\right.\right)=\frac{\exp(i\pi a_j)}{2\pi i}G_{p-1,q-1}^{m,n}\left(ze^{-i\pi}\left|\begin{matrix}a_1,\ldots,a_{j-1},a_{j+1},\ldots,a_p\\c_1,\ldots,c_{k-1},c_{k+1},\ldots,c_q\end{matrix}\right.\right)$$
$$-\frac{\exp(-i\pi a_j)}{2\pi i}G_{p-1,q-1}^{m,n}\left(ze^{i\pi}\left|\begin{matrix}a_1,\ldots,a_{j-1},a_{j+1},\ldots,a_p\\c_1,\ldots,c_{k-1},c_{k+1},\ldots,c_q\end{matrix}\right.\right).\tag{5.15}$$

The proof of this last identity starts with (replacing c_k with a_j)

$$G_{p,q}^{m,n}\left(z\left|\begin{matrix}a_p\\c_q\end{matrix}\right.\right)=\frac{1}{2\pi i}\int_C\frac{\prod_{\ell=1}^m\Gamma(c_\ell-s)\prod_{\ell=1}^n\Gamma(1-a_\ell+s)z^s}{\Gamma(1-a_j+s)\Gamma(a_j-s)\prod_{\substack{\ell=m+1\\ \ell\neq k}}^q\Gamma(1-c_\ell+s)\prod_{\substack{\ell=n+1\\ \ell\neq j}}^p\Gamma(a_\ell-s)}$$

$$=\frac{1}{2\pi i}\int_C\frac{\sin(a_j-s)\prod_{\ell=1}^m\Gamma(c_\ell-s)\prod_{\ell=1}^n\Gamma(1-a_\ell+s)z^s}{\pi\prod_{\substack{\ell=m+1\\ \ell\neq k}}^q\Gamma(1-c_\ell+s)\prod_{\substack{\ell=n+1\\ \ell\neq j}}^p\Gamma(a_\ell-s)},\tag{5.16}$$

where we applied ($G7$) in Appendix A. If we now make the replacement

$$\sin(a_j-s)=\frac{1}{2i}\left[e^{i\pi(a_j-s)}-e^{-i\pi(a_j-s)}\right],$$

then Eq. (5.16) becomes

$$G_{p,q}^{m,n}\left(z\middle|\begin{matrix}a_p\\c_q\end{matrix}\right)=\frac{1}{2\pi i}\int_C\frac{(e^{i\pi a_j}e^{-i\pi s}-e^{-i\pi a_j}e^{i\pi s})\prod_{\ell=1}^{m}\Gamma(c_\ell-s)\prod_{\ell=1}^{n}\Gamma(1-a_\ell+s)z^s}{2\pi i\prod_{\substack{\ell=m+1\\ \ell\neq k}}^{q}\Gamma(1-c_\ell+s)\prod_{\substack{\ell=n+1\\ \ell\neq j}}^{p}\Gamma(a_\ell-s)}$$
$$=\frac{\exp(i\pi a_j)}{2\pi i}G_{p-1,q-1}^{m,n}\left(ze^{-i\pi}\middle|\begin{matrix}a_1,\dots,a_{j-1},a_{j+1},\dots,a_p\\c_1,\dots,c_{k-1},c_{k+1},\dots,c_q\end{matrix}\right)$$
$$-\frac{\exp(-i\pi a_j)}{2\pi i}G_{p-1,q-1}^{m,n}\left(ze^{i\pi}\middle|\begin{matrix}a_1,\dots,a_{j-1},a_{j+1},\dots,a_p\\c_1,\dots,c_{k-1},c_{k+1},\dots,c_q\end{matrix}\right). \tag{5.17}$$

5.2.2 Integral formulas

Because of the generality of the Meijer G-function, any integral involving this function serves as a master formula for all of the specializations derived from the particular G-function. The dummy variable of integration is real, but some of the other parameters can be real or complex. The first integral is

$$\int_0^\infty e^{-\beta x}x^{-\lambda}G_{p,q}^{m,n}\left(\alpha x\middle|\begin{matrix}a_1,\dots,a_p\\c_1,\dots,c_q\end{matrix}\right)dx=\beta^{\lambda-1}G_{p+1,q}^{m,n+1}\left(\frac{\alpha}{\beta}\middle|\begin{matrix}\lambda,a_1,\dots,a_p\\c_1,\dots,c_q\end{matrix}\right), \tag{5.18}$$

where $\beta>0$, $p+q<2(m+n)$, $c_j-\lambda>-1$, $j=1,\dots,m$. To derive this relation, we replace the Meijer G-function with its Barnes integral representation [Eq. (5.4)] and interchange the order of integration. Then, based on (G19) in Appendix A, this action leads to

$$\int_0^\infty e^{-\beta x}x^{-\lambda}G_{p,q}^{m,n}\left(\alpha x\middle|\begin{matrix}a_1,\dots,a_p\\c_1,\dots,c_q\end{matrix}\right)dx$$
$$=\frac{1}{2\pi i}\int_C\frac{\prod_{\ell=1}^{m}\Gamma(c_\ell-s)\prod_{\ell=1}^{n}\Gamma(1-a_\ell+s)}{\prod_{\ell=m+1}^{q}\Gamma(1-c_\ell+s)\prod_{\ell=n+1}^{p}\Gamma(a_\ell-s)}\alpha^s\int_0^\infty e^{-\beta x}x^{s-\lambda}dx\,ds$$
$$=\frac{\beta^{\lambda-1}}{2\pi i}\int_C\frac{\Gamma(1-\lambda+s)\prod_{\ell=1}^{m}\Gamma(c_\ell-s)\prod_{\ell=1}^{n}\Gamma(1-a_\ell+s)}{\prod_{\ell=m+1}^{q}\Gamma(1-c_\ell+s)\prod_{\ell=n+1}^{p}\Gamma(a_\ell-s)}\left(\frac{\alpha}{\beta}\right)^s ds$$
$$=\frac{\beta^{\lambda-1}}{2\pi i}\int_C\frac{\prod_{\ell=1}^{m}\Gamma(c_\ell-s)[\Gamma(1-\lambda+s)\prod_{\ell=2}^{n+1}\Gamma(1-a_{\ell-1}+s)]}{\prod_{\ell=m+1}^{q}\Gamma(1-c_\ell+s)\prod_{\ell=n+2}^{p+1}\Gamma(a_{\ell-1}-s)}\left(\frac{\alpha}{\beta}\right)^s ds$$
$$=\beta^{\lambda-1}G_{p+1,q}^{m,n+1}\left(\frac{\alpha}{\beta}\middle|\begin{matrix}\lambda,a_1,\dots,a_p\\c_1,\dots,c_q\end{matrix}\right). \tag{5.19}$$

Equation (5.18) can be interpreted as the Laplace transform of $x^{-\lambda}$ times the Meijer G-function (see Section 7.4). Similar integration formulas involving other integral transforms are given below.

The following three integrals are all examples of a Mellin transform (see Sections 2.4 and 7.5). We start with

$$\int_0^1 x^{\mu-1}(1-x)^{\nu-1}\, G_{p,q}^{m,n}\left(\alpha x \middle| \begin{matrix} a_1,\ldots,a_p \\ c_1,\ldots,c_q \end{matrix}\right) dx$$
$$= \Gamma(\nu) G_{p+1,q+1}^{m,n+1}\left(\alpha \middle| \begin{matrix} 1-\mu,\ a_1,\ldots,a_p \\ c_1,\ldots,c_q,\ 1-\mu-\nu \end{matrix}\right), \tag{5.20}$$
$$p+q<2(m+n), \quad |\arg(\alpha)| < \left(m+n-\frac{p}{2}-\frac{q}{2}\right)\pi,$$
$$\mathrm{Re}(\mu+c_j)>0,\ j=1,2,\ldots,m; \quad \mathrm{Re}(\nu)>0.$$

Following along the same lines as in the proof of Eq. (5.18) leads to

$$\int_0^1 x^{\mu-1}(1-x)^{\nu-1}\, G_{p,q}^{m,n}\left(\alpha x \middle| \begin{matrix} a_1,\ldots,a_p \\ c_1,\ldots,c_q \end{matrix}\right) dx$$
$$=\frac{1}{2\pi i}\int_C \frac{\prod_{\ell=1}^m \Gamma(c_\ell - s)\prod_{\ell=1}^n \Gamma(1-a_\ell+s)}{\prod_{\ell=m+1}^q \Gamma(1-c_\ell+s)\prod_{\ell=n+1}^p \Gamma(a_\ell - s)}\alpha^s \int_0^1 x^{\mu+s-1}(1-x)^{\nu-1}dx\,ds$$
$$=\frac{\Gamma(\nu)}{2\pi i}\int_C \frac{\prod_{\ell=1}^m \Gamma(c_\ell - s)\left[\Gamma(\mu+s)\prod_{\ell=2}^{n+1}\Gamma(1-a_{\ell-1}+s)\right]}{\left[\prod_{\ell=m+1}^q \Gamma(1-c_\ell+s)\Gamma(\mu+\nu+s)\right]\prod_{\ell=n+2}^{p+1}\Gamma(a_{\ell-1}-s)}\alpha^s ds$$
$$=\Gamma(\nu) G_{p+1,q+1}^{m,n+1}\left(\alpha \middle| \begin{matrix} 1-\mu,\ a_1,\ldots,a_p \\ c_1,\ldots,c_q,\ 1-\mu-\nu \end{matrix}\right), \tag{5.21}$$

where we have used the beta function (G17) in Appendix A to evaluate the integral in the second line. The next Mellin transform integral is

$$\int_0^\infty x^{\mu-1}(x+\beta)^{-\nu}\, G_{p,q}^{m,n}\left(\alpha x \middle| \begin{matrix} a_1,\ldots,a_p \\ c_1,\ldots,c_q \end{matrix}\right) dx$$
$$=\frac{\beta^{\mu-\nu}}{\Gamma(\nu)} G_{p+1,q+1}^{m+1,n+1}\left(\alpha\beta \middle| \begin{matrix} 1-\mu,\ a_1,\ldots,a_p \\ \nu-\mu,\ c_1,\ldots,c_q \end{matrix}\right),$$
$$(p+q)<2(m+n), \quad |\arg(\alpha)| < \left(m+n-\frac{p}{2}-\frac{q}{2}\right)\pi, \quad |\arg(\beta)|<\pi,$$
$$\mathrm{Re}(\mu+c_j)>0,\ j=1,2,\ldots,m, \quad \mathrm{Re}(\mu-\nu+a_k)<1,\ k=1,2,\ldots,n. \tag{5.22}$$

The proof of this integral relation is nearly the same as that of Eq. (5.20) by simply recognizing that

$$\begin{aligned}\int_0^\infty \frac{x^{\mu+s-1}}{(x+\beta)^\nu}dx &= \beta^{\mu-\nu+s}\int_0^\infty \frac{t^{\mu+s-1}}{(t+1)^\nu}dt \\ &= \frac{\beta^{\mu-\nu+s}\Gamma(\mu+s)\Gamma(\nu-\mu-s)}{\Gamma(\nu)},\end{aligned} \tag{5.23}$$

where we made the change of variable $x=\beta t$ and used the beta function identity ($G18$) in Appendix A. Our final Mellin transform integral is

$$\begin{gathered}\int_0^\infty x^{\lambda-1}G_{p,q}^{m,n}\left(\alpha x\left|\begin{matrix}a_1,\ldots,a_p\\ c_1,\ldots,c_q\end{matrix}\right.\right)dx=\frac{1}{\alpha^\lambda}\frac{\prod_{j=1}^m\Gamma(c_j+\lambda)\prod_{j=1}^n\Gamma(1-a_j-\lambda)}{\prod_{j=m+1}^q\Gamma(1-c_j-\lambda)\prod_{j=n+1}^p\Gamma(a_j+\lambda)},\\ p+q<2(m+n),\quad |\arg(\alpha)|<\left(m+n-\frac{p}{2}-\frac{q}{2}\right)\pi,\\ -\min\mathrm{Re}(c_j)<\mathrm{Re}(\lambda)<1-\max\mathrm{Re}(a_k),\, j=1,2,\ldots,m,\, k=1,2,\ldots,n.\end{gathered} \tag{5.24}$$

To prove this integral identity, we first write

$$\begin{aligned}\int_0^\infty x^{\lambda-1}G_{p,q}^{m,n}\left(\alpha x\left|\begin{matrix}a_1,\ldots,a_p\\ c_1,\ldots,c_q\end{matrix}\right.\right)dx &= \lim_{\beta\to 0}\int_0^\infty \frac{x^\lambda}{(x+\beta)}G_{p,q}^{m,n}\left(\alpha x\left|\begin{matrix}a_1,\ldots,a_p\\ c_1,\ldots,c_q\end{matrix}\right.\right)dx \\ &= \lim_{\beta\to 0}\beta^\lambda G_{p+1,q+1}^{m+1,n+1}\left(\alpha\beta\left|\begin{matrix}\lambda,\, a_1,\ldots,a_p\\ -\lambda,\, c_1,\ldots,c_q\end{matrix}\right.\right),\end{aligned} \tag{5.25}$$

where we have used the result of Eq. (5.22). We next make use of the definition from Eq. (5.1) to obtain the limit

$$\begin{aligned}\lim_{\beta\to 0}\beta^\lambda G_{p+1,q+1}^{m+1,n+1}\left(\alpha\beta\left|\begin{matrix}\lambda,\, a_p\\ -\lambda,\, c_q\end{matrix}\right.\right) &= \lim_{\beta\to 0}\beta^\lambda\frac{\prod_{j=2}^{m+1}\Gamma(c_{j-1}+\lambda)\prod_{j=2}^{n+1}\Gamma(1-\lambda-a_{j-1})}{\prod_{j=m+2}^{q+1}\Gamma(1-\lambda-c_{j-1})\prod_{j=n+2}^{p+1}\Gamma(a_{j-1}+\lambda)}(\alpha\beta)^{-\lambda} \\ &= \frac{1}{\alpha^\lambda}\frac{\prod_{j=1}^m\Gamma(c_j+\lambda)\prod_{j=1}^n\Gamma(1-\lambda-a_j)}{\prod_{j=m+1}^q\Gamma(1-\lambda-c_j)\prod_{j=n+1}^p\Gamma(a_j+\lambda)}.\end{aligned} \tag{5.26}$$

For this last equation, it follows that all terms in the summation in Eq. (5.1) are zero except for $k=1$, and the remaining generalized hypergeometric function with argument zero has been set to unity.

A general Hankel transform integral of order ν (see Section 7.6) is given by

$$\int_0^\infty x^{2\lambda} J_\nu(\beta x) G_{p,q}^{m,n}\left(\alpha x^2 \left| \begin{matrix} a_1, \ldots, a_p \\ c_1, \ldots, c_q \end{matrix} \right.\right) dx$$
$$= \frac{2^{2\lambda}}{\beta^{2\lambda+1}} G_{p+2,q}^{m,n+1}\left(\frac{4\alpha}{\beta^2} \left| \begin{matrix} \mu,\ a_1, \ldots, a_p,\ \sigma \\ c_1, \ldots, c_q \end{matrix} \right.\right),$$
$$\beta > 0, \quad \mu = \frac{1-\nu}{2} - \lambda, \quad \sigma = \frac{1+\nu}{2} - \lambda, \quad p+q < 2(m+n),$$
$$c_j + \lambda + \frac{\nu}{2} > -\frac{1}{2}, \quad j = 1, \ldots, m, \quad a_k + \lambda < \frac{3}{4}, \quad k = 1, \ldots, n. \tag{5.27}$$

By again following the proof of Eq. (5.18), we first note that by interchanging the order of integration, we can write

$$\int_0^\infty x^{2\lambda} J_\nu(\beta x) G_{p,q}^{m,n}\left(\alpha x^2 \left| \begin{matrix} a_1, \ldots, a_p \\ c_1, \ldots, c_q \end{matrix} \right.\right) dx$$
$$= \frac{1}{2\pi i} \int_C \frac{\prod_{\ell=1}^m \Gamma(c_\ell - s) \prod_{\ell=1}^n \Gamma(1 - a_\ell + s)}{\prod_{\ell=m+1}^q \Gamma(1 - c_\ell + s) \prod_{\ell=n+1}^p \Gamma(a_\ell - s)} \alpha^s \int_0^\infty x^{2(\lambda+s)} J_\nu(\beta x)\, dx\, ds. \tag{5.28}$$

To evaluate the last integral in Eq. (5.28), we rely on ($C9$) in Appendix C and the integral result of Eq. (5.24) to obtain

$$\alpha^s \int_0^\infty x^{2(\lambda+s)} J_\nu(\beta x)\, dx = \alpha^s \int_0^\infty x^{2(\lambda+s)} G_{02}^{10}\left(\frac{\beta^2 x^2}{4} \left| \frac{\nu}{2}, -\frac{\nu}{2} \right.\right) dx$$
$$= \frac{\alpha^s}{2} \int_0^\infty t^{(\lambda+s-1/2)} G_{02}^{10}\left(\frac{\beta^2 t}{4} \left| \frac{\nu}{2}, -\frac{\nu}{2} \right.\right) dt$$
$$= \frac{2^{2\lambda}}{\beta^{2\lambda+1}} \left(\frac{4\alpha}{\beta^2}\right)^s \frac{\Gamma\left(\frac{1+\nu}{2} + \lambda + s\right)}{\Gamma\left(\frac{1+\nu}{2} - \lambda - s\right)}, \tag{5.29}$$

where we made the change of variable $t = x^2$. By substituting this last expression into Eq. (5.28), we get the result of Eq. (5.27), viz.,

$$\int_0^\infty x^{2\lambda} J_\nu(\beta x) G_{p,q}^{m,n}\left(\alpha x^2 \left| \begin{matrix} a_1, \ldots, a_p \\ c_1, \ldots, c_q \end{matrix} \right.\right) dx$$
$$= \frac{2^{2\lambda}}{\beta^{2\lambda+1}} \frac{1}{2\pi i} \int_C \frac{\prod_{\ell=1}^m \Gamma(c_\ell - s)[\Gamma(1-\mu+s) \prod_{\ell=2}^{n+1} \Gamma(1 - a_{\ell-1} + s)]}{\prod_{\ell=m+1}^q \Gamma(1 - c_\ell + s)[\prod_{\ell=n+2}^{p+1} \Gamma(a_{\ell-1} - s)\Gamma(\sigma - s)]} \left(\frac{4\alpha}{\beta^2}\right)^s ds$$
$$= \frac{2^{2\lambda}}{\beta^{2\lambda+1}} G_{p+2,q}^{m,n+1}\left(\frac{4\alpha}{\beta^2} \left| \begin{matrix} \mu,\ a_1, \ldots, a_p,\ \sigma \\ c_1, \ldots, c_q \end{matrix} \right.\right). \tag{5.30}$$

Another way to evaluate the integral in Eq. (5.27) is to use Mellin transform properties (pp. 262–263 of Ref. 4) to obtain

$$\mathcal{M}\{J_\nu(\beta x)\} = \frac{1}{\beta^s}\mathcal{M}\{J_\nu(x)\} = \frac{2^{s-1}\Gamma\left(\frac{\nu+s}{2}\right)}{\beta^s\Gamma\left(\frac{\nu-s}{2}+1\right)}, \tag{5.31}$$

coupled with the translation property $\mathcal{M}\{x^a f(x)\} = F(s+a)$. Then, by replacing $s \to 2s$ and $a = 2\lambda + 1$, we find that we can directly obtain the result of Eq. (5.27) without the use of Eq. (5.24).

The restrictions listed for the above integrals represent only one possible case. Most of the integrals are valid for several additional cases that we haven't listed (see, for example, pp. 157–166 of Ref. 5).

5.3 Asymptotic Relations for Large Arguments

From Eqs. (3.72) and (3.74), we know that the Barnes integral for the general ${}_pF_q$ function is proportional to

$$f(a_1, \ldots, a_p;\ c_1, \ldots, c_q;\ z) = \frac{1}{2\pi i}\int_{-i\infty}^{i\infty} \frac{\prod_{j=1}^{p}\Gamma(a_j+s)}{\prod_{j=1}^{q}\Gamma(c_j+s)}\Gamma(-s)z^s ds. \tag{5.32}$$

To see its connection with the Meijer G-function, we compare Eq. (5.32) with

$$G_{p,q}^{m,n}\left(z\middle| \begin{matrix}\hat{a}_1, \ldots, \hat{a}_p\\ \hat{c}_1, \ldots, \hat{c}_q\end{matrix}\right) = \frac{1}{2\pi i}\int_C \frac{\prod_{j=1}^{m}\Gamma(\hat{c}_j - s)\prod_{j=1}^{n}\Gamma(1-\hat{a}_j+s)}{\prod_{j=m+1}^{q}\Gamma(1-\hat{c}_j+s)\prod_{j=n+1}^{p}\Gamma(\hat{a}_j - s)} z^s ds. \tag{5.33}$$

- By choosing $m = 1$ and $\hat{c}_1 = 0$ in Eq. (5.33), the first product of gamma functions in the numerator reduces to $\Gamma(-s)$.
- By setting $n = p$ and $\hat{a}_j = 1 - a_j$, the remaining product in the numerator of Eq. (5.33) coincides with that in Eq. (5.32).
- With $n = p$, the second product in the denominator of Eq. (5.33) does not exist, so it is taken as unity.
- If we set $\hat{c}_j = 1 - c_{j-1}$ and make the change of index $j \to j + 1$, the first denominator product runs from $j = 1$ to $j = q + 1$. Thus, q in Eq. (5.33) becomes $q + 1$, and we find that

$$f(a_1, \ldots, a_p;\ c_1, \ldots, c_q;\ z) = G_{p,q+1}^{1,p}\left(z\middle| \begin{matrix}1-a_1, \ldots, 1-a_p\\ 0,\ 1-c_1, \ldots, 1-c_q\end{matrix}\right). \tag{5.34}$$

Hence, from Eq. (3.74) we now deduce the relationship

$$ {}_pF_q(a_1,\ldots,a_p;\ c_1,\ldots,c_q;\ -z) = \frac{\prod_{j=1}^{q}\Gamma(c_j)}{\prod_{j=1}^{p}\Gamma(a_j)} G_{p,q+1}^{1,p}\left(z\,\middle|\,\begin{matrix} 1-a_1,\ldots,1-a_p \\ 0,\ 1-c_1,\ldots,1-c_q \end{matrix}\right). \tag{5.35} $$

Last, by use of the property of Eq. (5.9), we also have the important relation

$$ {}_pF_q(a_1,\ldots,a_p;\ c_1,\ldots,c_q;\ -z) = \frac{\prod_{j=1}^{q}\Gamma(c_j)}{\prod_{j=1}^{p}\Gamma(a_j)} G_{q+1,p}^{p,1}\left(\frac{1}{z}\,\middle|\,\begin{matrix} 1,\ c_1,\ldots,c_q \\ a_1,\ldots,a_p \end{matrix}\right). \tag{5.36} $$

5.3.1 The function ${}_1F_1$

Equation (5.36) provides us with a direct way of finding asymptotic formulas for generalized hypergeometric functions with large argument without reintroducing the Barnes integral combined with residue theory (recall Chapter 3). To illustrate the method, let us start with the confluent hypergeometric function. From Eq. (5.36), we have

$$ {}_1F_1(a;\ c;\ -z) = \frac{\Gamma(c)}{\Gamma(a)} G_{21}^{11}\left(\frac{1}{z}\,\middle|\,\begin{matrix} 1,\ c \\ a \end{matrix}\right), \tag{5.37} $$

and by using Eq. (5.1), we find directly that

$$ \begin{aligned} {}_1F_1(a;\ c;\ -z) &\sim \frac{\Gamma(c)}{\Gamma(a)}\frac{\Gamma(a)}{\Gamma(c-a)}\left(\frac{1}{z}\right)^a {}_2F_0\left(a;\ 1+a-c;\ -;\ \frac{1}{z}\right) \\ &\sim \frac{\Gamma(c)}{\Gamma(c-a)} z^{-a}, \qquad z\to\infty. \end{aligned} \tag{5.38} $$

Thus, we produced the large-argument result by simply using Eq. (5.36) combined with Eq. (5.1).

5.3.2 The function ${}_1F_k$

Following the approach used for the confluent hypergeometric function, we first write ($k = 2, 3, \ldots$)

$$ {}_1F_k(a;\ c_1,\ldots,c_k;\ -z) = \frac{\prod_{j=1}^{k}\Gamma(c_j)}{\Gamma(a)} G_{k+1,1}^{1,1}\left(\frac{1}{z}\,\middle|\,\begin{matrix} 1,\ c_1,\ldots,c_k \\ a \end{matrix}\right). \tag{5.39} $$

We then use Eq. (5.1) to get

$$
\begin{aligned}
{}_1F_k(a;\, c_1, \ldots, c_k;\, -z) &\sim \frac{\prod_{j=1}^{k}\Gamma(c_j)}{\prod_{j=1}^{k}\Gamma(c_j - a)} z^{-a} \\
&\quad \times {}_{k+1}F_0\left(a,\, 1 - c_1 + a, \ldots, 1 - c_k + a;\, -;\, \frac{1}{z}\right), \\
&\sim \frac{\prod_{j=1}^{k}\Gamma(c_j)}{\prod_{j=1}^{k}\Gamma(c_j - a)} z^{-a}, \qquad z \to \infty.
\end{aligned} \tag{5.40}
$$

5.3.3 The function ${}_2F_1$

For the hypergeometric function, we again start with Eq. (5.36) to obtain

$$
{}_2F_1(a,\, b;\, c;\, -z) = \frac{\Gamma(c)}{\Gamma(a)\Gamma(b)} G_{22}^{21}\left(\frac{1}{z}\middle|\begin{matrix} 1,\, c \\ a,\, b \end{matrix}\right). \tag{5.41}
$$

Now, by using Eq. (5.1), we produce the following (noting that $m = 2$ in this case):

$$
\begin{aligned}
{}_2F_1(a, b; c; -z) &= \frac{\Gamma(c)}{\Gamma(a)\Gamma(b)} \frac{\Gamma(a)\Gamma(b-a)z^{-a}}{\Gamma(c-a)} {}_2F_1\left(a,\, 1-c+a;\, 1-b+a;\, -\frac{1}{z}\right) \\
&\quad + \frac{\Gamma(c)}{\Gamma(a)\Gamma(b)} \frac{\Gamma(b)\Gamma(a-b)z^{-b}}{\Gamma(c-b)} {}_2F_1\left(b,\, 1-c+b;\, 1-a+b;\, -\frac{1}{z}\right) \\
&\sim \frac{\Gamma(c)\Gamma(b-a)z^{-a}}{\Gamma(b)\Gamma(c-a)} + \frac{\Gamma(c)\Gamma(a-b)z^{-b}}{\Gamma(a)\Gamma(c-b)}, \qquad z \to \infty,
\end{aligned} \tag{5.42}
$$

where in the last line we have again retained only the leading terms.

5.3.4 The function ${}_3F_3$

The derivations given above illustrate the efficiency of using the Meijer G-function in obtaining large-argument asymptotic formulas for any of the generalized hypergeometric functions. Here we do the same for the particular function ${}_3F_3$. We choose this function because it appears in the development of some of the applications in later chapters.

Based on Eq. (5.36), we deduce that

$$
{}_3F_3(a_1,\, a_2,\, a_3;\, c_1,\, c_2,\, c_3;\, -z) = \frac{\prod_{j=1}^{3}\Gamma(c_j)}{\prod_{j=1}^{3}\Gamma(a_j)} G_{43}^{31}\left(\frac{1}{z}\middle|\begin{matrix} 1,\, c_1,\, c_2,\, c_3 \\ a_1,\, a_2,\, a_3 \end{matrix}\right), \tag{5.43}
$$

and by applying Eq. (5.1), we are led to (noting that $m = 3$ in this case)

$$
{}_3F_3(a_1, a_2, a_3; c_1, c_2, c_3; -z) \sim \frac{\prod_{j=1}^{3}\Gamma(c_j)}{\prod_{j=1}^{3}\Gamma(a_j)}\left[\frac{\Gamma(a_2-a_1)\Gamma(a_3-a_1)\Gamma(a_1)}{\Gamma(c_1-a_1)\Gamma(c_2-a_1)\Gamma(c_3-a_1)}z^{-a_1}\right.
$$
$$
+\frac{\Gamma(a_1-a_2)\Gamma(a_3-a_2)\Gamma(a_2)}{\Gamma(c_1-a_2)\Gamma(c_2-a_2)\Gamma(c_3-a_2)}z^{-a_2}
$$
$$
\left.+\frac{\Gamma(a_1-a_3)\Gamma(a_2-a_3)\Gamma(a_3)}{\Gamma(c_1-a_3)\Gamma(c_2-a_3)\Gamma(c_3-a_3)}z^{-a_3}\right], \; z \to \infty. \tag{5.44}
$$

As in the previous cases, we have retained only the lead terms in the sum on the right-hand side of Eq. (5.44).

5.4 MacRobert *E*-Function

In the late 1930s, T. M. MacRobert made an attempt to give meaning to the function ${}_pF_q$ when $p > q+1$. For $p \geq q+1$, MacRobert's original work led to the E-function definition,[6]

$$
E(a_1, \ldots, a_p;\; c_1, \ldots, c_q;\; x) = \sum_{n=1}^{p} \frac{\prod_{k=1}^{\prime p}\Gamma(a_k - a_n)}{\prod_{j=1}^{q}\Gamma(c_j - a_n)}\Gamma(a_n)x^{a_n}
$$
$$
\times\; {}_{q+1}F_{p-1}[a_n,\, 1+a_n-c_1, \ldots, 1+a_n-c_q;
$$
$$
1+a_n-a_1, \ldots, *, \ldots, 1+a_n-a_p;\; (-1)^{p+q}x], \tag{5.45}
$$

where $|x| < 1$ if $p = q+1$. In addition, he also derived

$$
E(a_1, \ldots, a_p;\; c_1, \ldots, c_q;\; x) = \frac{\prod_{k=1}^{p}\Gamma(a_k)}{\prod_{j=1}^{q}\Gamma(c_j)}\, {}_pF_q\left(a_1, \ldots, a_p;\; c_1, \ldots, c_q;\; -\frac{1}{x}\right), \tag{5.46}
$$

where $p < q$, $x \neq 0$, or $p = q+1$ and $|x| > 1$.

Although MacRobert's E-function analysis led to an asymptotic expansion for large argument, his E-function never gained wide recognition because it was found to be a special case of the Meijer G-function, i.e.,

$$
E(a_1, \ldots, a_p;\; c_1, \ldots, c_q;\; x) = G_{q+1,p}^{p,1}\left(x \,\middle|\, \begin{matrix} 1, c_1, \ldots, c_q \\ a_1, \ldots, a_p \end{matrix}\right). \tag{5.47}
$$

Hence, properties of the E-function are simple consequences of the properties of the Meijer G-function.

5.5 Comments on the G-Function

In this chapter we have presented some of the most useful properties associated with the Meijer *G*-function. These and other useful relations are contained in Appendices A, C, and D. It is the generality of the Meijer *G*-function in representing other special functions and elementary functions that contributes to its usefulness. That is, most special functions and even products of special functions can be written as a Meijer *G*-function. The same is true for many of the elementary functions. For this reason, it is possible to solve a wide variety of integrals that appear in physics and engineering applications by using relations given in Appendix C along with the integral formulas listed in Appendix D. Many examples of this integration technique are provided in later chapters of the text. Additional properties and background can be found in Meijer's original papers. Readers are also encouraged to examine the treatment of the Meijer *G*-function and other generalized hypergeometric functions in Refs. 3, 5, 7, and 8.

5.6 Historical Remarks

Thomas Murray MacRobert (1884–1962) was a Scottish mathematician who was first educated at Irving Royal Academy along with his identical twin brother, Alexander, but later moved on to the University of Glasgow to complete his degree. He was the author of three textbooks: *Functions of a Complex Variable*, *Spherical Harmonics*, and *Trigonometry*. Much of his research was conducted later in his career where his most important work concerned the discovery of the MacRobert *E*-function. His later research papers were primarily concerned with the *E*-function.

The Dutch mathematician Cornelis Simon Meijer (1904–1974) is probably most noted for his introduction of the Meijer *G*-function and the Meijer integral transform. His other scientific work involved asymptotic expansions with error estimates. Also, the integral representation of special functions of mathematical physics was his main field of interest in later years. During the late 1930s and early 1940s, he worked with the Hungarian-British mathematician Arthur D. Erdélyi (1908–1977) to develop the theory of integral representations of Whittaker functions and their products. Later, Erdélyi was instrumental in helping to complete Meijer's 1941 development of the *G*-function using the Barnes integral.

The English mathematician Harry Bateman (1882–1946) of Caltech, who was a prolific scientist and a student of Sir Edmund Whittaker (1873–1956), had collected a gigantic amount of material on all of the special functions known at that time. His goal was to publish this material in a form suitable for the scientific community to access. Upon Bateman's untimely death in 1946, Whittaker was asked to recommend a mathematician who could start the enormous task of finishing and publishing Bateman's manuscripts. Erdélyi was the natural choice for Whittaker, and after arriving at Caltech in 1947 as a Visiting Professor, Erdélyi headed a team of prominent mathematicians to complete the project. The *Bateman Manuscript Project* consists of three texts on properties of higher transcendental functions (published in 1953 and 1955) and two texts on integral transform tables (published in 1954).[3]

References

1. C. S. Meijer, "Einige integraldarstellungen aus der Theorie der Besselschen und der Whittaker funktionen," *Akad. Wet. Amst. Proc.* **39**, 394–403, 519–527 (1936).
2. C. S. Meijer, "Neue integraldarstellungen für Whittakersche funcktionen," *Proc. Ned. Akad. V. Wetensch* **XLIV**, 82–92, 186–194, 298–307, 442–451, 590–598 (1941).
3. A. Erdelyi, W. Magnus, F. Oberhettinger, and F. G. Tricomi, *Higher Transcendental Functions*, 3 Vols., Bateman Manuscript Project, McGraw Hill, New York (1953), (1955).
   A. Erdelyi, W. Magnus, F. Oberhettinger, and F. G. Tricomi, *Tables of Integral Transforms*, 2 Vols., Bateman Manuscript Project, McGraw Hill, New York (1954).
4. I. N. Sneddon, *The Use of Integral Transforms*, McGraw Hill, New York (1972).
5. Y. L. Luke, *The Special Functions and Their Applications*, 2 vols., Academic Press, New York (1969).
6. T. M. MacRobert, "Induction proofs of the relations between certain asymptotic expansions and corresponding generalized hypergeometric series," *Proc. Roy. Soc. Edinburgh* **58**, 1–13 (1938).
7. A. M. Mathai and R. K. Saxena, *Generalized Hypergeometric Functions with Applications in Statistics and Physical Sciences*, Springer-Verlag, New York, (1970).
8. I. S. Gradshteyn and I. M. Ryzhik, *Table of Integrals, Series, and Products*, Academic Press, New York (1980).

Chapter 6
Approximations

6.0 Introduction

In previous chapters we derived simple formulas for numerical evaluation of the hypergeometric-type functions and Bessel functions when the argument is either very small or very large. However, it can be useful to combine these asymptotic formulas by deriving simple algebraic expressions that bridge the asymptotic behavior between small and large arguments, and that may therefore be valid for all values of the argument. The method we illustrate here for combining the large- and small-argument approximations is not a universal technique for all ${}_pF_q$ functions but seems to work remarkably well for the case when the real or imaginary part of the function is either monotonically increasing or decreasing. Essentially, the main idea is to replace a given ${}_pF_q$ function with a simple binomial function.

6.1 Approximations for the ${}_2F_1$ Function

For the first illustrative examples presented here, let us consider some particular hypergeometric functions. In doing so, we will need to refer to asymptotic relations provided in Appendix A or E.

Example 1: ${}_2F_1(-\frac{5}{6}, \frac{1}{2}; \frac{3}{2}; -z) \approx (1 + 0.31z)^{5/6}$.

To derive the above relation, we note that for small arguments, the result ($HY9$) in Appendix A leads to

$$ {}_2F_1\left(-\frac{5}{6}, \frac{1}{2}; \frac{3}{2}; -z\right) \sim 1 + 0.278z, \qquad z \to 0, \tag{6.1} $$

whereas for large arguments, we can use ($HY7$) in Appendix A to write

$$
\begin{aligned}
{}_2F_1\left(-\frac{5}{6}, \frac{1}{2}; \frac{3}{2}; -z\right) &= (1+z)^{5/6}{}_2F_1\left(-\frac{5}{6}, 1; \frac{3}{2}; \frac{z}{1+z}\right) \\
&\sim z^{5/6}{}_2F_1\left(-\frac{5}{6}, 1; \frac{3}{2}; 1\right) \\
&\sim 0.375\, z^{5/6}, \quad z \to \infty.
\end{aligned} \tag{6.2}
$$

We chose the dominant Pochhammer parameter "$-\frac{5}{6}$" for the large-argument expression and ignored "$\frac{1}{2}$." We also used ($HY4$) of Appendix A in the last step. Now, based on the asymptotic expressions in Eqs. (6.1) and (6.2), we choose a constant α so that the hypergeometric function has the algebraic approximation of a binomial function with the proper asymptotic behavior; that is,

$$
{}_2F_1\left(-\frac{5}{6}, \frac{1}{2}; \frac{3}{2}; -z\right) \approx (1+\alpha z)^{5/6} \sim \begin{cases} 1 + \dfrac{5}{6}\alpha z, & z \to 0 \\ (\alpha z)^{5/6}, & z \to \infty. \end{cases} \tag{6.3}
$$

The trick is to select a value of α that provides a good approximation to both Eqs. (6.1) and (6.2). In the upper expression of Eq. (6.3) we should choose $\alpha = 0.334$ to match Eq. (6.1), whereas in the lower expression we should choose $\alpha = 0.308$. Fortunately, these values of α are fairly close to one another so we will choose a compromising value, $\alpha = 0.31$, leading to

$$
{}_2F_1\left(-\frac{5}{6}, \frac{1}{2}; \frac{3}{2}; -z\right) \approx (1+0.31z)^{5/6}. \tag{6.4}
$$

In Fig. 6.1, we plot both functions in Eq. (6.4) for real values $z = x$ over the interval $-2 \le x \le 10$, avoiding values of x such that $1 + 0.31x \le 0$.

If the two values of α that match each asymptotic expression are not sufficiently similar, an algebraic approximation like that in Eq. (6.4) will be less accurate. In such cases the value of α can be chosen to favor either the

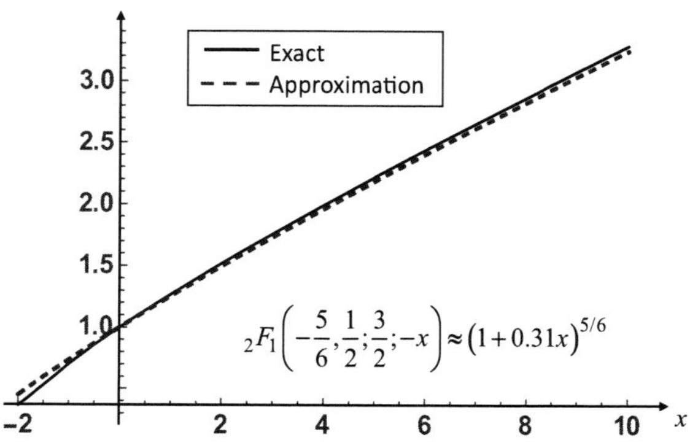

Figure 6.1 Plot of the hypergeometric function and its approximation given in Eq. (6.4) for real $z = x$.

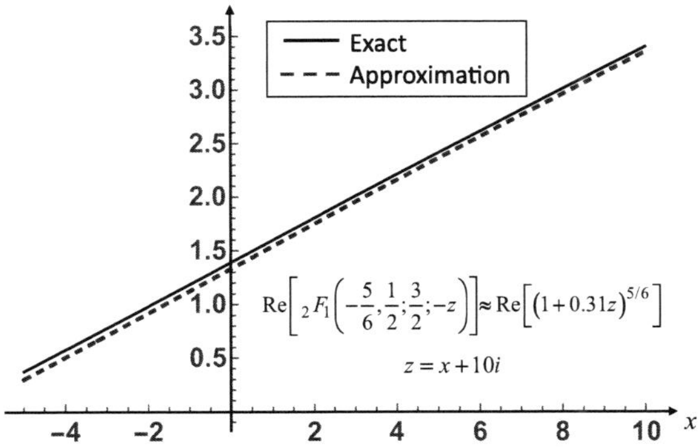

(a)

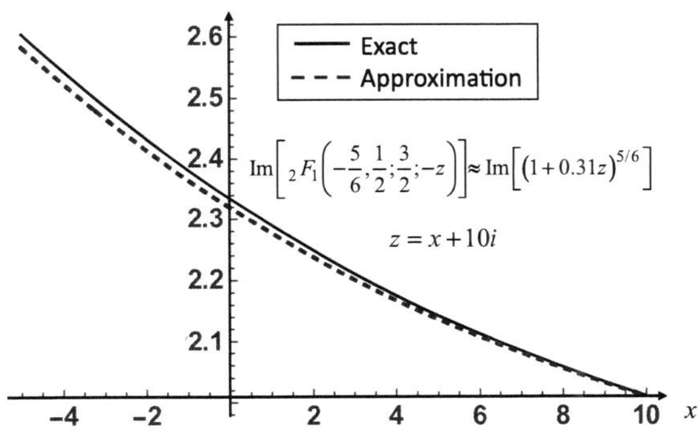

(b)

Figure 6.2 (a) Plot of the real parts of Eq. (6.4) and (b) plot of the imaginary parts of Eq. (6.4). Both (a) and (b) assume that $z = x + 10i$.

small-argument approximation or the large-argument approximation. Otherwise, we might select α somewhere in the middle.

Although in Fig. 6.1 we plotted both functions appearing in Eq. (6.4) for real values of z, the approximation (6.4) is also valid for complex values of z through analytic continuation. For example, in Fig. 6.2 we plot the real parts and imaginary parts of both functions in Eq. (6.4) for complex $z = x + 10i$, $-5 \le x \le 10$.

Example 2: ${}_2F_1(-\frac{5}{6}, \frac{11}{6}; \frac{17}{6}; z) \approx (1 - 0.66z)^{5/6}$.

As another example of the ${}_2F_1$ approximation, we now consider the function in Example 2 where we assume that z is once again complex. The series for small arguments leads to

$$ {}_2F_1\left(-\frac{5}{6}, \frac{11}{6}; \frac{17}{6}; z\right) \sim 1 - 0.539z, \quad z \to 0. \tag{6.5} $$

For large arguments, this time we will use the first term of ($HY10$) in Appendix A to get

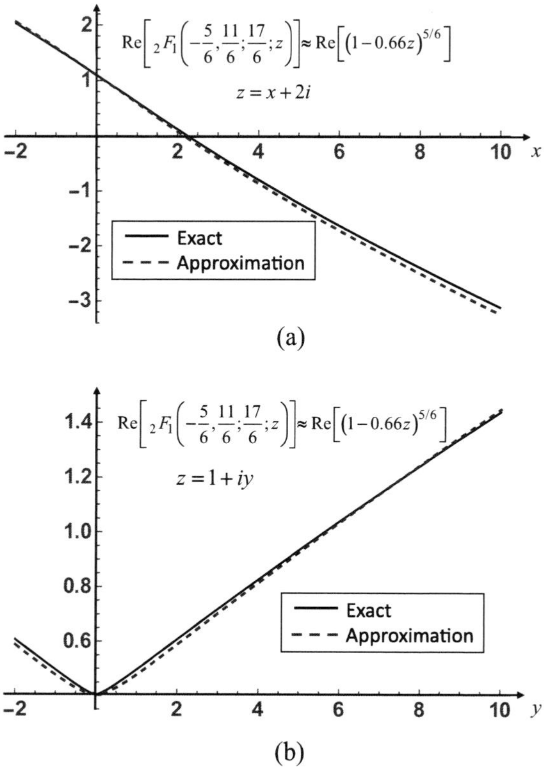

Figure 6.3 Plot of the real parts of Eq. (6.7) for (a) $z = x + 2i$ and for (b) $z = 1 + iy$.

$$
\begin{aligned}
{}_2F_1\left(-\frac{5}{6}, \frac{11}{6}; \frac{17}{6}; z\right) &\sim \frac{\Gamma(17/6)\Gamma(8/3)}{\Gamma(11/6)\Gamma(11/3)}(-z)^{5/6} \\
&\sim 0.688(-z)^{5/6}, \quad z \to \infty.
\end{aligned} \tag{6.6}
$$

By combining the results of Eqs. (6.5) and (6.6), we find a reasonably good approximation to be

$$
{}_2F_1\left(-\frac{5}{6}, \frac{11}{6}; \frac{17}{6}; z\right) \approx (1 - 0.66z)^{5/6}. \tag{6.7}
$$

In Fig. 6.3, we plot the real parts of both terms in Eq. (6.7) with $z = x + 2i$ in (a) and $z = 1 + iy$ in (b). Note that this produces a monotonically decreasing function in Fig. 6.3(a) and a monotonically increasing function in Fig. 6.3(b) for $y > 0$.

6.2 Approximations for the ${}_1F_1$ Function

We can use the same technique illustrated in Section 6.1 to develop similar algebraic approximations for the confluent hypergeometric function.

Example 3: ${}_1F_1(-\frac{5}{6};\,1;\,-z) \approx 1 + \frac{5z}{6}(1 + 0.232z)^{-1/6}$.

To begin, let us consider the asymptotic expressions from Appendix E,

$$ {}_1F_1\left(-\frac{5}{6};\,1;\,-z\right) \sim \begin{cases} 1 + 0.833z, & z \to 0 \\ 1.063z^{5/6}, & z \to \infty. \end{cases} \tag{6.8} $$

Thus, we might consider setting $\alpha = 1$, which provides the simple approximation

$$ {}_1F_1\left(-\frac{5}{6};\,1;\,-z\right) \approx (1 + z)^{5/6}. \tag{6.9} $$

In Fig. 6.4(a) we plot both functions in Eq. (6.9) for real values $z = x$, $0 \le x \le 10$. Although the approximation over this interval looks good for smaller values of x, we can improve on it for larger values in the following way.[1] Let us first write out the series for ${}_1F_1$ with the first term displayed and make the change of index $n \to n + 1$ in the remaining series to get

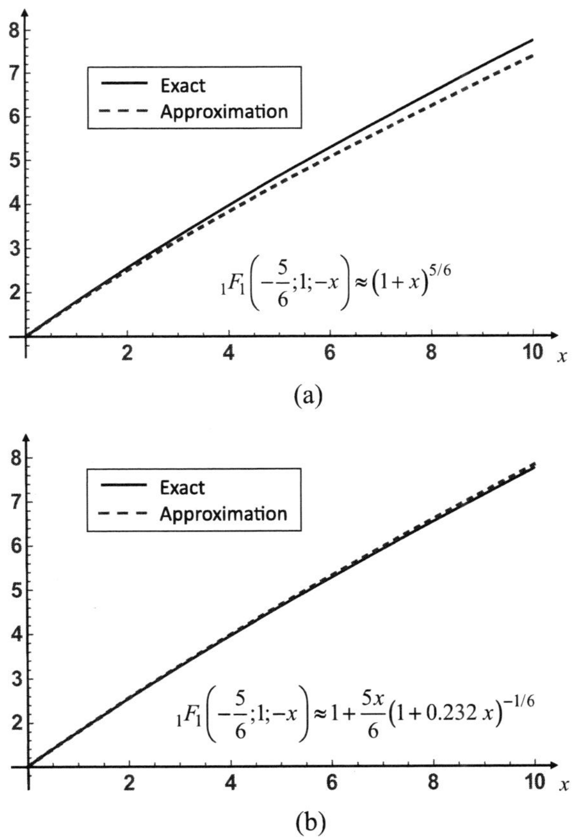

Figure 6.4 Plots of (a) Eq. (6.9) and (b) Eq. (6.13) for real arguments.

$$
\begin{aligned}
{}_1F_1\left(-\frac{5}{6};\ 1;\ -z\right) &= 1 + \underbrace{\sum_{n=1}^{\infty}}_{n\to n+1} (-1)^n \frac{\left(-\frac{5}{6}\right)_n}{(1)_n}\frac{z^n}{n!} \\
&= 1 - z\sum_{n=0}^{\infty}(-1)^n \frac{\left(-\frac{5}{6}\right)_{n+1}}{(1)_{n+1}}\frac{z^n}{(n+1)!}.
\end{aligned}
\tag{6.10}
$$

From the Pochhammer symbol identities in Appendix A,

$$
(a)_{n+1} = a(a+1)_n, \qquad \frac{1}{(n+1)!} = \frac{n!}{(n+1)!n!} = \frac{(1)_n}{(2)_n n!},
$$

we find that Eq. (6.10) becomes

$$
\begin{aligned}
{}_1F_1\left(-\frac{5}{6};\ 1;\ -z\right) &= 1 + \frac{5z}{6}\sum_{n=0}^{\infty}(-1)^n \frac{\left(\frac{1}{6}\right)_n (1)_n}{(2)_n (2)_n}\frac{z^n}{n!} \\
&= 1 + \frac{5z}{6}\,{}_2F_2\left(\frac{1}{6},\ 1;\ 2,\ 2;\ -z\right).
\end{aligned}
\tag{6.11}
$$

To continue, we calculate the asymptotic results

$$
{}_2F_2\left(\frac{1}{6},\ 1;\ 2,\ 2;\ -z\right) \sim \begin{cases} 1 - 0.042z, & z \to 0 \\ 1.276z^{-1/6}, & z \to \infty, \end{cases}
\tag{6.12}
$$

where we have used the small- and large-argument expressions for ${}_2F_2$ in Appendix E. Thus, combining Eqs. (6.11) and (6.12), we have the approximation

$$
{}_1F_1\left(-\frac{5}{6};\ 1;\ -z\right) \approx 1 + \frac{5z}{6}(1 + 0.232z)^{-1/6}.
\tag{6.13}
$$

In this example we used the value obtained from the large-argument asymptote in Eq. (6.12) to estimate α. We made this choice for α because for small values of z, the approximation in Eq. (6.13) agrees exactly with that in Eq. (6.8).

In Fig. 6.4(b) we plot both functions in Eq. (6.13) for real values of the argument, $z = x,\ 0 \le x \le 10$, for purposes of comparison with the simpler approximation [Eq. (6.9)] appearing in Fig. 6.4(a). Clearly, Eq. (6.13) is a better approximation compared to Eq. (6.9). This is just another technique that can sometimes be used when the first approximation is considered not sufficiently accurate over the range of interest—most often, when a single value of α doesn't match very well with both small and large arguments. In general, this secondary approach will be more accurate.

Example 4: ${}_1F_1(-\frac{11}{6};\ 1;\ -z) \approx (1 + 0.73z)^{11/6}$.

Once again, we refer to Appendix E to obtain the asymptotic results,

$${}_1F_1\left(-\frac{11}{6};\ 1;\ -z\right) \sim \begin{cases} 1 + 1.833z, & z \to 0 \\ 0.58z^{11/6}, & z \to \infty. \end{cases} \tag{6.14}$$

Approximating the ${}_1F_1$ function by $(1 + \alpha z)^{11/6}$, we select $\alpha = 0.73$, from which it follows that

$${}_1F_1\left(-\frac{11}{6};\ 1;\ -z\right) \approx (1 + 0.73z)^{11/6}. \tag{6.15}$$

In Fig. 6.5, we plot the real part of both functions in Eq. (6.15) for pure imaginary values of the argument, $z = iy$, $-4 \le y \le 5$.

Example 5: ${}_1F_1(\frac{1}{2};\ \frac{7}{3};\ -z) \approx (1 + 0.6z)^{-1/2}$.

Based on results in Appendix E, the asymptotic expressions are

$${}_1F_1\left(\frac{1}{2};\ \frac{7}{3};\ -z\right) \sim \begin{cases} 1 - 0.214z, & z \to 0 \\ 1.266z^{-1/2}, & z \to \infty, \end{cases} \tag{6.16}$$

from which we infer that

$${}_1F_1\left(\frac{1}{2};\ \frac{7}{3};\ -z\right) \approx (1 + 0.6z)^{-1/2}. \tag{6.17}$$

In Fig. 6.6 we plot the real parts of Eq. (6.17) for values $z = iy$, $0 \le y \le 30$. Notice that the confluent hypergeometric function in Fig. 6.6 has

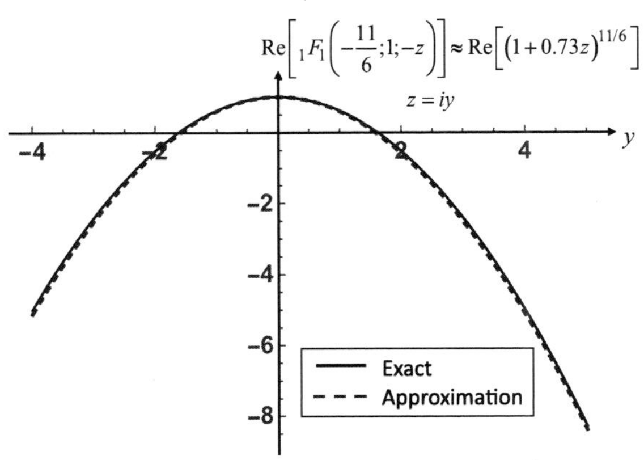

Figure 6.5 Plot of the real parts of Eq. (6.15) for $z = iy$.

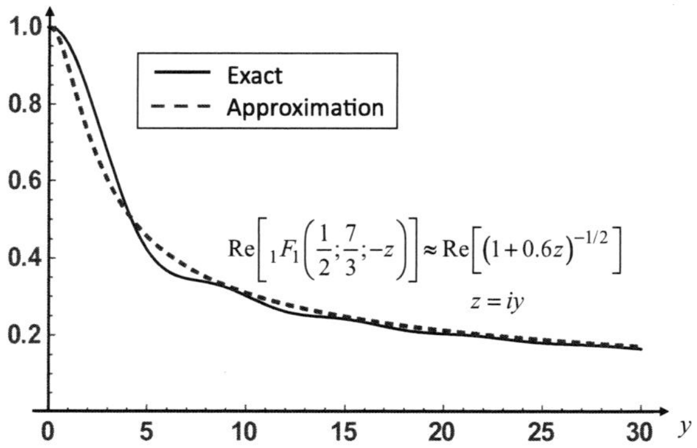

Figure 6.6 Plot of the real parts of Eq. (6.17) for $z = iy$.

a small oscillatory behavior for increasing values of y. This type of behavior cannot be approximated, in general, with a simple binomial function like that in Eq. (6.17). Nonetheless, the approximation shown in Fig. 6.6 may still be sufficiently accurate in some applications, particularly for larger arguments $y > 8$.

All of the previous examples had the argument given as $-z$ in the confluent hypergeometric function. For our last example, let us consider the confluent hypergeometric function ${}_1F_1(\frac{11}{6};\ \frac{7}{3};\ z)$ where the argument is z.

Example 6: ${}_1F_1(\frac{11}{6};\ \frac{7}{3};\ z) \approx e^z(1 + 0.6\,z)^{-1/2}$.

Here we start with the identity ($CH4$) in Appendix A, i.e.,

$$ {}_1F_1\left(\frac{11}{6};\ \frac{7}{3};\ z\right) = e^z{}_1F_1\left(\frac{1}{2};\ \frac{7}{3};\ -z\right). \tag{6.18} $$

The ${}_1F_1$ function on the right in Eq. (6.18) is that from the last example. Thus, our approximation this time is

$$ {}_1F_1\left(\frac{11}{6};\ \frac{7}{3};\ z\right) \approx e^z(1 + 0.6z)^{-1/2}. \tag{6.19} $$

Because the function grows exponentially for real positive arguments, we plot both sides of Eq. (6.19) with $z = x$ in Fig. 6.7(a) only over the interval $0 \le x \le 5$. And although the confluent hypergeometric function on the left is also defined for negative values of x, it is clear that the binomial function on the right is restricted to $x > -5/3$ in order to avoid the singularity at $x = -5/3$.

To illustrate that the approximations are not totally restricted to functions that do not oscillate, we plot the real parts of Eq. (6.19) in Fig. 6.7(b) for the complex argument $z = 2 + iy$. Here we see that, although the real part of the given ${}_1F_1$ function has an oscillatory behavior with a complex argument, the algebraic approximation still provides an excellent fit over all y. A similar fit

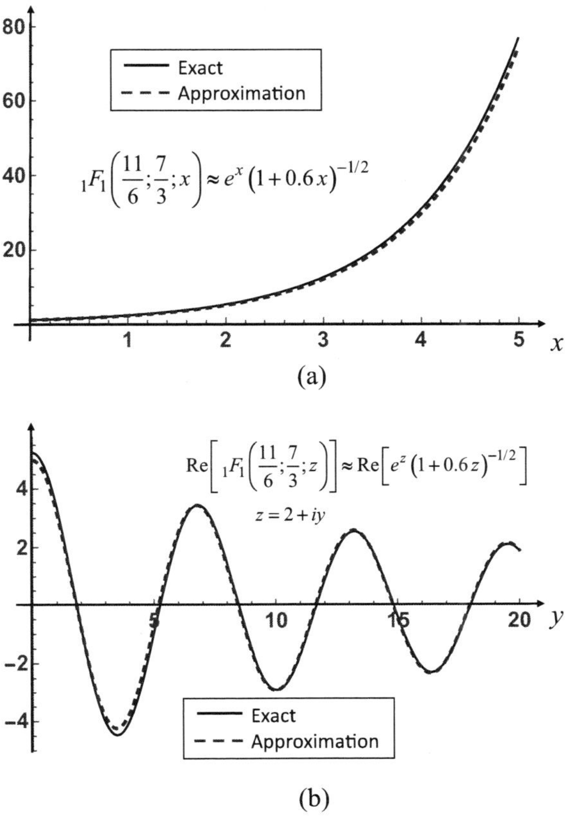

Figure 6.7 Plot of both functions in Eq. (6.19) with (a) $z = x$ and (b) the real parts of both functions with $z = 2 + iy$.

can also be obtained by plotting the imaginary parts of Eq. (6.19) with $z = 2 + iy$.

6.3 Other Functions

In this section we apply the techniques illustrated in Sections 6.1 and 6.2 to approximate some of the other special functions introduced in previous chapters.

6.3.1 Error function

In Section 3.10.1 we derived the error function relation

$$\mathrm{erf}(z) = \frac{2z}{\sqrt{\pi}}\,{}_1F_1\left(\frac{1}{2};\,\frac{3}{2};\,-z^2\right). \tag{6.20}$$

By applying the technique illustrated in this chapter with real $z = x$, we obtain

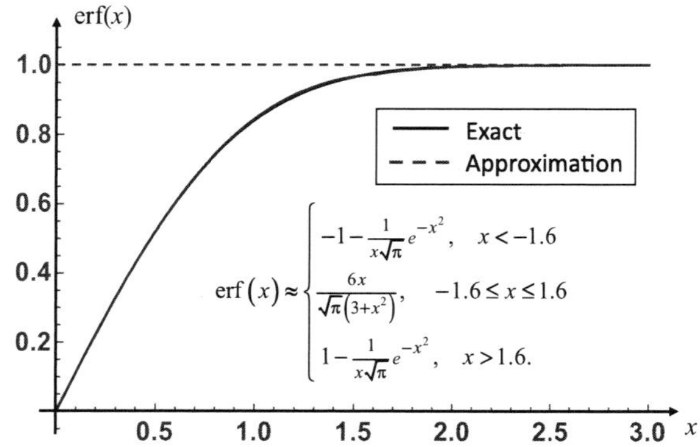

Figure 6.8 Plot of the error function and the approximation of Eq. (6.22).

$$\frac{2x}{\sqrt{\pi}}\,{}_1F_1\left(\frac{1}{2};\,\frac{3}{2};\,-x^2\right) \approx \frac{6x}{\sqrt{\pi}(3+x^2)}, \qquad -1.6 \le x \le 1.6. \tag{6.21}$$

Thus, if we also use Eq. (3.102) for $x < -1.6$ and $x > 1.6$, we arrive at

$$\mathrm{erf}(x) \approx \begin{cases} -1 - \dfrac{1}{x\sqrt{\pi}}e^{-x^2}, & x < -1.6 \\ \dfrac{6x}{\sqrt{\pi}\,(3+x^2)}, & -1.6 \le x \le 1.6 \\ 1 - \dfrac{1}{x\sqrt{\pi}}e^{-x^2}, & x > 1.6. \end{cases} \tag{6.22}$$

In Fig. 6.8 we plot the graph of the error function and the approximation in Eq. (6.22) over the interval $0 \le x \le 3$. With this approximation, the exact values are virtually indistinguishable from the approximate values for all x.

6.3.2 Bessel functions

The standard Bessel functions with real argument have an oscillatory behavior as illustrated in Fig. 4.1. Although the technique we are illustrating here usually works best on functions that are monotonically increasing or decreasing, the approximations can still follow the general behavior in other cases, as shown in Figs. 6.5, 6.7(b), and below in Fig. 6.9.

Based on Eq. (4.14), we have the relationship

$$J_p(z) = \frac{(z/2)^p}{\Gamma(p+1)}e^{iz}\,{}_1F_1\left(p+\frac{1}{2};\; 2p+1;\; -2iz\right). \tag{6.23}$$

We now apply the approximation technique we have been using on the confluent hypergeometric function in Eq. (6.23). This leads to the following approximations for real $z = x$ and $p = 0,\ 1$, and 2:

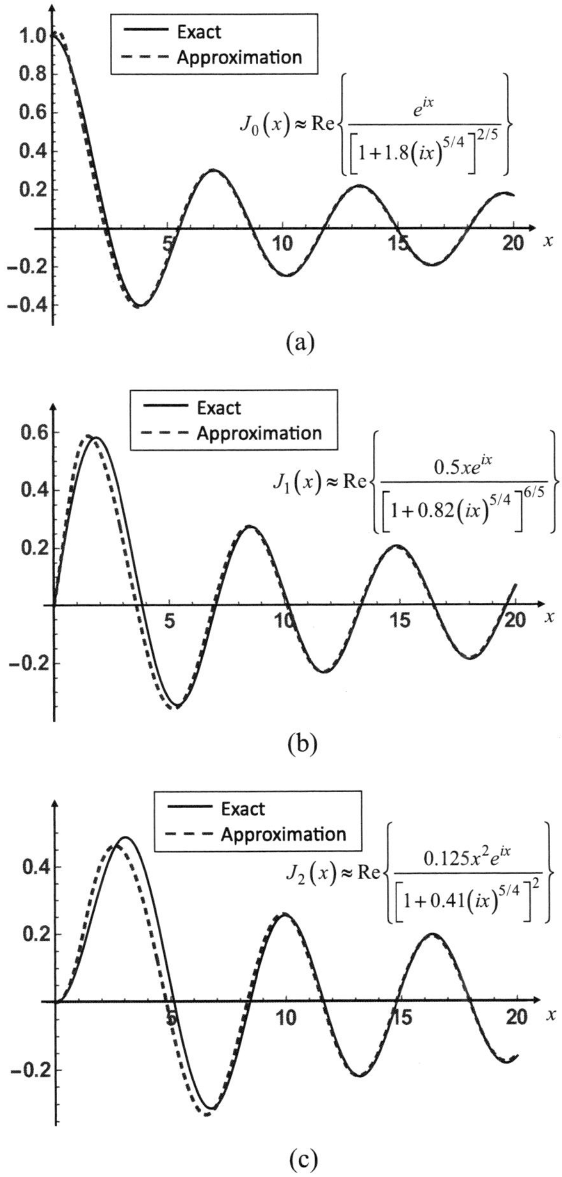

Figure 6.9 Plot of a standard Bessel function with the approximation of (a) Eq. (6.24), (b) Eq. (6.25), and (c) Eq. (6.26).

$$J_0(x) \approx \mathrm{Re}\{e^{ix}[1 + 1.8(ix)^{5/4}]^{-2/5}\}, \tag{6.24}$$

$$J_1(x) \approx \mathrm{Re}\left\{\frac{0.5xe^{ix}}{[1 + 0.82(ix)^{5/4}]^{6/5}}\right\}, \tag{6.25}$$

$$J_2(x) \approx \mathrm{Re}\left\{\frac{0.125x^2 e^{ix}}{[1+0.41(ix)^{5/4}]^2}\right\}. \tag{6.26}$$

Approximations like those in Eqs. (6.24) through (6.26) can also be found for other-order Bessel functions with real argument. We plot the graphs of the above three Bessel functions in Fig. 6.9 along with their approximations. Although the approximation given by Eq. (6.24) shown in Fig. 6.9(a) is quite good for all x, the approximations given by Eqs. (6.25) and (6.26) are not as good at small values of x; we chose the constants to provide a better fit as x increases. We make no claim that these represent the best approximations—only that they are reasonable. Other choices for the constants may improve the fit. Also, the method of Eqs. (6.10) through (6.13) might improve the fit.

6.3.3 Modified Bessel functions

From Eq. (4.41), we recall the modified Bessel function identity:

$$I_p(z) = \frac{(2z)^p \Gamma\left(p+\frac{1}{2}\right)}{\sqrt{\pi}\,\Gamma(2p+1)} e^z {}_1F_1\left(p+\frac{1}{2};\ 2p+1;\ -2z\right). \tag{6.27}$$

Hence, we can develop an algebraic approximation based on the confluent hypergeometric function. Doing so for general p leads to

$$I_p(z) \approx \frac{\Gamma\left(p+\frac{1}{2}\right)}{\sqrt{\pi}\,\Gamma(2p+1)} \frac{(2z)^p e^z}{(1+\alpha z)^{p+1/2}}, \quad \mathrm{Re}(z) \ge 0, \tag{6.28}$$

where

$$\alpha \approx 1.72\left[\frac{\Gamma(p+1/2)}{\Gamma(2p+1)}\right]^{1/(p+1/2)}, \quad 0 \le p \le 3. \tag{6.29}$$

In Fig. 6.10 we plot the modified Bessel functions corresponding to $p = 0$ and $p = 2$ along with the approximation [Eq. (6.28)] for real $z = x$. For values of $p > 3$, the choice of constant 1.72 in Eq. (6.29) may need to be adjusted slightly for a better fit.

Suppose we wish to find an approximation for the modified Bessel function $K_1(z)$. In the next example we will do so but will also approach the approximations a little differently than done thus far. Let us first start with the asymptotic behavior of the product,

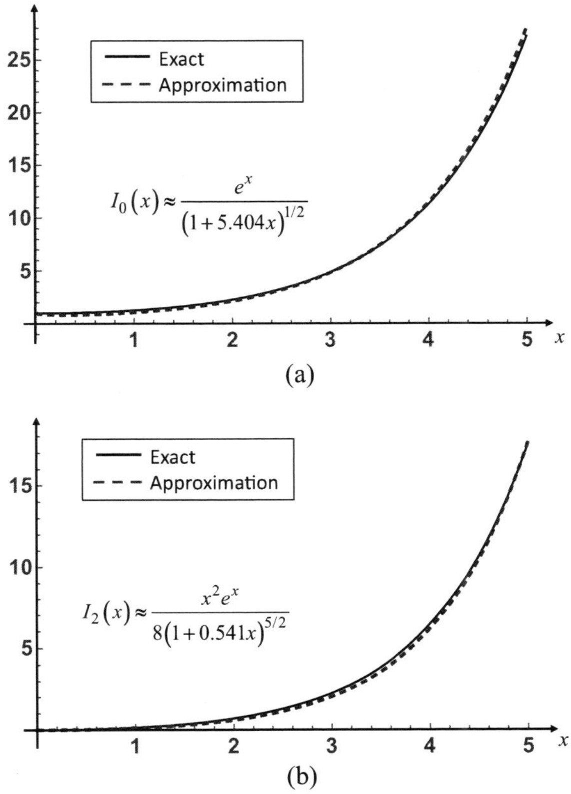

Figure 6.10 Plot of modified Bessel functions and the approximation of Eq. (6.28) for (a) $p = 0$ and (b) $p = 2$.

$$I_1(z)K_1(z) \sim \begin{cases} 1/2, & z \to 0 \\ 1/2z, & z \to \infty. \end{cases} \tag{6.30}$$

Based on the asymptotic results in Eq. (6.30), a reasonable approximation to this product of modified Bessel functions is

$$I_1(z)K_1(z) \approx \frac{1}{2(1 + z^{9/5})^{5/9}}. \tag{6.31}$$

Both sides of this last equation are plotted over the real interval $0 \le x \le 4$ in Fig. 6.11. Although the fit is also excellent for larger values of x, we chose a smaller interval to display the small-argument behavior.

For the function $K_1(z)$ alone, we use Eq. (6.28) for $I_1(z)$ along with Eq. (6.31) to get the approximation

$$K_1(z) \approx \frac{1}{2I_1(z)(1 + z^{9/5})^{5/9}} \approx \frac{(1 + z)^{3/2}e^{-z}}{z(1 + z^{7/5})^{5/7}}. \tag{6.32}$$

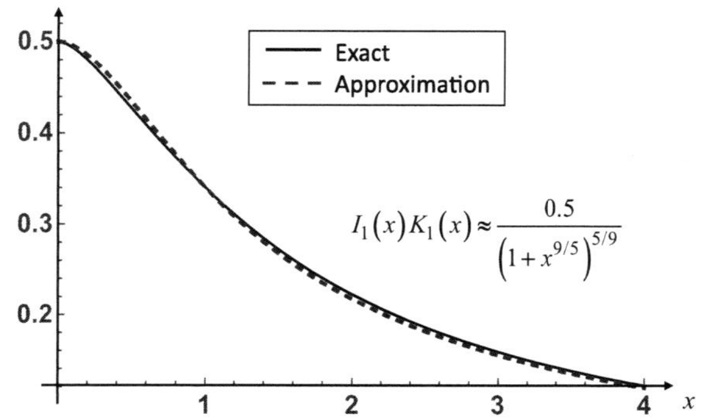

Figure 6.11 Plot of a product of modified Bessel functions with the approximation of Eq. (6.31).

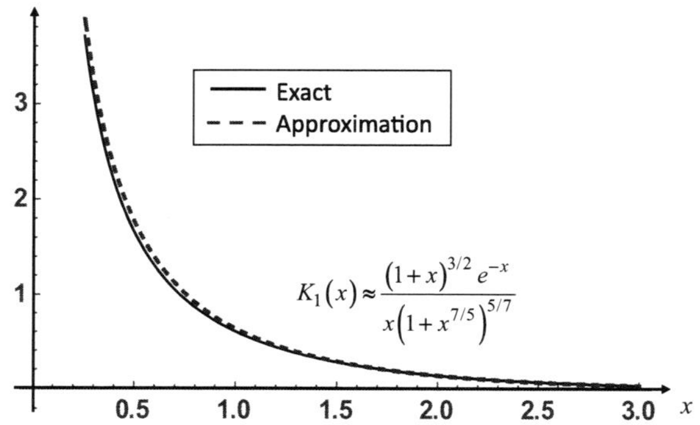

Figure 6.12 Plot of a modified Bessel function of the second kind with the approximation of Eq. (6.32).

Because of the approximation given by Eq. (6.28) for $I_1(z)$, we made a small adjustment on the term in the denominator of Eq. (6.32) from that in Eq. (6.31). The approximation given by Eq. (6.32) is shown in Fig. 6.12 over the real interval $0 < x \leq 3$.

6.3.4 Modified Struve function

The asymptotic form of the modified Struve function for large arguments involves the difference $I_p(z) - \mathbf{L}_p(z)$. Thus, we consider the approximation based on this difference. For example, when $p = 2$, we can obtain the small-argument approximation from Eq. (4.84) and the large-argument approximation from Eq. (4.85), or both from Appendix E. Then, by slightly adjusting the constants in these asymptotic relations, we obtain

$$I_2(z) - \mathbf{L}_2(z) \sim \begin{cases} z^2/8, & z \to 0 \\ z/4.7, & z \to \infty. \end{cases} \tag{6.33}$$

Next, by selecting the constant α for a best fit from the asymptotic results in Eq. (6.33), we arrive at the approximation

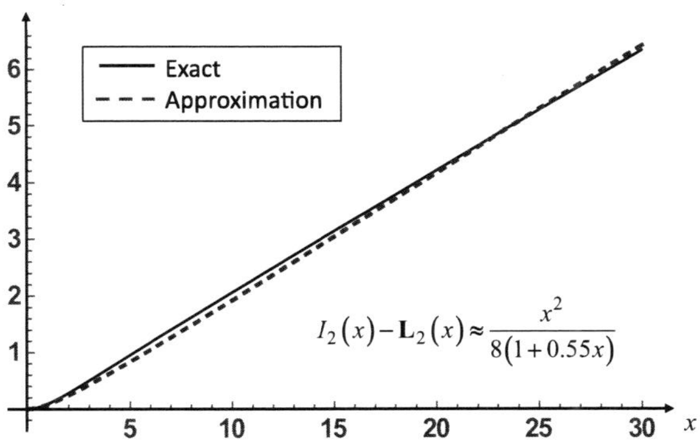

Figure 6.13 Plot of the difference of a modified Bessel function and the Struve function given by Eq. (6.34) for $p = 2$.

$$I_2(z) - \mathbf{L}_2(z) \approx \frac{z^2}{8(1 + 0.55z)}. \tag{6.34}$$

In Fig. 6.13 we plot both sides of Eq. (6.34) over the real interval $0 \le x \le 30$. We should point out that both the modified Bessel function and the modified Struve function exhibit exponential growth as the argument increases, but their difference causes a partial cancellation of such growth. Also, by using the approximation of Eq. (6.28) for $I_2(z)$, the following approximation can be generated for the modified Struve function alone:

$$\mathbf{L}_2(z) \approx \frac{z^2 e^z}{8(1 + 0.541z)^{5/2}} - \frac{z^2}{8(1 + 0.5z)}. \tag{6.35}$$

Once again, because of the approximation for the modified Bessel function $I_2(z)$, we have replaced the constant 0.55 in Eq. (6.34) with 0.5 for a better fit at smaller arguments. This approximation is shown in Fig. 6.14.

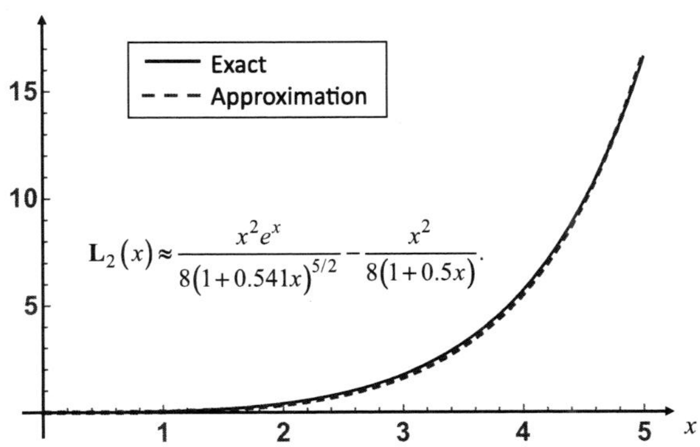

Figure 6.14 Plot of the Struve function given by Eq. (6.35) for $p = 2$.

6.3.5 Kelvin's functions

From the relationships given by Eqs. (4.90) and (6.28), we have the following approximations for real $z = x$:

$$\mathrm{ber}(x) \approx \mathrm{Re}\left[\frac{e^{x\sqrt{i}}}{(1 + 5.404x\sqrt{i})^{1/2}}\right] \tag{6.36}$$

and

$$\mathrm{bei}(x) \approx \mathrm{Im}\left[\frac{e^{x\sqrt{i}}}{(1 + 5.404x\sqrt{i})^{1/2}}\right]. \tag{6.37}$$

Plots of Eqs. (6.36) and (6.37) are shown in Fig. 6.15.

If we restrict values to $0 \le |z| \le 5$, we can approximate these functions by the first few terms of their series. Also, the large-argument approximations may be sufficiently accurate for $|z| > 5$ with an error on the order of 2–3%. Therefore, in this case we suggest the alternative approximations,

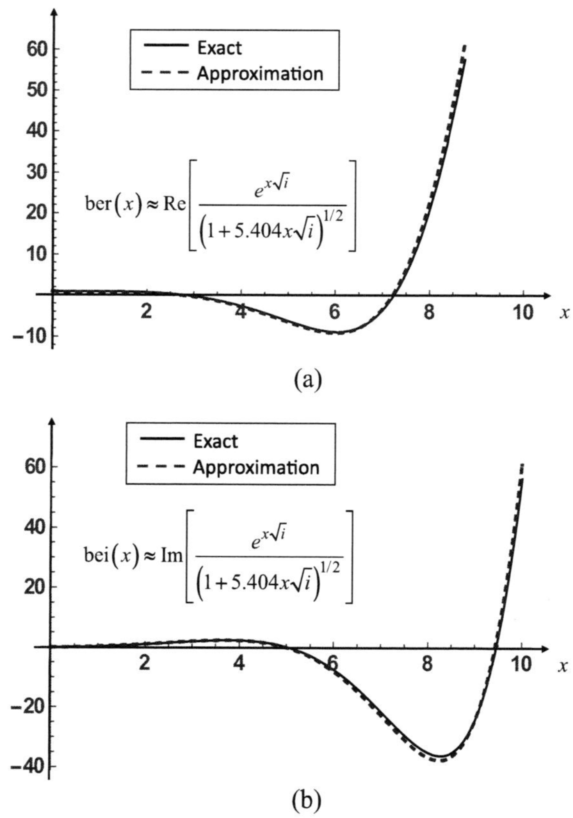

Figure 6.15 (a) Plot of Kelvin's function ber(*x*) and its approximation given by Eq. (6.36) and (b) plot of Kelvin's function bei(*x*) and its approximation given by Eq. (6.37).

$$\mathrm{ber}(z) \approx \begin{cases} 1 - \frac{1}{4}\left(\frac{z}{2}\right)^4\left[1 - 0.111\left(\frac{z}{4}\right)^4 + 0.00099\left(\frac{z}{4}\right)^8\right], & 0 \le |z| \le 5 \\ \dfrac{e^{z/\sqrt{2}}}{\sqrt{2\pi z}}\cos\left(\dfrac{z}{\sqrt{2}} - \dfrac{\pi}{8}\right), & |z| > 5 \end{cases} \tag{6.38}$$

and

$$\mathrm{bei}(z) \approx \begin{cases} \left(\frac{z}{2}\right)^2\left\{1 - \frac{1}{36}\left(\frac{z}{2}\right)^4\left[1 - 0.04\left(\frac{z}{4}\right)^4 + 0.00073\left(\frac{z}{4}\right)^8\right]\right\}, & 0 \le |z| \le 5 \\ \dfrac{e^{z/\sqrt{2}}}{\sqrt{2\pi z}}\sin\left(\dfrac{z}{\sqrt{2}} - \dfrac{\pi}{8}\right), & |z| > 5. \end{cases} \tag{6.39}$$

6.4 Discussion

The idea of finding an algebraic approximation to a complicated function has been around for many years. In particular, engineers often introduce in their analysis what is called an "engineering approximation" for such functions. Nonetheless, we believe the approximation technique we have illustrated in this chapter is a new and innovative approach. It is not a universal method and is generally limited to functions that are either monotonically increasing or decreasing over the interval of interest. However, the graphical plots in Figs. 6.5, 6.7(b), and 6.9 show that this is not a strict requirement. In both of the last two cases, the approximations in these figures closely match the oscillations of the function being approximated. The delicate part in this approach is to choose the constant α in the binomial approximation $(1 + \alpha z)^{\beta}$ to achieve the best approximation, based on a balance between the small-argument fit of the given function and the large-argument fit. We have found in our use of this technique that the best results usually come from choosing α to favor the large argument. Because the functions we are trying to approximate are analytic functions over all or much of the complex plane, we can use the principle of analytic continuation to extend the argument of the approximation to complex numbers, as we have illustrated with several examples. We should point out that the numerical constants that appear in the approximations presented here are not considered optimal—they are chosen on the basis of providing a reasonable approximation for illustration purposes. Other choices of constants might lead to a better approximation.

As a final comment, we also point out that the method illustrated in this chapter works best on approximating a generalized hypergeometric function that stands alone. If the generalized hypergeometric function is part of a sum (or a difference or a quotient) of terms in a given expression, we have found in some cases that the approximation may be poor when combined with other terms, even when the approximation for the generalized hypergeometric function by itself is quite good. Some examples of this are provided in Chapter 11.

Reference

1. L. C. Andrews, S. Vester, and C. E. Richardson, "Analytic expressions for the wave structure function based on a bump spectral model for refractive index," *J. Mod. Opt.* **40**, 931–938 (1993).

Part II
Applications

Chapter 7
Integrals and Products

7.0 Introduction

Although a variety of techniques exist for evaluating integrals, many of the integrals that commonly appear in applications are quite difficult to solve by conventional methods. Nonetheless, they can often be solved by using properties of the generalized hypergeometric functions, particularly the Meijer *G*-function. We illustrate this concept throughout the first section of this chapter by solving an assortment of definite integrals. Next, we turn to products of special functions that are quite commonplace in some areas of application. For example, in adaptive optics calculations involving Zernike filter functions, the product of two or more Bessel functions often occurs in the integral we wish to evaluate (cf. Chapters 9 and 10). We first show how to simplify such products and then solve some integrals involving these products. We end the chapter by adapting the integration methods featured here to a few Laplace, Mellin, and Hankel integral transforms,[1,2] followed by a few applications involving these transforms.

The short list of identities given below will be useful in some of the following analysis. Although they are listed in Appendix A or B, we find it convenient to also display them here for quick reference:

$$(a)_{n-k} = \frac{(-1)^k (a)_n}{(1-a-n)_k}, \quad k = 0, 1, \ldots, n, \tag{7.1}$$

$$\frac{1}{(n-k)!} = \frac{(-1)^k (-n)_k}{n!}, \quad k = 0, 1, \ldots, n, \tag{7.2}$$

$${}_2F_1(a, b; c; 1) = \frac{\Gamma(c)\Gamma(c-a-b)}{\Gamma(c-a)\Gamma(c-b)}. \tag{7.3}$$

Dixon's Theorem:

$$
\begin{aligned}
&{}_3F_2(a,\, b,\, c;\, 1+a-b,\, 1+a-c;\, 1) \\
&\quad = \frac{\Gamma(1+a/2)\Gamma(1+a-b)\Gamma(1+a-c)\Gamma(1+a/2-b-c)}{\Gamma(1+a)\Gamma(1+a/2-b)\Gamma(1+a/2-c)\Gamma(1+a-b-c)}.
\end{aligned} \tag{7.4}
$$

7.1 Miscellaneous Integrals

In this first section we present several different types of integrals to illustrate the technique of solving integrals using generalized hypergeometric functions. We start with some improper integrals involving elementary functions.

7.1.1 Elementary functions

Example 1: $\int_0^\infty e^{-a^2x^2-b^2/x^2}dx = \frac{\sqrt{\pi}}{2a}\, e^{-2ab}, \quad a>0,\, b>0.$

Based on $(C1)$ from Appendix C and $(MG2)$ from Appendix A, we find that

$$
e^{-b^2/x^2} = G_{01}^{10}\left(\frac{b^2}{x^2}\middle|\, 0\right) = G_{10}^{01}\left(\frac{x^2}{b^2}\middle|\, 1\right). \tag{7.5}
$$

The substitution of Eq. (7.5) into the integral in Example 1 gives us

$$
\begin{aligned}
\int_0^\infty e^{-a^2x^2-b^2/x^2}dx &= \int_0^\infty e^{-a^2x^2}\, G_{10}^{01}\left(\frac{x^2}{b^2}\middle|\, 1\right)dx \\
&= \frac{1}{2}\int_0^\infty t^{-1/2}e^{-a^2t}\, G_{10}^{01}\left(\frac{t}{b^2}\middle|\, 1\right)dt,
\end{aligned} \tag{7.6}
$$

where we introduced a change of variable $t = x^2$. Now, from $(D6)$ in Appendix D, $(MG2)$ in Appendix A, and $(C23)$ in Appendix C, we are led to

$$
\begin{aligned}
\int_0^\infty e^{-a^2x^2-b^2/x^2}dx &= \frac{1}{2a}\, G_{20}^{02}\left(\frac{1}{a^2b^2}\middle|\, \frac{1}{2},\, 1\right) \\
&= \frac{1}{2a}\, G_{02}^{20}\left(a^2b^2\middle|\, \frac{1}{2},\, 0\right) \\
&= \frac{1}{a}\sqrt{ab}\, K_{1/2}(2ab).
\end{aligned} \tag{7.7}
$$

Finally, by recalling $(BK9)$ in Appendix A, this last result becomes

$$
\int_0^\infty e^{-a^2x^2-b^2/x^2}dx = \frac{\sqrt{\pi}}{2a}\, e^{-2ab}. \tag{7.8}
$$

In an analogous manner, it can also be shown that (e.g., see Section 7.7.2)

$$\int_0^\infty x^{p-1}e^{-a/x-bx}dx = 2\left(\frac{a}{b}\right)^{p/2} K_p(2\sqrt{ab}), \quad \text{Re}(a) > 0, \text{Re}(b) > 0. \quad (7.9)$$

Example 2: $\int_0^\infty e^{-t^2} \frac{\sin(2xt)}{t}\, dt = \frac{\pi}{2}\text{erf}(x)$.

From the Meijer G-function identity ($C3$) in Appendix C,

$$\sin(2xt) = \sqrt{\pi}\, G_{02}^{10}\left(x^2t^2 \middle| \frac{1}{2},\ 0\right), \quad (7.10)$$

we obtain

$$\begin{aligned}\int_0^\infty e^{-t^2} \frac{\sin(2xt)}{t}\, dt &= \sqrt{\pi}\int_0^\infty e^{-t^2}t^{-1}G_{02}^{10}\left(x^2t^2 \middle| \frac{1}{2},\ 0\right) dt \\ &= \frac{\sqrt{\pi}}{2}\int_0^\infty e^{-u}u^{-1}G_{02}^{10}\left(x^2u \middle| \frac{1}{2},\ 0\right) du,\end{aligned} \quad (7.11)$$

where we made the change of variable $u = t^2$. By using the integral relation ($D6$) in Appendix D, we see that Eq. (7.11) reduces to

$$\begin{aligned}\int_0^\infty e^{-t^2} \frac{\sin(2xt)}{t}\, dt &= \frac{\sqrt{\pi}}{2} G_{12}^{11}\left(x^2 \middle| \begin{matrix} 1 \\ \frac{1}{2},\ 0 \end{matrix}\right) \\ &= \sqrt{\pi}\, x\; {}_1F_1\left(\frac{1}{2}; \frac{3}{2}; -x^2\right) \\ &= \frac{\pi}{2}\text{erf}(x).\end{aligned} \quad (7.12)$$

The final result comes from ($C14$) in Appendix C and ($B10$) in Appendix B.

7.1.2 Bessel functions

Because Bessel functions are so common in practice, we now consider several integrals involving them.

Example 3: $\int_0^\infty \frac{x^{p+1}J_p(bx)}{(x^2+a^2)^{\mu+1}}\, dx = \frac{a^{p-\mu}\, b^{\mu}}{2^{\mu}\Gamma(\mu+1)} K_{p-\mu}(ab)$,

$$-1 < \text{Re}(p) < \text{Re}\left(2\mu + \frac{3}{2}\right), \quad a, b > 0.$$

From ($C11$) in Appendix C, we first establish the relationship

$$\begin{aligned}\frac{1}{(x^2+a^2)^{\mu+1}} &= \frac{1}{a^{2(\mu+1)}(x^2/a^2+1)^{\mu+1}} \\ &= \frac{1}{a^{2(\mu+1)}\Gamma(\mu+1)}\, G_{11}^{11}\left(\frac{x^2}{a^2} \middle| \begin{matrix} -\mu \\ 0 \end{matrix}\right),\end{aligned} \quad (7.13)$$

which, based on ($D12$) in Appendix D followed by ($MG2$) and ($MG3$) in Appendix A, leads to

$$\begin{aligned}\int_0^\infty \frac{x^{p+1}J_p(bx)}{(x^2+a^2)^{\mu+1}}\,dx &= \frac{1}{a^{2(\mu+1)}\Gamma(\mu+1)}\int_0^\infty x^{p+1}J_p(bx)\,G_{11}^{11}\left(\frac{x^2}{a^2}\middle|\begin{matrix}-\mu\\0\end{matrix}\right)dx\\ &= \frac{2^{p+1}}{a^{2(\mu+1)}b^{p+2}\Gamma(\mu+1)}\,G_{31}^{12}\left(\frac{4}{a^2b^2}\middle|\begin{matrix}-p,\ -\mu,\ 0\\0\end{matrix}\right)\\ &= \frac{2^{p+1}}{a^{2(\mu+1)}b^{p+2}\Gamma(\mu+1)}\,G_{13}^{21}\left(\frac{a^2b^2}{4}\middle|\begin{matrix}1\\1+p,\ 1+\mu,\ 1\end{matrix}\right)\\ &= \frac{2^{p+1}}{a^{2(\mu+1)}b^{p+2}\Gamma(\mu+1)}\,G_{02}^{20}\left(\frac{a^2b^2}{4}\middle|\,1+p,\ 1+\mu\right).\end{aligned} \tag{7.14}$$

Now, by referring to ($C23$) in Appendix C, this last expression reduces to the identity

$$\int_0^\infty \frac{x^{p+1}J_p(bx)}{(x^2+a^2)^{\mu+1}}\,dx = \frac{a^{p-\mu}\,b^{\mu}}{2^{\mu}\Gamma(\mu+1)}\,K_{p-\mu}(ab). \tag{7.15}$$

We could also have solved this integral by first replacing the Bessel function with ($C9$) in Appendix C combined with ($D11$) in Appendix D.

Example 4: $\int_0^\infty x^{-1}J_0(tx)\sin x\,dx = \begin{cases}\pi/2, & 0<t<1\\ \csc^{-1} t, & t>1.\end{cases}$

We start with the sine function representation ($C3$) in Appendix C,

$$\sin x = \sqrt{\pi}\,G_{02}^{10}\left(\frac{x^2}{4}\middle|\,\frac{1}{2},\ 0\right), \tag{7.16}$$

from which we find that

$$\begin{aligned}\int_0^\infty x^{-1}J_0(tx)\sin x\,dx &= \sqrt{\pi}\int_0^\infty x^{-1}J_0(tx)G_{02}^{10}\left(\frac{x^2}{4}\middle|\,\frac{1}{2},\ 0\right)dx\\ &= \frac{\sqrt{\pi}}{2}G_{22}^{11}\left(\frac{1}{t^2}\middle|\begin{matrix}1,\ 1\\ \frac{1}{2},\ 0\end{matrix}\right)\\ &= \frac{1}{t}\,{}_2F_1\left(\frac{1}{2},\frac{1}{2};\frac{3}{2};\frac{1}{t^2}\right), \quad t>1.\end{aligned} \tag{7.17}$$

In arriving at this last relation, we have used ($D12$) in Appendix D and ($C18$) in Appendix C. From ($B33$) in Appendix B, it is recognized that the ${}_2F_1$ function in Eq. (7.17) is related to the inverse sine function; hence,

$$\int_0^\infty x^{-1}J_0(tx)\sin x\,dx = \sin^{-1}\left(\frac{1}{t}\right) = \csc^{-1} t, \quad t > 1. \tag{7.18}$$

For the interval $0 < t < 1$, we use ($MG2$) in Appendix A and ($C18$) in Appendix C to find from the middle line in Eq. (7.17),

$$\frac{\sqrt{\pi}}{2}G_{22}^{11}\left(\frac{1}{t^2}\middle|\begin{matrix}1,\,1\\ \frac{1}{2},\,0\end{matrix}\right) = \frac{\sqrt{\pi}}{2}G_{22}^{11}\left(t^2\middle|\begin{matrix}\frac{1}{2},\,1\\ 0,\,0\end{matrix}\right) = \frac{\pi}{2}, \quad 0 < t < 1. \tag{7.19}$$

Thus, by combining Eqs. (7.18) and (7.19), we have shown that

$$\int_0^\infty x^{-1}J_0(tx)\sin x\,dx = \begin{cases}\pi/2, & 0 < t < 1\\ \csc^{-1} t, & t > 1.\end{cases} \tag{7.20}$$

We continue this section on Bessel functions by considering several integrals that will arise in Chapter 11. First, we calculate the statistical moments associated with the probability density function (PDF),

$$p_{\mathbf{r}}(r) = \frac{r}{N}e^{-(r^2+A^2)/2N}I_0\left(\frac{Ar}{N}\right), \quad r > 0. \tag{7.21}$$

Here, N denotes a noise variance, and A is the amplitude of a sinusoidal signal. The statistical moments calculation requires evaluation of the integral

$$\langle r^\nu\rangle = \frac{e^{-A^2/2N}}{N}\int_0^\infty r^{\nu+1}e^{-r^2/2N}I_0\left(\frac{Ar}{N}\right)dr, \quad \nu > 0, \tag{7.22}$$

where the bracket notation $\langle\,\rangle$ denotes an ensemble average. From the relationships ($B4$) in Appendix B and ($C32$) in Appendix C, we have

$$I_0\left(\frac{Ar}{N}\right) = {}_0F_1\left(-;\,1;\,\frac{A^2r^2}{4N^2}\right) = G_{01}^{10}\left(-\frac{A^2r^2}{4N^2}\middle|\,0,\,0\right). \tag{7.23}$$

Next, by substituting the Meijer G-function in Eq. (7.23) into Eq. (7.22) and making the change of variable $x = r^2$, the statistical moments become

$$\begin{aligned}\langle r^\nu\rangle &= \frac{e^{-A^2/2N}}{2N}\int_0^\infty x^{\nu/2}e^{-x/2N}G_{02}^{10}\left(-\frac{A^2x}{4N^2}\middle|\,0,\,0\right)dx\\ &= \frac{e^{-A^2/2N}}{2N}(2N)^{\nu/2+1}G_{12}^{11}\left(-\frac{A^2}{2N}\middle|\begin{matrix}-\frac{\nu}{2}\\ 0,\,0\end{matrix}\right)\\ &= (2N)^{\nu/2}\Gamma\left(1+\frac{\nu}{2}\right){}_1F_1\left(1+\frac{\nu}{2};\,1;\,\frac{A^2}{2N}\right)e^{-A^2/2N}.\end{aligned} \tag{7.24}$$

Both (*D*6) in Appendix D and (*C*13) from Appendix C were used to obtain the last result. Now with the application of (*CH*4) in Appendix A, it follows that

$$\langle r^\nu \rangle = (2N)^{\nu/2}\Gamma\left(1+\frac{\nu}{2}\right) {}_1F_1\left(-\frac{\nu}{2};\, 1;\, -\frac{A^2}{2N}\right), \qquad \nu > 0. \tag{7.25}$$

Example 5: $\int_0^{2\pi} \cos\theta \exp(z\cos\theta)\, d\theta = 2\pi I_1(z)$.

To prove this result, we start with the identity (*BI*3) in Appendix A,

$$\exp(z\cos\theta) = I_0(z) + 2\sum_{n=1}^{\infty} I_n(z)\cos n\theta, \tag{7.26}$$

which, after multiplying by $\cos\theta$ and integrating the result termwise over $(0, 2\pi)$, leads to

$$\begin{aligned}\int_0^{2\pi} \cos\theta \exp(z\cos\theta)\, d\theta &= I_0(z)\int_0^{2\pi} \cos\theta\, d\theta \\ &\quad + 2\sum_{n=1}^{\infty} I_n(z)\int_0^{2\pi} \cos\theta\cos n\theta\, d\theta.\end{aligned} \tag{7.27}$$

Based on the orthogonality property of the cosine function, which causes each term on the right in Eq. (7.27) to vanish except for $n = 1$, we are left with

$$\int_0^{2\pi} \cos\theta \exp(z\cos\theta)\, d\theta = 2\pi I_1(z). \tag{7.28}$$

Example 6:

$$\int_0^\infty r_2^2 \exp\left[-\frac{r_2^2}{2N(1-\rho^2)}\right] I_1\left[\frac{r_1 r_2 \rho}{N(1-\rho^2)}\right] dr_2 = N\rho(1-\rho^2) r_1 \exp\left[\frac{\rho^2 r_1^2}{2N(1-\rho^2)}\right].$$

For notational simplicity, let us define the integral by

$$H = \int_0^\infty r_2^2 \exp\left[-\frac{r_2^2}{2N(1-\rho^2)}\right] I_1\left[\frac{r_1 r_2 \rho}{N(1-\rho^2)}\right] dr_2, \tag{7.29}$$

where ρ is a correlation coefficient and N again denotes a random noise variance. To evaluate this integral, we start with

$$I_1(ar_2) = \frac{1}{i} J_1(iar_2) = \frac{1}{i} G_{02}^{10}\left(-\frac{a^2 r_2^2}{4}\,\middle|\, \frac{1}{2}, -\frac{1}{2}\right), \tag{7.30}$$

where we used (*C*9) in Appendix C and set $a = \rho r_1 / N(1-\rho^2)$. Then the integral in Eq. (7.29) takes the form

$$H = \frac{1}{i}\int_0^\infty r_2^2 e^{-br_2^2} G_{02}^{10}\left(-\frac{a^2r_2^2}{4}\middle|\frac{1}{2}, -\frac{1}{2}\right)dr_2$$
$$= \frac{1}{2i}\int_0^\infty x^{1/2} e^{-bx} G_{02}^{10}\left(-\frac{a^2x}{4}\middle|\frac{1}{2}, -\frac{1}{2}\right)dx, \tag{7.31}$$

where $b = 1/2N(1-\rho^2)$ and we made the change of variable $x = r_2^2$. Using ($D6$) in Appendix D followed by application of ($MG3$) in Appendix A, we have

$$H = \frac{1}{2i}b^{-3/2}G_{12}^{11}\left(-\frac{a^2}{4b}\middle|\begin{matrix}-\frac{1}{2}\\ \frac{1}{2}, -\frac{1}{2}\end{matrix}\right) = \frac{1}{2i}b^{-3/2}G_{01}^{10}\left(-\frac{a^2}{4b}\middle|\frac{1}{2}\right). \tag{7.32}$$

From ($C1$) in Appendix C, we identify

$$G_{01}^{10}\left(-\frac{a^2}{4b}\middle|\frac{1}{2}\right) = \frac{ia}{2\sqrt{b}}\exp\left(\frac{a^2}{4b}\right); \tag{7.33}$$

thus, by replacing a and b with their definitions above, we arrive at

$$\int_0^\infty r_2^2 \exp\left[-\frac{r_2^2}{2N(1-\rho^2)}\right] I_1\left[\frac{r_1r_2\rho}{N(1-\rho^2)}\right]dr_2 = N\rho(1-\rho^2)r_1\exp\left[\frac{\rho^2r_1^2}{2N(1-\rho^2)}\right]. \tag{7.34}$$

7.1.3 Hypergeometric functions

A number of integrals involving hypergeometric functions are readily evaluated with the methods illustrated here. In doing so, in some cases it is straightforward to express the hypergeometric function by its power series and use termwise integration. In other cases we use the Meijer G-function.

Example 7:

$$\int_0^1 t^{s-1}(1-t)^{c-s-1}\,{}_2F_1(a, b; s; zt)\,dt = \frac{\Gamma(s)\Gamma(c-s)}{\Gamma(c)}\,{}_2F_1(a, b; c; z)$$
$$c - s > 0,\ z \neq 1,\ |\arg(1-z)| < \pi.$$

We start by defining the integral

$$I = \int_0^1 t^{s-1}(1-t)^{c-s-1}\,{}_2F_1(a, b; s; zt)\,dt. \tag{7.35}$$

Because the limits of integration are $0 \le t \le 1$, we have the option to express the ${}_2F_1$ function by its power series provided we temporarily assume that $|z| < 1$. Thus, under the assumption that $|z| < 1$, this action leads to

$$I = \sum_{n=0}^{\infty} \frac{(a)_n (b)_n}{(s)_n} \frac{z^n}{n!} \int_0^1 t^{n+s-1}(1-t)^{c-s-1}\, dt$$
$$= \sum_{n=0}^{\infty} \frac{(a)_n (b)_n}{(s)_n} \frac{z^n}{n!} \frac{\Gamma(n+s)\Gamma(c-s)}{\Gamma(n+c)}, \tag{7.36}$$

where we have recalled the beta function ($G17$) in Appendix A. Simplifying this last expression by converting to Pochhammer symbols, we obtain

$$\int_0^1 t^{s-1}(1-t)^{c-s-1}\, {}_2F_1(a, b; s; zt)\, dt = \frac{\Gamma(s)\Gamma(c-s)}{\Gamma(c)}\, {}_2F_1(a, b; c; z). \tag{7.37}$$

In the final result of Eq. (7.37), the restriction $|z| < 1$ can be removed.

Example 8:

$$\int_0^1 t^{s-1}(1-t)^{b-s-1}\, {}_2F_1(a, b; c; t)\, dt = \frac{\Gamma(c)\Gamma(s)\Gamma(b-s)\Gamma(c-a-s)}{\Gamma(b)\Gamma(c-a)\Gamma(c-s)}.$$

This integral is a small variation of that in Example 7. To prove the identity this time, we will first use ($HY1$) in Appendix A and replace the hypergeometric function with the Meijer G-function ($C19$) in Appendix C,

$${}_2F_1(a, b; c; t) \equiv {}_2F_1(b, a; c; t) = \frac{\Gamma(c)}{\Gamma(a)\Gamma(b)}\, G_{22}^{12}\left(-t \,\middle|\, \begin{matrix} 1-b, 1-a \\ 0, 1-c \end{matrix}\right), \tag{7.38}$$

followed by the use of ($D9$) in Appendix D and ($MG3$) in Appendix A, to find that

$$\frac{\Gamma(c)}{\Gamma(a)\Gamma(b)} \int_0^1 t^{s-1}(1-t)^{b-s-1}\, G_{22}^{12}\left(-t \,\middle|\, \begin{matrix} 1-b, 1-a \\ 0, 1-c \end{matrix}\right) dt$$
$$= \frac{\Gamma(c)\Gamma(b-s)}{\Gamma(a)\Gamma(b)}\, G_{33}^{13}\left(-1 \,\middle|\, \begin{matrix} 1-s, 1-b, 1-a \\ 0, 1-c, 1-b \end{matrix}\right) \tag{7.39}$$
$$= \frac{\Gamma(c)\Gamma(b-s)}{\Gamma(a)\Gamma(b)}\, G_{22}^{12}\left(-1 \,\middle|\, \begin{matrix} 1-s, 1-a \\ 0, 1-c \end{matrix}\right).$$

Thus, referring back to Eq. (7.38), it can be seen that

$$\int_0^1 t^{s-1}(1-t)^{b-s-1}\, {}_2F_1(a, b; c; t)\, dt = \frac{\Gamma(s)\Gamma(b-s)}{\Gamma(b)}\, {}_2F_1(s, a; c; 1), \tag{7.40}$$

and by applying Eq. (7.3), we get the intended result,

$$\int_0^1 t^{s-1}(1-t)^{b-s-1}\, {}_2F_1(a, b; c; t)\, dt = \frac{\Gamma(c)\Gamma(s)\Gamma(b-s)\Gamma(c-a-s)}{\Gamma(b)\Gamma(c-a)\Gamma(c-s)}. \tag{7.41}$$

Example 9: $\int_0^\infty e^{-xt}t^{c-1}\,{}_2F_1(a,\,b;\,c;\,-t)\,dt = \Gamma(c)x^{a-c}U(a;\,1+a-b;\,x)$.

The function $U(a;\,c;\,z)$ is the confluent hypergeometric function of the second kind. The final example in this section involves the integral that we denote by

$$I = \int_0^\infty e^{-xt}t^{c-1}\,{}_2F_1(a,\,b;\,c;\,-t)dt. \tag{7.42}$$

Although not necessary, we can interpret this integral as a Laplace transform (Section 7.4) of the function $f(t) = t^{c-1}\,{}_2F_1(a,\,b;\,c;\,-t)$.

Because of the infinite limit of integration, we choose to replace the hypergeometric function with the Meijer G-function using Eq. (7.38), which leads to

$$\begin{aligned} I &= \frac{\Gamma(c)}{\Gamma(a)\Gamma(b)}\int_0^\infty e^{-xt}t^{c-1}\,G_{22}^{12}\left(t\,\middle|\,\begin{matrix}1-a,\,1-b\\ 0,\,1-c\end{matrix}\right)dt \\ &= \frac{\Gamma(c)}{\Gamma(a)\Gamma(b)}x^{-c}\,G_{32}^{13}\left(\frac{1}{x}\,\middle|\,\begin{matrix}1-c,\,1-a,\,1-b\\ 0,\,1-c\end{matrix}\right) \\ &= \frac{\Gamma(c)}{\Gamma(a)\Gamma(b)}x^{-c}\,G_{21}^{12}\left(\frac{1}{x}\,\middle|\,\begin{matrix}1-a,\,1-b\\ 0\end{matrix}\right) \\ &= \frac{\Gamma(c)}{\Gamma(a)\Gamma(b)}x^{-c}\,G_{12}^{21}\left(x\,\middle|\,\begin{matrix}1\\ a,\,b\end{matrix}\right), \end{aligned} \tag{7.43}$$

where we have also used $(D6)$ in Appendix D as well as $(MG3)$ and $(MG2)$ in Appendix A. The last Meijer G-function can be related to the Whittaker function of the second kind by using $(C27)$ in Appendix C, which, after recalling $(B54)$ in Appendix B, yields

$$\begin{aligned} \int_0^\infty e^{-xt}t^{c-1}\,{}_2F_1(a,\,b;\,c;\,-t)\,dt &= \Gamma(c)x^{(a+b-2c-1)/2}e^{x/2}\,W_{\frac{1}{2}(1-a-b),\frac{1}{2}(a-b)}(x) \\ &= \Gamma(c)x^{a-c}U(a;\,1+a-b;\,x). \end{aligned} \tag{7.44}$$

7.2 Products of Special Functions

Various products of special functions can often be replaced by a single function. In particular, shown below are certain products of generalized hypergeometric functions and products of two and three Bessel functions.

7.2.1 ${}_0F_1$ functions

For the first product of generalized hypergeometric functions, we consider the product of two ${}_0F_1$ functions.

Example 10:

$$ {}_0F_1(-;\, a;\, z)\, {}_0F_1(-;\, b;\, z) = {}_2F_3\left(\frac{a+b-1}{2},\, \frac{a+b}{2};\, a,\, b,\, a+b-1;\, 4z\right). $$

To prove this identity, we start with the Cauchy product of two infinite series defined by Eq. (2.33),

$$ \begin{aligned} {}_0F_1(-;\, a;\, z)\, {}_0F_1(-;\, b;\, z) &= \sum_{m=0}^{\infty} \frac{z^m}{(a)_m m!} \sum_{k=0}^{\infty} \frac{z^k}{(b)_k k!} \\ &= \sum_{n=0}^{\infty} \sum_{k=0}^{n} \frac{z^n}{(a)_{n-k}(n-k)!(b)_k k!}, \end{aligned} \tag{7.45} $$

where in the second step we have made the index change $m = n - k$. Now based on the identities (7.1) and (7.2), we obtain

$$ \begin{aligned} {}_0F_1(-;\, a;\, z)\, {}_0F_1(-;\, b;\, z) &= \sum_{n=0}^{\infty} \sum_{k=0}^{n} \frac{(-n)_k (1-a-n)_k z^n}{(a)_n (b)_k k! n!} \\ &= \sum_{n=0}^{\infty} \frac{z^n}{(a)_n n!}\, {}_2F_1(-n,\, 1-a-n;\, b;\, 1). \end{aligned} \tag{7.46} $$

To finish the proof, we make use of Eq. (7.3) and properties of the Pochhammer symbol in Appendix A to arrive at

$$ {}_0F_1(-;\, a;\, z)\, {}_0F_1(-;\, b;\, z) = \sum_{n=0}^{\infty} \frac{\left(\frac{a+b-1}{2}\right)_n \left(\frac{a+b}{2}\right)_n (4z)^n}{(a)_n (b)_n (a+b-1)_n n!}, \tag{7.47} $$

which we recognize as

$$ {}_0F_1(-;\, a;\, z)\, {}_0F_1(-;\, b;\, z) = {}_2F_3\left(\frac{a+b-1}{2},\, \frac{a+b}{2};\, a,\, b,\, a+b-1;\, 4z\right). \tag{7.48} $$

When the arguments of the ${}_0F_1$ functions are not the same, we have a more general relationship for the product, as given in Example 11.

Example 11:

$$ {}_0F_1(-;\, a;\, pz)\, {}_0F_1(-;\, b;\, qz) = \sum_{n=0}^{\infty} \frac{(pz)^n}{(a)_n n!}\, {}_2F_1\left(-n,\, 1-a-n;\, b;\, \frac{p}{q}\right). $$

This identity can be proved similarly to Example 10 down to Eq. (7.46).

Because of the relation given by Eq. (4.3) between the Bessel function and the generalized hypergeometric function ${}_0F_1$, we can use the product identity provided in Example 10 to simplify the product of two Bessel functions of different order.

Example 12:

$$J_\mu(z)J_\nu(z) = \frac{(z/2)^{\mu+\nu}}{\Gamma(1+\mu)\Gamma(1+\nu)} \times {}_2F_3\left(\frac{\mu+\nu+1}{2}, \frac{\mu+\nu+2}{2}; \mu+1, \nu+1, \mu+\nu+1; -z^2\right).$$

The proof of this identity follows directly from the relationship in Eq. (4.3) and Example 10, i.e.,

$$\begin{aligned} J_\mu(z)J_\nu(z) &= \frac{(z/2)^{\mu+\nu}}{\Gamma(1+\mu)\Gamma(1+\nu)} {}_0F_1\left(-; 1+\mu; -\frac{z^2}{4}\right) {}_0F_1\left(-; 1+\nu; -\frac{z^2}{4}\right) \\ &= \frac{(z/2)^{\mu+\nu}}{\Gamma(1+\mu)\Gamma(1+\nu)} \\ &\quad \times {}_2F_3\left(\frac{\mu+\nu+1}{2}, \frac{\mu+\nu+2}{2}; \mu+1, \nu+1, \mu+\nu+1; -z^2\right). \end{aligned} \tag{7.49}$$

For the more general product $J_\mu(pz)J_\nu(qz)$, we can use Example 11 to get the result.

7.2.2 $_1F_1$ and $_2F_1$ functions

Certain products of interest involve the product of two confluent hypergeometric functions—the most famous is called the *Ramanujan theorem*.

Example 13: ${}_1F_1(a; b; z)\,{}_1F_1(a; b; -z) = {}_2F_3(a, b-a; b, \frac{b}{2}, \frac{b+1}{2}; \frac{z^2}{4})$.

Following the Cauchy product analysis in Example 10, we have

$$\begin{aligned} {}_1F_1(a; b; z)\,{}_1F_1(a; b; -z) &= \sum_{m=0}^{\infty} \frac{(a)_m z^m}{(b)_m m!} \sum_{k=0}^{\infty} \frac{(-1)^k (a)_k z^k}{(b)_k k!} \\ &= \sum_{n=0}^{\infty}\sum_{k=0}^{n} \frac{(-1)^k (a)_{n-k}(a)_k z^n}{(b)_{n-k}(n-k)!(b)_k k!}. \end{aligned} \tag{7.50}$$

Now applying the identities (7.1) and (7.2), this last result takes the form

$$\begin{aligned} {}_1F_1(a; b; z)\,{}_1F_1(a; b; -z) &= \sum_{n=0}^{\infty}\sum_{k=0}^{n} \frac{(-n)_k (a)_n (a)_k (1-b-n)_k z^n}{(b)_n (b)_k (1-a-n)_k k! n!} \\ &= \sum_{n=0}^{\infty} \frac{(a)_n z^n}{(b)_n n!} {}_3F_2(-n, a, 1-b-n; b, 1-a-n; 1). \end{aligned} \tag{7.51}$$

To simplify Eq. (7.51), we use *Dixon's theorem* defined by Eq. (7.4),

$$ {}_3F_2(-n,\, a,\, 1-b-n;\, b,\, 1-a-n;\, 1) = \frac{\Gamma(1-n/2)\Gamma(1-a-n)\Gamma(b)\Gamma(b-a+n/2)}{\Gamma(1-n)\Gamma(1-a-n/2)\Gamma(b+n/2)\Gamma(b-a)}. \tag{7.52} $$

However, based on ($G6$) in Appendix A, all terms in Eq. (7.52) with odd $n \geq 1$ are zero, so we make the change of index $n = 2m$, which leads to

$$ {}_1F_1(a;\, b;\, z)\, {}_1F_1(a;\, b;\, -z) = \sum_{m=0}^{\infty} \frac{(a)_{2m} z^{2m}}{(b)_{2m}(2m)!} \times \frac{\Gamma(1-m)}{\Gamma(1-2m)} \frac{\Gamma(1-a-2m)\Gamma(b)\Gamma(b-a+m)}{\Gamma(1-a-m)\Gamma(b+m)\Gamma(b-a)}. \tag{7.53} $$

Next, rewriting all gamma functions in Eq. (7.53) in terms of other gamma functions and Pochhammer symbols from identities in Appendix A, we find that

$$ \begin{aligned} \frac{\Gamma(1-m)}{\Gamma(1-2m)} &= \frac{(-1)^m \Gamma(2m+1)}{\Gamma(m+1)} = \frac{(-1)^m (2m)!}{m!}, \\ \Gamma(1-a-2m) &= \Gamma(1-a)(1-a)_{-2m} = \frac{\Gamma(1-a)}{2^{2m}\left(\frac{a}{2}\right)_m \left(\frac{a+1}{2}\right)_m}, \\ \Gamma(1-a-m) &= \Gamma(1-a)(1-a)_{-m} = \frac{(-1)^m \Gamma(1-a)}{(a)_m}, \\ \Gamma(b-a+m) &= \Gamma(b-a)(b-a)_m. \end{aligned} \tag{7.54} $$

Therefore, using the results in Eq. (7.54) and the Pochhammer identity for $(a)_{2m}$ and $(b)_{2m}$ from ($PS12$) in Appendix A, Eq. (7.53) becomes

$$ {}_1F_1(a;\, b;\, z)\, {}_1F_1(a;\, b;\, -z) = \sum_{m=0}^{\infty} \frac{(a)_m (b-a)_m}{(b)_m \left(\frac{b}{2}\right)_m \left(\frac{b+1}{2}\right)_m} \frac{(z/2)^{2m}}{m!}. \tag{7.55} $$

Thus, by recognizing the series on the right in Eq. (7.55) as a generalized hypergeometric series, it follows that

$$ {}_1F_1(a;\, b;\, z)\, {}_1F_1(a;\, b;\, -z) = {}_2F_3\left(a,\, b-a;\, b,\, \frac{b}{2},\, \frac{b+1}{2};\, \frac{z^2}{4}\right). \tag{7.56} $$

A special case of Example 13 with $a = 1$ and $b = 2$ leads to a simple but interesting result based on ($B13$) in Appendix B. Starting with

$$
\begin{aligned}
{}_2F_3\left(1,\ 1;\ 2,\ 1,\ \frac{3}{2};\ z\right) &= {}_1F_1(1;\ 2;\ 2\sqrt{z})\,{}_1F_1(1;\ 2;\ -2\sqrt{z}) \\
&= -\frac{1}{2\sqrt{z}}(1-e^{2\sqrt{z}})\frac{1}{2\sqrt{z}}\left(1-e^{-2\sqrt{z}}\right) \\
&= \frac{1}{4z}(e^{\sqrt{z}}-e^{-\sqrt{z}})(e^{\sqrt{z}}-e^{-\sqrt{z}}),
\end{aligned}
\tag{7.57}
$$

we deduce that (including the reduction of ${}_2F_3$ by canceling the common Pochhammer symbols in the numerator and denominator)

$$
{}_2F_3\left(1,\ 1;\ 2,\ 1,\ \frac{3}{2};\ z\right) = {}_1F_2\left(1;\ 2,\ \frac{3}{2};\ z\right) = \frac{1}{z}\sinh^2\sqrt{z}. \tag{7.58}
$$

Example 14:

$$
{}_1F_1(a;b;pz)\,{}_1F_1(c;d;qz) = \sum_{n=0}^{\infty}\frac{(a)_n(pz)^n}{(b)_n n!}\,{}_3F_2\left(-n,-c,1-b-n;d,1-a-n;-\frac{q}{p}\right).
$$

This result is a generalization of Example 13 with a similar proof down to Eq. (7.51).

Many special functions can be expressed in terms of the confluent hypergeometric function, e.g., the incomplete gamma function, Hermite and Laguerre functions, and various combinations of Bessel functions (cf. Appendix B). Consequently, the identities in Examples 13 and 14 may prove useful in representing certain products of these other functions. Also, there are additional identities involving products of hypergeometric functions, such as those below (see also Refs. 3 and 4).

A special product of hypergeometric functions is given by

$$
\left[{}_2F_1\left(a,\ b;\ a+b+\frac{1}{2};\ z\right)\right]^2 = {}_3F_2\left(2a,\ a+b,\ 2b;\ a+b+\frac{1}{2},\ a+2b;\ z\right). \tag{7.59}
$$

It turns out that this is the only case in which the square of a hypergeometric function can be expressed in terms of a ${}_3F_2$ generalized hypergeometric function. A similar product is

$$
\begin{aligned}
&{}_2F_1\left(a,\ b;\ a+b+\frac{1}{2};\ z\right){}_2F_1\left(a,\ b;\ a+b-\frac{1}{2};\ z\right) \\
&= {}_3F_2\left(2a,\ a+b,\ 2b;\ a+b+\frac{1}{2},\ 2a+2b-1;\ z\right),
\end{aligned}
\tag{7.60}
$$

and another such product is provided in the next example.

Example 15:

$$
{}_2F_1(a,\, b;\, c;\, pz)\,{}_2F_1(\alpha,\, \beta;\, \gamma;\, qz) = \sum_{n=0}^{\infty} \frac{(a)_n (b)_n (pz)^n}{(c)_n n!}\, {}_4F_3\!\left(-n,\, \alpha,\, \beta,\, 1-c-n;\, \gamma,\, 1-a-n,\, 1-b-n;\, \frac{q}{p}\right).
$$

To prove this identity, we take the product of the power series for the hypergeometric functions and make a change of index $m = n - k$, which, with the help of Eqs. (7.1) and (7.2), yields

$$
\begin{aligned}
{}_2F_1(a, b; c; pz)\,{}_2F_1(\alpha, \beta; \gamma; qz) &= \sum_{m=0}^{\infty} \frac{(a)_m (b)_m (pz)^m}{(c)_m m!} \sum_{k=0}^{\infty} \frac{(\alpha)_k (\beta)_k (qz)^k}{(\gamma)_k k!} \\
&= \sum_{n=0}^{\infty} \sum_{k=0}^{n} \frac{(a)_{n-k} (b)_{n-k} (pz)^{n-k}}{(c)_{n-k} (n-k)!} \frac{(\alpha)_k (\beta)_k (qz)^k}{(\gamma)_k k!} \\
&= \sum_{n=0}^{\infty} \frac{(a)_n (b)_n (pz)^n}{(c)_n n!} \sum_{k=0}^{n} \frac{(-n)_k (\alpha)_k (\beta)_k (1-c-n)_k}{(\gamma)_k (1-a-n)_k (1-b-n)_k k!} \left(\frac{q}{p}\right)^k.
\end{aligned}
\tag{7.61}
$$

The finite summation in this last expression is a ${}_4F_3$ function (polynomial); thus, we have shown that

$$
{}_2F_1(a,\, b;\, c;\, pz)\,{}_2F_1(\alpha,\, \beta;\, \gamma;\, qz) = \sum_{n=0}^{\infty} \frac{(a)_n (b)_n (pz)^n}{(c)_n n!}\, {}_4F_3\!\left(-n,\, \alpha,\, \beta,\, 1-c-n;\, \gamma,\, 1-a-n,\, 1-b-n;\, \frac{q}{p}\right).
\tag{7.62}
$$

7.2.3 Saalschütz's theorem

If n is a non-negative integer, then analogous to the identity known as Dixon's theorem is the identity known as *Saalschütz's theorem.*

Example 16: ${}_3F_2(-n,\, a,\, b;\, c,\, 1+a+b-c-n;\, 1) = \dfrac{(c-a)_n (c-b)_n}{(c)_n (c-a-b)_n}.$

To prove this result, we need the identity (*HY*8) in Appendix A,

$$
{}_2F_1(c-a,\, c-b;\, c;\, z) = (1-z)^{a+b-c}\,{}_2F_1(a,\, b;\, c;\, z),
\tag{7.63}
$$

which we proved in Section 3.5.3. Let us begin by expanding each term of Eq. (7.63) in a power series and making the index change $m = n - k$,

$$
\begin{aligned}
\sum_{n=0}^{\infty} \frac{(c-a)_n (c-b)_n}{(c)_n} \frac{z^n}{n!} &= \sum_{m=0}^{\infty} \frac{(c-a-b)_m}{m!} z^m \sum_{k=0}^{\infty} \frac{(a)_k (b)_k}{(c)_k} \frac{z^k}{k!} \\
&= \sum_{n=0}^{\infty} \sum_{k=0}^{n} \frac{(a)_k (b)_k}{(c)_k k!} \frac{(c-a-b)_{n-k}}{(n-k)!} z^n.
\end{aligned}
\tag{7.64}
$$

From the identities (7.1) and (7.2), we then obtain

$$\sum_{n=0}^{\infty}\frac{(c-a)_n(c-b)_n}{(c)_n}\frac{z^n}{n!}=\sum_{n=0}^{\infty}\frac{(c-a-b)_n}{n!}z^n\sum_{k=0}^{n}\frac{(-n)_k(a)_k(b)_k}{(c)_k(1+a+b-c-n)_k k!}$$
$$=\sum_{n=0}^{\infty}(c-a-b)_n\,{}_3F_2(-n,a,b;c,1+a+b-c-n;1)\frac{z^n}{n!}. \tag{7.65}$$

Finally, by comparing like coefficients of $z^n/n!$ in each series and rearranging terms, we get the result

$${}_3F_2(-n,\ a,\ b;\ c,\ 1+a+b-c-n;\ 1)=\frac{(c-a)_n(c-b)_n}{(c)_n(c-a-b)_n}. \tag{7.66}$$

7.3 Products and Integrals of Bessel Functions

Integrals that contain one or more special functions usually require sophisticated methods of solution. In this section we will solve integrals involving products of two and three Bessel functions that will prove useful in some of the material covered in Chapters 9 and 10.

7.3.1 Product of two Bessel functions

Equation (7.49) provides a relationship for the product of two Bessel functions of different order. However, our interest here involves the product of two Bessel functions of the same order.

Example 17: $t^{\beta-1}J_\mu(t)J_\mu(t)=\frac{1}{\sqrt{\pi}}G_{13}^{11}\left(t^2\,\middle|\,\begin{matrix}\frac{\beta}{2}\\ \mu+\frac{\beta-1}{2},\ \frac{\beta-1}{2},\ \frac{\beta-1}{2}-\mu\end{matrix}\right).$

To prove this relationship, we take the product of Bessel functions given in Example 12 and set $\nu=\mu$ to find that

$$J_\mu(t)J_\mu(t)=\frac{t^{2\mu}}{2^{2\mu}\Gamma^2(\mu+1)}\,{}_2F_3\left(\mu+\frac{1}{2},\ \mu+1;\ \mu+1,\ \mu+1,\ 2\mu+1;\ -t^2\right)$$
$$=\frac{t^{2\mu}\Gamma\left(\mu+\frac{1}{2}\right)}{\sqrt{\pi}\Gamma(2\mu+1)\Gamma(\mu+1)}\,{}_1F_2\left(\mu+\frac{1}{2};\ \mu+1,\ 2\mu+1;\ -t^2\right). \tag{7.67}$$

In the second step above we have canceled the common Pochhammer symbol and used the gamma function duplication formula (*G*10) in Appendix A. We now write the ${}_1F_2$ function as a Meijer *G*-function using (*C*32) in Appendix C and multiply by $t^{\beta-1}$, which yields

$$t^{\beta-1}J_\mu(t)J_\mu(t) = \frac{t^{2\mu+\beta-1}}{\sqrt{\pi}}G_{13}^{11}\left(t^2 \middle| \begin{matrix} \frac{1}{2}-\mu \\ 0,\ -\mu,\ -2\mu \end{matrix}\right)$$

$$= \frac{1}{\sqrt{\pi}}G_{13}^{11}\left(t^2 \middle| \begin{matrix} \frac{\beta}{2} \\ \mu+\frac{\beta-1}{2},\ \frac{\beta-1}{2},\ \frac{\beta-1}{2}-\mu \end{matrix}\right). \tag{7.68}$$

We used the relation ($MG5$) in Appendix A for the last step.

Example 18: $\int_0^\infty t^{\beta-1}J_\mu^2(t)\,dt = \dfrac{1}{2\sqrt{\pi}}\dfrac{\Gamma\left(\mu+\frac{\beta}{2}\right)}{\Gamma\left(\frac{1-\beta}{2}\right)\Gamma\left(1-\frac{\beta}{2}\right)\Gamma\left(1+\mu-\frac{\beta}{2}\right)}.$

Based on the product identity in Example 17, we obtain

$$\int_0^\infty t^{\beta-1}J_\mu^2(t)\,dt = \frac{1}{\sqrt{\pi}}\int_0^\infty G_{13}^{11}\left(t^2 \middle| \begin{matrix} \frac{\beta}{2} \\ \mu+\frac{\beta-1}{2},\ \frac{\beta-1}{2},\ \frac{\beta-1}{2}-\mu, \end{matrix}\right)dt$$

$$= \frac{1}{2\sqrt{\pi}}\int_0^\infty x^{-1/2}G_{13}^{11}\left(x \middle| \begin{matrix} \frac{\beta}{2} \\ \mu+\frac{\beta-1}{2},\ \frac{\beta-1}{2},\ \frac{\beta-1}{2}-\mu, \end{matrix}\right)dx, \tag{7.69}$$

where we made the change of variable $t^2 = x$. From the integral relation ($D8$) in Appendix D, we now deduce that

$$\int_0^\infty t^{\beta-1}J_\mu^2(t)dt = \frac{1}{2\sqrt{\pi}}\frac{\Gamma\left(\mu+\frac{\beta}{2}\right)\Gamma\left(\frac{1-\beta}{2}\right)}{\Gamma\left(1-\frac{\beta}{2}\right)\Gamma\left(1+\mu-\frac{\beta}{2}\right)}. \tag{7.70}$$

Using Eq. (7.49), ($C32$) in Appendix C, and ($D8$) in Appendix D, the evaluation of an integral with a product of two Bessel functions of different order leads to the similar result,

$$\int_0^\infty t^{\beta-1}J_\mu(t)J_\nu(t)\,dt$$

$$= \frac{2^{\beta-1}\Gamma(1-\beta)\Gamma\left(\frac{\beta+\mu+\nu}{2}\right)}{\Gamma\left(\frac{\nu-\mu-\beta}{2}+1\right)\Gamma\left(\frac{\mu-\nu-\beta}{2}+1\right)\Gamma\left(\frac{\mu+\nu-\beta}{2}+1\right)}. \tag{7.71}$$

Both of these last two integrals can be interpreted as a Mellin transform of the product of two Bessel functions with transform variable $s = \beta$. And although the result in Eq. (7.71) with $\nu = \mu$ looks different than the result in Eq. (7.70), the two are actually equal by virtue of the gamma function identity ($G10$) in Appendix A, which leads to

$$\frac{\Gamma\left(\frac{1-\beta}{2}\right)}{2\sqrt{\pi}} = \frac{2^{\beta-1}\Gamma(1-\beta)}{\Gamma\left(1-\frac{\beta}{2}\right)}, \quad \beta \neq 1, 2, 3, \ldots . \tag{7.72}$$

For another case concerning the product of two Bessel functions, let us consider the following integral.

Example 19:

$$\int_0^\infty te^{-ct^2} J_0(at)J_0(bt)\,dt = \frac{1}{2c}\exp\left[-\frac{(a^2+b^2)}{4c}\right] I_0\left(\frac{ab}{2c}\right), \quad a, b, c > 0.$$

We start here by introducing the shorthand notation for the integral

$$I = \int_0^\infty te^{-ct^2} J_0(at)J_0(bt)\,dt. \tag{7.73}$$

In this case we use ($C9$) in Appendix C,

$$J_0(at) = G_{02}^{10}\left(\frac{a^2t^2}{4}\middle|\, 0, 0\right), \tag{7.74}$$

and express $J_0(bt)$ in a power series to get

$$\begin{aligned} I &= \sum_{n=0}^\infty \frac{(-1)^n}{(n!)^2}\left(\frac{b^2}{4}\right)^n \int_0^\infty t^{2n+1}e^{-ct^2} G_{02}^{10}\left(\frac{a^2t^2}{4}\middle|\, 0, 0\right) dt \\ &= \sum_{n=0}^\infty \frac{(-1)^n}{(n!)^2}\left(\frac{b^2}{4}\right)^n \frac{1}{2c^{n+1}} G_{12}^{11}\left(\frac{a^2}{4c}\middle|\, \begin{matrix} -n \\ 0, 0 \end{matrix}\right), \end{aligned} \tag{7.75}$$

where we have made the change of variable $x = t^2$ and used ($D6$) in Appendix D. Next, we utilize the relation ($C13$) from Appendix C and the identity ($CH4$) in Appendix A to obtain

$$\begin{aligned} G_{12}^{11}\left(\frac{a^2}{4c}\middle|\, \begin{matrix} -n \\ 0, 0 \end{matrix}\right) &= \Gamma(n+1)\,{}_1F_1\left(n+1;\, 1;\, -\frac{a^2}{4c}\right) \\ &= n!e^{-a^2/4c}\,{}_1F_1\left(-n;\, 1;\, \frac{a^2}{4c}\right) \\ &= n!e^{-a^2/4c} L_n\left(\frac{a^2}{4c}\right), \end{aligned} \tag{7.76}$$

where $L_n(x)$ is the Laguerre polynomial [see ($B22$) in Appendix B]. Combining Eq. (7.76) with Eq. (7.75), we obtain

$$I = \frac{e^{-a^2/4c}}{2c} \sum_{n=0}^{\infty} \frac{(-1)^n}{n!} \left(\frac{b^2}{4c}\right)^n L_n\left(\frac{a^2}{4c}\right), \tag{7.77}$$

which can be further simplified by use of the identity (p. 247 of Ref. 6)

$$e^t (xt)^{-m/2} J_m(2\sqrt{xt}) = \sum_{n=0}^{\infty} L_n^{(m)}(x) \frac{t^n}{(n+m)!}, \quad m = 0, 1, 2, \ldots . \tag{7.78}$$

In particular, for $m = 0$ and $t \to -t$, the identity (7.78) becomes

$$e^{-t} I_0(2\sqrt{xt}) = \sum_{n=0}^{\infty} \frac{(-1)^n}{n!} t^n L_n(x), \tag{7.79}$$

so, upon using Eq. (7.79) to sum the series in Eq. (7.77), we are led to

$$\int_0^{\infty} t e^{-ct^2} J_0(at) J_0(bt)\, dt = \frac{1}{2c} \exp\left[-\frac{(a^2+b^2)}{4c}\right] I_0\left(\frac{ab}{2c}\right). \tag{7.80}$$

7.3.2 Product of three Bessel functions

Here we wish to pursue the evaluation of an integral involving the product of three Bessel functions. A more general form of the integral that we wish to solve is given by

$$I = \int_0^{\infty} t^{\beta-1} J_\mu(at) J_\nu(bt) J_p(ct)\, dt. \tag{7.81}$$

The integral in Eq. (7.81) has been previously solved through contour integration methods[7] and Mellin transform techniques (p. 164 of Ref. 8) and is tabulated in terms of a hypergeometric function of two variables (p. 694 of Ref. 5). An integral of this type evolved from an article by Hu et al.[9] concerning the application of Zernike polynomials to optical wave propagation through turbulence. In the next example, we derive the case where $\nu = \mu$ and $a = b = 1$.

Example 20:

$$\int_0^{\infty} t^{\beta-1} J_\mu^2(t) J_p(ct)\, dt = \frac{1}{c\sqrt{\pi}} G_{33}^{12}\left(\frac{4}{c^2}\left|\begin{matrix} \frac{1-p}{2}, \frac{\beta}{2}, \frac{1+p}{2} \\ \mu + \frac{\beta-1}{2}, \frac{\beta-1}{2}, \frac{\beta-1}{2} - \mu \end{matrix}\right.\right).$$

We present the proof of this specialization of Eq. (7.81) by first using the identity given in Example 17. From this result we obtain through the use of ($D12$) in Appendix D

$$
\begin{aligned}
I &= \int_0^\infty t^{\beta-1} J_\mu^2(t) J_p(ct)\,dt \\
&= \frac{1}{\sqrt{\pi}} \int_0^\infty J_p(ct) G_{13}^{11}\left(t^2 \middle| \begin{matrix} \frac{\beta}{2} \\ \mu + \frac{\beta-1}{2}, \frac{\beta-1}{2}, \frac{\beta-1}{2} - \mu \end{matrix} \right) dt \\
&= \frac{1}{c\sqrt{\pi}} G_{33}^{12}\left(\frac{4}{c^2} \middle| \begin{matrix} \frac{1-p}{2}, \frac{\beta}{2}, \frac{1+p}{2} \\ \mu + \frac{\beta-1}{2}, \frac{\beta-1}{2}, \frac{\beta-1}{2} - \mu \end{matrix} \right).
\end{aligned}
\tag{7.82}
$$

7.4 Laplace Transform

In Section 2.4 we introduced the *Fourier integral theorem* for piecewise continuous and absolutely integrable functions $g(x)$,

$$
g(t) = \frac{1}{2\pi} \int_{-\infty}^\infty e^{-i\xi t} \int_{-\infty}^\infty g(x) e^{i\xi x} dx\, d\xi, \tag{7.83}
$$

from which we developed the Mellin transform and its inversion formula. The Laplace transform and its inversion formula are also derivable from the Fourier integral theorem. For example, suppose that a given function $f(t)$ and its derivative $f'(t)$ are both piecewise continuous functions for all $t \geq 0$ and that $f(t)$ is zero for $t < 0$. While we don't require the function $f(t)$ to be absolutely integrable, we assume that the related function,

$$
g(t) = e^{-ct} f(t), \quad t \geq 0, \tag{7.84}
$$

where c is a positive constant, does have this property. Consequently, the function $g(t)$ satisfies the Fourier integral theorem and leads to

$$
f(t) = \frac{e^{ct}}{2\pi} \int_{-\infty}^\infty e^{-i\xi t} \int_0^\infty f(x) e^{-(c-i\xi)x} dx\, d\xi. \tag{7.85}
$$

If we now introduce the change of variable $s = c - i\xi$, then Eq. (7.85) becomes

$$
f(t) = \frac{1}{2\pi i} \int_{c-i\infty}^{c+i\infty} e^{st} \int_0^\infty e^{-sx} f(x)\, dx\, ds. \tag{7.86}
$$

Therefore, in this manner we can introduce the *Laplace transform*,

$$
\mathscr{L}\{f(t)\} \equiv F(s) = \int_0^\infty e^{-st} f(t)\, dt, \quad \text{Re}(s) > 0, \tag{7.87}
$$

and the *Laplace inversion formula*,

$$\mathscr{L}^{-1}\{F(s)\} \equiv f(t) = \frac{1}{2\pi i}\int_{c-i\infty}^{c+i\infty} e^{st}F(s)\,ds, \qquad t \geq 0. \tag{7.88}$$

The Laplace transform is one of the most widely used transforms in all engineering problems. Many of the routine Laplace transforms can be found by simple calculus operations, whereas others require more advanced methods. Many of the examples we present below are generally well known and tabulated in Laplace transform tables,[3] but we use them here for illustrative purposes.

7.4.1 Error function

To address the evaluation of Laplace transforms by the methods illustrated in this chapter, let us first consider the Laplace transform of the error function $f(t) = \operatorname{erf}(a\sqrt{t})$.

Example 21: $\mathscr{L}\{\operatorname{erf}(a\sqrt{t})\} = \dfrac{a}{s\sqrt{s+a^2}}, \quad a > 0.$

From (C12) in Appendix C, we have that

$$\operatorname{erf}(\sqrt{z}) = \sqrt{\frac{z}{\pi}}\, G_{12}^{11}\left(z \,\middle|\, \begin{matrix} \frac{1}{2} \\ 0, \ -\frac{1}{2} \end{matrix}\right), \tag{7.89}$$

and based on (D6) in Appendix D, the Laplace transform becomes

$$\begin{aligned}\int_0^\infty e^{-st}\operatorname{erf}(a\sqrt{t})\,dt &= \frac{a}{\sqrt{\pi}}\int_0^\infty e^{-st}t^{1/2}\, G_{12}^{11}\left(a^2 t \,\middle|\, \begin{matrix} \frac{1}{2} \\ 0, \ -\frac{1}{2} \end{matrix}\right)dt \\ &= \frac{a}{s\sqrt{\pi s}}\, G_{22}^{12}\left(\frac{a^2}{s} \,\middle|\, \begin{matrix} -\frac{1}{2}, \ \frac{1}{2} \\ 0, \ -\frac{1}{2} \end{matrix}\right).\end{aligned} \tag{7.90}$$

However, we can apply the reduction formula (MG3) in Appendix A to simplify Eq. (7.90) to

$$\int_0^\infty e^{-st}\operatorname{erf}(a\sqrt{t})\,dt = \frac{a}{s\sqrt{\pi s}}\, G_{11}^{11}\left(\frac{a^2}{s} \,\middle|\, \begin{matrix} \frac{1}{2} \\ 0 \end{matrix}\right) = \frac{a}{s\sqrt{s}}\left(1 + \frac{a^2}{s}\right)^{-1/2}, \tag{7.91}$$

where we have also used (C11) from Appendix C. Upon algebraic simplification of this last expression, we arrive at

$$\int_0^\infty e^{-st}\operatorname{erf}(a\sqrt{t})\,dt = \frac{a}{s\sqrt{s+a^2}}. \tag{7.92}$$

From Eq. (7.92) and the identity $\operatorname{erfc}(z) = 1 - \operatorname{erf}(z)$, we also get the bonus transform relation

$$\int_0^\infty e^{-st}\mathrm{erfc}(a\sqrt{t})\,dt = \frac{1}{s}\left(1 - \frac{a}{\sqrt{s+a^2}}\right). \tag{7.93}$$

Example 22: $\mathscr{L}\{\mathrm{erf}(t)\} = \frac{1}{s}e^{s^2/4}\mathrm{erfc}(\frac{s}{2})$.

To find the Laplace transform this time, we replace the error function with its integral representation [Eq. (3.93)],

$$\begin{aligned}\int_0^\infty e^{-st}\mathrm{erf}(t)\,dt &= \frac{2}{\sqrt{\pi}}\int_0^\infty e^{-st}\int_0^t e^{-x^2}dx\,dt\\ &= \frac{2}{\sqrt{\pi}}\int_0^\infty e^{-x^2}\int_x^\infty e^{-st}\,dt\,dx\\ &= \frac{2}{s\sqrt{\pi}}\int_0^\infty e^{-x^2-sx}dx\\ &= \frac{2}{s\sqrt{\pi}}e^{s^2/4}\int_0^\infty e^{-(x+s/2)^2}dx,\end{aligned} \tag{7.94}$$

where we have interchanged the order of integration in the second step. We also evaluated the inside integral, followed by writing $x^2+sx=(x+s/2)^2-s^2/4$. Last, by making the change of variable $u=x+s/2$, we obtain

$$\begin{aligned}\int_0^\infty e^{-st}\mathrm{erf}(t)\,dt &= \frac{2}{s\sqrt{\pi}}e^{s^2/4}\int_{s/2}^\infty e^{-u^2}du\\ &= \frac{1}{s}e^{s^2/4}\mathrm{erfc}\left(\frac{s}{2}\right).\end{aligned} \tag{7.95}$$

As an alternative to the method above, we can use ($C12$) in Appendix C, ($MG5$) in Appendix A, and ($D7$) in Appendix D to get

$$\begin{aligned}\int_0^\infty e^{-st}\mathrm{erf}(t)\,dt &= \frac{1}{\sqrt{\pi}}\int_0^\infty te^{-st}\,G_{12}^{11}\left(t^2\left|\begin{matrix}\frac{1}{2}\\ 0,\ -\frac{1}{2}\end{matrix}\right.\right)dt\\ &= \frac{1}{\sqrt{\pi}}\int_0^\infty e^{-st}G_{12}^{11}\left(t^2\left|\begin{matrix}1\\ \frac{1}{2},\ 0\end{matrix}\right.\right)dt\\ &= \frac{1}{\pi s}G_{32}^{13}\left(\frac{4}{s^2}\left|\begin{matrix}0,\ \frac{1}{2},\ 1\\ \frac{1}{2},\ 0\end{matrix}\right.\right).\end{aligned} \tag{7.96}$$

Then, applying ($MG3$) from Appendix A and ($C28$) in Appendix C, we find that

$$\int_0^\infty e^{-st}\operatorname{erf}(t)\,dt = \frac{1}{s}e^{s^2/4}\operatorname{erfc}\left(\frac{s}{2}\right). \tag{7.97}$$

7.4.2 Bessel functions

In various applications it is common to see integrals involving one or more Bessel functions along with elementary functions like the exponential function. For example, such integrals are common in solving problems in optical wave propagation through the atmosphere (cf. Chapters 8–10). We next examine the Laplace transform involving Bessel functions.

Example 23: $\mathscr{L}\{t^{p/2}J_p(2a\sqrt{t})\} = \dfrac{a^p}{s^{p+1}}\,e^{-a^2/s}, \quad a > 0, p > -\dfrac{1}{2}.$

Let us start by recalling the relation (*C*9) from Appendix C,

$$J_p(z) = G_{02}^{10}\left(\frac{z^2}{4}\,\middle|\,\frac{p}{2}, -\frac{p}{2}\right), \tag{7.98}$$

which, from (*D*6) in Appendix D, leads to

$$\begin{aligned}\int_0^\infty e^{-st}t^{p/2}J_p(2a\sqrt{t})\,dt &= \int_0^\infty e^{-st}t^{p/2}G_{02}^{10}\left(a^2t\,\middle|\,\frac{p}{2}, -\frac{p}{2}\right)dt\\ &= \frac{1}{s^{p/2+1}}\,G_{12}^{11}\left(\frac{a^2}{s}\,\middle|\,\begin{matrix}-\frac{p}{2}\\ \frac{p}{2}, -\frac{p}{2}\end{matrix}\right)\\ &= \frac{1}{s^{p/2+1}}\,G_{01}^{10}\left(\frac{a^2}{s}\,\middle|\,\frac{p}{2}\right).\end{aligned} \tag{7.99}$$

We have also used the reduction formula (*MG*3) in Appendix A to get the last step. From this result we deduce from (*C*1) in Appendix C that

$$\int_0^\infty e^{-st}t^{p/2}J_p(2a\sqrt{t})\,dt = \frac{a^p}{s^{p+1}}\,e^{-a^2/s}. \tag{7.100}$$

Example 24:

$$\mathscr{L}\{t^{n+m/2}J_m(2\sqrt{xt})\} = n!\,e^{-x/s}\frac{x^{m/2}}{s^{n+m+1}}L_n^{(m)}\left(\frac{x}{s}\right),\ m > 0,\ n = 0, 1, 2, \ldots.$$

The function $L_n^{(m)}(x/s)$ is the *associated Laguerre polynomial.* Although we could use Eq. (7.98) again to solve this integral, we choose to express the Bessel function this time in a power series and interchange the order of integration and summation. Also using (*G*19) in Appendix A, this action leads to

$$\int_0^\infty e^{-st} t^{n+m/2} J_m(2\sqrt{xt})\,dt = x^{m/2} \sum_{k=0}^\infty \frac{(-1)^k x^k}{\Gamma(k+m+1)k!} \int_0^\infty e^{-st} t^{k+n+m} dt$$
$$= \frac{x^{m/2}}{s^{n+m+1}} \sum_{k=0}^\infty \frac{(-1)^k (x/s)^k}{\Gamma(k+m+1)k!} \Gamma(k+n+m+1). \tag{7.101}$$

By writing each of the gamma functions in terms of the Pochhammer symbol from Appendix A, this last expression becomes

$$\int_0^\infty e^{-st} t^{n+m/2} J_m(2\sqrt{xt})\,dt = \frac{x^{m/2}}{s^{n+m+1}} \frac{\Gamma(n+m+1)}{\Gamma(m+1)} \times \sum_{k=0}^\infty \frac{(-1)^k (n+m+1)_k}{(m+1)_k k!} \left(\frac{x}{s}\right)^k$$
$$= \frac{x^{m/2}}{s^{n+m+1}} \frac{\Gamma(n+m+1)}{\Gamma(m+1)} {}_1F_1\left(n+m+1; m+1; -\frac{x}{s}\right)$$
$$= \frac{x^{m/2}}{s^{n+m+1}} \frac{\Gamma(n+m+1)}{\Gamma(m+1)} e^{-x/s} {}_1F_1\left(-n; m+1; \frac{x}{s}\right), \tag{7.102}$$

where the last step is based on ($CH4$) in Appendix A. However, it is known from ($B22$) in Appendix B that

$${}_1F_1\left(-n;\ m+1;\ \frac{x}{s}\right) = \frac{n!\Gamma(m+1)}{\Gamma(n+m+1)} L_n^{(m)}\left(\frac{x}{s}\right), \tag{7.103}$$

which inserted into Eq. (7.102) gives the result

$$\int_0^\infty e^{-st} t^{n+m/2} J_m(2\sqrt{xt})\,dt = n!\, e^{-x/s} \frac{x^{m/2}}{s^{n+m+1}} L_n^{(m)}\left(\frac{x}{s}\right). \tag{7.104}$$

As a third example involving Bessel functions, we consider the Laplace transform of $f(t) = t^p J_p(at)$, $a > 0$.

Example 25:

$$\mathscr{L}\{t^p J_p(at)\} = \frac{(2a)^p}{\sqrt{\pi}} \frac{\Gamma(p+\frac{1}{2})}{(s^2+a^2)^{p+1/2}}, \quad p > -\frac{1}{2}.$$

We start by recalling Eq. (4.13) for the Bessel function and the identity ($C13$) in Appendix C to arrive at

$$\begin{aligned} J_p(at) &= \frac{(at/2)^p 2^{2p}\Gamma(p+1/2)}{\sqrt{\pi}\,\Gamma(2p+1)} e^{-iat}\, {}_1F_1\left(p+\frac{1}{2};\, 2p+1;\, 2iat\right) \\ &= \frac{(2at)^p}{\sqrt{\pi}} e^{-iat}\, G_{12}^{11}\left(-2iat \,\middle|\, \begin{matrix} \frac{1}{2}-p \\ 0,\, -2p \end{matrix}\right). \end{aligned} \tag{7.105}$$

Although this integral could be solved using the Meijer G-function representation of Eq. (7.98) and integral ($D7$) in Appendix D, this time we choose to use the above Meijer G-function representation of the Bessel function for variety. Thus, the Laplace transform integral becomes

$$\begin{aligned} \int_0^\infty e^{-st} t^p J_p(at)\, dt &= \frac{(2a)^p}{\sqrt{\pi}} \int_0^\infty e^{-(s+ia)t} t^{2p} G_{12}^{11}\left(-2iat \,\middle|\, \begin{matrix} \frac{1}{2}-p \\ 0,\, -2p \end{matrix}\right) dt \\ &= \frac{(2a)^p}{\sqrt{\pi}} \frac{1}{(s+ia)^{2p+1}} G_{22}^{12}\left(-\frac{2ia}{s+ia} \,\middle|\, \begin{matrix} -2p,\, \frac{1}{2}-p \\ 0,\, -2p \end{matrix}\right) \\ &= \frac{(2a)^p}{\sqrt{\pi}} \frac{1}{(s+ia)^{2p+1}} G_{11}^{11}\left(-\frac{2ia}{s+ia} \,\middle|\, \begin{matrix} \frac{1}{2}-p \\ 0 \end{matrix}\right), \end{aligned} \tag{7.106}$$

where we have used both ($D6$) in Appendix D and ($MG3$) in Appendix A. Now from the relation ($C11$) in Appendix C, we obtain

$$G_{11}^{11}\left(-\frac{2ia}{s+ia} \,\middle|\, \begin{matrix} \frac{1}{2}-p \\ 0 \end{matrix}\right) = \Gamma\left(p+\frac{1}{2}\right)\left(1-\frac{2ia}{s+ia}\right)^{-(p+1/2)}, \tag{7.107}$$

which substituted into Eq. (7.106), and followed by algebraic simplification, leads to

$$\int_0^\infty e^{-st} t^p J_p(at)\, dt = \frac{(2a)^p}{\sqrt{\pi}} \frac{\Gamma\left(p+\frac{1}{2}\right)}{(s^2+a^2)^{p+1/2}}. \tag{7.108}$$

7.4.3 Generalized hypergeometric functions

As a wide-ranging example involving the Laplace transform, we wish to find the transform of the generalized hypergeometric function

$$f(t) = t^{\beta-1}{}_pF_q(a_1, \ldots, a_p;\, c_1, \ldots, c_q;\, t), \quad p \le q,\ \beta > 0. \tag{7.109}$$

Example 26:

$$\mathscr{L}\{t^{\beta-1}{}_pF_q(a_1,\ldots,a_p;\, c_1,\ldots,c_q;\, t)\} = \frac{\Gamma(\beta)}{s^\beta}\, {}_{p+1}F_q\left(\beta, a_1,\ldots,a_p;\, c_1,\ldots,c_q;\, \frac{1}{s}\right).$$

We could write the ${}_pF_q$ function in a convergent power series because of the restriction $p \le q$, but we choose to represent it as a Meijer G-function. Recalling (C32) in Appendix C, this approach yields

$$\begin{aligned}\int_0^\infty & e^{-st}\, t^{\beta-1}\, {}_pF_q(a_1, \ldots, a_p; c_1, \ldots, c_q; t)\, dt \\ &= \frac{\prod_{j=1}^q \Gamma(c_j)}{\prod_{j=1}^p \Gamma(a_j)} \int_0^\infty e^{-st}\, t^{\beta-1} G_{p,q+1}^{1,p}\left(-t \left| \begin{matrix} 1-a_1, \ldots, 1-a_p \\ 0, 1-c_1, \ldots, 1-c_q \end{matrix} \right.\right) dt \\ &= \frac{\prod_{j=1}^q \Gamma(c_j)}{\prod_{j=1}^p \Gamma(a_j)} \frac{1}{s^\beta} G_{p+1,q+1}^{1,p+1}\left(-\frac{1}{s} \left| \begin{matrix} 1-\beta, 1-a_1, \ldots, 1-a_p \\ 0, 1-c_1, \ldots, 1-c_q \end{matrix} \right.\right), \end{aligned} \tag{7.110}$$

where we have used the integral relation (D6) in Appendix D. Next, we use (C32) in Appendix C to express the G-function in terms of another generalized hypergeometric function,

$$\begin{aligned} G_{p+1,q+1}^{1,p+1}&\left(-\frac{1}{s} \left| \begin{matrix} 1-\beta, 1-a_1, \ldots, 1-a_p \\ 0, 1-c_1, \ldots, 1-c_q \end{matrix} \right.\right) \\ &= \frac{\prod_{j=1}^p \Gamma(a_j)}{\prod_{j=1}^q \Gamma(c_j)} \Gamma(\beta)\, {}_{p+1}F_q\left(\beta, a_1, \ldots, a_p; c_1, \ldots, c_q; \frac{1}{s}\right). \end{aligned} \tag{7.111}$$

Finally, by combining Eqs. (7.110) and (7.111), we obtain the required Laplace transform relation

$$\begin{aligned}\int_0^\infty & e^{-st}\, t^{\beta-1}\, {}_pF_q(a_1, \ldots, a_p; c_1, \ldots, c_q; t)\, dt \\ &= \frac{\Gamma(\beta)}{s^\beta}\, {}_{p+1}F_q\left(\beta, a_1, \ldots, a_p; c_1, \ldots, c_q; \frac{1}{s}\right), \quad p \le q. \end{aligned} \tag{7.112}$$

As a special case, let $p = q = 1$ in Eq. (7.112), from which we obtain

$$\int_0^\infty e^{-st}\, t^{\beta-1}\, {}_1F_1(a; c; t)\, dt = \frac{\Gamma(\beta)}{s^\beta}\, {}_2F_1\left(\beta, a; c; \frac{1}{s}\right). \tag{7.113}$$

The Laplace transform relation (7.112) represents a master formula for calculating the Laplace transform of any generalized hypergeometric function with $p \le q$, or any function that can be expressed entirely in terms of such a generalized hypergeometric function. For example, the function ${}_1F_1(-n; 1+m; z)$ is directly related to the associated Laguerre polynomial.

Example 27: $\mathscr{L}\{L_n(t)\} = \dfrac{(s-1)^n}{s^{n+1}}, \quad n = 1, 2, 3, \ldots.$

From the result of (B22) in Appendix B and Eq. (7.113), we first evaluate the more general integral,

$$\int_0^\infty e^{-st} t^{\beta-1} L_n^{(m)}(t)\,dt = \frac{\Gamma(n+m+1)}{\Gamma(m+1)n!} \int_0^\infty e^{-st} t^{\beta-1}\, {}_1F_1(-n;\, 1+m;\, t)\,dt$$
$$= \frac{\Gamma(n+m+1)\Gamma(\beta)}{\Gamma(m+1)n!\, s^\beta}\, {}_2F_1\left(\beta,\, -n;\, 1+m;\, \frac{1}{s}\right). \tag{7.114}$$

Because of $-n$ in the second position of the hypergeometric function, it follows that this function represents a polynomial in $1/s$ of degree n. Moreover, if $m = 0$ and $\beta = 1$, then canceling the common Pochhammer symbol, Eq. (7.114) reduces further to the much simpler result,

$$\int_0^\infty e^{-st} L_n(t)\,dt = \frac{1}{s}\, {}_1F_0\left(-n;\, -;\, \frac{1}{s}\right)$$
$$= \frac{(s-1)^n}{s^{n+1}}, \tag{7.115}$$

where we have also applied ($B8$) in Appendix B.

Example 28:
$$\mathscr{L}\{t^{\beta-1}{}_0F_1(-;a;t){}_0F_1(-;b;t)\} = \frac{\Gamma(\beta)}{s^\beta}{}_3F_3\left(\beta, \frac{a+b-1}{2}, \frac{a+b}{2}; a, b, a+b-1; \frac{4}{s}\right).$$
In this exercise we start with the result of Example 10, which leads directly to a specialization of Eq. (7.112) and finally to the identity above.

7.4.4 Miscellaneous functions

Example 29: $\mathscr{L}\{(a+t)^{-1/2}\} = \sqrt{\frac{\pi}{s}}\, e^{as}\,\mathrm{erfc}(\sqrt{as}), \quad a > 0.$
The Laplace transform of $f(t) = (a+t)^{-1/2}$ leads to

$$\int_0^\infty e^{-st}(a+t)^{-1/2}\,dt = \frac{1}{\sqrt{a\pi}} \int_0^\infty e^{-st} G_{11}^{11}\left(\frac{t}{a}\middle|\begin{matrix}\frac{1}{2}\\ 0\end{matrix}\right) dt$$
$$= \frac{1}{s\sqrt{a\pi}} G_{21}^{12}\left(\frac{1}{as}\middle|\begin{matrix}0,\ \frac{1}{2}\\ 0\end{matrix}\right) \tag{7.116}$$
$$= \frac{1}{s\sqrt{a\pi}} G_{12}^{21}\left(as\middle|\begin{matrix}1\\ 1,\ \frac{1}{2}\end{matrix}\right),$$

where we have used ($C11$) in Appendix C, ($D6$) in Appendix D, and ($MG2$) from Appendix A. Now by using ($C28$) in Appendix C, we obtain the relation

$$G_{12}^{21}\left(as\middle|\begin{matrix}1\\ 1,\ \frac{1}{2}\end{matrix}\right) = \pi\,\sqrt{as}\, e^{as}\mathrm{erfc}(\sqrt{as}),$$

and hence,

$$\int_0^\infty e^{-st}(a+t)^{-1/2}\,dt = \sqrt{\frac{\pi}{s}}\,e^{as}\,\mathrm{erfc}(\sqrt{as}). \tag{7.117}$$

Example 30: $\mathscr{L}\{e^{-t^2}\} = \frac{\sqrt{\pi}}{2}e^{s^2/4}\mathrm{erfc}\left(\frac{s}{2}\right).$

Using ($C6$) and ($C7$) in Appendix C, we start by replacing the exponential function e^{-st} with

$$\begin{aligned} e^{-st} &= \cosh st - \sinh st \\ &= \sqrt{\pi}\,G^{10}_{02}\left(-\frac{s^2t^2}{4}\middle|\,0,\,\frac{1}{2}\right) + i\sqrt{\pi}\,G^{10}_{02}\left(-\frac{s^2t^2}{4}\middle|\,\frac{1}{2},\,0\right). \end{aligned} \tag{7.118}$$

Thus, from the first Meijer G-function in Eq. (7.118) and ($D6$) in Appendix D, we get

$$\begin{aligned} \sqrt{\pi}\int_0^\infty e^{-t^2}G^{10}_{02}\left(-\frac{s^2t^2}{4}\middle|\,0,\,\frac{1}{2}\right)dt &= \frac{\sqrt{\pi}}{2}\int_0^\infty e^{-x}x^{-1/2}\,G^{10}_{02}\left(-\frac{s^2x}{4}\middle|\,0,\,\frac{1}{2}\right)dx \\ &= \frac{\sqrt{\pi}}{2}\,G^{11}_{12}\left(-\frac{s^2}{4}\middle|\,\begin{matrix}\frac{1}{2}\\ 0,\,\frac{1}{2}\end{matrix}\right). \end{aligned} \tag{7.119}$$

This result with the help of ($MG3$) in Appendix A and ($C1$) in Appendix C reduces to

$$\begin{aligned} \sqrt{\pi}\int_0^\infty e^{-t^2}G^{10}_{02}\left(-\frac{s^2t^2}{4}\middle|\,0,\,\frac{1}{2}\right)dt &= \frac{\sqrt{\pi}}{2}\,G^{10}_{01}\left(-\frac{s^2}{4}\middle|\,0\right) \\ &= \frac{\sqrt{\pi}}{2}\,e^{s^2/4}. \end{aligned} \tag{7.120}$$

In analogous fashion, we find for the second Meijer G-function in Eq. (7.118) that

$$\begin{aligned} i\sqrt{\pi}\int_0^\infty e^{-t^2}G^{10}_{02}\left(-\frac{s^2t^2}{4}\middle|\,\frac{1}{2},\,0\right)dt &= \frac{i\sqrt{\pi}}{2}\int_0^\infty e^{-x}x^{-1/2}G^{10}_{02}\left(-\frac{s^2x}{4}\middle|\,\frac{1}{2},\,0\right)dx \\ &= \frac{i\sqrt{\pi}}{2}\,G^{11}_{12}\left(-\frac{s^2}{4}\middle|\,\begin{matrix}\frac{1}{2}\\ \frac{1}{2},\,0\end{matrix}\right). \end{aligned} \tag{7.121}$$

By using ($C14$) in Appendix C, we can rewrite the Meijer G-function as

$$G_{12}^{11}\left(-\frac{s^2}{4}\middle|\begin{matrix}\frac{1}{2}\\ \frac{1}{2},\,0\end{matrix}\right)=\frac{2}{\sqrt{\pi}}\left(-\frac{s^2}{4}\right)^{1/2}{}_1F_1\left(1;\,\frac{3}{2};\,\frac{s^2}{4}\right)=i\,e^{s^2/4}\,\mathrm{erf}(s/2),\tag{7.122}$$

the last step of which follows ($CH4$) in Appendix A and ($B10$) in Appendix B. Now, by combining Eqs. (7.120), (7.121), and (7.122), we obtain the transform

$$\int_0^\infty e^{-st}e^{-t^2}\,dt=\frac{\sqrt{\pi}}{2}e^{s^2/4}\mathrm{erfc}\left(\frac{s}{2}\right).\tag{7.123}$$

A much shorter and more direct derivation of the transform relation given by Eq. (7.123) can be found by use of ($C1$) in Appendix C followed by use of ($D7$) in Appendix D, viz.,

$$\begin{aligned}\int_0^\infty e^{-st}e^{-t^2}\,dt&=\int_0^\infty e^{-st}G_{01}^{10}(t^2|\,0)\,dt\\&=\frac{1}{\sqrt{\pi}s}G_{21}^{12}\left(\frac{4}{s^2}\middle|\begin{matrix}0,\,\frac{1}{2}\\ 0\end{matrix}\right)\\&=\frac{1}{\sqrt{\pi}s}G_{12}^{21}\left(\frac{s^2}{4}\middle|\begin{matrix}1\\ 1,\,\frac{1}{2}\end{matrix}\right)\\&=\frac{\sqrt{\pi}}{2}e^{s^2/4}\mathrm{erfc}\left(\frac{s}{2}\right),\end{aligned}\tag{7.124}$$

where we have also used ($MG2$) in Appendix A and ($C28$) in Appendix C.

Example 31: $\mathscr{L}\{t^{b-1}\,\gamma(a,\,t)\}=\dfrac{\Gamma(a+b)}{a(1+s)^{a+b}}{}_2F_1\left(a+b,\,1;\,1+a;\,\dfrac{1}{1+s}\right).$

For the incomplete gamma function $\gamma(a,\,t)$, we recall ($B17$) in Appendix B and ($C13$) in Appendix C, which leads to

$$\gamma(a,\,t)=\frac{t^a}{a}{}_1F_1(a;\,a+1;\,-t)=t^aG_{12}^{11}\left(t\middle|\begin{matrix}1-a\\ 0,\,-a\end{matrix}\right).\tag{7.125}$$

Then, from ($D6$) in Appendix D and ($C19$) in Appendix C, we get

$$\begin{aligned}\int_0^\infty e^{-st}t^{b-1}\gamma(a,\,t)dt&=\int_0^\infty e^{-st}t^{a+b-1}G_{12}^{11}\left(t\middle|\begin{matrix}1-a\\ 0,\,-a\end{matrix}\right)dt\\&=\frac{1}{s^{a+b}}G_{22}^{12}\left(\frac{1}{s}\middle|\begin{matrix}1-a-b,\,1-a\\ 0,\,-a\end{matrix}\right)\\&=\frac{\Gamma(a+b)}{a\,s^{a+b}}{}_2F_1\left(a+b,\,a;\,1+a;\,-\frac{1}{s}\right).\end{aligned}\tag{7.126}$$

Finally, by use of the property ($HY7$) in Appendix A, we see that

$$ {}_2F_1\left(a+b,\, a;\, 1+a;\, -\frac{1}{s}\right) = \left(1+\frac{1}{s}\right)^{-(a+b)} {}_2F_1\left(a+b,\, 1;\, 1+a;\, \frac{1}{1+s}\right), \tag{7.127} $$

and by combining this last result with Eq. (7.126), we arrive at

$$ \int_0^\infty e^{-st} t^{b-1} \gamma(a,\, t)\, dt = \frac{\Gamma(a+b)}{a(1+s)^{a+b}}\, {}_2F_1\left(a+b,\, 1;\, 1+a;\, \frac{1}{1+s}\right). \tag{7.128} $$

7.5 Mellin Transform

The *Mellin transform* and *inversion formula* are defined by Eqs. (2.36) and (2.37). Similar to the above analysis concerning the Laplace transform, we can also use the Meijer G-function in calculating certain Mellin transforms. Again, most of the examples below are not considered new results but are found tabulated in many Mellin transform tables.[3]

7.5.1 Elementary functions

We start with the Mellin transform of a simple binomial function.

Example 32: $\mathcal{M}\{(1+x^a)^{-b}\} = \dfrac{\Gamma(s/a)\Gamma(b-s/a)}{a\Gamma(b)}, \quad 0 < \mathrm{Re}(s) < ab.$

From ($C11$) provided in Appendix C, we have

$$ (1+x^a)^{-b} = \frac{1}{\Gamma(b)}\, G_{11}^{11}\left(x^a \middle| \begin{matrix} 1-b \\ 0 \end{matrix}\right), \tag{7.129} $$

which leads to the Mellin transform integral,

$$ \begin{aligned} \int_0^\infty x^{s-1}(1+x^a)^{-b}\, dx &= \frac{1}{\Gamma(b)} \int_0^\infty x^{s-1}\, G_{11}^{11}\left(x^a \middle| \begin{matrix} 1-b \\ 0 \end{matrix}\right) dx \\ &= \frac{1}{a\Gamma(b)} \int_0^\infty t^{s/a-1}\, G_{11}^{11}\left(t \middle| \begin{matrix} 1-b \\ 0 \end{matrix}\right) dt. \end{aligned} \tag{7.130} $$

The second step above follows from setting $t = x^a$. Now recalling ($D8$) in Appendix D, we see that the last integral in Eq. (7.130) gives us

$$ \int_0^\infty x^{s-1}(1+x^a)^{-b}\, dx = \frac{\Gamma(s/a)\Gamma(b-s/a)}{a\Gamma(b)}. \tag{7.131} $$

We also could have solved this integral by using ($G18$) in Appendix A.

As a special interesting case, note that if $a = 2$ and $b = 1$, Eq. (7.131) with the use of ($G7$) in Appendix A reduces to [also recall Eq. (2.53)]

$$\int_0^\infty \frac{x^{s-1}}{1+x^2}\,dx = \frac{1}{2}\Gamma(s/2)\Gamma(1-s/2) \\ = \frac{\pi}{2\sin(\pi s/2)}, \quad 0 < \mathrm{Re}(s) < 2. \tag{7.132}$$

Example 33:

$$\mathcal{M}\left\{\left(\frac{1+\sqrt{1+x}}{2}\right)^{-2a}\right\} = \frac{\Gamma(2a+1)\Gamma(s)\Gamma(a-s)\Gamma\left(\frac{1}{2}+a-s\right)}{\Gamma(a)\Gamma\left(a+\frac{1}{2}\right)\Gamma(1+2a-s)}, \quad 0<\mathrm{Re}(s)<a.$$

Here we first recognize the relationships

$$\left(\frac{1+\sqrt{1+x}}{2}\right)^{-2a} = {}_2F_1\left(a,\, a+\frac{1}{2};\, 2a+1;\, -x\right) \\ = \frac{\Gamma(2a+1)}{\Gamma(a)\Gamma\left(a+\frac{1}{2}\right)} G_{22}^{12}\left(x\left|\begin{matrix}1-a,\, \frac{1}{2}-a\\ 0,\, -2a\end{matrix}\right.\right), \quad a>0, \tag{7.133}$$

where we used ($B29$) in Appendix B and ($C19$) in Appendix C. Therefore, based on ($D8$) in Appendix D, the Mellin transform becomes

$$\int_0^\infty x^{s-1}\left(\frac{1+\sqrt{1+x}}{2}\right)^{-2a} dx = \frac{\Gamma(2a+1)}{\Gamma(a)\Gamma\left(a+\frac{1}{2}\right)}\int_0^\infty x^{s-1} G_{22}^{12}\left(x\left|\begin{matrix}1-a,\, \frac{1}{2}-a\\ 0,\, -2a\end{matrix}\right.\right)dx \\ = \frac{\Gamma(2a+1)\Gamma(s)\Gamma(a-s)\Gamma\left(\frac{1}{2}+a-s\right)}{\Gamma(a)\Gamma\left(a+\frac{1}{2}\right)\Gamma(1+2a-s)}. \tag{7.134}$$

The next Mellin transform we consider in this section involves an important integral relation, variations of which are often required in dealing with optical wave propagation models.

Example 34:

$$\mathcal{M}\left\{\frac{\exp(-\kappa^2/\kappa_m^2)}{(\kappa^2+\kappa_0^2)^{11/6}}\right\} = \frac{1}{2}\kappa_0^{s-11/3}\Gamma\left(\frac{s}{2}\right)U\left(\frac{s}{2};\, \frac{s}{2}-\frac{5}{6};\, \frac{\kappa_0^2}{\kappa_m^2}\right).$$

By factoring out κ_0^2 in the denominator followed by the change of variable $t = \kappa^2/\kappa_0^2$, we find that

$$\int_0^\infty \kappa^{s-1} \frac{\exp(-\kappa^2/\kappa_m^2)}{(\kappa^2+\kappa_0^2)^{11/6}}\, d\kappa = \frac{1}{2}\kappa_0^{s-11/3} \int_0^\infty t^{s/2-1} \frac{\exp(-\kappa_0^2 t/\kappa_m^2)}{(t+1)^{11/6}}\, dt. \quad (7.135)$$

The integral on the right-hand side of Eq. (7.135) is now recognized as a special case of the confluent hypergeometric function of the second kind, defined by ($CH16$) in Appendix A with parameters $z = \kappa_0^2/\kappa_m^2$, $a = s/2$, and $c = s/2 - 5/6$. Making these associations leads to

$$\int_0^\infty \kappa^{s-1} \frac{\exp(-\kappa^2/\kappa_m^2)}{(\kappa^2+\kappa_0^2)^{11/6}}\, d\kappa = \frac{1}{2}\kappa_0^{s-11/3}\Gamma\left(\frac{s}{2}\right) U\left(\frac{s}{2}; \frac{s}{2} - \frac{5}{6}; \frac{\kappa_0^2}{\kappa_m^2}\right). \quad (7.136)$$

To find the Mellin transform of some trigonometric functions, we begin with the simple cosine function.

Example 35: $\mathcal{M}\{\cos x\} = \Gamma(s)\cos\left(\frac{\pi s}{2}\right), \quad 0 < \mathrm{Re}(s) < 1.$

We could write $\cos x = \mathrm{Re}(e^{-ix})$ and treat the integral as a Laplace transform. However, we choose to start with the relation ($C4$) in Appendix C,

$$\cos x = \sqrt{\pi}\, G_{02}^{10}\left(\frac{x^2}{4}\,\middle|\, 0, \frac{1}{2}\right), \quad (7.137)$$

and find through use of ($D8$) in Appendix D that

$$\begin{aligned} \int_0^\infty x^{s-1}\cos x\, dx &= \sqrt{\pi}\int_0^\infty x^{s-1} G_{02}^{10}\left(\frac{x^2}{4}\,\middle|\, 0, \frac{1}{2}\right) dx \\ &= 2^{s-1}\sqrt{\pi}\int_0^\infty t^{s/2-1} G_{02}^{10}\left(t\,\middle|\, 0, \frac{1}{2}\right) dt \\ &= 2^{s-1}\sqrt{\pi}\, \frac{\Gamma\left(\frac{s}{2}\right)}{\Gamma\left(\frac{1-s}{2}\right)}. \end{aligned} \quad (7.138)$$

Now, recalling the gamma function identities in Appendix A,

$$\Gamma\left(\frac{s}{2}\right) = \frac{\sqrt{\pi}\,\Gamma(s)}{2^{s-1}\Gamma\left(\frac{1+s}{2}\right)},$$

$$\Gamma\left(\frac{1+s}{2}\right)\Gamma\left(\frac{1-s}{2}\right) = \frac{\pi}{\cos(\pi s/2)},$$

the result in Eq. (7.138) reduces to

$$\int_0^\infty x^{s-1}\cos x\, dx = \Gamma(s)\cos\left(\frac{\pi s}{2}\right). \quad (7.139)$$

In a similar manner, it can be shown that

$$\int_0^\infty x^{s-1} \sin x \, dx = \Gamma(s) \sin\left(\frac{\pi s}{2}\right), \qquad 0 < \mathrm{Re}(s) < 1. \tag{7.140}$$

As a generalization of Eq. (7.139), it follows from ($C1$) in Appendix C and ($D8$) in Appendix D that

$$\begin{aligned}\int_0^\infty x^{s-1} e^{-x\cos\phi} \cos(x \sin\phi)\, dx &= \mathrm{Re} \int_0^\infty x^{s-1} \exp(-xe^{i\phi})\, dx \\ &= \mathrm{Re} \int_0^\infty x^{s-1} G_{01}^{10}(xe^{i\phi} \,|\, 0)\, dx \\ &= \Gamma(s) \cos(s\phi), \qquad -\frac{\pi}{2} < \phi \le \frac{\pi}{2}, \quad \mathrm{Re}(s) > 0.\end{aligned} \tag{7.141}$$

7.5.2 Bessel functions

We now consider the Mellin transform featuring Bessel functions. We start with $f(x) = x^{-p} J_p(x)$.

Example 36:

$$\mathcal{M}\{x^{-p} J_p(x)\} = 2^{s-p-1} \frac{\Gamma(s/2)}{\Gamma(1+p-s/2)}, \qquad 0 < \mathrm{Re}(s) < 1, \, p > -\frac{1}{2}.$$

From the Bessel function relationship of Eq. (7.98), we have

$$\begin{aligned}\int_0^\infty x^{s-1} x^{-p} J_p(x)\, dx &= \int_0^\infty x^{s-p-1}\, G_{02}^{10}\left(\frac{x^2}{4} \middle| \frac{p}{2}, -\frac{p}{2}\right) dx \\ &= 2^{s-p-1} \int_0^\infty t^{(s-p)/2-1}\, G_{02}^{10}\left(t \middle| \frac{p}{2}, -\frac{p}{2}\right) dt \\ &= 2^{s-p-1} \frac{\Gamma(s/2)}{\Gamma(1+p-s/2)}.\end{aligned} \tag{7.142}$$

The last step above follows from ($D8$) in Appendix D. If we modify the function in Example 36 by $f(x) = J_p(x)$, $p > -\frac{1}{2}$, it then follows directly from Eq. (7.142) that

$$\begin{aligned}\int_0^\infty x^{s-1} J_p(x)\, dx &= 2^{s-1} \int_0^\infty t^{s/2-1}\, G_{02}^{10}\left(t \middle| \frac{p}{2}, -\frac{p}{2}\right) dt \\ &= \frac{2^{s-1}\Gamma\left(\frac{s+p}{2}\right)}{\Gamma\left(1+\frac{p-s}{2}\right)}, \qquad -p < \mathrm{Re}(s) < p+2.\end{aligned} \tag{7.143}$$

Example 37:

$$\mathcal{M}\{e^{-\alpha x^2}J_p(\beta x)\} = \frac{\beta^p\Gamma\left(\frac{p+s}{2}\right)}{2^{p+1}\alpha^{(p+s)/2}\Gamma(1+p)}\,{}_1F_1\left(\frac{p+s}{2};\,1+p;\,-\frac{\beta^2}{4\alpha}\right).$$

Among other applications, this integral arises in an application of the Hankel transform concerning the vibrations of a thin plate (p. 513 of Ref. 10). By letting $x = \sqrt{t}$, we can also identify this integral as a Laplace transform integral.

To prove the integral identity in Example 37, we start with ($C1$) in Appendix C for the exponential function representation

$$e^{-\alpha x^2} = G_{01}^{10}(\alpha x^2|0), \tag{7.144}$$

from which it follows that

$$\begin{aligned}\int_0^\infty x^{s-1}e^{-\alpha x^2}J_p(\beta x)\,dx &= \int_0^\infty x^{s-1}J_p(\beta x)\,G_{01}^{10}(\alpha x^2|0)\,dx\\ &= \frac{2^{s-1}}{\beta^s}\,G_{21}^{11}\left(\frac{4\alpha}{\beta^2}\middle|\begin{matrix}1-\frac{p+s}{2},\;1+\frac{p-s}{2}\\ 0\end{matrix}\right)\\ &= \frac{2^{s-1}}{\beta^s}\,G_{12}^{11}\left(\frac{\beta^2}{4\alpha}\middle|\begin{matrix}1\\ \frac{p+s}{2},\;-\frac{p-s}{2}\end{matrix}\right).\end{aligned} \tag{7.145}$$

In deriving this result, we have used both ($D12$) in Appendix D and ($MG2$) in Appendix A. This last Meijer G-function is related to the confluent hypergeometric function via the identity ($C14$) in Appendix C,

$$G_{12}^{11}\left(z\middle|\begin{matrix}a\\ b,\,c\end{matrix}\right) = \frac{\Gamma(1-a+b)}{\Gamma(1-c+b)}\,z^b\,{}_1F_1(1-a+b;\,1-c+b;\,-z). \tag{7.146}$$

Hence, upon use of Eq. (7.146), we see that Eq. (7.145) takes the form

$$\int_0^\infty x^{s-1}e^{-\alpha x^2}J_p(\beta x)\,dx = \frac{\beta^p\Gamma\left(\frac{p+s}{2}\right)}{2^{p+1}\alpha^{(p+s)/2}\Gamma(1+p)}\,{}_1F_1\left(\frac{p+s}{2};\,1+p;\,-\frac{\beta^2}{4\alpha}\right). \tag{7.147}$$

The last Mellin transform example we derive concerns the product of a modified Bessel function and a standard Bessel function.

Example 38:

$$\mathcal{M}\{K_\mu(ax)J_\nu(bx)\} = \frac{2^{s-2}b^\nu}{a^{s+\nu}} \frac{\Gamma\left(\frac{s+\nu+\mu}{2}\right)\Gamma\left(\frac{s+\nu-\mu}{2}\right)}{\Gamma(1+\nu)}$$
$$\times {}_2F_1\left(\frac{s+\nu+\mu}{2}, \frac{s+\nu-\mu}{2}; 1+\nu; -\frac{b^2}{a^2}\right), \quad a, b > 0.$$

If we use ($C23$) in Appendix C,

$$K_\mu(ax) = \frac{1}{2} G_{02}^{20}\left(\frac{a^2x^2}{4} \middle| \frac{\mu}{2}, -\frac{\mu}{2}\right), \quad \mu > 0, \tag{7.148}$$

the transform integral becomes

$$\begin{aligned}\int_0^\infty x^{s-1}K_\mu(ax)J_\nu(bx)\,dx &= \frac{1}{2}\int_0^\infty x^{s-1}J_\nu(bx)G_{02}^{20}\left(\frac{a^2x^2}{4} \middle| \frac{\mu}{2}, -\frac{\mu}{2}\right)dx \\ &= \frac{2^{s-2}}{b^s} G_{22}^{21}\left(\frac{a^2}{b^2} \middle| \begin{matrix} 1-\frac{(\nu+s)}{2}, 1+\frac{(\nu-s)}{2} \\ \frac{\mu}{2}, -\frac{\mu}{2} \end{matrix}\right) \\ &= \frac{2^{s-2}}{b^s} G_{22}^{12}\left(\frac{b^2}{a^2} \middle| \begin{matrix} 1-\frac{\mu}{2}, 1+\frac{\mu}{2} \\ \frac{(\nu+s)}{2}, -\frac{(\nu-s)}{2} \end{matrix}\right),\end{aligned} \tag{7.149}$$

where we have used ($D12$) in Appendix D and ($MG2$) in Appendix A. If we now call on ($MG1$) in Appendix A, Eq. (7.149) reduces to

$$\int_0^\infty x^{s-1}K_\mu(ax)J_\nu(bx)\,dx = \frac{2^{s-2}b^\nu}{a^{s+\nu}} \frac{\Gamma\left(\frac{s+\nu+\mu}{2}\right)\Gamma\left(\frac{s+\nu-\mu}{2}\right)}{\Gamma(1+\nu)}$$
$$\times {}_2F_1\left(\frac{s+\nu+\mu}{2}, \frac{s+\nu-\mu}{2}; 1+\nu; -\frac{b^2}{a^2}\right). \tag{7.150}$$

We could also solve this integral by using Eq. (7.98) and ($D13$) in Appendix D.

7.6 Hankel Transform

Hankel transforms arise naturally in solving boundary-value problems in cylindrical coordinates, and they are used in diffraction theory to describe the image-forming process of a linear filtering operation for coherent and incoherent imaging systems.

The *Hankel transform* of order ν and its *inverse transform* are defined, respectively, by (pp. 274–290 of Ref. 2)

$$\mathcal{H}_\nu\{f(r)\} = \int_0^\infty rf(r)J_\nu(\rho r)\,dr = F(\rho), \quad \nu > -\frac{1}{2}, \tag{7.151}$$

$$\mathcal{H}_\nu^{-1}\{F(\rho)\} = \int_0^\infty \rho F(\rho)J_\nu(r\rho)\,d\rho = f(r), \quad \nu > -\frac{1}{2}. \tag{7.152}$$

In this case the interesting feature is that the Hankel transform and its inverse have exactly the same functional form. Therefore, either one can be called the transform and the other the inverse transform. The basic requirement for the existence of the Hankel transform is that the function $\sqrt{r}f(r)$ be piecewise continuous and absolutely integrable on the positive real axis. Last, we note that the two-dimensional Fourier transform with circular symmetry leads to the Hankel transform with $\nu = 0$ following a transformation to polar coordinates (pp. 274–275 of Ref. 2).

7.6.1 Elementary functions

Let us start by evaluating the Hankel transform of order zero for some elementary functions. The first function we consider is $f(r) = e^{-a^2r^2}$.

Example 39: $\mathcal{H}_0\{e^{-a^2r^2}\} = \dfrac{1}{2a^2}e^{-\rho^2/4a^2}$.

From ($C9$) in Appendix C, we find that

$$\begin{aligned}\mathcal{H}_0\{e^{-a^2r^2}\} &= \int_0^\infty re^{-a^2r^2}J_0(\rho r)\,dr \\ &= \int_0^\infty re^{-a^2r^2}G_{02}^{10}\left(\frac{\rho^2r^2}{4}\middle|\,0,\,0\right)dr \\ &= \frac{1}{2a^2}G_{12}^{11}\left(\frac{\rho^2}{4a^2}\middle|\begin{matrix}0\\0,\,0\end{matrix}\right),\end{aligned} \tag{7.153}$$

where we have set $x = r^2$ and used ($D6$) in Appendix D. Now applying ($MG3$) in Appendix A followed by ($C1$) in Appendix C, we are led to

$$\begin{aligned}\int_0^\infty re^{-a^2r^2}J_0(\rho r)\,dr &= \frac{1}{2a^2}G_{01}^{10}\left(\frac{\rho^2}{4a^2}\middle|\,0\right) \\ &= \frac{1}{2a^2}e^{-\rho^2/4a^2}.\end{aligned} \tag{7.154}$$

Example 40: $\mathcal{H}_0\{\frac{1}{r}e^{-a^2r^2/4}\} = \frac{\sqrt{\pi}}{a}\,{}_1F_1\left(\frac{1}{2};\,1;\,-\frac{\rho^2}{a^2}\right)$.

Let the exponential function be defined as in Eq. (7.144), from which we obtain

$$\begin{aligned}\int_0^\infty e^{-a^2r^2/4}J_0(\rho r)dr &= \int_0^\infty J_0(\rho r)G_{01}^{10}\left(\frac{a^2r^2}{4}\middle|0\right)dr\\ &= \frac{1}{\rho}G_{21}^{11}\left(\frac{a^2}{\rho^2}\middle|\begin{matrix}\frac{1}{2},\ \frac{1}{2}\\ 0\end{matrix}\right)\\ &= \frac{1}{\rho}G_{21}^{11}\left(\frac{\rho^2}{a^2}\middle|\begin{matrix}1\\ \frac{1}{2},\ \frac{1}{2}\end{matrix}\right),\end{aligned} \tag{7.155}$$

or

$$\int_0^\infty e^{-a^2r^2/4}J_0(\rho r)\,dr = \frac{\sqrt{\pi}}{a}\,{}_1F_1\left(\frac{1}{2};\ 1;\ -\frac{\rho^2}{a^2}\right). \tag{7.156}$$

We used ($D12$) in Appendix D, ($MG2$) in Appendix A, and ($C14$) from Appendix C to help reach the last step in Eq. (7.156). Furthermore, from identity ($B9$) in Appendix B for the ${}_1F_1$ function, we can also express this transform as

$$\mathcal{H}_0\left\{\frac{1}{r}e^{-a^2r^2/4}\right\} = \frac{\sqrt{\pi}}{a}e^{-\rho^2/a^2}I_0\left(\frac{\rho^2}{2a^2}\right). \tag{7.157}$$

7.6.2 Bessel functions

To begin, we calculate the Hankel transform of $f(r) = \frac{1}{r}J_0(ar)$.

Example 41: $\int_0^\infty J_0(ar)J_0(br)\,dr = \dfrac{2}{\pi b}\,K(a/b), \quad 0 < a < b.$

The function $K(m)$ is the complete elliptic integral from Section 3.10.6. In this integral we will interpret b as the transform variable ρ. Then, replacing $J_0(ar)$ with its Meijer G-function representation [Eq. (7.98)], we get

$$\begin{aligned}\int_0^\infty J_0(ar)\,J_0(br)\,dr &= \int_0^\infty J_0(br)\,G_{02}^{10}\left(\frac{a^2r^2}{4}\middle|0,\ 0\right)dr\\ &= \frac{1}{b}\,G_{22}^{11}\left(\frac{a^2}{b^2}\middle|\begin{matrix}\frac{1}{2},\ \frac{1}{2}\\ 0,\ 0\end{matrix}\right),\end{aligned} \tag{7.158}$$

where we have used ($D12$) in Appendix D. Now recalling ($C18$) in Appendix C and also ($B34$) in Appendix B, we obtain

$$\begin{aligned}\int_0^\infty J_0(ar)J_0(br)\,dr &= \frac{1}{b}\,{}_2F_1\left(\frac{1}{2},\ \frac{1}{2};\ 1;\ \frac{a^2}{b^2}\right)\\ &= \frac{2}{\pi b}\,K(a/b).\end{aligned} \tag{7.159}$$

Example 42:

$$\int_0^\infty r^{1+\mu-\nu} J_\mu(ar) J_\nu(br)\, dr = \frac{a^\mu (b^2 - a^2)^{\nu-\mu-1}}{2^{\nu-\mu-1} b^\nu \Gamma(\nu - \mu)}, \qquad \mu < \nu,\ a < b.$$

We can interpret this integral as a Hankel transform of order ν with transform variable $\rho = b$. Using the relation in Eq. (7.98), we have

$$\begin{aligned}\int_0^\infty r^{1+\mu-\nu} J_\mu(ar) J_\nu(br)\, dr &= \int_0^\infty r^{1+\mu-\nu} J_\nu(br) G_{02}^{10}\left(\frac{a^2 r^2}{4} \middle| \frac{\mu}{2}, -\frac{\mu}{2}\right) dr \\ &= \frac{2^{1+\mu-\nu}}{b^{2+\mu-\nu}} G_{22}^{11}\left(\frac{a^2}{b^2} \middle| \begin{matrix} -\frac{\mu}{2}, \nu - \frac{\mu}{2} \\ \frac{\mu}{2}, -\frac{\mu}{2} \end{matrix}\right),\end{aligned} \tag{7.160}$$

from which we deduce that

$$\begin{aligned}\int_0^\infty r^{1+\mu-\nu} J_\mu(ar) J_\nu(br)\, dr &= \frac{2^{1+\mu-\nu}}{b^{2+\nu-\mu}} G_{11}^{10}\left(\frac{a^2}{b^2} \middle| \begin{matrix} \nu - \frac{\mu}{2} \\ \frac{\mu}{2} \end{matrix}\right) \\ &= \frac{a^\mu (b^2 - a^2)^{\nu-\mu-1}}{2^{\nu-\mu-1} b^\nu \Gamma(\nu - \mu)}.\end{aligned} \tag{7.161}$$

In reaching these last steps, we have utilized ($MG3$) in Appendix A and ($C2$) from Appendix C.

For the last example we will use a different approach to evaluating the Hankel transform.

Example 43: $\int_0^\infty x^{\mu-1} e^{-ax} J_\nu(bx)\, dx = \frac{2^{\mu/2-1}}{a^\mu \sqrt{\pi}} G_{22}^{12}\left(\frac{b^2}{a^2} \middle| \begin{matrix} 1 - \frac{\mu}{2}, \frac{1-\mu}{2} \\ \frac{\nu}{2}, -\frac{\nu}{2} \end{matrix}\right).$

In this case we replace both the exponential function and Bessel function by using ($C1$) and ($C9$) in Appendix C followed by expressing ($C9$) in its Barnes integral representation to find that

$$\begin{aligned}\int_0^\infty x^{\mu-1} e^{-ax} J_\nu(bx)\, dx &= \int_0^\infty x^{\mu-1}\, G_{01}^{10}(ax|\,0)\, G_{02}^{10}\left(\frac{b^2 x^2}{4} \middle| \frac{\nu}{2}, -\frac{\nu}{2}\right) dx \\ &= \frac{1}{2\pi i} \int_C \frac{\Gamma\left(\frac{\nu}{2} - s\right)}{\Gamma\left(1 + \frac{\nu}{2} + s\right)} \left(\frac{b^2}{4}\right)^s \int_0^\infty x^{2s+\mu-1} G_{01}^{10}(ax|\,0)\, dx\, ds \\ &= \frac{1}{2\pi i} \int_C \frac{\Gamma\left(\frac{\nu}{2} - s\right) \Gamma(\mu + 2s)}{\Gamma\left(1 + \frac{\nu}{2} + s\right)} \left(\frac{b^2}{4a^2}\right)^s \frac{1}{a^\mu}\, ds.\end{aligned}$$

(7.162)

In the second step above we have interchanged the order of integration followed by ($D8$) in Appendix D. Now, recalling ($G10$) in Appendix A written as

$$\Gamma(\mu + 2s) = \frac{1}{\sqrt{\pi}} 2^{2s+\mu/2-1}\Gamma\left(\frac{\mu}{2} + s\right)\Gamma\left(\frac{\mu+1}{2} + s\right),$$

we are led to

$$\begin{aligned}
&\int_0^\infty x^{\mu-1}e^{-ax}J_\nu(bx)\,dx \\
&= \frac{2^{\mu/2-1}}{a^\mu\sqrt{\pi}}\frac{1}{2\pi i}\int_C \frac{\Gamma\left(\frac{\nu}{2} - s\right)\Gamma\left(\frac{\mu}{2} + s\right)\Gamma\left(\frac{\mu+1}{2} + s\right)}{\Gamma\left(1 + \frac{\nu}{2} + s\right)}\left(\frac{b^2}{a^2}\right)^s ds \qquad (7.163) \\
&= \frac{2^{\mu/2-1}}{a^\mu\sqrt{\pi}} G_{22}^{12}\left(\frac{b^2}{a^2}\middle|\begin{matrix} 1-\frac{\mu}{2}, \frac{1-\mu}{2} \\ \frac{\nu}{2}, -\frac{\nu}{2}\end{matrix}\right).
\end{aligned}$$

Note that by employing ($MG1$) in Appendix A, this last G-function can be expressed in terms of the hypergeometric function,

$$\begin{aligned}
&G_{22}^{12}\left(\frac{b^2}{a^2}\middle|\begin{matrix} 1-\frac{\mu}{2}, \frac{1-\mu}{2} \\ \frac{\nu}{2}, -\frac{\nu}{2}\end{matrix}\right) \\
&= \frac{\Gamma\left(\frac{\nu+\mu}{2}\right)\Gamma\left(\frac{1+\nu+\mu}{2}\right)}{\Gamma(1+\nu)}\left(\frac{b}{a}\right)^\nu {}_2F_1\left(\frac{\nu+\mu}{2}, \frac{1+\nu+\mu}{2}; 1+\nu; -\frac{b^2}{a^2}\right).
\end{aligned}$$

7.7 Applications of Integral Transforms

Applications of the Laplace transform are generally well known to most engineers. Less well known are applications involving the Mellin and Hankel transforms. In this section we will present applications involving several integral transforms.

7.7.1 Heat conduction in a long rod

A classic problem of heat flow concerns a long homogeneous rod, one end of which is exposed to a time-varying heat reservoir. It is assumed that the initial temperature distribution is 0°C along the rod and the lateral surface is insulated.

Example 44: The temperature $u(x, t)$ at any position in the rod at time t is a solution of the boundary value problem

$$\begin{aligned} &\frac{\partial^2 u}{\partial x^2} = a^{-2}\frac{\partial u}{\partial t}, \qquad 0 < x < \infty,\ t > 0 \\ &u(0,\ t) = f(t), \quad u(x,\ t) \to 0 \text{ as } x \to \infty \\ &u(x,\ 0) = 0, \quad 0 < x < \infty. \end{aligned} \tag{7.164}$$

The function $f(t)$ is the prescribed temperature at the finite boundary of the rod, and a^2 is a physical constant called the *diffusivity*. We wish to first find a solution formula for the heat conduction problem described by Eq. (7.164), and then complete the solution by prescribing two cases for $f(t)$.

By applying the Laplace transform to the boundary value problem defined by Eq. (7.164), the transformed problem becomes

$$\begin{aligned} &\frac{\partial^2 U}{\partial x^2} - \frac{s}{a^2}U = 0, \qquad 0 < x < \infty \\ &U(0,\ s) = F(s), \qquad U(x,\ s) \to 0 \text{ as } x \to \infty. \end{aligned} \tag{7.165}$$

In deriving the terms appearing in Eq. (7.165), we have made use of the Laplace transform relations

$$\mathscr{L}\{u(x,\ t)\} = U(x,\ s),\ \mathscr{L}\{f(t)\} = F(s),$$

$$\mathscr{L}\left\{\frac{\partial^2 u}{\partial x^2}\right\} = \frac{\partial^2 U}{\partial x^2}, \qquad \mathscr{L}\left\{\frac{\partial u}{\partial t}\right\} = sU(x,\ s) - u(x,\ 0).$$

The transformed problem (7.165) involves a second-order DE with a solution satisfying the prescribed boundary conditions given by

$$U(x,\ s) = F(s)e^{-x\sqrt{s}/a}. \tag{7.166}$$

To invert this solution back to the time domain, we use the *convolution theorem* of the Laplace transform, defined by (pp. 194–196 of Ref. 2)

$$\mathscr{L}^{-1}\{F(s)G(s)\} = \int_0^t f(\tau)g(t-\tau)\,d\tau. \tag{7.167}$$

The function $g(t)$ is the inverse transform of $G(s) = e^{-x\sqrt{s}/a}$ (p. 346 of Ref. 2),

$$g(t) = \frac{1}{2\pi i}\int_{c-i\infty}^{c+i\infty} e^{st}e^{-x\sqrt{s}/a}ds = \frac{x}{2a\sqrt{\pi}\,t^{3/2}}\exp\left(-\frac{x^2}{4a^2t}\right).$$

This relation can also be validated by taking the Laplace transform of the right-hand side, using the methods of integration illustrated in this chapter (e.g., see Example 1). Hence, the general solution formula we seek is described by the integral

$$u(x, t) = \frac{x}{2a\sqrt{\pi}} \int_0^t \frac{f(\tau)}{(t-\tau)^{3/2}} \exp\left[-\frac{x^2}{4a^2(t-\tau)}\right] d\tau. \tag{7.168}$$

For the first case involving $f(t)$, let us assume that $f(t) = T_0$ (constant). After the change of variable $z = x/2a\sqrt{t-\tau}$, Eq. (7.168) leads to the solution

$$\begin{aligned} u(x, t) &= \frac{2T_0}{\sqrt{\pi}} \int_{x/(2a\sqrt{t})}^{\infty} e^{-z^2} dz \\ &= T_0 \operatorname{erfc}\left(\frac{x}{2a\sqrt{t}}\right). \end{aligned} \tag{7.169}$$

For the second case, we impose the slightly different boundary condition

$$f(t) = \begin{cases} T_1, & 0 < t < b \\ 0, & t > b. \end{cases} \tag{7.170}$$

Here we start with the same substitution as in the first case. If $0 < t < b$, then the resulting solution is the same as that in Eq. (7.169), but with T_0 replaced by T_1. For $t > b$, we find that

$$\begin{aligned} u(x, t) &= \frac{2T_1}{\sqrt{\pi}} \int_{x/(2a\sqrt{t})}^{x/(2a\sqrt{t-b})} e^{-z^2} dz \\ &= T_1\left[\frac{2}{\sqrt{\pi}} \int_0^{x/(2a\sqrt{t-b})} e^{-z^2} dz - \frac{2}{\sqrt{\pi}} \int_0^{x/(2a\sqrt{t})} e^{-z^2} dz\right], \end{aligned} \tag{7.171}$$

and, thus, by combining solutions over each interval, we have

$$u(x, t) = \begin{cases} T_1 \operatorname{erfc}\left(\dfrac{x}{2a\sqrt{t}}\right), & 0 < t < b \\ T_1\left[\operatorname{erf}\left(\dfrac{x}{2a\sqrt{t-b}}\right) - \operatorname{erf}\left(\dfrac{x}{2a\sqrt{t}}\right)\right], & t > b. \end{cases} \tag{7.172}$$

7.7.2 Probability and statistics

Products of statistically independent random variables are common in certain engineering applications. Here we wish to find the probability density function (PDF) of the product $\mathbf{z} = \mathbf{xy}$, where $\mathbf{x}$ and $\mathbf{y}$ are statistically independent positive random variables. In Section 2.7.2 we derived the PDF for such a product by using the convolution integral of the Mellin transform resulting from the relationship

$$p_{\mathbf{z}}(z) = \int_0^{\infty} p_{\mathbf{x}}(x) p_{\mathbf{y}}(z/x) \frac{dx}{x}, \tag{7.173}$$

where $p_{\mathbf{x}}(x)$ and $p_{\mathbf{y}}(y)$ are PDFs of **x** and **y**, respectively. However, rather than calculating the Mellin transform of each PDF and finding the inverse Mellin transform of their product like we did in Section 2.7.2, here we will directly evaluate the integral in Eq. (7.173) by use of the Meijer *G*-function.

Example 45: Given that α and β are constants, find the PDF for $\mathbf{z} = \mathbf{xy}$, where both **x** and **y** are governed by gamma distributions,

$$p_{\mathbf{x}}(x) = \frac{\alpha(\alpha x)^{\alpha-1}}{\Gamma(\alpha)} \exp(-\alpha x), \quad \alpha > 0,\ x > 0,$$

$$p_{\mathbf{y}}(y) = \frac{\beta(\beta y)^{\beta-1}}{\Gamma(\beta)} \exp(-\beta y), \quad \beta > 0,\ y > 0.$$

By substituting both PDFs above into Eq. (7.173), we obtain

$$\begin{aligned} p_{\mathbf{z}}(z) &= \int_0^\infty \left[\frac{\alpha(\alpha x)^{\alpha-1}}{\Gamma(\alpha)} e^{-\alpha x}\right] \left[\frac{\beta(\beta z/x)^{\beta-1}}{\Gamma(\beta)} e^{-\beta z/x}\right] \frac{dx}{x} \\ &= \frac{\alpha^\alpha \beta^\beta z^{\beta-1}}{\Gamma(\alpha)\Gamma(\beta)} \int_0^\infty x^{\alpha-\beta-1} e^{-\alpha x - \beta z/x} dx. \end{aligned} \tag{7.174}$$

At this point we recognize the last integral being similar to that in Example 1. Hence, we use property (*C*1) from Appendix C and (*MG*2) from Appendix A to find that

$$\begin{aligned} \int_0^\infty x^{\alpha-\beta-1} e^{-\alpha x - \beta z/x} dx &= \int_0^\infty x^{\alpha-\beta-1} e^{-\alpha x} G_{10}^{01}\left(\frac{x}{\beta z}\middle| 1\right) dx \\ &= \alpha^{\beta-\alpha} G_{02}^{20}(\alpha\beta z | \alpha-\beta, 0) \\ &= 2\left(\frac{\alpha z}{\beta}\right)^{(\alpha-\beta)/2} K_{\alpha-\beta}(2\sqrt{\alpha\beta z}). \end{aligned} \tag{7.175}$$

Here we have also used (*D*6) in Appendix D followed by (*C*23) in Appendix C. Combining the results of Eqs. (7.174) and (7.175) leads to

$$p_{\mathbf{z}}(z) = \frac{2(\alpha\beta)^{(\alpha+\beta)/2}}{\Gamma(\alpha)\Gamma(\beta)} z^{(\alpha+\beta)/2-1} K_{\alpha-\beta}(2\sqrt{\alpha\beta z}), \quad z > 0, \tag{7.176}$$

which is called the *gamma-gamma PDF.*[11]

Example 46: Calculate the statistical moments of the gamma-gamma PDF,

$$p_{\mathbf{z}}(z) = \frac{2(\alpha\beta)^{(\alpha+\beta)/2}}{\Gamma(\alpha)\Gamma(\beta)} z^{(\alpha+\beta)/2-1} K_{\alpha-\beta}(2\sqrt{\alpha\beta z}), \quad z > 0.$$

We start by writing the gamma-gamma PDF in terms of the Meijer G-function using (C23) in Appendix C, i.e.,

$$p_{\mathbf{z}}(z) = \frac{1}{\Gamma(\alpha)\Gamma(\beta)z} G_{02}^{20}(\alpha\beta z|\alpha,\, \beta). \tag{7.177}$$

The statistical moments of $\mathbf{z}$ are defined by the ensemble average

$$\begin{aligned}\langle z^n \rangle &= \frac{1}{\Gamma(\alpha)\Gamma(\beta)} \int_0^\infty z^{n-1} G_{02}^{20}(\alpha\beta z|\alpha,\, \beta)\, dz \\ &= \frac{1}{\Gamma(\alpha)\Gamma(\beta)} \frac{\Gamma(\alpha+n)\Gamma(\beta+n)}{(\alpha\beta)^n}, \qquad n = 1, 2, 3, \ldots .\end{aligned} \tag{7.178}$$

In reaching the above result, we used (D8) in Appendix D. Last, by setting $n = 1$ and $n = 2$, we get the first two moments: $\langle z \rangle = 1$ and $\langle z^2 \rangle = (1 + 1/\alpha) \times (1 + 1/\beta)$.

7.7.3 Summation of series

The Mellin transform can be an excellent tool for the summation of certain series.[12] To see this, we start with the inversion formula [Eq. (2.37)],

$$f(x) = \frac{1}{2\pi i} \int_{c-i\infty}^{c+i\infty} x^{-s} F(s) ds, \qquad x > 0. \tag{7.179}$$

For this application, we may also need the Mellin transform *scaling property* and *translation property* defined, respectively, by

$$\begin{aligned}\mathcal{M}\{f(ax)\} &= a^{-s} F(s), \\ \mathcal{M}\{x^m f(x)\} &= F(s+m).\end{aligned}$$

The scaling property follows by making the substitution $x = u/a$ in the Mellin transform integral for $f(ax)$, whereas the translation property arises from making the replacement $s \to s+m$ in the Mellin transform integral.

If we replace x with nx, where $n = 1, 2, 3, \ldots$, the inversion formula given by Eq. (7.179) becomes

$$f(nx) = \frac{1}{2\pi i} \int_{c-i\infty}^{c+i\infty} n^{-s} x^{-s} F(s) ds, \qquad n = 1, 2, 3, \ldots . \tag{7.180}$$

Let us sum both sides of Eq. (7.180) over all positive integers of n, interchanging the order of summation and integration. Doing so yields

$$\sum_{n=1}^{\infty} f(nx) = \frac{1}{2\pi i} \int_{c-i\infty}^{c+i\infty} x^{-s} \zeta(s) F(s) ds, \tag{7.181}$$

where $\zeta(s)$ is the *Riemann zeta function* defined by (pp. 32–35 of Ref. 3, Vol. I)

$$\zeta(s) = \sum_{n=1}^{\infty} \frac{1}{n^s}, \qquad \mathrm{Re}(s) > 1. \tag{7.182}$$

Therefore, from Eq. (7.181), we have produced the result

$$\sum_{n=1}^{\infty} f(n) = \mathcal{M}^{-1}\{\zeta(s)F(s)\}|_{x=1}. \tag{7.183}$$

Example 47: Use Eq. (7.183) to sum the series $\sum_{n=1}^{\infty} \frac{\cos(an)}{n^2}$.

Based on the given series, we first define the function $f(x) = \cos x$, which has the Mellin transform (recall Example 35)

$$\mathcal{M}\{\cos x\} = \Gamma(s)\cos\left(\frac{1}{2}\pi s\right). \tag{7.184}$$

From the scaling and translation properties, it follows that

$$\begin{aligned}\mathcal{M}\{\cos(ax)\} &= \frac{\Gamma(s)}{a^s}\cos\left(\frac{1}{2}\pi s\right),\\ \mathcal{M}\left\{\frac{\cos(ax)}{x^2}\right\} &= -\frac{\Gamma(s-2)}{a^{s-2}}\cos\left(\frac{1}{2}\pi s\right),\end{aligned} \tag{7.185}$$

and, thus, based on the result in Eq. (7.183), we deduce that

$$\begin{aligned}\sum_{n=1}^{\infty}\frac{\cos(an)}{n^2} &= -\mathcal{M}^{-1}\left\{\frac{\Gamma(s-2)}{a^{s-2}}\zeta(s)\cos\left(\frac{1}{2}\pi s\right)\right\}\Bigg|_{x=1}\\ &= -\frac{a^2}{2}\mathcal{M}^{-1}\left\{\left(\frac{2\pi}{a}\right)^s\frac{\zeta(1-s)\Gamma(s-2)}{\Gamma(s)}\right\}\Bigg|_{x=1},\end{aligned} \tag{7.186}$$

the last step coming from the zeta function identity

$$\pi^s\zeta(1-s) = 2^{1-s}\Gamma(s)\zeta(s)\cos\left(\frac{1}{2}\pi s\right).$$

Useful values of the zeta function based on this last identity are

$$\zeta(0) = -\frac{1}{2}, \quad \zeta(1) = \infty, \quad \zeta(2) = \frac{\pi^2}{6}, \quad \zeta(-1) = -\frac{1}{12},$$
$$\zeta(-2n) = 0,\ n = 1, 2, 3, \ldots.$$

Next, from application of the recurrence formula ($G3$) in Appendix A, the inverse transform relation becomes

$$\sum_{n=1}^{\infty} \frac{\cos(an)}{n^2} = -\frac{a^2}{2} \mathcal{M}^{-1} \left\{ \left(\frac{2\pi}{a} \right)^s \frac{\zeta(1-s)}{(s-1)(s-2)} \right\} \bigg|_{x=1}. \tag{7.187}$$

The zeta function has a simple pole at $s = 0$, and there are additional poles coming from the terms in the denominator at $s = 1, 2$. Hence, by calculating the residues at each pole, we obtain

$$\mathrm{Res}(0) = \zeta(0) = -\frac{1}{2},$$
$$\mathrm{Res}(1) = -\frac{2\pi}{a}\zeta(0) = \frac{\pi}{a},$$
$$\mathrm{Res}(2) = \frac{4\pi^2}{a^2}\zeta(-1) = -\frac{\pi^2}{3a^2},$$

and finally, the sum of the residues gives us

$$\begin{aligned} \sum_{n=1}^{\infty} \frac{\cos(an)}{n^2} &= -\frac{a^2}{2}[\mathrm{Res}(0) + \mathrm{Res}(1) + \mathrm{Res}(2)] \\ &= \frac{a^2}{4} - \frac{\pi a}{2} + \frac{\pi^2}{6}. \end{aligned} \tag{7.188}$$

7.7.4 Optical wave propagation

In Chapter 8 we will introduce a variety of statistical relations associated with the lowest-order Gaussian-beam wave at the exit aperture of a laser transmitter, defined by

$$U(r, 0) = A_0 \exp\left(-\frac{1}{2} \alpha_0 k r^2 \right). \tag{7.189}$$

Here, A_0 is maximum amplitude, $k = 2\pi/\lambda$ is optical wave number corresponding to wavelength λ, and α_0 is a complex parameter describing the beam radius W_0 and phase front radius of curvature F_0 according to

$$\alpha_0 = \frac{2}{kW_0^2} + i\frac{1}{F_0}. \tag{7.190}$$

Finding the optical field $U(r, z) = e^{ikz} V(r, z)$ in free space at distance z from the transmitter can be done by solving either the *Huygens–Fresnel integral* (see Section 8.2.1) or the *paraxial wave equation* for $V(r, z,)$ (p. 86 of Ref. 13),

$$\frac{1}{r}\frac{\partial}{\partial r}\left(r \frac{\partial V}{\partial r} \right) + 2ik\frac{\partial V}{\partial z} = 0. \tag{7.191}$$

Example 48: Use the Hankel transform to solve the boundary value problem involving the paraxial wave equation

$$\frac{1}{r}\frac{\partial}{\partial r}\left(r\frac{\partial V}{\partial r}\right) + 2ik\frac{\partial V}{\partial z} = 0, \quad V(r, 0) = A_0 \exp\left(-\frac{1}{2}\alpha_0 kr^2\right). \tag{7.192}$$

By introducing the zero-order Hankel transform relations (pp. 516–517 of Ref. 14)

$$\mathcal{H}_0\{V(r, z)\} = \hat{V}(\rho, z), \quad \mathcal{H}_0\left\{\frac{\partial V}{\partial z}\right\} = \frac{\partial \hat{V}}{\partial z},$$
$$\mathcal{H}_0\left\{\frac{1}{r}\frac{\partial}{\partial r}\left(r\frac{\partial V}{\partial r}\right)\right\} = -\rho^2\hat{V}(\rho, z),$$

the transformed problem becomes

$$\frac{\partial \hat{V}}{\partial z} = \frac{\rho^2}{2ik}\hat{V}(\rho, z), \quad \hat{V}(\rho, 0) = \frac{A_0}{k\alpha_0}\exp\left(-\frac{\rho^2}{2k\alpha_0}\right). \tag{7.193}$$

The transformed boundary condition in Eq. (7.193) comes from the Hankel transform relation in Example 39. The problem described by Eq. (7.193) represents a first-order linear DE with a boundary condition that can be solved through separation of variables. Doing so yields

$$\begin{aligned}\hat{V}(\rho, z) &= \frac{A_0}{k\alpha_0}\exp\left(-\frac{\rho^2}{2k\alpha_0}\right)\exp\left(-\frac{iz\rho^2}{2k}\right)\\ &= \frac{A_0}{k\alpha_0}\exp\left[-\frac{1}{2k\alpha_0}(1 + i\alpha_0 z)\rho^2\right],\end{aligned} \tag{7.194}$$

where we have applied the boundary condition. Relying on Example 39 again for the inverse Hankel transform this time, we obtain the solution

$$V(r, z) = \frac{A_0}{1 + i\alpha_0 z}\exp\left[-\frac{k\alpha_0 r^2}{2(1 + i\alpha_0 z)}\right]. \tag{7.195}$$

Finally, multiplying the solution (7.195) by e^{ikz}, we produce the desired Gaussian-beam wave solution $U(r, z) = e^{ikz}V(r, z)$ at position z along the propagation path.

7.7.5 Steady-state heat conduction in a cylinder

For another example featuring the Hankel transform, let us consider finding the steady-state temperature $u(r, z)$ in a large solid cylinder made of homogeneous material. We assume that the cylinder occupies the half-space $z \geq 0$.

Example 49: The problem we wish to solve is characterized by

$$\begin{aligned} \frac{\partial^2 u}{\partial r^2} + \frac{1}{r}\frac{\partial u}{\partial r} + \frac{\partial^2 u}{\partial z^2} &= 0, \quad 0 < r < \infty,\ z > 0 \\ u(r,0) &= f(r), \quad 0 < r < \infty \\ u(r,z) &\to 0 \text{ as } \sqrt{r^2+z^2} \to \infty. \end{aligned} \tag{7.196}$$

The zero-order Hankel transform of the above problem leads to

$$\frac{\partial^2 U}{\partial z^2} - \rho^2 U = 0, \quad U(p,0) = F(\rho). \tag{7.197}$$

In obtaining the transformed problem, we are using the properties

$$\begin{aligned} \mathcal{H}_0\{u(r,z)\} &= U(\rho,z), \quad \mathcal{H}_0\{f(r)\} = F(\rho), \\ \mathcal{H}_0\left\{\frac{\partial^2 u}{\partial z^2}\right\} &= \frac{\partial^2 U}{\partial z^2}, \end{aligned}$$

and relying on a transform relation given in Example 48 by recognizing that

$$\mathcal{H}_0\left\{\frac{\partial^2 u}{\partial r^2} + \frac{1}{r}\frac{\partial u}{\partial r}\right\} = \mathcal{H}_0\left\{\frac{1}{r}\frac{\partial}{\partial r}\left(r\frac{\partial u}{\partial r}\right)\right\} = \rho^2 U(\rho,z).$$

The solution of the transformed problem (7.197) is elementary and leads to

$$U(\rho,z) = F(\rho)e^{-\rho z}. \tag{7.198}$$

Therefore, upon taking the inverse Hankel transform, we arrive at the solution formula

$$\begin{aligned} u(r,z) &= \mathcal{H}_0^{-1}\{F(\rho)e^{-\rho z}\} \\ &= \int_0^\infty \rho F(\rho)\, e^{-\rho z} J_0(\rho r)\, d\rho. \end{aligned} \tag{7.199}$$

Suppose we solve the problem described by Eq. (7.196) for the special case where the boundary condition at $z = 0$ is given by

$$f(r) = \frac{1}{(r^2+a^2)^{3/2}}, \quad a > 0. \tag{7.200}$$

By using (*C*9) in Appendix C, we first find the Hankel transform of the function in Eq. (7.200),

$$
\begin{aligned}
F(\rho) &= \int_0^\infty r f(r)\, J_0(\rho r)\, dr \\
&= \int_0^\infty \frac{r}{(r^2+a^2)^{3/2}} G_{02}^{10}\left(\frac{\rho^2 r^2}{4}\middle|\, 0,\, 0\right) dr \\
&= \frac{1}{a\sqrt{\pi}} G_{13}^{21}\left(\frac{a^2\rho^2}{4}\middle|\, \begin{matrix} 0 \\ \frac{1}{2},\, 0,\, 0 \end{matrix}\right) \\
&= \frac{1}{a\sqrt{\pi}} G_{02}^{20}\left(\frac{a^2\rho^2}{4}\middle|\, \frac{1}{2},\, 0\right).
\end{aligned}
\tag{7.201}
$$

To arrive at this result, we have used (*D*11) in Appendix D and (*MG*3) in Appendix A. Now from (*C*23) in Appendix C, we obtain

$$
F(\rho) = \frac{2}{\sqrt{\pi}}\sqrt{\frac{a\rho}{2}} K_{1/2}(a\rho) = \frac{1}{a} e^{-a\rho}. \tag{7.202}
$$

The last step in Eq. (7.202) follows by using the identity (*BK*9) in Appendix A. Therefore, returning to Eq. (7.199), we write the Bessel function in a power series and integrate termwise. Consequently,

$$
\begin{aligned}
u(r, z) &= \frac{1}{a}\int_0^\infty \rho\, e^{-(z+a)\rho} J_0(\rho r)\, d\rho \\
&= \frac{1}{a}\sum_{n=0}^\infty \frac{(-1)^n (r/2)^{2n}}{(n!)^2}\int_0^\infty \rho^{2n+1}\, e^{-(z+a)\rho}\, d\rho \\
&= \frac{1}{a(z+a)^2}\sum_{n=0}^\infty \frac{(-1)^n \Gamma(2+2n)}{(n!)^2}\left(\frac{1}{2}\right)^{2n}\left(\frac{r}{z+a}\right)^{2n} \\
&= \frac{1}{a(z+a)^2}\sum_{n=0}^\infty \frac{(-1)^n \left(\frac{3}{2}\right)_n}{n!}\left(\frac{r}{z+a}\right)^{2n}.
\end{aligned}
\tag{7.203}
$$

Upon summing the series as a generalized hypergeometric series and simplifying with (*B*8) from Appendix B, we get the final result

$$
\begin{aligned}
u(r, z) &= \frac{1}{a(z+a)^2}\, {}_1F_0\left[\frac{3}{2};\, -;\, -\left(\frac{r}{z+a}\right)^2\right] \\
&= \frac{z+a}{a[(z+a)^2 + r^2]^{3/2}}.
\end{aligned}
\tag{7.204}
$$

7.7.6 Fourier transform

Although we didn't give separate treatment of the Fourier transform in this chapter, like the Laplace transform, it can be derived from the Fourier integral theorem, written as

$$f(t) = \frac{1}{2\pi}\int_{-\infty}^{\infty} e^{-ist}\int_{-\infty}^{\infty} f(x)e^{isx}dx\,ds. \tag{7.205}$$

Thus, from this relation we define the *Fourier transform* by

$$\mathcal{F}\{f(x)\} = \int_{-\infty}^{\infty} e^{isx}f(x)\,dx = F(s) \tag{7.206}$$

and the *inverse Fourier transform* by

$$\mathcal{F}^{-1}\{F(s)\} = \frac{1}{2\pi}\int_{-\infty}^{\infty} e^{-isx}F(s)\,ds = f(x). \tag{7.207}$$

In some cases the constant $1/2\pi$ appears in the transform integral but not in the inverse, and the exponential functions e^{isx} and e^{-isx} may be switched.

Among other applications of the Fourier transform, we can use it to calculate the characteristic function of a random variable $\mathbf{x}$ through the transform relation

$$C_{\mathbf{x}}(s) = \int_{-\infty}^{\infty} e^{isx}p_{\mathbf{x}}(x)\,dx. \tag{7.208}$$

Example 50: Find the PDF for $\mathbf{z} = \mathbf{y}/\mathbf{x}$ given that $\mathbf{x}$ and $\mathbf{y}$ are statistically independent Gaussian random variables defined by the joint PDF

$$p_{\mathbf{xy}}(x, y) = \frac{1}{2\pi\sigma^2}e^{-(x^2+y^2)/2\sigma^2}, \quad -\infty < x < \infty, \ -\infty < y < \infty.$$

Thus, the characteristic function for $\mathbf{z}$ can be derived from the two-dimensional Fourier transform

$$\begin{aligned} C_{\mathbf{z}}(s) &= \int_{-\infty}^{\infty}\int_{-\infty}^{\infty} e^{isy/x}p_{\mathbf{xy}}(x, y)\,dy\,dx \\ &= \frac{1}{2\pi\sigma^2}\int_{-\infty}^{\infty} e^{-x^2/2\sigma^2}\int_{-\infty}^{\infty} e^{isy/x-y^2/2\sigma^2}dy\,dx. \end{aligned} \tag{7.209}$$

The inside integral in Eq. (7.209) leads to

$$\begin{aligned} \int_{-\infty}^{\infty} e^{isy/x-y^2/2\sigma^2}dy &= 2\int_0^{\infty}\cos\left(\frac{sy}{x}\right)e^{-y^2/2\sigma^2}dy \\ &= \sqrt{\pi}\int_0^{\infty} t^{-1/2}e^{-t/2\sigma^2}G_{02}^{10}\left(\frac{ts^2}{4x^2}\middle|\, 0, \frac{1}{2}\right)dt \\ &= \sigma\sqrt{2\pi}\,G_{12}^{11}\left(\frac{\sigma^2 s^2}{2x^2}\middle|\begin{matrix}\frac{1}{2}\\ 0, \frac{1}{2}\end{matrix}\right), \end{aligned} \tag{7.210}$$

where we have used ($C4$) in Appendix C, set $t = y^2$, and used ($D6$) in Appendix D. By employing ($MG3$) in Appendix A followed by ($C1$) in Appendix C, we obtain

$$\int_{-\infty}^{\infty} e^{isy/x - y^2/2\sigma^2} dy = \sigma\sqrt{2\pi}\, e^{-\sigma^2 s^2/2x^2}. \tag{7.211}$$

Returning to Eq. (7.209), we have

$$\begin{aligned} C_{\mathbf{z}}(s) &= \frac{1}{\sigma\sqrt{2\pi}} \int_{-\infty}^{\infty} e^{-x^2/2\sigma^2 - \sigma^2 s^2/2x^2} dx \\ &= \frac{1}{\sigma}\sqrt{\frac{2}{\pi}} \int_{0}^{\infty} e^{-x^2/2\sigma^2 - \sigma^2 s^2/2x^2} dx \\ &= e^{-|s|}, \end{aligned} \tag{7.212}$$

where we are recalling Example 1.

If we now invert the characteristic function, we arrive at

$$p_{\mathbf{z}}(z) = \frac{1}{2\pi} \int_{-\infty}^{\infty} e^{-izs} e^{-|s|}\, ds = \frac{1}{\sqrt{\pi}} \int_{0}^{\infty} e^{-s} G_{02}^{10}\left(\frac{z^2 s^2}{4} \middle| 0, \frac{1}{2} \right) ds, \tag{7.213}$$

from which it follows, using ($D7$) in Appendix D and ($MG3$) in Appendix A, that

$$p_{\mathbf{z}}(z) = \frac{1}{\pi} G_{22}^{12}\left(z^2 \middle| \begin{matrix} 0, \frac{1}{2} \\ 0, \frac{1}{2} \end{matrix} \right) = \frac{1}{\pi} G_{11}^{11}\left(z^2 \middle| \begin{matrix} 0 \\ 0 \end{matrix} \right). \tag{7.214}$$

Finally, by recognizing ($C11$) in Appendix C, we obtain the *Cauchy distribution*,

$$p_{\mathbf{z}}(z) = \frac{1}{\pi(1+z^2)}. \tag{7.215}$$

In imaging applications, it is customary to call upon the two-dimensional Fourier transform (cf. Section 9.1). For example, the point spread function (PSF) associated with an incident spherical wave in free space is defined by the intensity distribution in the focal plane of an incoherent imaging system. The normalized two-dimensional Fourier transform of the PSF leads to the optical transfer function (OTF). In most of the literature, the development of the OTF is based on computing the area common to two overlapping circles (e.g., see pp. 171–173 of Ref. 15), but here we will directly use the two-dimensional Fourier transform.

Example 51: Given that the PSF of an incoherent imaging system in free space is defined by the *Airy disk*,

$$\mathrm{PSF}(\mathbf{r}) = \left[\frac{2J_1(\pi Dr/\lambda F_{Rx})}{\pi Dr/\lambda F_{Rx}}\right]^2,$$

find the OTF by directly calculating the normalized two-dimensional Fourier transform of the PSF.

Starting with the Fourier transform of the PSF in polar coordinates, we obtain

$$\begin{aligned} I(\nu) &= \iint_{-\infty}^{\infty} e^{-2\pi i \mathbf{r}\cdot\boldsymbol{\nu}} \left[\frac{2J_1(\pi Dr/\lambda F_{Rx})}{\pi Dr/\lambda F_{Rx}}\right]^2 d^2 r \\ &= 8\pi \int_0^{\infty} rJ_0(2\pi\nu r)\left[\frac{J_1(\alpha r)}{\alpha r}\right]^2 dr, \end{aligned} \tag{7.216}$$

where ν is magnitude of the two-dimensional spatial frequency. In obtaining the second step in Eq. (7.216), we have integrated over the angle variable in $\mathbf{r}\cdot\boldsymbol{\nu} = r\nu\cos\phi$ and introduced the notation

$$\alpha = \frac{\pi D}{\lambda F_{Rx}}. \tag{7.217}$$

By recalling (C16) in Appendix C and employing the integral relation (D12) in Appendix D, we obtain

$$\begin{aligned} I(\nu) &= \frac{8\pi}{\alpha^2\sqrt{\pi}} \int_0^{\infty} r^{-1} J_0(2\pi\nu r)\, G_{13}^{11}\left(\alpha^2 r^2 \middle|\, \begin{matrix} \frac{1}{2} \\ 1,\ 0,\ -1 \end{matrix}\right) dr \\ &= \frac{4\sqrt{\pi}}{\pi^2\beta^2}\, G_{33}^{12}\left(\frac{\beta^2}{\nu^2}\middle|\, \begin{matrix} 1,\ \frac{1}{2},\ 1 \\ 1,\ 0,\ -1 \end{matrix}\right) \\ &= \frac{4\sqrt{\pi}}{\pi^2\beta^2}\, G_{33}^{21}\left(\frac{\nu^2}{\beta^2}\middle|\, \begin{matrix} 0,\ 1,\ 2 \\ 0,\ \frac{1}{2},\ 0 \end{matrix}\right). \end{aligned} \tag{7.218}$$

We have used (MG2) from Appendix A in the last step and introduced

$$\beta^2 = \frac{\alpha^2}{\pi^2} = \left(\frac{D}{\lambda F_{Rx}}\right)^2. \tag{7.219}$$

Now we apply (MG3) followed by (MG1) in Appendix A to get

$$\begin{aligned} G_{33}^{21}\left(\frac{\nu^2}{\beta^2}\middle|\, \begin{matrix} 0,\ 1,\ 2 \\ 0,\ \frac{1}{2},\ 0 \end{matrix}\right) &= G_{22}^{20}\left(\frac{\nu^2}{\beta^2}\middle|\, \begin{matrix} 1,\ 2 \\ 0,\ \frac{1}{2} \end{matrix}\right) \\ &= \sqrt{\pi} - \frac{4}{\sqrt{\pi}}\left(\frac{\nu}{\beta}\right) {}_2F_1\left(\frac{1}{2}, -\frac{1}{2}; \frac{3}{2}; \frac{\nu^2}{\beta^2}\right) \\ &= \frac{2}{\sqrt{\pi}}\left[\frac{\pi}{2} - 2\left(\frac{\nu}{\beta}\right) {}_2F_1\left(\frac{1}{2}, -\frac{1}{2}; \frac{3}{2}; \frac{\nu^2}{\beta^2}\right)\right]. \end{aligned} \tag{7.220}$$

At this point we make use of the identity ($HY14$) in Appendix A,

$$(a-b)\,{}_2F_1(a,b;c;z) = a\,{}_2F_1(a+1,b;c;z) - b\,{}_2F_1(a,b+1;c;z).$$

Thus, the hypergeometric function in the last line of Eq. (7.220) becomes

$$\begin{aligned}2\left(\frac{\nu}{\beta}\right){}_2F_1\left(\frac{1}{2},-\frac{1}{2};\frac{3}{2};\frac{\nu^2}{\beta^2}\right) &= \left(\frac{\nu}{\beta}\right){}_2F_1\left(\frac{3}{2},-\frac{1}{2};\frac{3}{2};\frac{\nu^2}{\beta^2}\right)\\ &\quad+\left(\frac{\nu}{\beta}\right){}_2F_1\left(\frac{1}{2},\frac{1}{2};\frac{3}{2};\frac{\nu^2}{\beta^2}\right)\\ &= \left(\frac{\nu}{\beta}\right){}_1F_0\left(-\frac{1}{2};-;\frac{\nu^2}{\beta^2}\right)+\sin^{-1}\left(\frac{\nu}{\beta}\right)\\ &= \frac{\nu}{\beta}\sqrt{1-\frac{\nu^2}{\beta^2}}+\frac{\pi}{2}-\cos^{-1}\left(\frac{\nu}{\beta}\right),\end{aligned} \tag{7.221}$$

where we have used both ($B33$) and ($B8$) in Appendix B. Combining this last result with Eqs. (7.218) and (7.220) leads to

$$I(\nu) = \frac{8}{\alpha^2}\left\{\cos^{-1}\left(\frac{\lambda F_{Rx}\nu}{D}\right)-\left(\frac{\lambda F_{Rx}\nu}{D}\right)\left[1-\left(\frac{\lambda F_{Rx}\nu}{D}\right)^2\right]^{1/2}\right\}, \quad \nu \le \frac{D}{\lambda F_{Rx}}, \tag{7.222}$$

and $I(\nu) = 0,\ \nu > D/\lambda F_{Rx}$.

Based on Eq. (7.222), it follows that

$$I(0) = \frac{4\pi}{\alpha^2}, \tag{7.223}$$

from which we deduce the normalized OTF given by

$$\begin{aligned}\mathrm{OTF}(\nu) &= \frac{I(\nu)}{I(0)}\\ &= \frac{2}{\pi}\left\{\cos^{-1}\left(\frac{\lambda F_{Rx}\nu}{D}\right)-\left(\frac{\lambda F_{Rx}\nu}{D}\right)\left[1-\left(\frac{\lambda F_{Rx}\nu}{D}\right)^2\right]^{1/2}\right\}, \quad \nu \le \frac{D}{\lambda F_{Rx}}.\end{aligned} \tag{7.224}$$

7.7.7 Inverse transforms

We previously used the Meijer G-function only in calculating integral transforms, but the Meijer G-function can also be used in evaluating some inverse transforms by use of ($D16$) through ($D18$) in Appendix D. Integrals marked as ($D17$) and ($D18$) can be interpreted as either cosine and sine transforms, or inverse cosine and sine transforms. That is, like the Hankel transform, the cosine and sine integral transforms and their inverses look the same. However, these integrals as transforms or inverse transforms are normally multiplied by $\sqrt{2/\pi}$.

Example 52: Find the inverse Laplace transform of $F(s) = \frac{a}{s\sqrt{s+a^2}}$, where $a > 0$.

Let us start with

$$\mathscr{L}^{-1}\left\{\frac{a}{s\sqrt{s+a^2}},\right\} = \frac{1}{2\pi i}\int_{c-i\infty}^{c+i\infty} e^{st}s^{-1}(1+s/a^2)^{-1/2}ds. \tag{7.225}$$

By the use of (C11) in Appendix C followed by (MG2) in Appendix A and (D16) in Appendix D, we have

$$\begin{aligned}\frac{1}{2\pi i}\int_{c-i\infty}^{c+i\infty} e^{st}s^{-1}(1+s/a^2)^{-1/2}ds &= \frac{1}{\sqrt{\pi}}\frac{1}{2\pi i}\int_{c-i\infty}^{c+i\infty} e^{st}s^{-1}G_{11}^{11}\left(\frac{a^2}{s}\middle|\begin{matrix}1\\ \frac{1}{2}\end{matrix}\right)ds \\ &= \frac{1}{\sqrt{\pi}}G_{12}^{11}\left(a^2t\middle|\begin{matrix}1\\ \frac{1}{2},0\end{matrix}\right) \\ &= 2a\sqrt{\frac{t}{\pi}}\,{}_1F_1\left(\frac{1}{2};\,\frac{3}{2};\,-a^2t\right).\end{aligned} \tag{7.226}$$

To obtain the last step, we referred to (C14) in Appendix C. Finally, from (B10) in Appendix B, we obtain (also recall Example 21)

$$\mathscr{L}^{-1}\left\{\frac{a}{s\sqrt{s+a^2}},\right\} = \operatorname{erf}(a\sqrt{t}), \quad t > 0. \tag{7.227}$$

Example 53: Find the inverse cosine transform of $F(s) = \frac{1}{\sqrt{1+s^2}}$.

Based on (D17) in Appendix D along with (C11) in Appendix C, we have

$$\begin{aligned}\int_0^\infty \frac{\cos(st)}{\sqrt{1+s^2}}ds &= \frac{1}{\sqrt{\pi}}\int_0^\infty \cos(st)G_{11}^{11}\left(s^2\middle|\begin{matrix}\frac{1}{2}\\ 0\end{matrix}\right)ds \\ &= t^{-1}G_{31}^{12}\left(\frac{4}{t^2}\middle|\begin{matrix}\frac{1}{2},\frac{1}{2},0\\ 0\end{matrix}\right) \\ &= t^{-1}G_{13}^{21}\left(\frac{t^2}{4}\middle|\begin{matrix}1\\ \frac{1}{2},\frac{1}{2},1\end{matrix}\right),\end{aligned} \tag{7.228}$$

where the last step is based on (MG2) in Appendix A. If we now apply (MG3) in Appendix A followed by (C23) in Appendix C, we arrive at

$$\int_0^\infty \frac{\cos(st)}{\sqrt{1+s^2}}\, ds = t^{-1} G_{02}^{20}\left(\frac{t^2}{4}\,\middle|\, \frac{1}{2}, \frac{1}{2}\right) \\ = K_0(t), \quad t > 0, \tag{7.229}$$

where $K_0(t)$ is the modified Bessel function of the second kind. To obtain the inverse cosine transform, we multiply both sides of the last result by $\sqrt{2/\pi}$.

7.8 Discussion

Throughout much of this chapter, we have illustrated how to use hypergeometric-type functions to solve an assortment of improper integrals. The properties of these functions listed in the first four Appendices at the end of the text are sufficiently detailed to assist in the evaluation of integrals that commonly arise in a variety of applications. The basic idea employed in most examples is to replace one of the functions under the integral with a Meijer *G*-function found in Appendix C in order to match one of the integral forms given in Appendix D. Functions under the integral that do not have a Meijer *G*-function listed in Appendix C may have another representation in terms of a ${}_pF_q$ function found in Appendix B. This function can then be expressed as a Meijer *G*-function. Once the integral has been evaluated by this method, we usually replace the new Meijer *G*-function with a more familiar function (when possible). Finally, we should point out that the particular choice for *G*-function we made in most integrals is not necessarily the only choice or best choice. In this regard, we have offered an alternative approach in some of the examples.

7.9 Historical Remarks

Numerous integral transforms have been developed over the years, many of which are highly specialized. The one best known to engineers and physicists is perhaps the Fourier transform. Joseph Fourier (1768–1830) was one of several famous French mathematicians who flourished during the time of Napoleon. The theory widely known as Fourier series is credited to Fourier, who came across such representations in his classic studies of heat conduction. His basic papers, presented to the Academy of Sciences in Paris in 1807 and 1811, were criticized by the referees (most strongly by Lagrange and Laplace) for a lack of rigor and consequently were not then published. Fourier was called to Paris by the Academy of Sciences in 1816, whereupon he subsequently became *secrétaire perpétuel* of the Ecole Polytechnique. When publishing his classic *Théorie Analytique de la Chaleur* in 1822, he also incorporated his earlier work that was previously rejected, almost without change.

The Laplace transform can be traced back to the work of Oliver Heaviside (1850–1925), who used operational calculus to solve circuit problems with discontinuous input functions. However, the term Laplace transform dates back to the French mathematician Pierre-Simon Laplace (1749–1827), who made use of the transform integral in his work on probability theory in the 1780s. The French mathematician Siméon Denis Poisson (1781–1840) also knew of the Laplace transform integral in the 1820s, and the integral appeared in Fourier's famous 1811 paper on heat conduction. However, it was Heaviside who generated many of its properties that are used today. Heaviside was an English electrical engineer who set himself apart from the established scientists by studying mostly alone. One of his greatest achievements was in reducing Maxwell's equations from 20 equations with 20 unknowns to just 4 vector equations with two unknowns. When his fellow engineers and mathematicians were perplexed by his lack of rigor and new mathematical techniques involving operational calculus, he alienated himself from them by calling them names such as "woodenheaded." Consequently, mostly ignored by his colleagues, much of his work did not receive the credit he deserved.

The Mellin transform is attributed to Hjalmar Mellin (1854–1933), who introduced it in a paper in 1897 in *Acta Societatis Scientiarum Fennicae*. It is closely related to the Laplace transform and the Fourier transform, although it is most often used in number theory, in mathematical statistics, and in the theory of asymptotic expansions. In physics and engineering, the most famous application is the computation of the solution of a potential problem in a wedge-shaped region where the unknown function (e.g., temperature or electrostatic potential) satisfies Laplace's equation with given boundary conditions on the edges. It is also used in electromagnetic wave propagation through turbulence.[8]

The Hankel transform is known as the Fourier–Bessel transform in certain settings. The Hankel transform was first developed by the German mathematician Hermann Hankel (1839–1873), who was a student of Riemann (who himself was a student of Gauss). Hankel also worked on the theory of complex numbers, the theory of functions, and the history of mathematics.

References

1. I. N. Sneddon, *The Use of Integral Transforms*, McGraw Hill, New York (1972).
2. L. C. Andrews and B. K. Shivamoggi, *Integral Transforms for Engineers*, SPIE Press, Bellingham, Washington (1999) [doi: 10.1117/3.339204].
3. A. Erdelyi, W. Magnus, F. Oberhettinger, and F. G. Tricomi, *Higher Transcendental Functions*, 3 Vols., Bateman Manuscript Project, McGraw Hill, New York (1953), (1955).

A. Erdelyi, W. Magnus, F. Oberhettinger, and F. G. Tricomi, *Tables of Integral Transforms*, 2 Vols., Bateman Manuscript Project, McGraw Hill, New York (1954).

4. E. D. Rainville, *Special Functions*, Academic Press, New York (1960).
5. I. S. Gradshteyn and I. M. Ryzhik, *Table of Integrals, Series, and Products*, Academic Press, New York (1980).
6. L. C. Andrews, *Special Functions of Mathematics for Engineers*, 2nd Edition, SPIE Press, Bellingham, Washington (1998) [doi: 10.1117/3.270709].
7. G. A. Tyler, "Analysis of propagation through turbulence: evaluation of an integral involving the product of three Bessel functions," *J. Opt. Soc. Am. A* **7**(7), 1218–1223 (1990).
8. R. J. Sasiela, *Electromagnetic Wave Propagation in Turbulence: Evaluation and Application of Mellin Transforms*, 2nd Edition, SPIE Press, Bellingham, Washington (2007) [doi: 10.1117/3.741372].
9. P. H. Hu, J. Stone, and J. S. Stanley, "Applications of Zernike polynomials to atmospheric propagation problems," *J. Opt. Soc. Am. A* **6**(10), 1595–1608 (1989).
10. I. N. Sneddon, *Fourier Transforms*, McGraw Hill, New York (1951).
11. M. A. Al-Habash, L. C. Andrews, and R. L. Phillips, "Mathematical model for the irradiance PDF of a laser beam propagating through turbulent media," *Opt. Eng.* **40**(8), 1554–1562 (2001) [doi: 10.1117/1.1386641].
12. G. G. MacFarlane, "The application of Mellin transforms to the summation of slowly convergent series," *The London, Edinburgh, and Dublin Philosophical Magazine and Journal of Science* **40**(301), 188–197 (1949).
13. L. C. Andrews and R. L. Phillips, *Laser Beam Propagation through Random Media*, 2nd Edition, SPIE Press, Bellingham, Washington (2005) [doi: 10.1117/3.626196].
14. L. C. Andrews and R. L. Phillips, *Mathematical Techniques for Engineers and Scientists*, SPIE Press, Bellingham, Washington (2003) [doi: 10.1117/3.467443].
15. L. B. Stotts, *Free Space Optical Systems Engineering: Design and Analysis*, John Wiley & Sons, Hoboken, New Jersey (2017).

Chapter 8
Electromagnetic Wave Propagation

8.0 Introduction

Electromagnetic waves are different from mechanical waves in that they do not require a medium to propagate. Types of electromagnetic (EM) waves include radio waves, microwaves, millimeter, infrared, visible, and ultraviolet rays, X-rays, and gamma rays, all of which are part of the electromagnetic spectrum illustrated in Fig. 8.1. In this figure the position of an EM wave in the spectrum can be characterized by either its frequency of oscillation or its wavelength. Common units for the frequency bands range from kilohertz (kHz) to megahertz (MHz) to gigahertz (GHz) to terahertz (THz). The

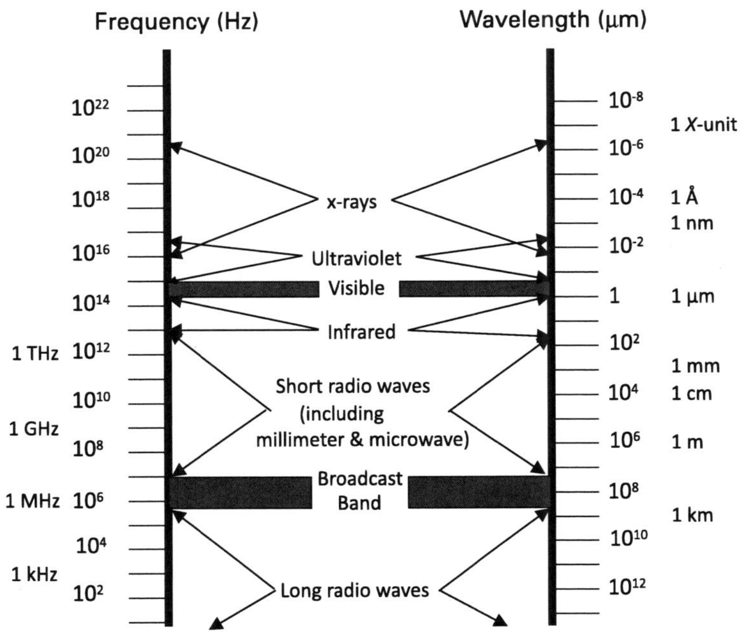

Figure 8.1 Electromagnet spectrum (adapted from Ref. 1).

standard units of measurement for the various wavelengths include the kilometer (km), meter (m), centimeter (cm), millimeter (mm), micrometer (μm), nanometer (nm), angstrom (Å), and X-unit (XU).

Radiofrequency (RF) waves have wavelengths that typically extend from roughly 20 km down to 1–2 mm—microwaves exist between 1 m and 10 mm. At wavelengths shorter than 10 mm lie the millimeter and infrared (IR) bands. The visible spectrum exists from 0.4 to 0.7 μm, beyond which are the ultraviolet (UV) bands from roughly 10 to 400 nm. Between 0.01 and 10 nm are the X-rays, and gamma rays occur at even shorter wavelengths measured in X-units.

Optical/IR wave propagation through a random medium is a subset of general electromagnetic wave propagation. By *random medium*, we mean one whose basic properties are random functions of space and time—common examples include the atmosphere and ocean. In general, an optical wave propagating through the atmosphere experiences distortions caused by small temperature variations related to the sun's heating of the Earth and the turbulent motion of air due to wind and convection. The first theoretical treatments of optical wave propagation through the atmosphere involved the mathematical analysis of starlight, which acts like an infinite *plane wave* when received by a ground-based telescope. In addition to the infinite plane wave model, some of the initial investigations included a small-diameter wave model that acts like a *spherical wave* or *point source*. With the demonstration of the first laser in 1960, theoretical investigations of optical wave propagation went beyond the interest of astronomers as scientists began to examine the basic characteristics associated with a finite-diameter laser beam. This caused much interest in new application areas like *free-space optical communications* (FSOC), *remote sensing*, *laser ranging*, *target designation*, and *laser radar*, among others. In such applications it was recognized that the simple plane wave and spherical wave models that had been previously studied were not adequate to describe the focusing and finite beam size commonly associated with a laser source. To model a propagating laser beam, researchers began to study the lowest-order transverse electromagnetic (TEM_{00}) *Gaussian-beam wave* as a reasonable alternative to plane waves and spherical waves. Moreover, with a fundamental Gaussian-beam wave model, infinite plane waves and spherical waves can always be treated as limiting cases.

Fairly complete mathematical models of laser beam propagation through atmospheric turbulence have now been developed for second-order and fourth-order field moments of a Gaussian-beam wave.[1,2] Nonetheless, many of the mathematical models that have previously been established are missing critical details in their derivations. The presentations often consist of displaying the defining integrals followed by the resulting model itself. To fill the void, we use the mathematical techniques developed in earlier chapters in this text to provide some of those critical details along with the development of several new mathematical models. The mathematical techniques we use here are not always the original methods that researchers used. However, these innovative techniques may be superior to other methods and therefore may be helpful to future researchers.

Because many of the mathematical models associated with optical/IR wave propagation often involve one or more special functions of the generalized hypergeometric type, they can be intimidating for engineers and scientists who are not well-versed in the properties of these functions. Fortunately, several software packages are available today to help with numerical calculations. Moreover, these models can often be simplified by use of the approximation techniques illustrated in Chapter 6, producing more tractable solution forms in terms of algebraic functions. In many cases examined here we present exact results first followed by simple algebraic approximations.

8.1 Atmospheric Turbulence

Classical studies of turbulence were concerned mostly with velocity fluctuations of a viscous fluid. Experience has dictated that a viscous fluid has two distinct states of motion—laminar and turbulent. Although the flow characteristics in a *laminar flow* are uniform, the velocity field in a *turbulent flow* loses its uniform characteristics due to dynamic mixing and acquires random sub-flows called turbulent eddies. The theory of turbulence as developed by Kolmogorov in 1941 for velocity fluctuations was presented in terms of a set of hypotheses based heavily on physical insight. These basic ideas were later applied to conservative passive scalars such as temperature and index of refraction—both of which are considered passive because they do not exchange energy with velocity turbulence.

Atmospheric turbulence is the random fluctuations in the fluid motion of air superimposed on its aggregate flow. The random flow fluctuations develop swirls or "eddies" generated by wind shear, convection, and surface resistance. The eddy swirls break up into smaller eddies due to inertial forces in what is known as an energy *cascade process*. This creates a continuum of eddy sizes over a large volume of space from the macro-effect size called the *outer scale* L_0 to the micro-effect size called the *inner scale* l_0, after which viscous forces result in *dissipation* of the energy into heat. The range of eddy sizes between l_0 and L_0 is commonly called the *inertial range* of turbulence. Up to a few meters above ground, the inner scale is on the order of millimeters and the outer scale is on the order of meters, but both l_0 and L_0 tend to increase with height above ground.

The index of refraction is the most significant parameter for optical wave propagation studies. Random fluctuations in the refractive index of the atmosphere, called *optical turbulence*, are directly associated with temperature, humidity, and pressure fluctuations. Although humidity and pressure fluctuations can usually be neglected for a transmitted optical wave through clear-air turbulence over land, humidity and temperature cross-correlation can still have an impact under certain conditions. All of the analysis in this chapter is based on a horizontal path for the optical wave. In such cases, the refractive-index structure parameter C_n^2 [$\mathrm{m}^{-2/3}$], which is a measure of the strength of the atmospheric turbulence, is taken to be a constant. Variations in C_n^2 along the path due to elevation changes will be addressed in Chapter 9.

When developing theoretical models of atmospheric turbulence effects on an optical wave, we generally partition the analysis into regimes called *weak irradiance fluctuations* and *strong irradiance fluctuations*. The usual way these regimes are identified is through the *Rytov variance* $\sigma_R^2 = 1.23C_n^2 k^{7/6} L^{11/6}$ or the *Rytov number* $\sigma_\chi^2 = \frac{1}{4}\sigma_R^2$. Here, $k = 2\pi/\lambda$ [rad/m] is the optical wave number with wavelength λ [m], and L [m] is the propagation distance. We identify weak irradiance fluctuations by $\sigma_R^2 < 1$ and moderate-to-strong irradiance fluctuations by $\sigma_R^2 > 1$.

8.1.1 Random fields

A *random process* is a collection of time functions and an associated probability description. A *random field* is a three-dimensional analog of a random process; it is a function of a vector spatial variable $\mathbf{R} = (x, y, z)$ and also of time t in some cases. Random fluctuations of velocity, temperature, and index of refraction have all been studied in the context of a random field. Knowledge of the fluctuation characteristics of these quantities is important in areas of application that involve optical/IR wave propagation.

We say that a random field $u(\mathbf{R})$ is *statistically homogeneous* if its mean value is constant, i.e., $\langle u(\mathbf{R})\rangle = m$, where the brackets $\langle\,\rangle$ denote an ensemble average, and its covariance function depends only on the difference $\mathbf{R}_2 - \mathbf{R}_1$, where $\mathbf{R}_1$ and $\mathbf{R}_2$ are vectors representing two points in space. If the random field is *statistically homogeneous* and *isotropic*, the covariance function further reduces to a function depending only on the scalar distance $R = |\mathbf{R}_2 - \mathbf{R}_1|$. Because of its complexity for a more complete description, it has become customary to describe a random process or random field primarily in terms of its lower-order statistical moments.

Atmospheric parameters like wind velocity and temperature fluctuations are not strictly stationary because their mean values are constant only over relatively short time periods and distances. In fact, because most random processes and random fields are not strictly stationary nor statistically homogeneous, they are sometimes studied through stationary increments, which are also called locally homogeneous increments, and a statistical quantity called the *structure function*. However, for a homogeneous random process or a random field that is statistically homogeneous and isotropic, the structure function of interest is directly related to the covariance function (see Section 8.1.3).

8.1.2 Power spectrum models

To model atmospheric turbulence, we use power spectrum models of the refractive index. The most commonly used model for optical/IR wave propagation is the three-dimensional *Kolmogorov spectrum*,

$$\Phi_n(\kappa) = 0.033C_n^2\kappa^{-11/3}, \quad 1/L_0 \ll \kappa \ll 1/l_0, \tag{8.1}$$

where κ is the atmospheric wave number or *spatial frequency*. The Kolmogorov spectrum is the most widely used model in theoretical studies because of its relatively simple algebraic form. This spectrum model is theoretically valid only over the spatial wave number inertial range $1/L_0 \ll \kappa \ll 1/l_0$, but is often extended over all wave numbers by assuming that the inner scale is negligibly small and the outer scale is infinite. However, this can often lead to integrals that do not converge, so caution is called for in extending this wave number domain.

Effects of inner scale and outer scale are taken into account in the frequently used spectrum model called the *modified von Kármán spectrum*,

$$\begin{aligned} \Phi_n(\kappa) &= 0.033C_n^2\frac{\exp(-\kappa^2/\kappa_m^2)}{(\kappa^2+\kappa_0^2)^{11/6}}, \quad 0 \le \kappa < \infty; \\ \kappa_m &= 5.92/l_0, \; \kappa_0 = 2\pi/L_0. \end{aligned} \tag{8.2}$$

In some of our analyses, we will use Eq. (8.2) but may ignore the outer scale in certain cases by setting $\kappa_0 = 0$, which reduces Eq. (8.2) to the *Tatarskii spectrum*. In other cases, we ignore the inner scale by allowing $\kappa_m \to \infty$. Without the inner scale term, the spectrum is simply the original form of the *von Kármán spectrum.*

Another spectrum model that includes both inner-scale and outer-scale parameters and exhibits a high wave number rise (or "bump") just prior to the dissipation range is also used in some analyses. Such a model should be used when greater accuracy is needed for comparing theory with actual outdoor experimental data involving small-scale effects like scintillation. Among other proposed bump spectrum models is the *modified atmospheric spectrum* described by[3]

$$\begin{gathered} \Phi_n(\kappa) = 0.033C_n^2\underbrace{\left[1+1.802\left(\frac{\kappa}{\kappa_l}\right)-0.254\left(\frac{\kappa}{\kappa_l}\right)^{7/6}\right]}_{\text{Bump term}}\frac{\exp(-\kappa^2/\kappa_l^2)}{(\kappa^2+\kappa_0^2)^{11/6}}, \\ 0 \le \kappa < \infty; \quad \kappa_l = 3.3/l_0, \; \kappa_0 = 2\pi/L_0. \end{gathered} \tag{8.3}$$

The bump in the spectrum was originally discovered in the temperature spectrum data of Champagne et al.[4] and in that of Williams and Paulson.[5] Noting this, Hill[6] later performed a hydrodynamic analysis that led to a numerical version of the bump spectrum (minus the outer scale) for both temperature and refractive index.

Outer-scale effects in the mathematical analysis of optical/IR wave propagation have generally been introduced by either the modified von Kármán spectrum (8.2) or the modified atmospheric spectrum (8.3). Some of the analysis in this text will still be based on these models. Nonetheless, the outer-scale term in those spectrum models is sometimes replaced by the exponential form,

$$\left(\kappa^2+\kappa_0^2\right)^{-11/6} \to \kappa^{-11/3}\left(1-e^{-\kappa^2/\kappa_0^2}\right).$$

In such cases the modified von Kármán spectrum (8.2), for example, is replaced by what is called the *exponential spectrum model*,

$$\begin{aligned}\Phi_n(\kappa) &= 0.033C_n^2\kappa^{-11/3}\exp\left(-\frac{\kappa^2}{\kappa_m^2}\right)\left[1-\exp\left(-\frac{\kappa^2}{\kappa_0^2}\right)\right]\\ &= 0.033C_n^2\kappa^{-11/3}\left[\exp\left(-\frac{\kappa^2}{\kappa_m^2}\right)-\exp\left(-\frac{\kappa^2}{\kappa_{m0}^2}\right)\right],\end{aligned} \tag{8.4}$$

where we assume that $\kappa_0 = c/L_0$ and define $\kappa_{m0}^2 = \kappa_m^2\kappa_0^2/(\kappa_m^2+\kappa_0^2)$. In the analysis here we will generally assume that $c = 2\pi$ when numerical calculations are made. Because of its greater tractability, we will use the exponential spectrum model in much of the analysis where inner scale and/or outer scale are taken into account.

8.1.3 Index-of-refraction structure function

Along near-ground horizontal propagation paths, it is customary, for mathematical simplicity, to assume that optical turbulence is statistically homogeneous and isotropic. Also, average fluctuations in the refractive-index structure parameter C_n^2 over such propagation paths are taken to remain constant over the entire propagation path, ignoring any small elevation or surface changes along the path. Slant path propagation to or from an airborne target, however, requires a refractive-index structure parameter C_n^2 that includes changes as a function of altitude (e.g., see Section 9.6).

The index of refraction in air is defined by $n(\mathbf{R}) \approx 1 + n_1(\mathbf{R})$, where $n_1(\mathbf{R})$ is the random component satisfying $\langle n_1(\mathbf{R})\rangle = 0$. The associated covariance function of the refractive index is then defined by the ensemble average $B_n(\mathbf{R}_1, \mathbf{R}_2) = \langle n_1(\mathbf{R}_1)n_1(\mathbf{R}_2)\rangle$. For statistically homogeneous and isotropic index-of-refraction fluctuations, the covariance function and power spectral density $\Phi_n(\kappa)$ are related by the Fourier transform relation

$$\begin{aligned} B_n(R) &= \iiint_{-\infty}^{\infty} \Phi_n(\kappa) \exp(i\mathbf{K}\cdot\mathbf{R}) d^3\kappa \\ &= \frac{4\pi}{R}\int_0^\infty \kappa\Phi_n(\kappa)\sin(\kappa R)d\kappa, \end{aligned}$$

where spherical symmetry has been taken into account in the second line. The corresponding index-of-refraction structure function is therefore described by

$$\begin{aligned} D_n(R) &= \langle [n_1(\mathbf{R}_1) - n_1(\mathbf{R}_2)]^2 \rangle \\ &= 2[B_n(0) - B_n(R)] \\ &= 8\pi \int_0^\infty \kappa^2 \Phi_n(\kappa)\left(1 - \frac{\sin \kappa R}{\kappa R}\right) d\kappa. \end{aligned} \tag{8.5}$$

Confined to the inertial range, the refractive-index structure function defined by Eq. (8.5) takes the simple algebraic form

$$D_n(R) = C_n^2 R^{2/3}, \quad l_0 \ll R \ll L_0. \tag{8.6}$$

Thus, we see that C_n^2 is actually the proportionality constant in the index-of-refraction structure function over the inertial range; i.e., $C_n^2 = D_n(R)/R^{2/3}$.

To derive a more inclusive model of $D_n(R)$, we begin by expanding the term $1 - (\sin \kappa R)/\kappa R$ in Eq. (8.5) in a power series and integrating the series termwise. Doing so, we find that the structure function based on spectrum model (8.2) can be expressed as

$$\begin{aligned} D_n(R) &= 0.829 C_n^2 \sum_{k=1}^{\infty} \frac{(-1)^{k-1} R^{2k}}{(2k+1)!} \int_0^\infty \kappa^{2k+2} \frac{\exp(-\kappa^2/\kappa_m^2)}{(\kappa^2+\kappa_0^2)^{11/6}} d\kappa \\ &= 0.415 C_n^2 \kappa_0^{-2/3} \sum_{k=1}^{\infty} \frac{(-1)^{k-1}(\kappa_0 R)^{2k}}{(2k+1)!} \Gamma\left(k+\frac{3}{2}\right) U\left(k+\frac{3}{2}; k+\frac{2}{3}; \frac{\kappa_0^2}{\kappa_m^2}\right), \end{aligned} \tag{8.7}$$

where in the second step we used ($D5$) in Appendix D. To proceed, we know that $\kappa_0^2/\kappa_m^2 \approx l_0^2/L_0^2 \ll 1$, which permits us to replace the confluent hypergeometric function $U(a;c;z)$ in Eq. (8.7) with its small argument approximation ($CH19$) found in Appendix A. This action leads to

$$\begin{aligned} D_n(R) &= 0.415 C_n^2 \kappa_0^{-2/3} \sum_{k=1}^{\infty} \frac{(-1)^{k-1}(\kappa_0 R)^{2k}}{(2k+1)!} \Gamma\left(k+\frac{3}{2}\right) \\ &\quad \times \left[\frac{\Gamma\left(\frac{1}{3}-k\right)}{\Gamma\left(\frac{11}{6}\right)} + \frac{\Gamma\left(k-\frac{1}{3}\right)}{\Gamma\left(k+\frac{3}{2}\right)} \left(\frac{\kappa_0^2}{\kappa_m^2}\right)^{\frac{1}{3}-k} \right]. \end{aligned} \tag{8.8}$$

Now, converting all gamma functions and factorials involving the index k in Eq. (8.8) to Pochhammer symbols based on identities from Appendix A, we have

$$\begin{aligned}(2k+1)! &= 2^{2k}\left(\frac{3}{2}\right)_k k!,\\ \Gamma\left(k+\frac{3}{2}\right) &= \Gamma\left(\frac{3}{2}\right)\left(\frac{3}{2}\right)_k,\\ \Gamma\left(\frac{1}{3}-k\right) &= \Gamma\left(\frac{1}{3}\right)\left(\frac{1}{3}\right)_{-k} = \frac{\Gamma\left(\frac{1}{3}\right)(-1)^k}{\left(\frac{2}{3}\right)_k},\\ \Gamma\left(k-\frac{1}{3}\right) &= \Gamma\left(-\frac{1}{3}\right)\left(-\frac{1}{3}\right)_k.\end{aligned}$$

After inserting these Pochhammer symbol identities into Eq. (8.8) and simplifying the resulting expressions, we obtain two infinite series:

$$\begin{aligned}D_n(R) = &-1.047C_n^2\kappa_0^{-2/3}\sum_{k=1}^{\infty}\frac{(\kappa_0 R/2)^{2k}}{\left(\frac{2}{3}\right)_k k!}\\ &+1.685C_n^2\kappa_m^{-2/3}\sum_{k=1}^{\infty}\frac{(-1)^k\left(-\frac{1}{3}\right)_k(\kappa_m R/2)^{2k}}{\left(\frac{3}{2}\right)_k k!}.\end{aligned} \tag{8.9}$$

Finally, by adding and subtracting unity to each summation and summing each resulting series as a generalized hypergeometric series, we get the expression

$$\begin{aligned}D_n(R) = &\,1.685C_n^2\kappa_m^{-2/3}\left[{}_1F_1\left(-\frac{1}{3};\frac{3}{2};-\frac{\kappa_m^2R^2}{4}\right)-1\right]\\ &+1.047C_n^2\kappa_0^{-2/3}\left[1-{}_0F_1\left(-;\frac{2}{3};\frac{\kappa_0^2R^2}{4}\right)\right].\end{aligned} \tag{8.10}$$

Note that Eq. (8.10) reduces to the inertial-range form of Eq. (8.6) by setting $\kappa_0 = 0$ and $\kappa_m \to \infty$.

Equation (8.10) involving hypergeometric-type functions is an exact result based on the small-argument approximation for $U(a;c;z)$. Nonetheless, we can follow the technique illustrated in Chapter 6 to develop a more tractable algebraic approximation for this structure function. We start with the asymptotic expressions [see ($CH6$) in Appendix A]

$$ {}_1F_1\left(-\frac{1}{3};\frac{3}{2};-x\right) \sim \begin{cases} 1+0.222x, & x\to 0 \\ 0.942x^{1/3}, & x\to\infty, \end{cases} $$

from which we deduce that

$$ {}_1F_1\left(-\frac{1}{3};\frac{3}{2};-x\right) \approx (1+0.84x)^{1/3}. \tag{8.11} $$

Also, by assuming that $\kappa_0 R < 2$ such that $1 - {}_0F_1(-;2/3;\kappa_0^2R^2/4) \approx -3\kappa_0^2R^2/8$, the structure function in Eq. (8.10) becomes

$$ D_n(R) \approx 1.685C_n^2\kappa_m^{-2/3}\left[(1+0.210\kappa_m^2R^2)^{1/3}-1\right] - 0.393C_n^2\kappa_0^{4/3}R^2. \tag{8.12} $$

In Fig. 8.2 we plot the scaled structure function $D_n(R)/1.685C_n^2\kappa_m^{-2/3}$ as defined by Eqs. (8.10) and (8.12) but setting $\kappa_0 = 0$ in both expressions.

If we use the modified atmospheric power spectrum model (8.3), the resulting index-of-refraction structure function can be derived in the same manner as above, which in this case takes the form(with $\kappa_0 = 0$)

$$ \begin{aligned} D_n(R) = 1.685C_n^2\kappa_l^{-2/3}\Bigg\{ & {}_1F_1\left(-\frac{1}{3};\frac{3}{2};-\frac{\kappa_l^2R^2}{4}\right) - 1 \\ & + 2.470\left[1 - {}_1F_1\left(\frac{1}{6};\frac{3}{2};-\frac{\kappa_l^2R^2}{4}\right)\right] - 0.227\left[1 - {}_1F_1\left(\frac{1}{4};\frac{3}{2};-\frac{\kappa_l^2R^2}{4}\right)\right]\Bigg\}. \end{aligned} \tag{8.13} $$

Although left to the reader, simple algebraic approximations like Eq. (8.11) can be derived for all three confluent hypergeometric functions in Eq. (8.13) by using the method illustrated in Chapter 6.

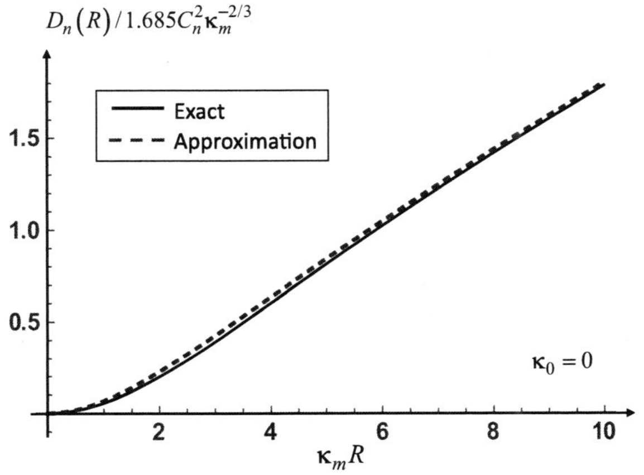

Figure 8.2 Scaled refractive-index structure function based on Eqs. (8.10) and (8.12).

Last, we should point out that the structure function for temperature has an identical form as the index-of-refraction structure function given by Eq. (8.10), but with C_n^2 replaced by the temperature structure constant C_T^2. Assuming negligible effects from humidity fluctuations as well as from temperature and humidity cross-correlation, the relation between the two structure constants is

$$C_n^2 = \left(79.2 \times 10^{-6} \frac{\overline{P}}{\overline{T}^2}\right)^2 C_T^2,$$

where $\overline{P}$ denotes mean pressure and $\overline{T}$ is mean temperature.

8.2 Optical/IR Wave Propagation

Several basic geometries are commonly used to describe various optical/IR wave models. Among others, these include the infinite *plane wave*, *spherical wave*, and *Gaussian-beam wave*. Limiting cases based on the TEM_{00} Gaussian-beam wave lead to the plane wave and spherical wave models.

8.2.1 Propagation in free space

Let us start with the propagation of a Gaussian-beam wave in free space (no atmospheric turbulence). Propagation is along the positive z-axis with the transmitter (Tx) aperture located in the plane $z = 0$. The amplitude distribution in this plane has beam radius (spot size) W_0 [m], where W_0 denotes the radius at which the field amplitude falls to $1/e$ of that on the beam axis. In addition, the phase front radius of curvature is F_0 [m], where the specific cases $F_0 > 0$, $F_0 < 0$, and $F_0 = \infty$ correspond to *convergent*, *divergent*, and *collimated* beam forms, respectively. If the Gaussian-beam wave has maximum amplitude A_0 at the output aperture of the Tx, then its complex optical field can be written as

$$z = 0\colon\ U_0(\mathbf{r}, 0) = A_0 \exp\left(-\frac{1}{2}\alpha_0 k r^2\right), \tag{8.14}$$

where $r = |\mathbf{r}| = \sqrt{x^2 + y^2}$ and

$$\alpha_0 = \frac{2}{kW_0^2} + i\frac{1}{F_0}. \tag{8.15}$$

Also, λ is the optical wavelength and $k = 2\pi/\lambda$ is the corresponding wave number.

The optical field at distance $z = L$ from the Tx along the positive z-axis is a solution of the *Huygens–Fresnel integral*,

$$z = L:\ U_0(\mathbf{r}, L) = -2ik \iint_{-\infty}^{\infty} G(\mathbf{s}, \mathbf{r}; L) U_0(\mathbf{s}, 0) d^2 s, \tag{8.16}$$

where $G(\mathbf{s}, \mathbf{r}; L)$ is the free-space *Green's function* under a paraxial approximation defined by

$$G(\mathbf{s}, \mathbf{r}; L) = \frac{1}{4\pi L} \exp\left(ikL + \frac{ik}{2L} |\mathbf{s} - \mathbf{r}|^2 \right). \tag{8.17}$$

By inserting Eq. (8.17) into Eq. (8.16) and evaluating the resulting integrals, the optical wave can be expressed as another Gaussian-beam wave (see pp. 90–91 of Ref. 1):

$$z = L:\ U_0(\mathbf{r}, L) = \frac{A_0}{1 + i\alpha_0 L} \exp\left[ikL - \frac{1}{2} \left(\frac{\alpha_0 k r^2}{1 + i\alpha_0 L} \right) \right]. \tag{8.18}$$

This same expression was also derived in Section 7.7.4 by solving the paraxial wave equation with application of the Hankel transform.

To facilitate the analysis in free space and in optical turbulence, it is useful to introduce two sets of *nondimensional Gaussian-beam parameters* that characterize the Gaussian beam at the Tx and receiver (Rx) pupil plane, respectively. Based on the above Tx beam characteristics, the nondimensional beam parameters at the Tx plane are defined by

$$z = 0: \begin{cases} \Theta_0 = \mathrm{Re}(1 + i\alpha_0 L) = 1 - \dfrac{L}{F_0}, \\[2ex] \Lambda_0 = \mathrm{Im}(1 + i\alpha_0 L) = \dfrac{2L}{kW_0^2}. \end{cases} \tag{8.19}$$

The complementary set of Rx beam parameters at range $z = L$ arise from the inverse mapping $\Theta - i\Lambda = (1 + i\alpha_0 L)^{-1} = (\Theta_0 + i\Lambda_0)^{-1}$, from which we deduce that

$$z = L: \begin{cases} \Theta = \dfrac{\Theta_0}{\Theta_0^2 + \Lambda_0^2} = 1 + \dfrac{L}{F}, \\[2ex] \overline{\Theta} = 1 - \Theta = -\dfrac{L}{F}, \\[2ex] \Lambda = \dfrac{\Lambda_0}{\Theta_0^2 + \Lambda_0^2} = \dfrac{2L}{kW^2}. \end{cases} \tag{8.20}$$

Here, F and W denote the phase front radius of curvature and spot radius of the beam at the Rx pupil plane, respectively. The spot size and phase front radius of curvature at the Rx pupil plane can be defined by any of the following relations:

$$W = W_0\sqrt{\Theta_0^2 + \Lambda_0^2} = \frac{W_0}{\sqrt{\Theta^2 + \Lambda^2}} = \sqrt{\frac{2L}{k\Lambda}},$$
$$F = \frac{F_0(\Theta_0^2 + \Lambda_0^2)(\Theta_0 - 1)}{(\Theta_0^2 + \Lambda_0^2 - \Theta_0)} = \frac{F_0(\Theta^2 + \Lambda^2 - \Theta)}{(\Theta - 1)(\Theta^2 + \Lambda^2)} = \frac{L}{\Theta - 1}. \tag{8.21}$$

The beam parameters defined by Eqs. (8.19) and (8.20) are useful when describing the beam in free space but are also useful in the presence of atmospheric turbulence. In all statistical models for a Gaussian-beam wave, the limiting case of an infinite plane wave is obtained by setting $\Theta = 1$, $\Lambda = 0$, whereas the limiting case of a spherical wave is obtained by setting $\Theta = \Lambda = 0$.

8.2.2 Rytov approximation

The *Rytov approximation* has been the most successful method of analysis in the development of mathematical models of optical/IR wave propagation through atmospheric turbulence under weak irradiance fluctuations.[1,2,7] The fundamental idea in the Rytov approximation is to write the optical field at the Rx as

$$U(\mathbf{r}, L) = U_0(\mathbf{r}, L)\exp\left[\psi_1(\mathbf{r}, L) + \psi_2(\mathbf{r}, L) + \cdots\right], \tag{8.22}$$

where $\psi_1(\mathbf{r}, L), \psi_2(\mathbf{r}, L), \ldots$ are complex phase perturbations and $U_0(\mathbf{r}, L)$ is the optical field in free space. Based on Rytov theory, it was previously shown that all of the statistical math models associated with second-order and fourth-order moments of the optical field in line-of-sight homogeneous and isotropic turbulence involve various linear combinations of three integrals[7] (see also Chapter 5 of Ref. 1):

$$\begin{aligned} E_1(0,0) &= \langle \psi_2(\mathbf{r}, L)\rangle + \frac{1}{2}\langle \psi_1^2(\mathbf{r}, L)\rangle \\ &= -2\pi^2 k^2 L \int_0^1\int_0^\infty \kappa\Phi_n(\kappa)\, d\kappa\, d\xi, \end{aligned} \tag{8.23}$$

$$\begin{aligned} E_2(\mathbf{r}_1, \mathbf{r}_2) &= \langle \psi_1(\mathbf{r}_1, L)\psi_1^*(\mathbf{r}_2, L)\rangle \\ &= 4\pi^2 k^2 L \int_0^1\int_0^\infty \kappa\Phi_n(\kappa)\exp\left(-\frac{\Lambda L\kappa^2\xi^2}{k}\right) \\ &\quad\times J_0[|(1-\overline{\Theta}\xi)\mathbf{p} - 2i\Lambda\xi\mathbf{r}|\,\kappa]\, d\kappa\, d\xi, \end{aligned} \tag{8.24}$$

$$\begin{aligned} E_3(\mathbf{r}_1, \mathbf{r}_2) &= \langle \psi_1(\mathbf{r}_1, L)\psi_1(\mathbf{r}_2, L)\rangle \\ &= -4\pi^2 k^2 L \int_0^1\int_0^\infty \kappa\Phi_n(\kappa) J_0\{[1 - (\overline{\Theta} + i\Lambda)\xi]\kappa\rho\} \\ &\quad\times \exp\left(-\frac{\Lambda L\kappa^2\xi^2}{k}\right)\exp\left[-\frac{iL\kappa^2}{k}\xi(1-\overline{\Theta}\xi)\right] d\kappa\, d\xi, \end{aligned} \tag{8.25}$$

where the notation * means complex conjugate, || denotes magnitude of the vector quantity, the parameter $\xi = 1 - z/L$ is a normalized distance variable along the positive z-axis, $\mathbf{r}_1$ and $\mathbf{r}_2$ represent two points on the wavefront at distance $z = L$, and we have introduced the transverse spatial vectors and magnitudes

$$\mathbf{r} = \frac{1}{2}(\mathbf{r}_1 + \mathbf{r}_2), \quad \mathbf{p} = \mathbf{r}_1 - \mathbf{r}_2; \quad r = |\mathbf{r}|, \quad \rho = |\mathbf{p}|.$$

Our concentration in this chapter will be on evaluating each of these integrals as they apply to various statistical quantities of the received waveform. To start, we note that the inside integral over wave number κ in Eqs. (8.24) and (8.25) under the spectrum model (8.2) with $\kappa_0 = 0$ are both of the general form

$$I = \int_0^\infty \kappa^\mu J_0(b\kappa) \exp(-a\kappa^2)\, d\kappa. \tag{8.26}$$

For the integral $E_2(\mathbf{r}_1, \mathbf{r}_2)$ defined by Eq. (8.24), the parameters a and b are

$$E_2(\mathbf{r}_1, \mathbf{r}_2): \begin{cases} a = \dfrac{1}{\kappa_m^2}(1 + \Lambda Q_m \xi^2), \\ b = |\,(1 - \overline{\Theta}\xi)\mathbf{p} - 2i\Lambda\xi\mathbf{r}\,|, \end{cases} \tag{8.27}$$

where $Q_m = L\kappa_m^2/k$. For $E_3(\mathbf{r}_1, \mathbf{r}_2)$ defined by Eq. (8.25), the corresponding parameters are

$$E_3(\mathbf{r}_1, \mathbf{r}_2): \begin{cases} a = \dfrac{1}{\kappa_m^2}\{1 + iQ_m\xi[1 - (\overline{\Theta} + i\Lambda)\xi]\}, \\ b = [1 - (\overline{\Theta} + i\Lambda)\xi]\rho. \end{cases} \tag{8.28}$$

The parameter μ in Eq. (8.26) takes on the value $\mu = -\frac{8}{3}$ for spectrum model (8.2) with $\kappa_0 = 0$, but we choose to not define it as such at this stage of the analysis to avoid a singularity in Eq. (8.26) at $\kappa = 0$. However, this singularity is eliminated once the integral is combined with other integrals from Eqs. (8.23), (8.24), or (8.25) in calculating the statistical quantities of interest. Only at this point can we take the limit $\mu \to -\frac{8}{3}$. Then the remaining integration on ξ can be handled in various ways depending on whether the optical wave is a plane wave, spherical wave, or Gaussian-beam wave.

To solve the integral in Eq. (8.26), we start with ($C1$) in Appendix C,

$$\exp(-a\kappa^2) = G_{01}^{10}(a\kappa^2|\,0), \tag{8.29}$$

which, temporarily assuming that $\mu > -1$, leads to

$$
\begin{aligned}
I &= \int_0^\infty \kappa^\mu J_0(b\kappa) G_{01}^{10}(a\kappa^2 | \, 0) d\kappa \\
&= \frac{2^\mu}{b^{\mu+1}} G_{21}^{11}\left(\frac{4a}{b^2} \middle| \begin{matrix} \frac{1-\mu}{2}, \frac{1-\mu}{2} \\ 0 \end{matrix} \right) \\
&= \frac{2^\mu}{b^{\mu+1}} G_{12}^{11}\left(\frac{b^2}{4a} \middle| \begin{matrix} 1 \\ \frac{1+\mu}{2}, \frac{1+\mu}{2} \end{matrix} \right).
\end{aligned}
\tag{8.30}
$$

Here we have used $(D12)$ in Appendix D followed by $(MG2)$ in Appendix A. Last, based on $(C14)$ in Appendix C, it follows that

$$
\int_0^\infty \kappa^\mu J_0(b\kappa) \exp\left(-a\kappa^2\right) d\kappa = \frac{1}{2a^{(1+\mu)/2}} \Gamma\left(\frac{1+\mu}{2}\right) {}_1F_1\left(\frac{1+\mu}{2}; 1; -\frac{b^2}{4a}\right).
\tag{8.31}
$$

We note here that Eq. (8.31) applies to both Eqs. (8.24) and (8.25) with parameters a and b defined, respectively, by Eqs. (8.27) and (8.28). To complete the analysis of the three integrals over wave number κ, we see that Eq. (8.23) leads to

$$
\int_0^\infty \kappa^\mu \exp\left(-\frac{\kappa^2}{\kappa_m^2}\right) d\kappa = \frac{1}{2} \kappa_m^{1+\mu} \Gamma\left(\frac{1+\mu}{2}\right),
\tag{8.32}
$$

where we have simply set $b = 0$ and $a = 1/\kappa_m^2$ in Eq. (8.31).

8.2.3 Comments on the exponential spectrum model

Although we have neglected the outer scale in deriving the integrals (8.31) and (8.32), we can still introduce the outer scale by relying on the exponential spectrum model. The exponential spectrum (8.4) is useful when incorporating outer-scale effects into the analysis because of the greater tractability it offers compared to the traditional spectrum models [Eqs. (8.2) and (8.3)], particularly when using the approach such as that introduced in Section 8.2.2. That is, because the outer-scale filter function in the exponential spectrum has the same mathematical form as the inner-scale filter function, the integral relations in Eqs. (8.31) and (8.32) can still be used for the outer-scale term, replacing κ_m with κ_{m0}. We illustrate the idea below in deriving expressions for the *wave structure function* (WSF) that include the outer scale. The scaling constant in the exponential spectrum will generally be chosen as $c = 2\pi$ but may be chosen differently depending on the application. Because the outer scale is still not well defined in the literature, it is difficult to proclaim the proper scaling constant for the outer scale in any spectrum model. Last, the outer-scale term in Eq. (8.4) can also be used in the same fashion with the modified atmospheric spectrum (8.3).

8.3 Mutual Coherence Function

One of the most basic statistical quantities associated with the propagation of a laser beam through optical turbulence is the second-order field moment called the *mutual coherence function* (MCF). The MCF is a two-point field moment that is used to describe the mean irradiance, beam spreading beyond diffraction, beam wander, angle-of-arrival fluctuations, and the loss of spatial coherence of the optical wave. An optical wave emitted from a laser source is spatially coherent as it exits the laser Tx aperture. As the beam travels along a horizontal or slant path through the atmosphere, it continually loses some of this spatial coherence, which can be determined from knowledge of the WSF derived from the MCF. The loss of spatial coherence of the optical wave is perhaps the most significant result derived from the MCF because it affects all of the other statistical quantities derived from the random optical field.

Let us assume that the optical field at the Tx is a Gaussian-beam wave as described by Eq. (8.14). If $U(\mathbf{r}, L)$ denotes the optical field at the Rx pupil plane, the MCF is defined by the ensemble average $\Gamma_2(\mathbf{r}_1, \mathbf{r}_2, L) = \langle U(\mathbf{r}_1, L)U^*(\mathbf{r}_2, L)\rangle$, which takes the mathematical form[8,9]

$$\begin{aligned}\Gamma_2(\mathbf{r}_1, \mathbf{r}_2, L) &= U_0(\mathbf{r}_1, L)U_0^*(\mathbf{r}_2, L)\exp\left[2E_1(0,0) + E_2(\mathbf{r}_1, \mathbf{r}_2)\right]\\ &= U_0(\mathbf{r}_1, L)U_0^*(\mathbf{r}_2, L)\exp\left[\left[-4\pi^2k^2L\int_0^1\int_0^\infty \kappa\Phi_n(\kappa)\right.\right.\\ &\quad\times\left.\left.\left\{1-\exp\left(-\frac{\Lambda L\kappa^2\xi^2}{k}\right)J_0\left[\left|(1-\overline{\Theta}\xi)\mathbf{p} - 2i\Lambda\xi\mathbf{r}\right|\kappa\right]\right\}d\kappa\, d\xi\right]\right]\\ &= U_0(\mathbf{r}_1, L)U_0^*(\mathbf{r}_2, L)\exp\left[\sigma_r^2(\mathbf{r}_1, L) + \sigma_r^2(\mathbf{r}_2, L) - T\right]\\ &\quad\times\exp\left[-\frac{1}{2}\Delta(\mathbf{r}_1, \mathbf{r}_2, L)\right].\end{aligned} \tag{8.33}$$

The terms in the first exponential function of the last line of Eq. (8.33) are

$$\begin{aligned}\sigma_r^2(\mathbf{r}, L) &= \frac{1}{2}\left[E_2(\mathbf{r}, \mathbf{r}) - E_2(0,0)\right]\\ &= 2\pi^2k^2L\int_0^1\int_0^\infty \kappa\Phi_n(\kappa)\exp\left(-\frac{\Lambda L\kappa^2\xi^2}{k}\right)\left[I_0(2\Lambda r\xi\kappa) - 1\right]d\kappa\, d\xi,\\ &\quad\mathbf{r} = \mathbf{r}_1, \mathbf{r}_2,\end{aligned} \tag{8.34}$$

$$\begin{aligned}T &= -2E_1(0,0) - E_2(0,0)\\ &= 4\pi^2k^2L\int_0^1\int_0^\infty \kappa\Phi_n(\kappa)\left[1-\exp\left(-\frac{\Lambda L\kappa^2\xi^2}{k}\right)\right]d\kappa\, d\xi.\end{aligned} \tag{8.35}$$

Physically, the quantity $\sigma_r^2(\mathbf{r}, L)$ describes the atmospherically induced change of the mean irradiance profile, and T describes the change in the on-axis mean irradiance. We use the on-axis mean irradiance to predict beam size in isotropic atmospheric turbulence (Sections 8.3.5 and 10.11.1). Also, $\exp\left[-\frac{1}{2}\Delta(\mathbf{r}_1, \mathbf{r}_2, L)\right]$ is the *complex degree of coherence* for which

$$\begin{aligned}\mathrm{Re}[\Delta(\mathbf{r}_1, \mathbf{r}_2, L)] &\equiv D(\mathbf{r}_1, \mathbf{r}_2, L)\\ &= \mathrm{Re}[E_2(\mathbf{r}_1, \mathbf{r}_1) + E_2(\mathbf{r}_2, \mathbf{r}_2) - 2E_2(\mathbf{r}_1, \mathbf{r}_2)],\end{aligned} \tag{8.36}$$

where $D(\mathbf{r}_1, \mathbf{r}_2, L)$ is the WSF. Last, the free-space optical field $U_0(\mathbf{r}, L)$ at the Rx pupil plane leads to the expression

$$U_0(\mathbf{r}_1, L)U_0^*(\mathbf{r}_2, L) = \frac{A_0^2 W_0^2}{W^2} \exp\left(-\frac{2r^2}{W^2} - \frac{\rho^2}{2W^2} - \frac{ik}{F}\mathbf{p}\cdot\mathbf{r}\right),$$

which represents the MCF in free space.

The *spatial coherence radius* ρ_0 of the optical wave is defined by the $1/e$ point of the normalized magnitude of the MCF, which is also called the *modulus of the complex degree of coherence* (DOC), defined by

$$\begin{aligned}\mathrm{DOC}(\mathbf{r}_1, \mathbf{r}_2, L) &= \frac{|\Gamma_2(\mathbf{r}_1, \mathbf{r}_2, L)|}{\sqrt{\Gamma_2(\mathbf{r}_1, \mathbf{r}_1, L)\Gamma_2(\mathbf{r}_2, \mathbf{r}_2, L)}}\\ &= \exp\left[-\frac{1}{2}D(\mathbf{r}_1, \mathbf{r}_2, L)\right].\end{aligned} \tag{8.37}$$

8.3.1 WSF for a plane wave

In the case of an infinite plane wave, we set $\Theta = 1$, $\Lambda = 0$ in Eq. (8.33) and find that $\sigma_r^2(r, L) = 0$ and $T = 0$. Also, because only $\rho = |\mathbf{r}_1 - \mathbf{r}_2|$ appears in Eq. (8.33) for a plane wave, the corresponding MCF is defined by

$$\Gamma_2(\rho, L) = A_0^2 \exp\left[-\frac{1}{2}D_{pl}(\rho, L)\right]. \tag{8.38}$$

For an infinite plane wave, therefore, the DOC and MCF are the same (with amplitude $A_0 = 1$), but this is not so for other optical waves.

For a plane wave, it can be shown that $E_2(\mathbf{r}_1, \mathbf{r}_1) + E_2(\mathbf{r}_2, \mathbf{r}_2) = -4E_1(0,0)$ in Eq. (8.36). Also, by making the substitution $\Theta = 1$, $\Lambda = 0$ in Eq. (8.27), we see that (accounting for both inner and outer scales)

$$a = \frac{1}{\kappa_m^2},\ b = \rho \quad \text{and} \quad a = \frac{1}{\kappa_{m0}^2},\ b = \rho.$$

In the following analysis we use the exponential spectrum model (8.4), from which we obtain

$$\begin{aligned}
D_{pl}(\rho, L) &= -4E_1(0,0) - 2E_2(\mathbf{r}_1, \mathbf{r}_2) \\
&= 2.606C_n^2k^2L\int_0^1\int_0^\infty \kappa^{-8/3}\left[\exp\left(-\frac{\kappa^2}{\kappa_m^2}\right) - \exp\left(-\frac{\kappa^2}{\kappa_{m0}^2}\right)\right] \\
&\quad\times[1 - J_0(\kappa\rho)]d\kappa\, d\xi \\
&= \lim_{\mu\to-8/3}\left[\left[1.303C_n^2k^2L\Gamma\left(\frac{1+\mu}{2}\right)\right.\right. \\
&\quad\times\left\{\kappa_m^{-(1+\mu)}\left[{}_1F_1\left(\frac{1+\mu}{2}; 1; -\frac{\kappa_m^2\rho^2}{4}\right) - 1\right]\right. \\
&\quad\left.\left.\left.- \kappa_{m0}^{-(1+\mu)}\left[{}_1F_1\left(\frac{1+\mu}{2}; 1; -\frac{\kappa_{m0}^2\rho^2}{4}\right) - 1\right]\right\}\right]\right].
\end{aligned} \tag{8.39}$$

In deriving Eq. (8.39) we used the results of ($D1$) and ($D2$) in Appendix D. Also, the WSF does not depend on the normalized distance variable ξ for the infinite plane wave, so now upon taking the limit $\mu \to -\frac{8}{3}$, Eq. (8.39) reduces to

$$\begin{aligned}
D_{pl}(\rho, L) = 8.702C_n^2k^2L&\left\{\kappa_m^{-5/3}\left[{}_1F_1\left(-\frac{5}{6}; 1; -\frac{\kappa_m^2\rho^2}{4}\right) - 1\right]\right. \\
&\left.- \kappa_0^{-5/3}\left[{}_1F_1\left(-\frac{5}{6}; 1; -\frac{\kappa_0^2\rho^2}{4}\right) - 1\right]\right\}.
\end{aligned} \tag{8.40}$$

Because $\kappa_m^2 \gg \kappa_0^2$, we have set $\kappa_{m0} \approx \kappa_0$ in Eq. (8.40). Moreover, if we also restrict $\kappa_0\rho < 1$, then Eq. (8.40) can be further simplified to

$$D_{pl}(\rho, L) = 8.702C_n^2k^2L\left\{\kappa_m^{-5/3}\left[{}_1F_1\left(-\frac{5}{6}; 1; -\frac{\kappa_m^2\rho^2}{4}\right) - 1\right] - 0.208\kappa_0^{1/3}\rho^2\right\}, \tag{8.41}$$

where we have used a small-argument approximation in the second confluent hypergeometric function in Eq. (8.40). By further setting $\kappa_0 = 0$ and $\kappa_m \to \infty$, both Eqs. (8.40) and (8.41) reduce to the well-known inertial-range result,

$$D_{pl}(\rho, L) = 2.914C_n^2k^2L\rho^{5/3}, \quad l_0 \ll \rho \ll L_0. \tag{8.42}$$

In order to obtain the spatial coherence radius ρ_0 for an infinite plane wave from Eq. (8.42), we solve $\mathrm{DOC}(\rho_0, L) = e^{-1}$. This leads to $D_{pl}(\rho_0, L) = 2$, from which we deduce the inertial range transverse coherence radius

$$\rho_0 \equiv \rho_{pl} = (1.457C_n^2k^2L)^{-3/5}, \quad l_0 \ll \rho_0 \ll L_0. \tag{8.43}$$

The expressions given by Eqs. (8.40) and (8.41) for the WSF are very similar to those that arise from use of the modified von Kármán spectrum.

For comparison, the WSF based on the modified von Kármán spectrum model (8.2) is

$$\begin{aligned} D_{pl}(\rho, L) = 8.702 C_n^2 k^2 L \Bigg\{ \kappa_m^{-5/3} \left[{}_1F_1\left(-\frac{5}{6}; 1; -\frac{\kappa_m^2\rho^2}{4}\right) - 1 \right] \\ - 0.180 \kappa_0^{-5/3} \left[{}_0F_1\left(-; \frac{1}{6}; \frac{\kappa_0^2\rho^2}{4}\right) - 1 \right] \Bigg\}, \end{aligned} \tag{8.44}$$

which can be derived in a manner similar to that used in obtaining the index-of-refraction structure function [Eq. (8.10)] in Section 8.1.3. Now imposing the small-argument approximation for the outer-scale term in Eq. (8.44) under the restriction $\kappa_0\rho < 1$, we find that Eq. (8.44) takes the form

$$D_{pl}(\rho, L) = 8.702 C_n^2 k^2 L \left\{ \kappa_m^{-5/3} \left[{}_1F_1\left(-\frac{5}{6}; 1; -\frac{\kappa_m^2\rho^2}{4}\right) - 1 \right] - 0.269 \kappa_0^{1/3} \rho^2 \right\}. \tag{8.45}$$

To compare the outer-scale behavior between the von Kármán spectrum and the exponential spectrum, we plot them both in Fig. 8.3(a) with vanishing inner scale and $L_0 = 1$ m (see also Fig. 3.8 in Ref. 1). We show the exponential spectrum with three values of the scaling constant: $\kappa_0 = 2\pi/L_0$ and $\kappa_0 \equiv \kappa_{00} = 2.5\pi/L_0$, $\kappa_0 \equiv \kappa_{00} = 3.5\pi/L_0$. Thus, increasing the scaling constant in the exponential spectrum moves it closer to the von Kármán spectrum. This is further illustrated in Fig. 8.3(b), where we show the behavior of the outer-scale terms in Eqs. (8.40) and (8.44) for increasing values of $\kappa_0\rho$ with the same scaling constant c for κ_0 in both equations. Although the expressions in Eqs. (8.41) and (8.45) are very similar in the outer-scale terms, the general behavior of the outer-scale terms in Eqs. (8.40) and (8.44) for $\kappa_0\rho > 1$ begin to differ more because the ${}_0F_1$ function in Eq. (8.44) increases more rapidly than the ${}_1F_1$ function in Eq. (8.40). By increasing the scaling constant so that $\kappa_0 \equiv \kappa_{00} = 2.5\pi/L_0$ in the ${}_1F_1$ function in Eq. (8.40), the resulting dashed curve in Fig. 8.3(b) is closer to that corresponding to the outer-scale term in Eq. (8.44). Nonetheless, because of the uncertainty of the scaling constant for κ_0 in various spectrum models, we will limit the outer-scale effect to small values of $\kappa_0\rho$ in some of the following analyses.

An algebraic approximation for the confluent hypergeometric function in Eq. (8.41) was previously provided by Eq. (6.13). If we now insert this approximation into the expression in Eq. (8.41), we obtain the simpler algebraic form

$$D_{pl}(\rho, L) \approx 3.280 C_n^2 k^2 L \rho^2 l_0^{-1/3} \left[\frac{1}{(1 + 2.033\,\rho^2/l_0^2)^{1/6}} - 0.553 (\kappa_0 l_0)^{1/3} \right]. \tag{8.46}$$

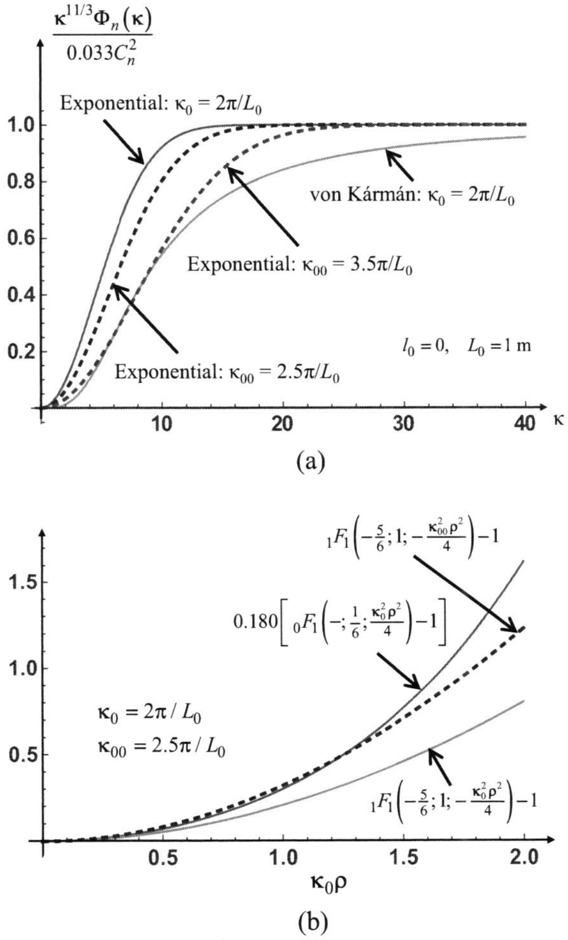

Figure 8.3 (a) Plot of the von Kármán and exponential spectrums with zero inner scale. (b) Plot of the outer-scale terms in Eqs. (8.40) and (8.44), both with $\kappa_0 = 2\pi/L_0$. Also shown by the dashed line is the outer-scale term in Eq. (8.40) with $\kappa_{00} = 2.5\pi/L_0$.

Based on the modified atmospheric spectrum model (8.3), the WSF for a plane wave takes the form[10]

$$\begin{aligned} D_{pl}(\rho, L) = 8.702 C_n^2 k^2 L \kappa_l^{-5/3} \Bigg\{ & {}_1F_1\left(-\frac{5}{6}; 1; -\frac{\kappa_l^2\rho^2}{4}\right) - 1 \\ & + 1.096\left[{}_1F_1\left(-\frac{1}{3}; 1; -\frac{\kappa_l^2\rho^2}{4}\right) - 1\right] \\ & - 0.186\left[{}_1F_1\left(-\frac{1}{4}; 1; -\frac{\kappa_l^2\rho^2}{4}\right) - 1\right] - 0.269 \kappa_l^{5/3} \kappa_0^{1/3} \rho^2 \Bigg\}. \end{aligned} \tag{8.47}$$

By following the approach from Chapter 6, we can write the ${}_1F_1$ functions in Eq. (8.47) in terms of ${}_2F_2$ functions and then approximate those to obtain the algebraic approximation given by

$$\begin{aligned}
D_{pl}(\rho, L) &= 2.700 C_n^2 k^2 L l_0^{-1/3} \rho^2 \left[{}_2F_2\left(\frac{1}{6}, 1; 2, 2; -\frac{\kappa_l^2 \rho^2}{4}\right) \right. \\
&\quad + 0.438\, {}_2F_2\left(\frac{2}{3}, 1; 2, 2; -\frac{\kappa_l^2 \rho^2}{4}\right) \\
&\quad \left. - 0.056\, {}_2F_2\left(\frac{3}{4}, 1; 2, 2; -\frac{\kappa_l^2 \rho^2}{4}\right) - 0.868(\kappa_0 l_0)^{1/3} \right] \\
&\approx 2.700 C_n^2 k^2 L l_0^{-1/3} \rho^2 \left[\frac{1}{(1 + 0.632\rho^2/l_0^2)^{1/6}} \right. \\
&\quad \left. + \frac{0.438}{(1 + 0.442\rho^2/l_0^2)^{2/3}} - \frac{0.056}{(1 + 0.376\rho^2/l_0^2)^{3/4}} - 0.868(\kappa_0 l_0)^{1/3} \right].
\end{aligned} \tag{8.48}$$

8.3.2 Outer-scale effect

Scale sizes larger than L_0 are not considered isotropic. Nonetheless, to better understand the effect of an outer scale on the WSF, we will use the modified von Kármán spectrum model (8.2) with $\kappa_m \to \infty$ to emphasize only the outer scale effect. Thus, in this case the WSF is now defined by

$$D_{pl}(\rho, L) = 2.606 C_n^2 k^2 L \int_0^\infty \frac{\kappa}{(\kappa^2 + \kappa_0^2)^{11/6}} [1 - J_0(\kappa\rho)]\, d\kappa, \qquad \rho \gg l_0. \tag{8.49}$$

If we represent the Bessel function by ($C9$) in Appendix C,

$$J_0(z) = G_{02}^{10}\left(\frac{z^2}{4}\middle|\, 0, 0\right), \tag{8.50}$$

we then obtain with the use of ($D11$) in Appendix D and ($MG3$) in Appendix A

$$\begin{aligned}
\int_0^\infty \frac{\kappa[1 - J_0(\kappa\rho)]}{(\kappa^2 + \kappa_0^2)^{11/6}}\, d\kappa &= \int_0^\infty \frac{\kappa}{(\kappa^2 + \kappa_0^2)^{11/6}} \left[1 - G_{02}^{10}\left(\frac{\kappa^2\rho^2}{4}\middle|\, 0, 0\right)\right] d\kappa \\
&= \frac{3}{5}\kappa_0^{-5/3} - \frac{\kappa_0^{-5/3}}{2\Gamma\left(\frac{11}{6}\right)} G_{02}^{20}\left(\frac{\kappa_0^2\rho^2}{4}\middle|\, \frac{5}{6}, 0\right),
\end{aligned} \tag{8.51}$$

which, from ($C23$) in Appendix C, reduces further to

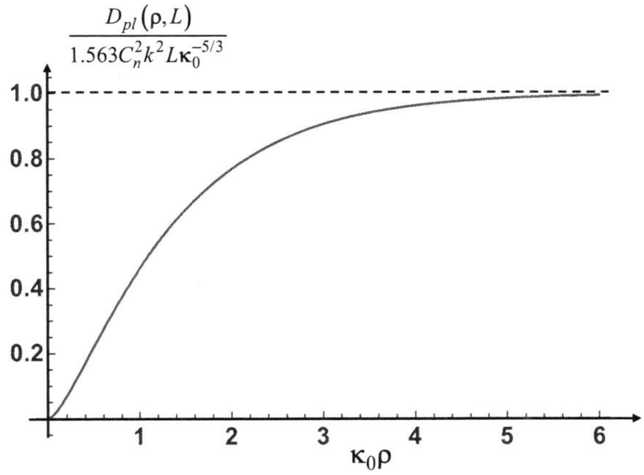

Figure 8.4 Scaled WSF [Eq. (8.53)] showing the effects of the outer scale.

$$\int_0^\infty \frac{\kappa\left[1 - J_0(\kappa\rho)\right]}{(\kappa^2 + \kappa_0^2)^{11/6}} d\kappa = \frac{3}{5}\kappa_0^{-5/3} - \frac{\rho^{5/6}K_{5/6}(\kappa_0\rho)}{2^{5/6}\Gamma\left(\frac{11}{6}\right)\kappa_0^{5/6}}. \tag{8.52}$$

Combining the result of Eq. (8.52) with the WSF (8.49) gives us the alternative expression for the plane wave structure function with zero inner scale,

$$D_{pl}(\rho, L) = 1.563C_n^2k^2L\kappa_0^{-5/3}\left[1 - 0.994(\kappa_0\rho)^{5/6}K_{5/6}(\kappa_0\rho)\right]. \tag{8.53}$$

We plot the scaled WSF (8.53) as a function of $\kappa_0\rho$ in Fig. 8.4. Notice that the outer-scale effect forces the scaled WSF to roll over, approaching a constant value of unity for increasing values of $\kappa_0\rho$. We will show in Section 8.8.3 that the scaling constant is actually $1.563C_n^2k^2L\kappa_0^{-5/3} = 2\sigma_{S,pl}^2$, where $\sigma_{S,pl}^2$ is the phase variance of a plane wave under a geometrical optics approximation.

8.3.3 WSF for a spherical wave

For a spherical wave ($\Theta = \Lambda = 0$), we see that the parameters in Eq. (8.27) reduce to

$$a = \frac{1}{\kappa_m^2},\ b = \rho\xi \quad \text{and} \quad a = \frac{1}{\kappa_{m0}^2},\ b = \rho\xi,$$

where we have also made the change of variable $\xi \rightarrow 1 - \xi$. Here, we again note that $E_2(\mathbf{r}_1, \mathbf{r}_1) + E_2(\mathbf{r}_2, \mathbf{r}_2) = -4E_1(0,0)$. Thus, with the use of (*D*1) and (*D*2) in Appendix D, we find that the corresponding WSF is defined by

$$\begin{aligned}
D_{sph}(\rho,\, L) &= -\,4E_1(0,0) - 2E_2(\mathbf{r}_1,\, \mathbf{r}_2)\\
&= 2.606C_n^2k^2L\int_0^1\int_0^\infty \kappa^{-8/3}\left[\exp\left(-\frac{\kappa^2}{\kappa_m^2}\right) - \exp\left(-\frac{\kappa^2}{\kappa_{m0}^2}\right)\right]\\
&\quad\times[1 - J(\kappa\rho\xi)]d\kappa\, d\xi\\
&= \lim_{\mu\to -8/3}\left[\left[-1.303C_n^2k^2L\,\Gamma\left(\frac{1+\mu}{2}\right)\right.\right.\\
&\quad\times\left\{\kappa_m^{-(1+\mu)}\int_0^1\left[1 - {}_1F_1\left(\frac{1+\mu}{2};\, 1;\, -\frac{\kappa_m^2\rho^2\xi^2}{4}\right)\right]d\xi\right.\\
&\quad\left.\left.\left.-\,\kappa_{m0}^{-(1+\mu)}\int_0^1\left[1 - {}_1F_1\left(\frac{1+\mu}{2};\, 1;\, -\frac{\kappa_{m0}^2\rho^2\xi^2}{4}\right)\right]d\xi\right\}\right]\right].
\end{aligned} \tag{8.54}$$

After taking the limit $\mu \to -\frac{8}{3}$, we complete the integration in Eq. (8.54) with the help of (C13) in Appendix C,

$$ {}_1F_1\left(-\frac{5}{6};\, 1;\, -\frac{\kappa_m^2\rho^2\xi^2}{4}\right) = \frac{1}{\Gamma\left(-\frac{5}{6}\right)}G_{12}^{11}\left(\frac{\kappa_m^2\rho^2\xi^2}{4}\middle|\begin{matrix}\frac{11}{6}\\ 0,\, 0\end{matrix}\right), \tag{8.55}$$

to find that

$$\begin{aligned}
&\int_0^1\left[1 - {}_1F_1\left(-\frac{5}{6};\, 1;\, -\frac{\kappa_m^2\rho^2\xi^2}{4}\right)\right]d\xi\\
&= \int_0^1\left[1 - \frac{1}{\Gamma\left(-\frac{5}{6}\right)}G_{12}^{11}\left(\frac{\kappa_m^2\rho^2\xi^2}{4}\middle|\begin{matrix}\frac{11}{6}\\ 0,\, 0\end{matrix}\right)\right]d\xi\\
&= 1 - \frac{1}{2\Gamma\left(-\frac{5}{6}\right)}G_{23}^{12}\left(\frac{\kappa_m^2\rho^2}{4}\middle|\begin{matrix}\frac{1}{2},\, \frac{11}{6}\\ 0,\, 0,\, -\frac{1}{2}\end{matrix}\right)\\
&= 1 - {}_2F_2\left(-\frac{5}{6},\, \frac{1}{2};\, 1,\, \frac{3}{2};\, -\frac{\kappa_m^2\rho^2}{4}\right).
\end{aligned} \tag{8.56}$$

In deriving the last step in Eq. (8.56), we applied (D9) in Appendix D and (C32) in Appendix C. Returning to Eq. (8.54), we see that the second ${}_1F_1$ function leads to the same result with κ_m replaced by κ_{m0} in Eq. (8.56); hence,

$$\begin{aligned} D_{sph}(\rho, L) &= 8.702 C_n^2 k^2 L \left\{ \kappa_m^{-5/3} \left[{}_2F_2\left(-\frac{5}{6}, \frac{1}{2}; 1, \frac{3}{2}; -\frac{\kappa_m^2\rho^2}{4}\right) - 1 \right] \right. \\ &\quad \left. - \kappa_0^{-5/3} \left[{}_2F_2\left(-\frac{5}{6}, \frac{1}{2}; 1, \frac{3}{2}; -\frac{\kappa_0^2\rho^2}{4}\right) - 1 \right] \right\} \\ &= 8.702 C_n^2 k^2 L \left\{ \kappa_m^{-5/3} \left[{}_2F_2\left(-\frac{5}{6}, \frac{1}{2}; 1, \frac{3}{2}; -\frac{\kappa_m^2\rho^2}{4}\right) - 1 \right] \right. \\ &\quad \left. - 0.069 \kappa_0^{1/3} \rho^2 \right\}. \end{aligned} \tag{8.57}$$

Once again, we replaced κ_{m0} with κ_0 in Eq. (8.57) and we have also imposed the outer-scale restriction $\kappa_0\rho < 1$. Finally, by setting $\kappa_0 = 0$ and $\kappa_m \to \infty$, this last expression reduces to

$$D_{sph}(\rho, L) = 1.093 C_n^2 k^2 L \rho^{5/3}, \quad l_0 \ll \rho \ll L_0. \tag{8.58}$$

The corresponding transverse spatial coherence radius deduced from Eq. (8.58) is

$$\rho_0 \equiv \rho_{sph} = (0.547 C_n^2 k^2 L)^{-3/5}, \quad l_0 \ll \rho_0 \ll L_0. \tag{8.59}$$

Following the approach outlined in Eqs. (6.10) through (6.13), we can find an algebraic approximation for the generalized hypergeometric function in Eq. (8.57). First note that

$${}_2F_2\left(-\frac{5}{6}, \frac{1}{2}; 1, \frac{3}{2}; -x\right) - 1 = \frac{5x}{18}\, {}_3F_3\left(\frac{1}{6}, \frac{3}{2}, 1; 2, 2, \frac{5}{2}; -x\right). \tag{8.60}$$

Although the large-argument expression in Appendix E for ${}_3F_3$ contains three terms on the right-hand side, we will retain only the most significant term of the three. Then, by imposing the asymptotic approximations on the ${}_3F_3$ function, we have

$${}_3F_3\left(\frac{1}{6}, \frac{3}{2}, 1; 2, 2, \frac{5}{2}; -x\right) \sim \begin{cases} -0.025x, & x \to 0 \\ 1.439x^{-1/6}, & x \to \infty, \end{cases}$$

from which we deduce that

$${}_3F_3\left(\frac{1}{6}, \frac{3}{2}, 1; 2, 2, \frac{5}{2}; -x\right) \approx (1 + 0.114x)^{-1/6}. \tag{8.61}$$

Now based on this approximation, Eq. (8.60) with $x = \kappa_m^2\rho^2/4 = 8.762(\rho^2/l_0^2)$ and the result of Eq. (8.61) lead to

$$ {}_2F_2\left(-\frac{5}{6}, \frac{1}{2}; 1, \frac{3}{2}; -\frac{\kappa_m^2\rho^2}{4}\right) - 1 \approx 2.434\left(\frac{\rho^2}{l_0^2}\right)\left(1+\frac{\rho^2}{l_0^2}\right)^{-1/6}. \tag{8.62} $$

Therefore, the WSF approximation of Eq. (8.57) for a spherical wave becomes

$$ D_{sph}(\rho, L) \approx 1.093 C_n^2 k^2 L l_0^{-1/3} \rho^2 \left[\frac{1}{(1+\rho^2/l_0^2)^{1/6}} - 0.553(\kappa_0 l_0)^{1/3}\right]. \tag{8.63} $$

If we now use the power spectrum model (8.3), it can be shown that the WSF with $\kappa_0 = 0$ takes the form[10]

$$ \begin{aligned} D_{sph}(\rho, L) = 8.702 C_n^2 k^2 L \kappa_l^{-5/3} \Bigg\{ & {}_2F_2\left(-\frac{5}{6}, \frac{1}{2}; 1, \frac{3}{2}; -\frac{\kappa_l^2\rho^2}{4}\right) - 1 \\ & + 1.096\left[{}_2F_2\left(-\frac{1}{3}, \frac{1}{2}; 1, \frac{3}{2}; -\frac{\kappa_l^2\rho^2}{4}\right) - 1\right] \\ & - 0.186\left[{}_2F_2\left(-\frac{1}{4}, \frac{1}{2}; 1, \frac{3}{2}; -\frac{\kappa_l^2\rho^2}{4}\right) - 1\right]\Bigg\}, \end{aligned} \tag{8.64} $$

which, following the same approach as presented in Eqs. (8.60) to (8.62), leads to the algebraic approximation

$$ \begin{aligned} D_{sph}(\rho, L) = 0.90 C_n^2 k^2 L l_0^{-1/3} \rho^2 \Bigg[& {}_3F_3\left(\frac{1}{6}, \frac{3}{2}, 1; 2, 2, \frac{5}{2}; -\frac{\kappa_l^2\rho^2}{4}\right) \\ & + 0.438\, {}_3F_3\left(\frac{2}{3}, \frac{3}{2}, 1; 2, 2, \frac{5}{2}; -\frac{\kappa_l^2\rho^2}{4}\right) \\ & - 0.056\, {}_3F_3\left(\frac{3}{4}, \frac{3}{2}, 1; 2, 2, \frac{5}{2}; -\frac{\kappa_l^2\rho^2}{4}\right)\Bigg] \\ \approx 0.90 C_n^2 k^2 L l_0^{-1/3} \rho^2 \Bigg[& \frac{1}{(1+0.311\rho^2/l_0^2)^{1/6}} \\ & + \frac{0.438}{(1+0.183\rho^2/l_0^2)^{2/3}} - \frac{0.056}{(1+0.149\rho^2/l_0^2)^{3/4}}\Bigg]. \end{aligned} \tag{8.65} $$

8.3.4 WSF for a Gaussian-beam wave

By taking two points on the wavefront such that $\mathbf{r}_2 = -\mathbf{r}_1$, the resulting WSF for a Gaussian-beam wave can be written as

$$\begin{aligned} D(\rho, L) &= 8\pi^2 k^2 L \int_0^1 \int_0^\infty \kappa\, \Phi_n(\kappa) \exp\left(-\frac{\Lambda L \kappa^2 \xi^2}{k}\right) \\ &\quad \times \{I_0(\Lambda\, \rho\, \xi\kappa) - J_0[(1-\overline{\Theta}\xi)\kappa\rho]\} d\kappa d\xi \\ &= I_1 + I_2, \end{aligned} \tag{8.66}$$

where

$$I_1 = 8\pi^2 k^2 L \int_0^1 \int_0^\infty \kappa\, \Phi_n(\kappa) \exp\left(-\frac{\Lambda L \kappa^2 \xi^2}{k}\right) [I_0(\Lambda\, \rho\, \xi\kappa) - 1]\, d\kappa d\xi, \tag{8.67}$$

$$I_2 = 8\pi^2 k^2 L \int_0^1 \int_0^\infty \kappa\, \Phi_n(\kappa) \exp\left(-\frac{\Lambda L \kappa^2 \xi^2}{k}\right) \{1 - J_0[(1-\overline{\Theta}\xi)\kappa\rho]\}\, d\kappa d\xi. \tag{8.68}$$

An approximation for the WSF of a Gaussian-beam based on the spectrum models of Eqs. (8.2) and (8.3) is provided in Refs. 8 and 9. Here, however, we will derive the WSF of a Gaussian-beam wave based on the exponential spectrum model (8.4). Because we are using the exponential spectrum and applying a different type of analysis, the final approximation we derive will not be quite the same as that given in Refs. 8 and 9, except over the inertial range where they agree.

Starting with Eq. (8.67) and the exponential spectrum model, we have

$$\begin{aligned} I_1 &= 2.606 C_n^2 k^2 L \int_0^1 \int_0^\infty \kappa^{-8/3} \left\{ \exp\left[-\frac{\kappa^2}{\kappa_m^2}(1 + \Lambda Q_m \xi^2)\right] \right. \\ &\quad \left. - \exp\left[-\frac{\kappa^2}{\kappa_0^2}(1 + \Lambda Q_0 \xi^2)\right] \right\} [I_0(\Lambda\, \rho\, \xi\kappa) - 1]\, d\kappa d\xi, \end{aligned} \tag{8.69}$$

where $Q_m = L\kappa_m^2/k$. Also, in the second exponential term, we have set $\kappa_{m0} \approx \kappa_0$ and $Q_0 = L\kappa_0^2/k$. Next, we introduce the small-argument approximation of the modified Bessel function $I_0(x) - 1 \sim x^2/4$, which for the inside integral leads to

$$\begin{aligned} &\frac{(\Lambda\rho\xi)^2}{4} \int_0^\infty \kappa^{-2/3} \left\{ \exp\left[-\frac{\kappa^2}{\kappa_m^2}(1 + \Lambda Q_m \xi^2)\right] - \exp\left[-\frac{\kappa^2}{\kappa_0^2}(1 + \Lambda Q_0 \xi^2)\right] \right\} d\kappa \\ &\qquad = \frac{(\Lambda\rho)^2}{8} \Gamma\left(\frac{1}{6}\right) \left[\frac{\kappa_m^{1/3}\xi^2}{(1 + \Lambda Q_m \xi^2)^{1/6}} - \frac{\kappa_0^{1/3}\xi^2}{(1 + \Lambda Q_0 \xi^2)^{1/6}} \right]. \end{aligned} \tag{8.70}$$

Finally, finishing up the integration on ξ by using the integral relations

$$\int_0^1 \frac{\xi^2}{(1+\Lambda Q_m\xi^2)^{1/6}}d\xi = \frac{1}{3}\,{}_2F_1\left(\frac{3}{2},\frac{1}{6};\frac{5}{2};-\Lambda Q_m\right),$$
$$\int_0^1 \frac{\xi^2}{(1+\Lambda Q_0\xi^2)^{1/6}}d\xi = \frac{1}{3}\,{}_2F_1\left(\frac{3}{2},\frac{1}{6};\frac{5}{2};-\Lambda Q_0\right),$$

we arrive at

$$\begin{aligned} I_1 &= 0.4914\sigma_R^2\Lambda^2\left(\frac{k\rho^2}{L}\right)\left[Q_m^{1/6}\,{}_2F_1\left(\frac{3}{2},\frac{1}{6};\frac{5}{2};-\Lambda Q_m\right)\right. \\ &\quad \left. - Q_0^{1/6}\,{}_2F_1\left(\frac{3}{2},\frac{1}{6};\frac{5}{2};-\Lambda Q_0\right)\right], \\ &\cong 0.4914\sigma_R^2\Lambda^2\left(\frac{k\rho^2}{L}\right)\left[Q_m^{1/6}\,{}_2F_1\left(\frac{3}{2},\frac{1}{6};\frac{5}{2};-\Lambda Q_m\right) - Q_0^{1/6}\right]. \end{aligned} \tag{8.71}$$

where $\sigma_R^2 = 1.23C_n^2k^{7/6}L^{11/6}$ is the Rytov variance and we used the small argument approximation $\Lambda Q_0 \ll 1$ in the last step of Eq. (8.71).

For the integral I_2 defined by Eq. (8.68), we begin with

$$\begin{aligned} I_2 &= 2.606C_n^2k^2L\int_0^1\int_0^\infty \kappa^{-8/3}\left\{\exp\left[-\frac{\kappa^2}{\kappa_m^2}(1+\Lambda Q_m\xi^2)\right]\right. \\ &\quad \left. - \exp\left[-\frac{\kappa^2}{\kappa_0^2}(1+\Lambda Q_0\xi^2)\right]\right\}\{1 - J_0[(1-\overline{\Theta}\xi)\kappa\rho]\}d\kappa d\xi. \end{aligned} \tag{8.72}$$

Upon evaluating the inside integral with the help of integrals ($D1$) and ($D2$) in Appendix D, we have

$$\begin{aligned} I_2 &= 1.303C_n^2k^2L\int_0^1\left[\left[\kappa_m^{-5/3}\Gamma\left(-\frac{5}{6}\right)(1+\Lambda Q_m\xi^2)^{5/6}\right.\right. \\ &\quad \left.\left. \times\left\{1 - {}_1F_1\left[-\frac{5}{6};1;-\frac{(1-\overline{\Theta}\xi)^2\kappa_m^2\rho^2}{4(1+\Lambda Q_m\xi^2)}\right]\right\}\right]\right]d\xi \\ &\quad - 1.303C_n^2k^2L\int_0^1\left[\left[\kappa_0^{-5/3}\Gamma\left(-\frac{5}{6}\right)(1+\Lambda Q_0\xi^2)^{5/6}\right.\right. \\ &\quad \left.\left. \times\left\{1 - {}_1F_1\left[-\frac{5}{6};1;-\frac{(1-\overline{\Theta}\xi)^2\kappa_0^2\rho^2}{4(1+\Lambda Q_0\xi^2)}\right]\right\}\right]\right]d\xi. \end{aligned} \tag{8.73}$$

To continue, we make use of the large-argument approximation for the first confluent hypergeometric function and the small-argument approximation for the second confluent hypergeometric function in Eq. (8.73). Combined with the integral relations

$$\int_0^1 (1+\Lambda Q_m \xi^2)^{5/6} d\xi = {}_2F_1\left(-\frac{5}{6}, \frac{1}{2}; \frac{3}{2}; -\Lambda Q_m\right),$$

$$\int_0^1 \frac{(1-\overline{\Theta}\xi)^2}{(1+\Lambda Q_0 \xi^2)^{1/6}} d\xi = \frac{1}{40\Lambda Q_0}\left[48\overline{\Theta} + (15\overline{\Theta}^2 - 48\overline{\Theta})(1+\Lambda Q_0)^{5/6} \right.$$
$$\left. + (40\Lambda Q_0 - 15\overline{\Theta}^2)\,{}_2F_1\left(\frac{1}{6}, \frac{1}{2}; \frac{3}{2}; -\Lambda Q_0\right)\right],$$

this action leads to

$$\begin{aligned} I_2 \approx 8.702 C_n^2 k^2 L \Bigg\{ & \frac{1.063\rho^{5/3}}{4^{5/6}} \int_0^1 (1-\overline{\Theta}\xi)^{5/3} d\xi \\ & - \kappa_m^{-5/3}\,{}_2F_1\left(-\frac{5}{6}, \frac{1}{2}; \frac{3}{2}; -\Lambda Q_m\right) \\ & - \frac{5\kappa_0^{1/3}\rho^2}{960\Lambda Q_0}\left[48\overline{\Theta} + (15\overline{\Theta}^2 - 48\overline{\Theta})(1+\Lambda Q_0)^{5/6} \right. \\ & \left. + (40\Lambda Q_0 - 15\overline{\Theta}^2)\,{}_2F_1\left(\frac{1}{6}, \frac{1}{2}; \frac{3}{2}; -\Lambda Q_0\right)\right]\Bigg\}. \end{aligned} \tag{8.74}$$

After the final integration on ξ, the last expression simplifies to

$$\begin{aligned} I_2 \approx\ & 0.889\sigma_R^2 a\left(\frac{k\rho^2}{L}\right)^{5/6} - 6.563\sigma_R^2 Q_m^{-5/3}\,{}_2F_1\left(-\frac{5}{6}, \frac{1}{2}; \frac{3}{2}; -\Lambda Q_m\right) \\ & - 1.474\sigma_R^2(\kappa_0\rho)^{1/3}\left(\frac{k\rho^2}{L}\right)^{5/6}\left\{\frac{1}{40\Lambda Q_0}\left[48\overline{\Theta} + (15\overline{\Theta}^2 - 48\overline{\Theta})(1+\Lambda Q_0)^{5/6} \right.\right. \\ & \left.\left. + (40\Lambda Q_0 - 15\overline{\Theta}^2)\,{}_2F_1\left(\frac{1}{6}, \frac{1}{2}; \frac{3}{2}; -\Lambda Q_0\right)\right]\right\} \\ \approx\ & 0.889\sigma_R^2 a\left(\frac{k\rho^2}{L}\right)^{5/6} - 1.474\sigma_R^2(\kappa_0\rho)^{1/3}\left(\frac{k\rho^2}{L}\right)^{5/6}. \end{aligned} \tag{8.75}$$

In the last step above, we have once again dropped the less significant terms that sum to nearly zero. The constant a is defined by

$$a = \begin{cases} \dfrac{1-\Theta^{8/3}}{1-\Theta}, & \Theta \geq 0 \\[2ex] \dfrac{1+|\Theta|^{8/3}}{1-\Theta}, & \Theta < 0. \end{cases} \tag{8.76}$$

Finally, by combining Eqs. (8.71) and (8.75), the total WSF is

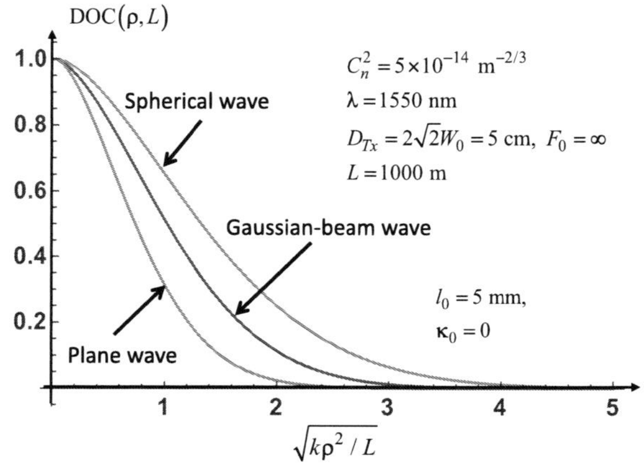

Figure 8.5 Plots of the modulus of the complex DOC for a Gaussian-beam wave, plane wave, and spherical wave where the WSF is defined by Eqs. (8.77), (8.41), and (8.57), respectively. The inner scale is taken to be 5 mm, and the outer scale is infinite.

$$\begin{aligned} D(\rho, L) &= I_1 + I_2 \\ &\approx \sigma_R^2 \left\{ 0.889a\left(\frac{k\rho^2}{L}\right)^{5/6} - 1.474Q_0^{1/6}\left(\frac{k\rho^2}{L}\right) \right. \\ &\quad \left. + 0.4914\Lambda^2\left[Q_m^{1/6}\,{}_2F_1\left(\frac{3}{2}, \frac{1}{6}; \frac{5}{2}; -\Lambda Q_m\right) - Q_0^{1/6}\right]\left(\frac{k\rho^2}{L}\right)\right\}. \end{aligned} \tag{8.77}$$

In Fig. 8.5, we plot the expression $\mathrm{DOC}(\rho, L) = \exp\left[-\frac{1}{2}D(\rho, L)\right]$ as a function of $\sqrt{k\rho^2/L}$ defined by Eq. (8.77) for a Gaussian-beam wave, a plane wave defined by Eq. (8.41), and a spherical wave defined by Eq. (8.57). The path length is 1000 m, the refractive-index structure constant is $C_n^2 = 5 \times 10^{-14}\ \mathrm{m}^{-2/3}$, the inner scale is 5 mm, and the outer scale is assumed to be infinite. The wavelength is 1550 nm.

By setting $l_0 = 0$ and $\kappa_0 = 0$, the WSF in Eq. (8.77) reduces to the inertial-range form

$$D(p, L) \approx 0.889\sigma_R^2\left[a\left(\frac{k\rho^2}{L}\right)^{5/6} + 0.622\Lambda^{11/6}\left(\frac{k\rho^2}{L}\right)\right], \tag{8.78}$$

which agrees with the previously derived expression.[8] In order to obtain the spatial coherence radius as provided in Eqs. (8.43) and (8.59), we first introduce the approximation $(k\rho^2/L) \approx (k\rho^2/L)^{5/6}$ in the last term of Eq. (8.78). Doing so, we find that the spatial coherence radius for a Gaussian-beam wave can be approximated by

$$\rho_0 \approx \left[\frac{8}{3(a+0.622\Lambda^{11/6})}\right]^{3/5}(1.457C_n^2k^2L)^{-3/5}, \qquad l_0 \ll \rho_0 \ll L_0. \tag{8.79}$$

In the near field of the Tx ($\Lambda_0 \ll 1$), the spatial coherence radius (8.79) for a collimated beam approaches that of a plane wave, whereas in the far field ($\Lambda_0 \gg 1$), it approaches that of a spherical wave (cf. Fig. 2.3 of Ref. 2).

Although the derived expressions for the WSF and spatial coherence radius for an unbounded plane wave and spherical wave are believed to be valid under both weak and strong irradiance fluctuations, the same is not true in general for the Gaussian-beam wave. However, as mentioned above, the Gaussian-beam wave in the far field approaches a spherical wave model and, in that case, has essentially the same coherence radius as that of a spherical wave.

8.3.5 Long-term beam radius

The notion of spot size is meaningful only in the case of a Gaussian-beam wave, and under isotropic turbulence it is normally determined from the on-axis mean irradiance calculated from the MCF by setting $\mathbf{r}_1 = \mathbf{r}_2 = \mathbf{0}$. The *long-term beam radius* at the pupil plane of a Rx located distance L from the Tx in the *near field* or *far field* range is defined by (p. 31 of Ref. 2)

$$W_{LT} = \begin{cases} W(1+T)^{1/2}, & \text{(near field: } \Lambda_0 \ll 1) \\ W(1+T)^{3/5}, & \text{(far field: } \Lambda_0 \gg 1), \end{cases} \tag{8.80}$$

where $W = W_0\sqrt{\Theta_0^2 + \Lambda_0^2}$ is the free-space spot radius at the Rx and T is defined by Eq. (8.35).

By expanding the exponential function in Eq. (8.35) in a power series and integrating termwise with spectrum model (8.2), we get

$$\begin{aligned} T &= 1.303C_n^2k^2L\sum_{n=1}^{\infty}\frac{(-1)^{n-1}(\Lambda L/k)^n}{n!}\int_0^{\infty}\kappa^{2n+1}\frac{\exp(-\kappa^2/\kappa_m^2)}{(\kappa^2+\kappa_0^2)^{11/6}}\,d\kappa\int_0^1 \xi^{2n}\,d\xi \\ &= 0.651\,C_n^2k^2L\,\kappa_0^{-5/3}\sum_{n=1}^{\infty}\frac{(-1)^{n-1}(\Lambda L\kappa_0^2/k)^n}{(2n+1)n!}\Gamma(n+1)U\left(n+1;n+\frac{1}{6};\frac{\kappa_0^2}{\kappa_m^2}\right), \end{aligned} \tag{8.81}$$

where we have recognized from (D5) in Appendix D the integral over spatial wave number κ as that of the confluent hypergeometric function of the second kind. Based on $\kappa_0^2 \ll \kappa_m^2$, we can use the small-argument approximation for the confluent hypergeometric function $U(a;\,c;\,z)$, which leads to

$$T = -\,0.782 C_n^2 k^2 L \kappa_0^{-5/3} \sum_{n=1}^{\infty} \frac{\left(\frac{1}{2}\right)_n (1)_n \left(\Lambda L \kappa_0^2/k\right)^n}{\left(\frac{3}{2}\right)_n \left(\frac{1}{6}\right)_n n!}$$
$$+\,4.351 C_n^2 k^2 L \kappa_m^{-5/3} \sum_{n=1}^{\infty} \frac{(-1)^n \left(-\frac{5}{6}\right)_n \left(\frac{1}{2}\right)_n \left(\Lambda L \kappa_m^2/k\right)^n}{\left(\frac{3}{2}\right)_n n!}. \tag{8.82}$$

In obtaining this last result, we followed the analysis provided in Section 8.1.3. By identifying each series in Eq. (8.82) as related to a generalized hypergeometric function, we arrive at the far-field long-term beam radius

$$W_{LT} = W\left\{1 + 4.351 C_n^2 k^2 L \kappa_m^{-5/3}\left[{}_2F_1\left(-\frac{5}{6}, \frac{1}{2}; \frac{3}{2}; -\Lambda Q_m\right) - 1\right]\right.$$
$$\left. -\,0.782 C_n^2 k^2 L \kappa_0^{-5/3}\left[{}_2F_2\left(\frac{1}{2}, 1; \frac{3}{2}, \frac{1}{6}; \Lambda Q_0\right) - 1\right]\right\}^{3/5}, \tag{8.83}$$

where $Q_m = L\kappa_m^2/k$ and $Q_0 = L\kappa_0^2/k$. The small-argument approximation for the ${}_2F_2$ function and the approximation method from Chapter 6 [recall Eq. (6.4)] for the hypergeometric function ${}_2F_1$ now permit us to approximate both hypergeometric-type functions in Eq. (8.83), from which we obtain

$$W_{LT} \approx W\left\{1 + 3.537\sigma_R^2 \Lambda^{5/6}\left[\frac{(1 + 0.31\Lambda Q_m)^{5/6} - 1}{(\Lambda Q_m)^{5/6}} - 0.358(\Lambda Q_0)^{1/6}\right]\right\}^{3/5}. \tag{8.84}$$

If we use the modified atmospheric spectrum model (8.3), the term T then takes the approximate form[10]

$$T \approx 3.537\sigma_R^2 \Lambda^{5/6}\left\{\frac{(1 + 0.31\Lambda Q_l)^{5/6} - 1.909}{(\Lambda Q_l)^{5/6}} + 1.096\frac{(1 + 0.22\Lambda Q_l)^{1/3}}{(\Lambda Q_l)^{5/6}}\right.$$
$$\left. -\,0.187\frac{(1 + 0.012\Lambda Q_l)^{1/4}}{(\Lambda Q_l)^{5/6}} - 0.358(\Lambda Q_0)^{1/6}\right\}, \tag{8.85}$$

where $Q_l = L\kappa_l^2/k$. If the outer scale is ignored or $L_0 \gg 1$ m, then the inner scale alone generally has a small effect on the long-term spot size W_{LT}. Thus, it is customary to base the spot radius primarily on the Kolmogorov spectrum (8.1) for which $T = 1.33\sigma_R^2 \Lambda^{5/6}$. However, if the outer scale is on the order of

1–2 m or less, then the spot size can be significantly reduced from that based on spectrum model (8.1).

8.4 Scintillation Index

The fourth-order field moment involves irradiance fluctuations of the optical wave leading to the *covariance function* and *scintillation index*. The covariance function is used to calculate the correlation width of the irradiance fluctuations, whereas knowledge of the scintillation index is critical for determining system performance in a laser communication system or laser radar link.

In general, the scintillation index of a Gaussian-beam wave under weak irradiance fluctuations is defined by

$$\begin{aligned}\sigma_I^2(r, L) &= 2\mathrm{Re}[E_2(\mathbf{r}, \mathbf{r}) + E_3(\mathbf{r}, \mathbf{r})] \\ &= 8\pi^2 k^2 L \int_0^1\int_0^\infty \kappa\Phi_n(\kappa)\exp\left(-\frac{\Lambda L\kappa^2\xi^2}{k}\right) \\ &\quad\times\left\{I_0(2\Lambda r\kappa\xi) - \cos\left[\frac{L\kappa^2}{k}\xi(1-\overline{\Theta}\xi)\right]\right\}d\kappa\, d\xi.\end{aligned} \tag{8.86}$$

In the following analysis we separately consider plane waves, spherical waves, and Gaussian-beam waves. We don't include the outer-scale parameter in the analysis because it was previously shown that the outer scale L_0 does not affect the scintillation index of an optical wave under weak irradiance fluctuations except at off-axis points in a Gaussian-beam wave.[11]

8.4.1 Plane wave

To begin our analysis, let us start with the scintillation index of an infinite plane wave based on Eq. (8.2) with $\kappa_0 = 0$. We first observe that the modified Bessel function and exponential function in Eq. (8.86) become unity when $\Theta = 1$, $\Lambda = 0$. Thus, by also setting $\cos x = \mathrm{Re}(e^{-ix})$, Eq. (8.86) reduces to

$$\begin{aligned}\sigma_{I,pl}^2(L) &= 8\pi^2 k^2 L\,\mathrm{Re}\int_0^1\int_0^\infty \kappa\Phi_n(\kappa)\left[1-\exp\left(-\frac{iL\kappa^2\xi}{k}\right)\right]d\kappa\, d\xi \\ &= 2.606 C_n^2 k^2 L \\ &\quad\times\lim_{\mu\to-8/3}\mathrm{Re}\int_0^1\int_0^\infty \kappa^\mu\left\{\exp\left(-\frac{\kappa^2}{\kappa_m^2}\right) - \exp\left[-\frac{\kappa^2}{\kappa_m^2}(1+iQ_m\xi)\right]\right\}d\kappa\, d\xi.\end{aligned} \tag{8.87}$$

For this case we use the result of ($D2$) in Appendix D to obtain

$$\begin{aligned}\sigma^2_{I,pl}(L) &= 1.303C_n^2k^2L \\ &\times \lim_{\mu\to -8/3}\text{Re}\left\{\kappa_m^{-(1+\mu)}\Gamma\left(\frac{1+\mu}{2}\right)\left[1-\int_0^1 (1+iQ_m\xi)^{(-1-\mu)/2}\,d\xi\right]\right\} \\ &= 7.07\sigma_R^2 Q_m^{-5/6}\left(\frac{6}{11}\right)\text{Re}\left[\frac{(1+iQ_m)^{11/6}}{iQ_m}-\frac{1}{iQ_m}-\frac{11}{6}\right].\end{aligned} \tag{8.88}$$

Upon calculating the real part, this last expression takes the form

$$\sigma^2_{I,pl}(L) = 3.86\sigma_R^2\left[\left(1+\frac{1}{Q_m^2}\right)^{11/12}\sin\left(\frac{11}{6}\tan^{-1}Q_m\right)-\frac{11}{6}Q_m^{-5/6}\right]. \tag{8.89}$$

If the inner scale is zero, then $Q_m \to \infty$ and Eq. (8.89) reduces to the Rytov variance $\sigma^2_{I,pl}(L)|_{l_0=0} = \sigma^2_R$.

Last, the scintillation index for a plane wave based on the modified atmospheric spectrum (8.3) is given by the similar expression,[11]

$$\begin{aligned}\sigma^2_{I,pl}(L) = 3.86\sigma_R^2\Bigg\{&\left(1+\frac{1}{Q_l^2}\right)^{11/12}\left[\sin\left(\frac{11}{6}\tan^{-1}Q_l\right)\right. \\ &+\frac{1.507}{(1+Q_l^2)^{1/4}}\sin\left(\frac{4}{3}\tan^{-1}Q_l\right) \\ &\left.-\frac{0.273}{(1+Q_l^2)^{7/24}}\sin\left(\frac{5}{4}\tan^{-1}Q_l\right)\right]-\frac{3.50}{Q_l^{5/6}}\Bigg\}.\end{aligned} \tag{8.90}$$

8.4.2 Spherical wave

For the case of a spherical wave, the scintillation index with $\Theta = \Lambda = 0$ and $\cos x = \text{Re}(e^{-ix})$ is defined by

$$\begin{aligned}\sigma^2_{I,sph}(L) &= 8\pi^2k^2L\text{Re}\int_0^1\int_0^\infty \kappa\Phi_n(\kappa)\left\{1-\exp\left[-\frac{iL\kappa^2\xi(1-\xi)}{k}\right]\right\}d\kappa\,d\xi \\ &= 2.606C_n^2k^2L\lim_{\mu\to -8/3}\text{Re}\int_0^1\int_0^\infty \kappa^\mu\left[\left[\exp\left(-\frac{\kappa^2}{\kappa_m^2}\right)\right.\right. \\ &\left.\left.-\exp\left\{-\frac{\kappa^2}{\kappa_m^2}[1+iQ_m\xi(1-\xi)]\right\}\right]\right]d\kappa\,d\xi,\end{aligned} \tag{8.91}$$

where we are again using the spectrum model (8.2) with $\kappa_0 = 0$. After calculating the inside integral using ($D2$) in Appendix D, taking the limit and performing some algebraic simplification, we get

$$\sigma_{I,sph}^2(L) = 7.07\sigma_R^2\, Q_m^{-5/6}\mathrm{Re}\int_0^1 \{[1 + iQ_m\xi(1-\xi)]^{5/6} - 1\}\, d\xi. \tag{8.92}$$

To continue, we impose the temporary assumption that $Q_m < 1$; thus, the binomial function within the integral in Eq. (8.92) has the following power series representation [recall Eq. (3.6)]:

$$\begin{aligned}\int_0^1 [1 + iQ_m\xi(1-\xi)]^{5/6} d\xi &= \sum_{n=0}^{\infty} (-1)^n\left(-\frac{5}{6}\right)_n \frac{(iQ_m)^n}{n!}\int_0^1 \xi^n(1-\xi)^n d\xi \\ &= \sum_{n=0}^{\infty} (-1)^n\left(-\frac{5}{6}\right)_n \frac{(iQ_m)^n}{n!}\frac{\Gamma^2(1+n)}{\Gamma(2+2n)}, \quad Q_m < 1,\end{aligned} \tag{8.93}$$

where we have used termwise integration and the beta function (*G*17) in Appendix A. After converting to Pochhammer symbols,

$$\frac{\Gamma^2(1+n)}{\Gamma(2+2n)} = \frac{(1)_n(1)_n}{2^{2n}(1)_n\left(\frac{3}{2}\right)_n} = \frac{(1)_n}{4^n\left(\frac{3}{2}\right)_n},$$

and summing the resulting series, we obtain

$$\sigma_{I,sph}^2(L) = 7.07\sigma_R^2 Q_m^{-5/6}\mathrm{Re}\left[{}_2F_1\left(-\frac{5}{6}, 1; \frac{3}{2}; -\frac{iQ_m}{4}\right) - 1\right], \tag{8.94}$$

thus removing the restriction that $Q_m < 1$. Although Eq. (8.94) is an exact result, we can produce a simple algebraic approximation to it by methods introduced in Chapter 6. In this case, we find that

$${}_2F_1\left(-\frac{5}{6}, 1; \frac{3}{2}; -\frac{iQ_m}{4}\right) \sim \begin{cases} 1 + 0.139\, iQ_m, & Q_m \to 0 \\ 0.221(iQ_m)^{5/6}, & Q_m \to \infty, \end{cases}$$

which leads to

$${}_2F_1\left(-\frac{5}{6}, 1; \frac{3}{2}; -\frac{iQ_m}{4}\right) \approx (1 + 0.162\, iQ_m)^{5/6}. \tag{8.95}$$

Taking the real part of Eq. (8.95), it follows that

$$\sigma_{I,sph}^2(L) \approx 7.07\sigma_R^2 Q_m^{-5/6}\left\{\left(1 + 0.0262\, Q_m^2\right)^{5/12}\cos\left[\frac{5}{6}\tan^{-1}(0.162Q_m)\right] - 1\right\}. \tag{8.96}$$

A similar solution for the scintillation index based on the spectrum model (8.3) that was previously derived is given by[11]

$$\sigma_{I,sph}^2(L) \approx 9.65\beta_0^2\left\{0.40\left(1+\frac{9}{Q_l^2}\right)^{11/12}\left[\sin\left(\frac{11}{6}\tan^{-1}\frac{Q_l}{3}\right)\right.\right.$$
$$+\frac{2.610}{(9+Q_l^2)^{1/4}}\sin\left(\frac{4}{3}\tan^{-1}\frac{Q_l}{3}\right) \tag{8.97}$$
$$\left.\left.-\frac{0.518}{(9+Q_l^2)^{7/24}}\sin\left(\frac{5}{4}\tan^{-1}\frac{Q_l}{3}\right)\right]-\frac{3.50}{Q_l^{5/6}}\right\},$$

where $\beta_0^2 = 0.496C_n^2k^{7/6}L^{11/6} = 0.40\,\sigma_R^2$. Both Eqs. (8.96) and (8.97) reduce to β_0^2 with $l_0 = 0$.

8.4.3 Gaussian-beam wave

Although it is known that beam wander can induce additional scintillation effects on a Gaussian-beam wave, we ignore beam-wander-induced scintillation in the analysis here by assuming that the beam is perfectly tracked (see Chapter 3 of Ref. 2 for a development of mathematical models concerning beam-wander-induced scintillation).

The scintillation index of a Gaussian-beam wave as defined by Eq. (8.86) can be written as a sum of *radial* and *longitudinal components*. The radial component based on Eq. (8.1), $I_0(x) = J_0(ix)$, and ($D1$) and ($D2$) in Appendix D leads to the result

$$\begin{aligned}\sigma_{I,r}^2(r, L) &= 8\pi^2k^2L\int_0^1\int_0^\infty \kappa\Phi_n(\kappa)\exp\left(-\frac{\Lambda L\kappa^2\xi^2}{k}\right)[I_0(2\Lambda r\kappa\xi)-1]d\kappa\, d\xi\\ &= 2.65\sigma_R^2\Lambda^{5/6}\left[1-{}_1F_1\left(-\frac{5}{6};1;\frac{2r^2}{W^2}\right)\right].\end{aligned} \tag{8.98}$$

To calculate the longitudinal component, we start by setting $r = 0$ and making the replacement $\cos x = \mathrm{Re}(e^{-ix})$ in Eq. (8.86). In this case, we will first use the Kolmogorov spectrum model (8.1), which leads to (after some algebraic rearranging)

$$\begin{aligned}\sigma_I^2(0, L) = 1.303C_n^2k^2L\mathrm{Re}\int_0^1\int_0^\infty \kappa^{-8/3}\\ \times\left[\left[\exp\left(-\frac{\Lambda L\kappa^2\xi^2}{k}\right)-\exp\left\{-\frac{iL\kappa^2\xi}{k}[1-(\overline{\Theta}+i\Lambda)\xi]\right\}\right]\right]d\kappa\, d\xi.\end{aligned} \tag{8.99}$$

Although we could rely on the technique used for the plane wave and spherical wave scintillation index, here we wish to illustrate a slightly different approach to removing the singularity at $\kappa = 0$.

The inside integral in Eq. (8.99) has the algebraic form

$$I = \int_0^\infty \kappa^{-8/3}[\exp(-\alpha\kappa^2) - \exp(-\beta\kappa^2)]\, d\kappa, \tag{8.100}$$

where

$$\alpha = \frac{\Lambda L \xi^2}{k}, \quad \beta = \frac{iL\xi}{k}[1 - (\overline{\Theta} + i\Lambda)\xi]. \tag{8.101}$$

By expressing each exponential function in Eq. (8.100) in a power series, we find that the $n = 0$ term from each series cancels, leaving

$$\begin{aligned}\exp(-\alpha\kappa^2) - \exp(-\beta\kappa^2) &= \sum_{\substack{n=1\\ n\to n+1}}^{\infty} \frac{(-1)^n \alpha^n \kappa^{2n}}{n!} - \sum_{\substack{n=1\\ n\to n+1}}^{\infty} \frac{(-1)^n \beta^n \kappa^{2n}}{n!} \\ &= -\alpha\kappa^2 \sum_{n=0}^{\infty} \frac{(-1)^n (1)_n \alpha^n \kappa^{2n}}{(2)_n n!} + \beta\kappa^2 \sum_{n=0}^{\infty} \frac{(-1)^n (1)_n \beta^n \kappa^{2n}}{(2)_n n!} \\ &= -\alpha\kappa^2\, {}_1F_1(1;2;-\alpha\kappa^2) + \beta\kappa^2\, {}_1F_1(1;2;-\beta\kappa^2).\end{aligned} \tag{8.102}$$

By this procedure, which is a variation of that presented in Section 8.2.2, we have removed the common singularity in each term of Eq. (8.99). Thus, each term in Eq. (8.102) inserted back into Eq. (8.100) can be integrated independently without introducing the limit operation as suggested in Section 8.2.2. In this case we use ($C13$) in Appendix C for both confluent hypergeometric functions in Eq. (8.102). For the first confluent hypergeometric function, this leads to

$$\begin{aligned}-\alpha \int_0^\infty \kappa^{-2/3} G_{12}^{11}\left(\alpha\kappa^2 \middle| \begin{matrix} 0 \\ 0,\ -1 \end{matrix}\right) d\kappa &= -\frac{\Gamma\left(\frac{1}{6}\right)\Gamma\left(\frac{5}{6}\right)}{2\Gamma\left(\frac{11}{6}\right)} \alpha^{5/6} \\ &= -3.34\alpha^{5/6},\end{aligned} \tag{8.103}$$

where we have made the change of variable $t = \kappa^2$ and applied ($D8$) in Appendix D. The remaining integral involving β leads to the same result with β replacing α. Next, by completing the integration in Eq. (8.99) on the variable ξ for the first term in that equation, we arrive at

$$\int_0^1 \alpha^{5/6} d\xi = \left(\frac{\Lambda L}{k}\right)^{5/6} \int_0^1 \xi^{5/3}\, d\xi = \frac{3}{8}\left(\frac{\Lambda L}{k}\right)^{5/6}. \tag{8.104}$$

For the remaining integral, we start with ($C2$) in Appendix C followed by ($D9$) in Appendix D to find that

$$\begin{aligned}\int_0^1 \xi^{5/6}[1-(\overline{\Theta}+i\Lambda)\xi]^{5/6}\, d\xi &= \Gamma\left(\frac{11}{6}\right)\int_0^1 \xi^{5/6} G_{11}^{10}\left[(\overline{\Theta}+i\Lambda)\xi\left|\begin{matrix}\frac{11}{6}\\ 0\end{matrix}\right.\right] d\xi\\ &= \Gamma\left(\frac{11}{6}\right) G_{22}^{11}\left[(\overline{\Theta}+i\Lambda)\left|\begin{matrix}-\frac{5}{6},\ \frac{11}{6}\\ 0,\ -\frac{11}{6}\end{matrix}\right.\right]\\ &= \frac{6}{11}\,{}_2F_1\left(-\frac{5}{6},\frac{11}{6};\frac{17}{6};\overline{\Theta}+i\Lambda\right).\end{aligned} \tag{8.105}$$

The last step in Eq. (8.105) is based on applying ($C18$) in Appendix C. Collecting the relevant expressions (8.99) through (8.105) leads to the result

$$\sigma_I^2(0,L) = 3.86\,\sigma_R^2 \mathrm{Re}\left[i^{5/6}\,{}_2F_1\left(-\frac{5}{6},\frac{11}{6};\frac{17}{6};\overline{\Theta}+i\Lambda\right)-\frac{11}{16}\Lambda^{5/6}\right]. \tag{8.106}$$

Based on Eq. (6.7) for the above hypergeometric function, we can approximate the scintillation index (8.106) by the expression

$$\begin{aligned}\sigma_I^2(0,L) &\approx 3.86\sigma_R^2 \mathrm{Re}\left\{i^{5/6}[1-0.65(\overline{\Theta}+i\Lambda)]^{5/6}-\frac{11}{16}\Lambda^{5/6}\right\}\\ &\approx 3.86\sigma_R^2[(1-0.65\overline{\Theta})^2+0.42\Lambda^2]^{5/12}\\ &\quad\times\cos\left[\frac{5}{6}\tan^{-1}\left(\frac{-0.65\Lambda}{1-0.65\overline{\Theta}}\right)+\frac{5\pi}{12}\right]-2.65\sigma_R^2\Lambda^{5/6}.\end{aligned} \tag{8.107}$$

In Fig. 8.6 we plot both Eqs. (8.106) and (8.107) as a function of range up to a distance of 1200 m for a collimated beam with $W_0 = 3$ cm. Because the two plots are virtually indistinguishable, we have not labeled them as approximation and exact as we have done in previous plots. For smaller beams with $W_0 < 3$ cm, the approximation (8.107) may be a little less accurate, depending on range and wavelength.

To include the effects of inner scale in the longitudinal component, we again use spectrum model (8.2), but with $\kappa_0 = 0$. By inserting $e^{-\kappa^2/\kappa_m^2}$ in Eq. (8.99) and then using ($D2$) in Appendix D to solve the integral over wave number κ, we see that

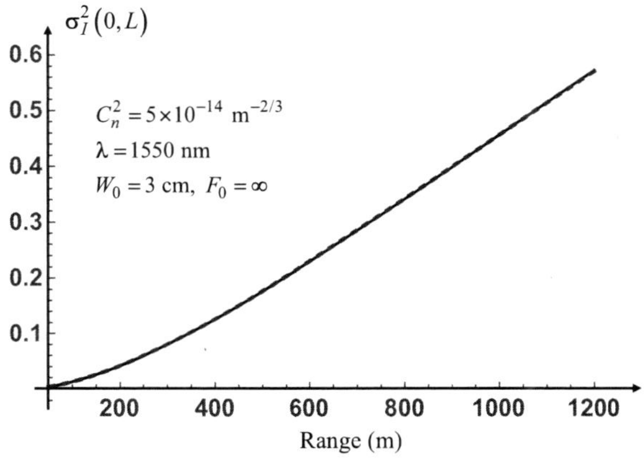

Figure 8.6 Plot of the on-axis scintillation index of a collimated Gaussian-beam wave given by Eq. (8.106)] and the approximation given by Eq. (8.107).

$$\int_0^\infty \kappa^{-8/3}\left\{\exp\left[-\frac{L\kappa^2}{k}\left(\Lambda\xi^2+\frac{1}{Q_m}\right)\right]-\exp\left[-\frac{L\kappa^2}{k}\left(\Lambda\xi^2+\frac{1}{Q_m}+i\xi(1-\overline{\Theta}\xi)\right)\right]\right\}d\kappa$$
$$=\frac{1}{2}\Gamma\left(-\frac{5}{6}\right)\left(\frac{L}{k}\right)^{5/6}Q_m^{-5/6}\{(1+\Lambda Q_m\xi^2)^{5/6}-[1+i\xi Q_m(1-\overline{\Theta}\xi-i\Lambda\xi)]^{5/6}\}. \tag{8.108}$$

Next, under the temporary assumption that $Q_m < 1$, we expand the binomial expressions in Eq. (8.108) in their respective power series as we did in Eq. (8.93), and then perform the integration on ξ. Doing so, we find that the longitudinal component of the scintillation index becomes

$$\begin{aligned}\sigma_I^2(0,\,L) = 7.08\sigma_R^2 Q_m^{-5/6}\Bigg\{&\mathrm{Re}\left[\sum_{n=0}^{\infty}\frac{(-5/6)_n(1)_n}{(2)_n n!}(-iQ_m)^n \right.\\ &\left.\times\,{}_2F_1(-n,\,n+1;\,n+2;\,\overline{\Theta}+i\Lambda)\right] \\ &-{}_2F_1\left(-\frac{5}{6},\,\frac{1}{2};\,\frac{3}{2};\,-\Lambda Q_m\right)\Bigg\},\qquad Q_m<1.\end{aligned} \tag{8.109}$$

The restriction $Q_m < 1$ is necessary for the infinite series in Eq. (8.109) to converge, but the restriction can be removed after summation of the series. The procedure we use for summing the series is to make the approximation

$${}_2F_1(-n,\,n+1;\,n+2;\,z)\approx\left(1-\frac{2z}{3}\right)^n, \tag{8.110}$$

which is exact only when $n = 0$ and $n = 1$. However, the first two terms of the series in Eq. (8.109) are the dominant terms and so, upon summing the series with the use of approximation (8.110), we obtain

$$\sum_{n=0}^{\infty} \frac{(-5/6)_n (1)_n}{(2)_n n!} (-iQ_m)^n \left[1 - \frac{2(\overline{\Theta} + i\Lambda)}{3}\right]^n$$
$$= {}_2F_1\left\{-\frac{5}{6}, 1; 2; -Q_m\left[\frac{2\Lambda}{3} + i\left(1 + \frac{2\overline{\Theta}}{3}\right)\right]\right\} \tag{8.111}$$
$$= \frac{6}{11}\left[\left[\frac{\{1 + Q_m[2\Lambda/3 + i(1 + 2\overline{\Theta}/3)]\}^{11/6} - 1}{Q_m[2\Lambda/3 + i(1 + 2\overline{\Theta}/3)]}\right]\right],$$

the right-hand side of which is now valid for all Q_m. In arriving at Eq. (8.111), we have summed the series as a hypergeometric function followed by use of the identity (B28) in Appendix B,

$${}_2F_1(1 - a, 1; 2; -z) = \frac{(1 + z)^a - 1}{az}. \tag{8.112}$$

The remaining hypergeometric function in Eq. (8.109) leads to the approximation

$${}_2F_1\left(-\frac{5}{6}, \frac{1}{2}; \frac{3}{2}; -\Lambda Q_m\right) \approx (1 + 0.31\Lambda Q_m)^{5/6}, \tag{8.113}$$

which follows from Eq. (6.4) and, thus, we obtain

$$\sigma_I^2(0, L) \approx 3.86\sigma_R^2 Q_m^{-5/6} \mathrm{Re}\left[\left[\frac{\{1 + Q_m[2\Lambda/3 + i(1 + 2\overline{\Theta}/3)]\}^{11/6} - 1}{Q_m[2\Lambda/3 + i(1 + 2\overline{\Theta}/3)]} - \frac{11}{6}\left(\frac{1 + 0.31\Lambda Q_m}{Q_m}\right)^{5/6}\right]\right]. \tag{8.114}$$

Selecting the real part of Eq. (8.114) leads to

$$\sigma_I^2(0, L) \approx 3.86\sigma_R^2\left\{0.40\frac{[(1 + 2\Theta)^2 + (2\Lambda + 3/Q_m)^2]^{11/12}}{[(1 + 2\Theta)^2 + 4\Lambda^2]^{1/2}}\sin\left(\frac{11}{6}\varphi_1 + \varphi_2\right) - \frac{6\Lambda}{Q_m^{11/6}[(1 + 2\Theta)^2 + 4\Lambda^2]} - \frac{11}{6}\left(\frac{1 + 0.31\Lambda Q_m}{Q_m}\right)^{5/6}\right\}, \tag{8.115}$$

where

$$\varphi_1 = \tan^{-1}\left[\frac{(1+2\Theta)Q_m}{3+2\Lambda Q_m}\right], \quad \varphi_2 = \tan^{-1}\left(\frac{2\Lambda}{1+2\Theta}\right). \tag{8.116}$$

An expression for the scintillation index of both the radial and the on-axis components of a Gaussian-beam wave based on the modified atmospheric spectrum model (8.3) is provided in Ref. 11 and on p. 41 of Ref. 2.

8.5 Irradiance Covariance Function: Weak Fluctuations

The scintillation index describes irradiance fluctuations at a single point in the beam. For an infinite plane wave or spherical wave, there is no distinction about which point we select, but for a Gaussian-beam wave only a point chosen on the optical axis leads to the longitudinal component—if the point is off the optical axis, it is part of the radial component. The *irradiance covariance function* on the other hand describes how the irradiance fluctuations at one point on the optical wavefront are related to those at a different point. Knowing the maximum distance between two points at which the irradiance fluctuations are still correlated gives us what is called the *irradiance correlation width* ρ_c. Rx aperture sizes smaller than ρ_c act like a point aperture. Thus, with increasing aperture size, ρ_c is an estimate of the smallest Rx aperture diameter at which aperture averaging (i.e., a reduction in the scintillation index) of the optical wave begins to take effect. This reduction due to aperture averaging is also called the *flux variance* or *power scintillation.*

8.5.1 Plane wave

If $\rho = |\mathbf{r}_1 - \mathbf{r}_2|$ denotes the distance between two points on the wavefront of a plane wave, the irradiance covariance function is defined by

$$\begin{aligned} B_{I,pl}(\rho, L) &= 2\mathrm{Re}[E_2(\mathbf{r}_1, \mathbf{r}_2) + E_3(\mathbf{r}_1, \mathbf{r}_2)] \\ &= 8\pi^2 k^2 L \int_0^1 \int_0^\infty \kappa \Phi_n(\kappa) J_0(\kappa\rho)\left[1 - \cos\left(\frac{L\kappa^2 \xi}{k}\right)\right] d\kappa\, d\xi. \end{aligned} \tag{8.117}$$

Here we will perform the integration on ξ first which, based on spectrum model (8.2) with $\kappa_0 = 0$, leads to

$$B_{I,pl}(\rho, L) = 2.606 C_n^2 k^2 L \int_0^\infty \kappa^{-8/3} e^{-\kappa^2/\kappa_m^2} J_0(\kappa\rho)\left[1 - \frac{k}{L\kappa^2}\sin\left(\frac{L\kappa^2}{k}\right)\right] d\kappa. \tag{8.118}$$

For the term in Eq. (8.118) involving the sine function, we make the replacement $-\sin x = \mathrm{Im}(e^{-ix})$. At this point we basically follow the steps like those illustrated in Section 8.2.2. That is, we temporarily replace the exponent

$-\frac{8}{3}$ in the term $\kappa^{-8/3}$ with a parameter μ ($\mu > -1$) so each individual integral term between the brackets in Eq. (8.118) converges.

For the first term between the brackets in Eq. (8.118), we use ($D1$) in Appendix D, which leads to

$$\int_0^\infty \kappa^{\mu} e^{-\kappa^2/\kappa_m^2} J_0(\kappa\rho)\, d\kappa = \frac{1}{2}\left(\frac{L}{k}\right)^{-(\mu+1)/2} Q_m^{(\mu+1)/2} \Gamma\left(\frac{\mu+1}{2}\right) {}_1F_1\left(\frac{\mu+1}{2};\ 1;\ -\frac{Q_m k\rho^2}{4L}\right). \tag{8.119}$$

For the second term in brackets in Eq. (8.118), we again use ($D1$) in Appendix D to obtain the similar integral result

$$\frac{k}{L}\int_0^\infty \kappa^{\mu-2} \exp\left[-\frac{\kappa^2}{\kappa_m^2}(1+iQ_m)\right] J_0(\kappa\rho)\, d\kappa = \frac{1}{2}\left(\frac{L}{k}\right)^{-(\mu+1)/2} \Gamma\left(\frac{\mu-1}{2}\right)\left(\frac{1}{Q_m}+i\right)^{(1-\mu)/2} {}_1F_1\left[\frac{\mu-1}{2};\ 1;\ -\frac{Q_m k\rho^2}{4L(1+iQ_m)}\right]. \tag{8.120}$$

By combining Eqs. (8.119) and (8.120) and then taking the limit $\mu \to -\frac{8}{3}$, the common singularity vanishes, and we get the exact expression

$$B_{I,pl}(\rho, L) = 3.86\sigma_R^2 \mathrm{Im}\left\{\left(\frac{1}{Q_m}+i\right)^{11/6} {}_1F_1\left[-\frac{11}{6};\ 1;\ -\frac{Q_m k\rho^2}{4L(1+iQ_m)}\right]\right\} - 7.07\sigma_R^2 Q_m^{-5/6}\, {}_1F_1\left(-\frac{5}{6};\ 1;\ -\frac{Q_m k\rho^2}{4L}\right). \tag{8.121}$$

Equation (8.121) represents the plane wave covariance function of irradiance based on the modified von Kármán spectrum model (8.2) with inner scale but setting $\kappa_0 = 0$. Also, we see that for negligible inner scale ($Q_m \to \infty$), Eq. (8.121) reduces to the result derived years ago by Fried and Cloud,[12]

$$B_{I,pl}(\rho, L) = 3.86\sigma_R^2\, \mathrm{Im}\left[i^{11/6}\, {}_1F_1\left(-\frac{11}{6};\ 1;\ \frac{ik\rho^2}{4L}\right)\right] - 7.52\sigma_R^2\left(\frac{k\rho^2}{4L}\right)^{5/6}. \tag{8.122}$$

As an observation here, we see that setting $\rho = 0$ in Eq. (8.121) and taking the imaginary part of the first term reduces Eq. (8.121) to the

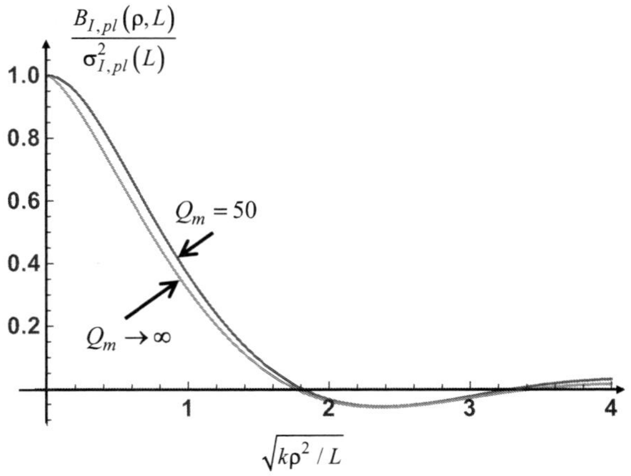

Figure 8.7 Normalized irradiance covariance function of a plane wave with inner scale defined by Eq. (8.121) and zero inner scale defined by Eq. (8.122).

previously derived plane wave scintillation index given by Eq. (8.89); i.e., $B_{I,pl}(0, L) = \sigma^2_{I,pl}(L)$. More importantly, it should be noted that the presence of inner scale has only a modest effect on the irradiance covariance function scaled by the scintillation index. This is shown in Fig. 8.7 where we plot scaled versions of both Eqs. (8.121) and (8.122) as a function of $\sqrt{k\rho^2/L}$ and $Q_m = 50$. For larger values of the inner-scale parameter $Q_m = 35.05L/kl_0^2$ (i.e., smaller values of inner scale l_0), the curves in Fig. 8.7 are even closer. Thus, unless $Q_m < 50$, which corresponds to very large inner-scale values and/or very short path lengths, there is only a small difference between Eqs. (8.121) and (8.122), a consequence of the fact that inner-scale effects are small and essentially cancel in the scaled covariance function. Last, based on the zero crossing of the covariance function in Fig. 8.7, it follows that the irradiance correlation width is $\rho_c = 1.756\sqrt{L/k} = 0.70\sqrt{\lambda L}$.

In many cases the modified atmospheric spectrum model (8.3) leads to an expression for certain statistical quantities that, because of the bump, separate from those based on the modified von Kármán spectrum model (8.2). To see if that is the case here, we basically follow the same steps that led to Eq. (8.121) but now using the modified atmospheric spectrum model (8.3) with $\kappa_0 = 0$ to derive the covariance function. To do so, we start with the integral

$$\begin{aligned} I = \lim_{\mu\to -8/3} \int_0^\infty \kappa^\mu \left[1 + 1.802\frac{\kappa}{\kappa_l} - 0.254\left(\frac{\kappa}{\kappa_l}\right)^{7/6}\right] \\ \times \left\{\exp\left(-\frac{\kappa^2}{\kappa_l^2}\right) J_0(\kappa\rho) - \frac{k}{L\kappa^2}\exp\left[-\frac{\kappa^2}{\kappa_l^2}(1 + iQ_l)\right] J_0(\kappa\rho)\right\} d\kappa, \end{aligned} \tag{8.123}$$

which, upon using ($D1$) in Appendix D and taking the limit $\mu \to -\frac{8}{3}$, leads to

$$\begin{aligned}
I = &\frac{1}{2}\left(\frac{L}{k}\right)^{5/6}\left\{Q_l^{-5/6}\Gamma\left(-\frac{5}{6}\right)\,{}_1F_1\left(-\frac{5}{6};\,1;\,-\frac{Q_l k\rho^2}{4L}\right)\right.\\
&+1.802Q_l^{-5/6}\Gamma\left(-\frac{1}{3}\right)\,{}_1F_1\left(-\frac{1}{3};\,1;\,-\frac{Q_l k\rho^2}{4L}\right)\\
&\left.-0.254Q_l^{-5/6}\Gamma\left(-\frac{1}{4}\right)\,{}_1F_1\left(-\frac{1}{4};\,1;\,-\frac{Q_l k\rho^2}{4L}\right)\right\}\\
&-\frac{1}{2}\left(\frac{L}{k}\right)^{5/6}\left\{\Gamma\left(-\frac{11}{6}\right)\left(\frac{1}{Q_l}+i\right)^{11/6}\,{}_1F_1\left[-\frac{11}{6};\,1;\,-\frac{Q_l k\rho^2}{4L(1+iQ_l)}\right]\right.\\
&+1.802\Gamma\left(-\frac{4}{3}\right)Q_l^{-11/6}(1+iQ_l)^{4/3}\,{}_1F_1\left[-\frac{4}{3};\,1;\,-\frac{Q_l k\rho^2}{4L(1+iQ_l)}\right]\\
&\left.-0.254\Gamma\left(-\frac{5}{4}\right)Q_l^{-11/6}(1+iQ_l)^{5/4}\,{}_1F_1\left[-\frac{5}{4}\,1;\,-\frac{Q_l k\rho^2}{4L(1+iQ_l)}\right]\right\}.
\end{aligned} \tag{8.124}$$

The combination of these results has once again canceled the singularity at $\kappa = 0$ in each term between double brackets of Eq. (8.123) and provides us with an expression for the plane wave irradiance covariance function given by

$$\begin{aligned}
B_{I,pl}(\rho, L) = &3.86\sigma_R^2 \mathrm{Im}\left[\left[\left(\frac{1}{Q_l}+i\right)^{11/6}\left\{{}_1F_1\left[-\frac{11}{6};1;-\frac{Q_l k\rho^2}{4L(1+iQ_l)}\right]\right.\right.\right.\\
&+\frac{1.507}{(1+iQ_l)^{1/2}}\,{}_1F_1\left[-\frac{4}{3};1;-\frac{Q_l k\rho^2}{4L(1+iQ_l)}\right]\\
&\left.\left.\left.-\frac{0.273}{(1+iQ_l)^{7/12}}\,{}_1F_1\left[-\frac{5}{4}\,1;-\frac{Q_l k\rho^2}{4L(1+iQ_l)}\right]\right\}\right]\right]\\
&-7.07\sigma_R^2 Q_l^{-5/6}\left\{{}_1F_1\left(-\frac{5}{6};1;-\frac{Q_l k\rho^2}{4L}\right)\right.\\
&\left.+1.096\,{}_1F_1\left(-\frac{1}{3};1;-\frac{Q_l k\rho^2}{4L}\right)-0.186\,{}_1F_1\left(-\frac{1}{4};1;-\frac{Q_l k\rho^2}{4L}\right)\right\}.
\end{aligned} \tag{8.125}$$

Equation (8.125) is an exact expression based on the modified atmospheric spectrum model (8.3) with $\kappa_0 = 0$. However, we found that the normalized covariance function $B_{I,pl}(\rho, L)/\sigma_{I,pl}^2(L)$ is basically indistinguishable from that deduced from Eq. (8.121). Hence, except for very large inner scale and/or very short ranges, we conclude once again that the normalized irradiance covariance function is essentially independent of inner scale.

8.5.2 Plane wave: aperture averaging

The plane wave irradiance covariance function derived in the last section involves analysis made in the Rx pupil plane. We now wish to take the Rx aperture diameter D into account (see p. 412 of Ref. 1). Based on spectrum model (8.1), this leads to the covariance function defined by

$$B_{I,pl}(\rho, L, D) = 2.606C_n^2k^2L\int_0^\infty \kappa^{-8/3}e^{-D^2\kappa^2/16}J_0(\kappa\rho)\left[1-\frac{k}{L\kappa^2}\sin\left(\frac{L\kappa^2}{k}\right)\right]d\kappa, \tag{8.126}$$

which has the same functional form as Eq. (8.118) with $1/\kappa_m^2$ replaced by $D^2/16$. Consequently, we can simply make this substitution in Eq. (8.121) to get

$$\begin{aligned} B_{I,pl}(\rho, L, D) = {} & 3.86\sigma_R^2 \mathrm{Im}\left\{\left(\frac{kD^2}{16L}+i\right)^{11/6} {}_1F_1\left[-\frac{11}{6};1;-\frac{4\rho^2}{D^2(1+16iL/kD^2)}\right]\right\} \\ & -7.07\sigma_R^2\left(\frac{kD^2}{16L}\right)^{5/6} {}_1F_1\left(-\frac{5}{6};1;-\frac{4\rho^2}{D^2}\right). \end{aligned} \tag{8.127}$$

By setting $\rho = 0$, we obtain the flux variance

$$\sigma_{I,pl}^2(L, D) = 3.86\sigma_R^2 \mathrm{Im}\left[\left(\frac{kD^2}{16L}+i\right)^{11/6}\right] - 7.07\sigma_R^2\left(\frac{kD^2}{16L}\right)^{5/6}. \tag{8.128}$$

This last expression is equivalent to that provided on p. 124 of Ref. 2.

In Fig. 8.8(a) we plot the aperture average plane wave irradiance covariance function [Eq. (8.127)] scaled by the flux variance [Eq. (8.128)]. Here we see that with a Rx aperture leading to $d = \sqrt{kD^2/4L} = 1$, the correlation width is widened over that with $d = 0$. This plot is quite similar to one shown in Fig. 4.7 of Wheelon[13] obtained through numerical integration. In Fig. 8.8(b) we plot the normalized irradiance covariance function of a plane wave as a function of separation distance over aperture diameter.

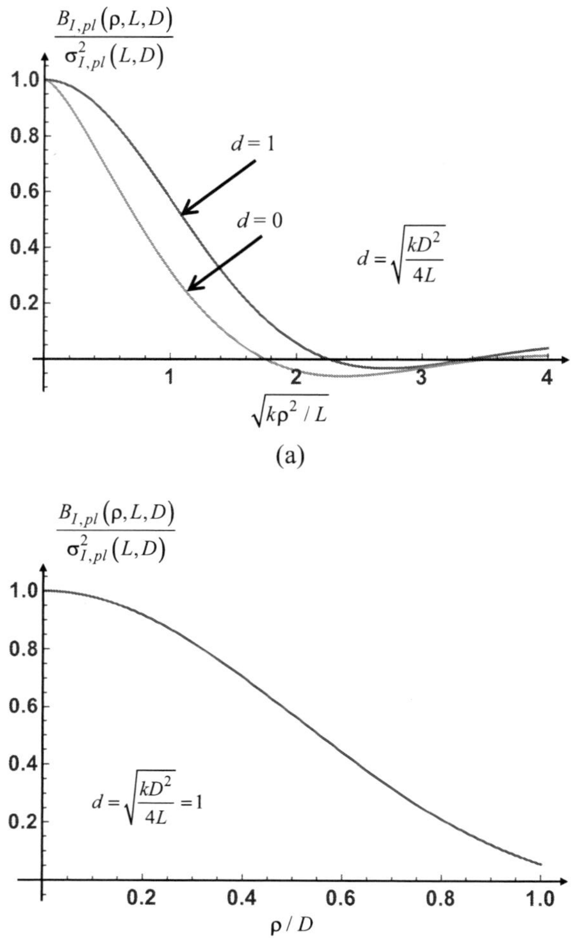

Figure 8.8 Normalized irradiance covariance function of a plane wave with aperture averaging [Eq. (8.127)] (a) as a function of separation distance over the Fresnel zone size and (b) as a function of separation distance over Rx diameter size.

8.5.3 Spherical wave

Following the analysis for the plane wave covariance function, we now consider the irradiance covariance function of a spherical wave,

$$B_{I,sph}(\rho, L) = 8\pi^2 k^2 L \int_0^1 \int_0^\infty \kappa \Phi_n(\kappa) J_0(\rho\xi\kappa) \left\{ 1 - \cos\left[\frac{L\kappa^2 \xi(1-\xi)}{k} \right] \right\} d\kappa \, d\xi. \tag{8.129}$$

Because the plane wave normalized covariance function is only weakly dependent on the inner scale of turbulence, we assume that the same is true for a spherical wave.

We start by replacing the cosine function with $\cos x = \mathrm{Re}(e^{-ix})$ as well as replacing the Bessel function with the Meijer G-function ($C9$) in Appendix C. Thus, based on the spectrum model (8.1), we have

$$\begin{aligned}B_{I,sph}(\rho,\,L) &= 2.60558 C_n^2 k^2 L \lim_{\mu\to -8/3} \mathrm{Re} \int_0^1 \int_0^\infty \left\{ \kappa^\mu G_{02}^{10}\left(\frac{\rho^2\xi^2\kappa^2}{4}\middle|\, 0,\,0\right)\right.\\ &\quad \left. - \kappa^\mu\, G_{02}^{10}\left(\frac{\rho^2\xi^2\kappa^2}{4}\middle|\, 0,\,0\right) \exp\left[-\frac{iL\kappa^2}{k}\xi(1-\xi)\right]\right\} d\kappa\, d\xi \\ &= 1.30279 C_n^2 k^2 L \mathrm{Re} \int_0^1 \left\{ \frac{\Gamma\left(-\frac{5}{6}\right)}{\Gamma\left(\frac{11}{6}\right)} \left(\frac{\rho}{2}\right)^{5/3} \xi^{5/3} \right. \\ &\quad \left. - \left(\frac{iL}{k}\right)^{5/6} [\xi(1-\xi)]^{5/6}\, G_{12}^{11}\left[\frac{k\rho^2\xi}{4iL(1-\xi)}\middle|\, \begin{matrix}\frac{11}{6}\\ 0,\,0\end{matrix}\right] \right\} d\xi,\end{aligned} \tag{8.130}$$

where we have taken the limit $\mu \to -\frac{8}{3}$ in the second step above after using ($D8$) and ($D6$) in Appendix D to evaluate the inside integral. The first term in the remaining integral in Eq. (8.130) involving only $\xi^{5/3}$ can be easily integrated over ξ, but in the second term we make a change of variable $x = \xi/(1-\xi)$, or equivalently, $\xi = x/(1+x)$. After making this change of variable and using ($D11$) in Appendix D, we are led to

$$\begin{aligned}\int_0^1 [\xi(1-\xi)]^{5/6} G_{12}^{11}\left[\frac{k\rho^2\xi}{4iL(1-\xi)}\middle|\, \begin{matrix}\frac{11}{6}\\ 0,0\end{matrix}\right] d\xi &= \int_0^\infty \frac{x^{5/6}}{(1+x)^{11/3}} G_{12}^{11}\left(\frac{k\rho^2 x}{4iL}\middle|\, \begin{matrix}\frac{11}{6}\\ 0,0\end{matrix}\right) dx \\ &= \frac{1}{\Gamma\left(\frac{11}{3}\right)} G_{23}^{22}\left(-\frac{ik\rho^2}{4L}\middle|\, \begin{matrix}-\frac{5}{6},\frac{11}{6}\\ \frac{11}{6},0,0\end{matrix}\right).\end{aligned} \tag{8.131}$$

Upon simplification, we obtain from Eqs. (8.130) and (8.131)

$$B_{I,sph}(\rho,L) = -6.99426\beta_0^2\left(\frac{k\rho^2}{4L}\right)^{5/6} - 0.65465\beta_0^2 \mathrm{Re}\left[i^{5/6} G_{23}^{22}\left(-\frac{ik\rho^2}{4L}\middle|\, \begin{matrix}-\frac{5}{6},\frac{11}{6}\\ \frac{11}{6},0,0\end{matrix}\right)\right]. \tag{8.132}$$

From the definition ($MG1$) in Appendix A, we can express this last Meijer G-function in terms of generalized hypergeometric functions,

$$G_{23}^{22}\left(-\frac{ik\rho^2}{4L}\middle|\begin{matrix}-\frac{5}{6},\frac{11}{6}\\ \frac{11}{6},0,0\end{matrix}\right)=\Gamma^2\left(\frac{11}{6}\right)\Gamma\left(-\frac{5}{6}\right){}_1F_1\left(\frac{11}{6};1;-\frac{ik\rho^2}{4L}\right)$$
$$+\frac{\Gamma\left(-\frac{11}{6}\right)\Gamma\left(\frac{11}{3}\right)}{\Gamma\left(\frac{17}{6}\right)}\left(-\frac{ik\rho^2}{4L}\right)^{11/6}{}_2F_2\left(\frac{11}{3},1;\frac{17}{6},\frac{17}{6};-\frac{ik\rho^2}{4L}\right).\tag{8.133}$$

Therefore, by combining the results of Eqs. (8.132) and (8.133), we obtain[14]

$$B_{I,sph}(\rho,\ L)=3.869\,\beta_0^2\mathrm{Re}\left[i^{5/6}\,{}_1F_1\left(\frac{11}{6};\ 1;\ -\frac{ik\rho^2}{4L}\right)\right]-6.994\,\beta_0^2\left(\frac{k\rho^2}{4L}\right)^{5/6}$$
$$-5.550\,\beta_0^2\left(\frac{k\rho^2}{4L}\right)^{11/6}\mathrm{Im}\left[{}_2F_2\left(\frac{11}{3},\ 1;\ \frac{17}{6},\ \frac{17}{6};\ -\frac{ik\rho^2}{4L}\right)\right].\tag{8.134}$$

In Fig. 8.9 we plot the normalized covariance function $B_I(\rho,\ L)/\sigma_I^2(L)$ [defined by either Eq. (8.132) or (8.134)] for a spherical wave as a function of $\sqrt{k\rho^2/L}$. For comparison, we also plot the normalized plane wave covariance function [Eq. (8.122)]. Note that the plot of the spherical wave normalized irradiance covariance function crosses the horizontal axis at a somewhat larger value of $\sqrt{k\rho^2/L}$ than that of the plane wave. In this case, the irradiance correlation width of a spherical wave is $\rho_c=4.539\sqrt{L/k}=1.81\sqrt{\lambda L}$.

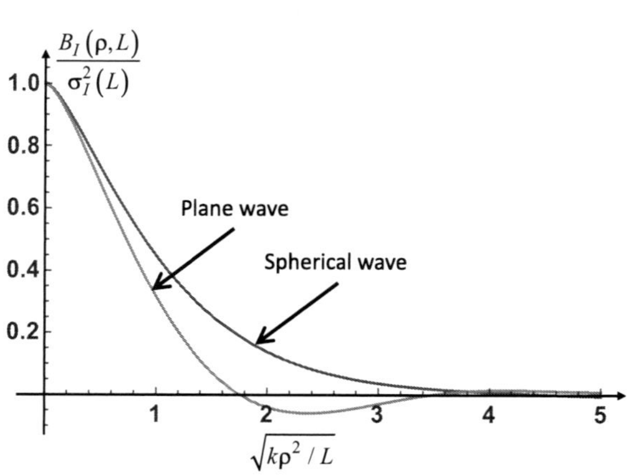

Figure 8.9 Normalized irradiance covariance function of a spherical wave defined by Eq. (8.132) and that of a plane wave defined by Eq. (8.122).

8.5.4 Spherical wave: aperture averaging

By following the plane wave aperture averaging analysis in Section 8.5.2, we can do the same for a spherical wave. With aperture averaging, the spherical wave irradiance covariance function takes the form (see p. 415 of Ref. 1)

$$\begin{aligned} B_{I,sph}(\rho, L, D) &= 2.606 C_n^2 k^2 L \lim_{\mu \to -8/3} \int_0^1 \int_0^\infty \kappa^\mu \exp\left(-\frac{D^2\xi^2\kappa^2}{16}\right) \\ &\quad \times G_{02}^{10}\left(\frac{\rho^2\xi^2\kappa^2}{4}\middle| 0, 0\right) \mathrm{Re}\left\{1 - \exp\left[-\frac{iL\kappa^2\xi(1-\xi)}{k}\right]\right\} d\kappa\, d\xi \\ &= 2.606\beta_0^2 \int_0^1 (I_1 - I_2)\, d\xi, \end{aligned} \tag{8.135}$$

where we have relied on ($D6$) in Appendix D and where

$$I_1 = \left(\frac{d\xi}{2}\right)^{5/3} G_{12}^{11}\left(\frac{k\rho^2}{Ld^2}\middle| \begin{matrix} \frac{11}{6} \\ 0, 0 \end{matrix}\right), \tag{8.136}$$

$$I_2 = \mathrm{Re}\left[b^{5/6} G_{12}^{11}\left(\frac{k\rho^2\xi^2}{4Lb}\middle| \begin{matrix} \frac{11}{6} \\ 0, 0 \end{matrix}\right)\right]. \tag{8.137}$$

In the last two equations, we have introduced the simplifying notation

$$d = \sqrt{\frac{kD^2}{4L}}, \quad b = i\xi\left[1 - \left(1 + \frac{id^2}{4}\right)\xi\right]. \tag{8.138}$$

At this point numerical integration is used to finish the evaluation in Eq. (8.135). However, we can find an approximation by replacing the atmosphere along the path with a thin random phase screen (cf. Chapter 15 in Ref. 1). To do so we replace ξ with $\varepsilon = 0.16$ (among other choices). This value of ε places the phase screen closer to the Rx than to the Tx. Hence, Eq. (8.135) takes the form

$$B_{I,sph}(\rho, L, D) \approx 2.606\beta_0^2 \mathrm{Re}\left\{\left(\frac{d\varepsilon}{2}\right)^{5/3} G_{12}^{11}\left(\frac{k\rho^2}{Ld^2}\middle| \begin{matrix} \frac{11}{6} \\ 0,0 \end{matrix}\right) - b^{5/6} G_{12}^{11}\left(\frac{k\rho^2\varepsilon^2}{4Lb}\middle| \begin{matrix} \frac{11}{6} \\ 0,0 \end{matrix}\right)\right\}, \tag{8.139}$$

where we now define $b = i\varepsilon[1 - (1 + id^2/4)\varepsilon]$, $\varepsilon = 0.16$.

In Fig. 8.10 we plot Eq. (8.139) with $d = 3$, scaled by the flux variance obtained from Eq. (8.139) with $\rho = 0$. Also shown by the dots are exact numerical integration results from Eq. (8.135). Last, plotted for comparison is the scaled irradiance covariance function ($d = 0$) described by Eq. (8.132).

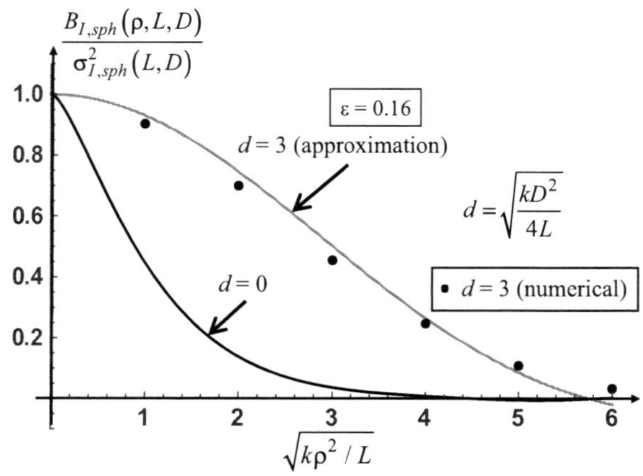

Figure 8.10 Scaled irradiance covariance function of a spherical wave with aperture averaging [Eq. (8.139)]. The dots are numerical integration results from Eq. (8.135).

8.5.5 Gaussian-beam wave

For a collimated Gaussian-beam wave, we choose $\mathbf{r}_2 = -\mathbf{r}_1$, and in this case the irradiance covariance function is defined by

$$B_I(\rho, L) = 2.606 C_n^2 k^2 L \lim_{\mu \to -8/3} \mathrm{Re} \int_0^1 \int_0^\infty \kappa^\mu \exp\left(-\frac{\Lambda L \xi^2 \kappa^2}{k}\right) \left\{ J_0[(1-\overline{\Theta}\xi)\kappa\rho] - \exp\left[-\frac{iL\kappa^2}{k}\xi(1-\overline{\Theta}\xi)\right] J_0[(1-\overline{\Theta}\xi - i\Lambda\xi)\kappa\rho] \right\} d\kappa\, d\xi. \tag{8.140}$$

Here it is convenient to split the inside integral into two separate integrals, viz.,

$$I_1 = \int_0^\infty \kappa^\mu \exp(-a_1\kappa^2) J_0(b_1\kappa)\, d\kappa, \tag{8.141}$$

where

$$a_1 = \frac{\Lambda L \xi^2}{k}, \quad b_1 = (1-\overline{\Theta}\xi)\rho, \tag{8.142}$$

and

$$I_2 = \int_0^\infty \kappa^\mu \exp(-a_2\kappa^2) J_0(b_2\kappa)\, d\kappa, \tag{8.143}$$

where

$$a_2 = \frac{iL\xi}{k}[1-(\overline{\Theta}+i\Lambda)\xi], \quad b_2 = [1-(\overline{\Theta}+i\Lambda)\xi]\rho. \tag{8.144}$$

Both Eqs. (8.141) and (8.143) are of the particular type ($D1$) in Appendix D. Hence, by evaluating the integrals for I_1 and I_2, we obtain

$$I_1 - I_2 = \frac{1}{2}\Gamma\left(-\frac{5}{6}\right)\left(\frac{\Lambda L}{k}\right)^{5/6} \xi^{5/3}{}_1F_1\left[-\frac{5}{6}; 1; -\frac{(1-\overline{\Theta}\xi)^2 k\rho^2}{4\Lambda L\xi^2}\right]$$
$$-\frac{1}{2}\Gamma\left(-\frac{5}{6}\right)\left(\frac{iL}{k}\right)^{5/6} \xi^{5/6}[1-(\overline{\Theta}+i\Lambda)\xi]^{5/6}{}_1F_1\left\{-\frac{5}{6}; 1; -\frac{[1-(\overline{\Theta}+i\Lambda)\xi]k\rho^2}{4iL\xi}\right\}, \tag{8.145}$$

where we have also taken the limit $\mu \to -\frac{8}{3}$. Returning to Eq. (8.140), we have

$$B_I(\rho, L) = 7.075\sigma_R^2\Bigg[\Bigg[-\Lambda^{5/6}\int_0^1 \xi^{5/3}{}_1F_1\left[-\frac{5}{6}; 1; -\frac{(1-\overline{\Theta}\xi)^2 k\rho^2}{4\Lambda L\xi^2}\right]d\xi$$
$$+\,\mathrm{Re}\int_0^1 (i\xi)^{5/6}[1-(\overline{\Theta}+i\Lambda)\xi]^{5/6}{}_1F_1\left\{-\frac{5}{6}; 1; -\frac{[1-(\overline{\Theta}+i\Lambda)\xi]k\rho^2}{4iL\xi}\right\}d\xi\Bigg]\Bigg]. \tag{8.146}$$

Once again, we have reached a point where numerical integration is called for to complete the evaluation in Eq. (8.146).

To avoid numerical integration of Eq. (8.146), we can use the same idea that we used in Section 8.5.4 for the irradiance covariance function of a spherical wave with aperture averaging. That is, a reasonable approximation for the near-field beams ($\Lambda_0 \le 1$) can be obtained by replacing the random medium between Tx and Rx with a single thin random phase screen located by setting $\varepsilon = 0.7$ in place of ξ, placing the phase screen closer to the optical source than to the receiver. With $\rho = 0$ we can also calculate the on-axis scintillation index $B_I(0, L)$. The result of introducing the thin random phase screen is

$$B_I(\rho, L) \approx 7.075\sigma_R^2\left\{-(\Lambda\varepsilon^2)^{5/6}{}_1F_1\left[-\frac{5}{6}; 1; -\frac{(1-\overline{\Theta}\varepsilon)^2 k\rho^2}{4\Lambda L\varepsilon^2}\right]\right\}$$
$$+\,7.075\sigma_R^2\,\mathrm{Re}\Bigg[\Bigg[(i\varepsilon)^{5/6}[1-(\overline{\Theta}+i\Lambda)\varepsilon]^{5/6}{}_1F_1\left\{-\frac{5}{6}; 1; -\frac{[1-(\overline{\Theta}+i\Lambda)\varepsilon]k\rho^2}{4iL\varepsilon}\right\}\Bigg]\Bigg]. \tag{8.147}$$

We plot the normalized approximation of Eq. (8.147) in Fig. 8.11 for a collimated Gaussian-beam wave with two values of Λ_0. We also perform numerical integration shown by the open and closed dots. We find that for far-field beams ($\Lambda_0 > 1$), the accuracy of this approximation is generally not as good.

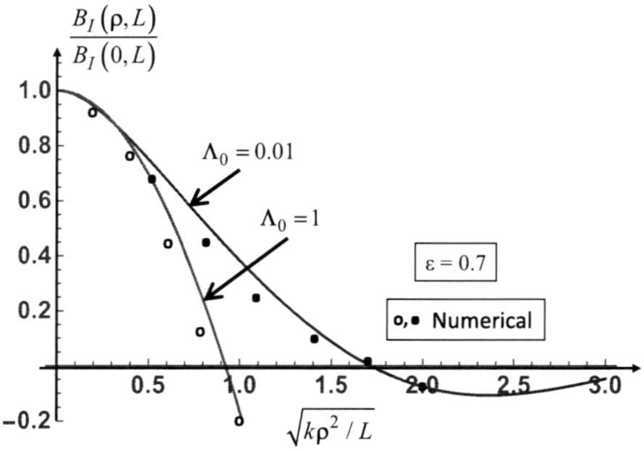

Figure 8.11 Approximations of the normalized covariance function for two collimated Gaussian-beam waves [Eq. (8.147)] corresponding to two values of Λ_0. The open and closed circles (dots) are numerical results from Eq. (8.146).

8.6 Irradiance Covariance Function: Strong Fluctuations

Up to this point we have concentrated on the irradiance covariance function under weak irradiance fluctuations. We now wish to consider the case of strong irradiance fluctuations. To do so, we will use the *extended Rytov theory*.[15] The treatment here will cover both an infinite plane wave and a spherical wave for which the covariance function takes the form

$$B_I(\rho, L) = \exp\left[B_{\ln X}(\rho, L) + B_{\ln Y}(\rho, L)\right] - 1, \tag{8.148}$$

where $B_{\ln X}(\rho, L)$ and $B_{\ln Y}(\rho, L)$ are the large-scale and small-scale covariances of log-irradiance.

8.6.1 Plane wave

We start by making the substitution $\eta = L\kappa^2/L$ in Eq. (8.117), which, based on the Kolmogorov spectrum (8.1), leads to the large-scale component

$$B_{\ln X}(\rho, L) = 1.059\sigma_R^2 \int_0^1 \int_0^\infty \eta^{-11/6} \exp\left(-\frac{\eta}{\eta_X}\right) J_0\left(\rho\sqrt{\frac{k\eta}{L}}\right)(1 - \cos\xi\eta)\, d\eta\, d\xi. \tag{8.149}$$

The function $\exp(-\eta/\eta_X)$ is the large-scale filter function with

$$\eta_X = \frac{2.61}{1 + 1.11\sigma_R^{12/5}}. \tag{8.150}$$

By use of ($C9$) in Appendix C and the small-argument approximation for the cosine function, we get

$$\begin{aligned} B_{\ln X}(\rho, L) &= 0.53\sigma_R^2 \int_0^1 \xi^2 \, d\xi \int_0^\infty \eta^{1/6} \exp\left(-\frac{\eta}{\eta_X}\right) G_{02}^{10}\left(\frac{k\rho^2\eta}{4L}\middle| 0, 0\right) d\eta \\ &= 0.16\sigma_R^2 \eta_X^{7/6} {}_1F_1\left(\frac{7}{6}; 1; -\frac{k\rho^2\eta_X}{4L}\right), \end{aligned} \tag{8.151}$$

where we have used ($D6$) in Appendix D and ($C13$) in Appendix C. The corresponding log-irradiance covariance for the small-scale fluctuations is

$$\begin{aligned} B_{\ln Y}(\rho, L) &= 1.059\sigma_R^2 \int_0^1 \int_0^\infty \frac{1}{(\eta + \eta_Y)^{11/6}} J_0\left(\rho\sqrt{\frac{k\eta}{L}}\right)(1 - \cos\xi\eta) \, d\eta \, d\xi \\ &= 1.059\sigma_R^2 \int_0^1 \int_0^\infty \frac{1}{(\eta + \eta_Y)^{11/6}} J_0\left(\rho\sqrt{\frac{k\eta}{L}}\right) d\eta \, d\xi, \end{aligned} \tag{8.152}$$

where we set $1 - \cos\xi\eta \approx 1$. The function $(\eta + \eta_Y)^{-11/6}$ is the small-scale filter function with

$$\eta_Y = 3(1 + 0.69\sigma_R^{12/5}). \tag{8.153}$$

By using ($C9$) in Appendix C once again and ($D11$) in Appendix D, we are led to

$$B_{\ln Y}(\rho, L) = 1.27\sigma_R^2 \left(\frac{k\rho^2}{L\eta_Y}\right)^{5/12} K_{5/6}\left(\sqrt{\frac{k\rho^2\eta_Y}{L}}\right). \tag{8.154}$$

The final result in Eq. (8.154) is based on ($MG3$) in Appendix A and ($C23$) in Appendix C.

Upon combining Eqs. (8.151) and (8.154) with Eq. (8.148) along with some algebraic simplification, we arrive at

$$\begin{aligned} B_{I,pl}(\rho, L) = \exp\Bigg[&\frac{0.49\sigma_R^2}{(1 + 1.11\sigma_R^{12/5})^{7/6}} {}_1F_1\left(\frac{7}{6}; 1; -\frac{k\rho^2\eta_X}{4L}\right) \\ &+ \frac{0.51\sigma_R^2}{(1 + 0.69\sigma_R^{12/5})^{5/6}} \left(\frac{k\rho^2\eta_Y}{L}\right)^{5/12} K_{5/6}\left(\sqrt{\frac{k\rho^2\eta_Y}{L}}\right)\Bigg] - 1. \end{aligned} \tag{8.155}$$

In Fig. 8.12 we plot a scaled version of Eq. (8.155) for scintillation values $\sigma_R^2(L) = 0.1$ and $\sigma_R^2(L) = 20$. In the strong irradiance fluctuation case defined by $\sigma_R^2(L) = 20$, note the quick drop in the scaled covariance function from unity (small-scale effect) followed by a long tail (large-scale effect), depicting a two-scale phenomenon.

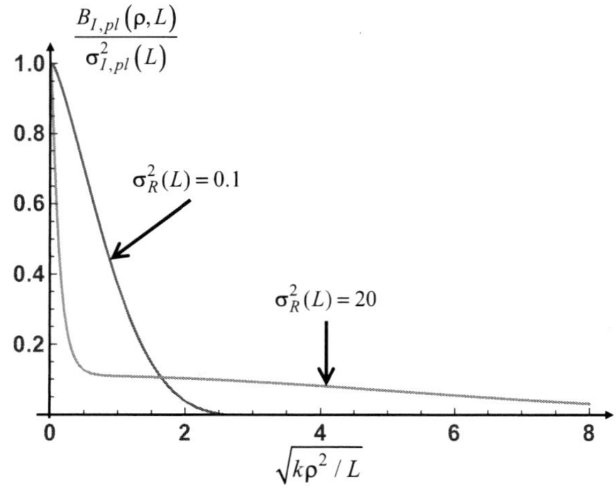

Figure 8.12 Normalized irradiance covariance function of a plane wave defined by Eq. (8.155). Each plot is based on the Kolmogorov spectrum (8.1) showing both weak and strong irradiance fluctuations.

8.6.2 Spherical wave

For a spherical wave, the large-scale covariance function is defined by

$$B_{\ln X}(\rho,\, L) = 2.62659\,\beta_0^2 \int_0^1\!\!\int_0^\infty \eta^{-11/6} \exp\left(-\frac{\eta}{\eta_X}\right) J_0\left(\rho\xi\sqrt{\frac{k\eta}{L}}\right) \times \{1 - \cos[\eta\xi(1-\xi)]\}\, d\eta\, d\xi, \tag{8.156}$$

where

$$\eta_X = \frac{8.56}{1 + 0.56\beta_0^{12/5}}. \tag{8.157}$$

Once again, we refer to ($C9$) in Appendix C and the small-argument approximation for the cosine function. We then use ($D6$) in Appendix D to obtain

$$\begin{aligned} B_{\ln X}(\rho, L) &= 1.3133\beta_0^2 \int_0^1 \xi^2(1-\xi)^2 \int_0^\infty \eta^{1/6} \exp\left(-\frac{\eta}{\eta_X}\right) G_{02}^{10}\left(\frac{k\rho^2\xi^2\eta}{4L}\middle|\, 0,0\right) d\eta\, d\xi \\ &= 1.3133\,\beta_0^2\eta_X^{7/6} \int_0^1 \xi^2(1-\xi)^2 G_{12}^{11}\left(\frac{k\rho^2\xi^2\eta_X}{4L}\middle|\, {-\tfrac{1}{6} \atop 0,0}\right) d\xi \\ &= 1.2183\beta_0^2\eta_X^{7/6} \int_0^1 \xi^2(1-\xi)^2 {}_1F_1\left(\frac{7}{6}; 1; -\frac{k\rho^2\xi^2\eta_X}{4L}\right) d\xi. \end{aligned} \tag{8.158}$$

By expressing the ${}_1F_1$ function in a power series and integrating termwise over ξ using the beta function ($G17$) in Appendix A, we find that

$$\begin{aligned} B_{\ln X}(\rho, L) &= 0.04061\ \beta_0^2 \eta_X^{7/6} \sum_{n=0}^{\infty} \frac{(-1)^n \left(\frac{7}{6}\right)_n \left(\frac{3}{2}\right)_n (2)_n}{(1)_n (3)_n \left(\frac{7}{2}\right)_n n!} \left(\frac{k\rho^2 \eta_X}{4L}\right)^n \\ &= 0.04061\ \beta_0^2 \eta_X^{7/6}\ {}_3F_3\left(\frac{7}{6}, \frac{3}{2}, 2; 1, 3, \frac{7}{2}; -\frac{k\rho^2\eta_X}{4L}\right). \end{aligned} \tag{8.159}$$

For the small-scale fluctuations, we have

$$B_{\ln Y}(\rho, L) = 2.62659\ \beta_0^2 \int_0^1 \int_0^\infty \frac{1}{(\eta + \eta_Y)^{11/6}} J_0\left(\rho\xi \sqrt{\frac{k\eta}{L}}\right) d\eta\, d\xi, \tag{8.160}$$

where we set $1 - \cos[\eta\xi(1-\xi)] \approx 1$ and define

$$\eta_Y = 9(1 + 0.69\beta_0^{12/5}). \tag{8.161}$$

From ($C9$) in Appendix C and ($D11$) in Appendix D, we are led to

$$\begin{aligned} B_{\ln Y}(\rho, L) &= \frac{2.7923}{\eta_Y^{5/6}} \beta_0^2 \int_0^1 G_{13}^{21}\left(\frac{k\rho^2\xi^2\eta_Y}{4L} \middle| \begin{matrix} 0 \\ \frac{5}{6}, 0, 0 \end{matrix}\right) d\xi \\ &= \frac{1.3962}{\eta_Y^{5/6}} \beta_0^2\, G_{13}^{21}\left(\frac{k\rho^2\eta_Y}{4L} \middle| \begin{matrix} \frac{1}{2} \\ \frac{5}{6}, 0, -\frac{1}{2} \end{matrix}\right), \end{aligned} \tag{8.162}$$

which follows from a change of variable $t = \xi^2$, ($MG3$) in Appendix A, and application of ($D9$) in Appendix D. However, we found that under moderate-to-strong irradiance fluctuations, erratic behavior begins to form in the calculation of Eq. (8.162) for $\sqrt{k\rho^2/L} > 3$. To avoid this behavior, we can replace the integral in Eq. (8.162) with the model of a thin phase screen, thus replacing ξ in the integral with $\varepsilon = 0.16$ as we did in Section 8.5.4. The combination of Eqs. (8.159) and (8.162), the latter with the phase screen approximation, provides the solution

$$\begin{aligned} B_{I,sph}(\rho, L) \approx \exp\Bigg[&\frac{0.49\,\beta_0^2}{(1 + 0.56\beta_0^{12/5})^{7/6}}\ {}_3F_3\left(\frac{7}{6}, \frac{3}{2}, 2; 1, 3, \frac{7}{2}; -\frac{k\rho^2\eta_X}{4L}\right) \\ &+ \frac{0.51\,\beta_0^2}{(1 + 0.69\beta_0^{12/5})^{5/6}} \left(\frac{k\rho^2\varepsilon^2\eta_Y}{L}\right)^{5/12} K_{5/6}\left(\sqrt{\frac{k\rho^2\varepsilon^2\eta_Y}{L}}\right)\Bigg] - 1. \end{aligned} \tag{8.163}$$

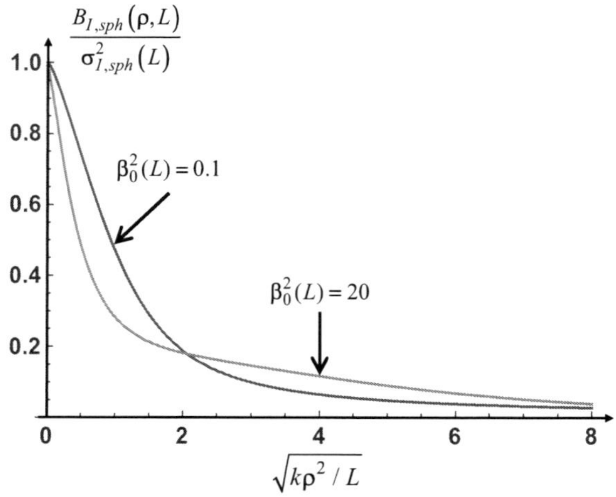

Figure 8.13 Normalized irradiance covariance function of a spherical wave defined by Eq. (8.163). Each plot is based on the Kolmogorov spectrum (8.1) showing both weak and strong irradiance fluctuations.

The last step leading to Eq. (8.163) arises by using (*MG*3) in Appendix A followed by (*C*23) in Appendix C.

We plot Eq. (8.163), scaled by the scintillation index, in Fig. 8.13 for values $\beta_0^2 = 0.1$ and $\beta_0^2 = 20$. In this case we found that the irradiance covariance function under weak irradiance fluctuations continually decreases similarly to that in Fig. 8.9 but does not cross the horizontal axis as it does in Fig. 8.9. However, like the plane wave case in Fig. 8.12, the two-scale behavior for a spherical wave under strong fluctuations is clearly visible for the case $\beta_0^2 = 20$.

8.7 Temporal Spectrum of Irradiance

The analysis thus far has described the effect of optical turbulence on an optical wave in terms of spatial statistics. In many applications, however, the *temporal statistics* are also of importance. Most propagation studies rely on *Taylor's frozen turbulence hypothesis* to describe temporal changes in irradiance and phase. This hypothesis implies that temporal variations of meteorological quantities at a given point are produced through advection of these quantities by the mean wind velocity flow and not by the quantities themselves. Thus, from knowledge of the mean wind velocity $\mathbf{V}_\perp$ transverse to the direction of propagation, we can convert spatial statistics to temporal statistics. For example, a turbulence variable $u(\mathbf{R}, t)$ in this case has time dependence given by $u(\mathbf{R}, t+\tau) = u(\mathbf{R} - \mathbf{V}_\perp\tau, t)$.

Our interest here is in the *temporal spectrum of irradiance fluctuations.* The temporal spectrum, also called the *power spectral density* (PSD), is defined by the Fourier transform relation

$$S_I(\omega) = 2\int_{-\infty}^{\infty} B_I(\tau, L)e^{-i\omega\tau}\,d\tau = 4\int_0^{\infty} B_I(\tau, L)\cos(\omega\tau)\,d\tau, \tag{8.164}$$

where ω is angular frequency and $B_I(\tau, L)$ is the *temporal covariance function* of irradiance. The extra factor of "2" in the transform integral is a consequence of considering only positive frequencies.

8.7.1 Plane wave

The temporal power spectrum of an infinite plane wave was first calculated by Tatarskii (pp. 261–263 of Ref. 16). The plane wave irradiance covariance function is defined by Eq. (8.117). To convert this expression from spatial statistics to temporal statistics, the frozen turbulence hypothesis allows us to make the association $\mathbf{p} = \mathbf{V}_\perp\tau$, where $\mathbf{V}_\perp$ is the mean transverse wind velocity. Hence, based on the Kolmogorov spectrum of Eq. (8.1), the temporal covariance function takes the form

$$\begin{aligned} B_{I,pl}(\tau, L) &= 8\pi^2k^2L\int_0^1\int_0^{\infty}\kappa\Phi_n(\kappa)J_0(\kappa V_\perp\tau)\left[1-\cos\left(\frac{L\kappa^2\xi}{k}\right)\right]d\kappa\,d\xi \\ &= 2.606C_n^2k^2L\int_0^{\infty}\kappa^{-8/3}J_0(\kappa V_\perp\tau)\left[1-\frac{k}{L\kappa^2}\sin\left(\frac{L\kappa^2}{k}\right)\right]d\kappa, \end{aligned} \tag{8.165}$$

where we have performed the integration on the normalized distance variable ξ. If we substitute this last expression into Eq. (8.164) and exchange the order of integration, we have

$$\begin{aligned} S_{I,pl}(\omega) &= 10.4223C_n^2k^2L\int_0^{\infty}\kappa^{-8/3}\int_0^{\infty}J_0(\kappa V_\perp\tau)\cos(\omega\tau)\,d\tau \\ &\quad\times\left[1-\frac{k}{L\kappa^2}\sin\left(\frac{L\kappa^2}{k}\right)\right]d\kappa. \end{aligned} \tag{8.166}$$

From the inside integral over time τ in Eq. (8.166), we obtain

$$\begin{aligned} \int_0^{\infty}J_0(\kappa V_\perp\tau)\cos(\omega\tau)\,d\tau &= \sqrt{\pi}\int_0^{\infty}J_0(\kappa V_\perp\tau)\,G_{02}^{10}\left(\frac{\omega^2\tau^2}{4}\,\middle|\,0, \frac{1}{2}\right)d\tau \\ &= \frac{\sqrt{\pi}}{\kappa V_\perp}G_{22}^{11}\left(\frac{\omega^2}{\kappa^2V_\perp^2}\,\middle|\,\begin{matrix}\frac{1}{2}, \frac{1}{2}\\ 0, \frac{1}{2}\end{matrix}\right) \\ &= \frac{\sqrt{\pi}}{\kappa V_\perp}G_{11}^{10}\left(\frac{\omega^2}{\kappa^2V_\perp^2}\,\middle|\,\begin{matrix}\frac{1}{2}\\ 0\end{matrix}\right), \end{aligned} \tag{8.167}$$

where we have applied ($C4$) in Appendix C followed by ($D12$) in Appendix D and ($MG3$) in Appendix A. Next, by imposing ($C2$) in Appendix C and performing some algebraic simplification, we arrive at

$$\int_0^\infty J_0(\kappa V_\perp \tau)\cos(\omega\tau)\,d\tau = \begin{cases} [(\kappa V_\perp)^2 - \omega^2]^{-1/2}, & \kappa > \omega/V_\perp \\ 0, & \kappa < \omega/V_\perp. \end{cases} \tag{8.168}$$

Thus, Eq. (8.166) becomes

$$\begin{aligned} S_{I,pl}(\omega) &= \frac{10.4223C_n^2k^2L}{V_\perp}\int_a^\infty \frac{\kappa^{-8/3}}{(\kappa^2-a^2)^{1/2}}\left[1-\frac{k}{L\kappa^2}\sin\left(\frac{L\kappa^2}{k}\right)\right]d\kappa \\ &= \frac{10.4223C_n^2k^2L}{V_\perp}(I_1+I_2), \end{aligned} \tag{8.169}$$

where we have set $a = \omega/V_\perp$. In the last line we partitioned Eq. (8.169) into a sum of two integrals with the first defined by

$$\begin{aligned} I_1 = \int_a^\infty \frac{\kappa^{-8/3}}{(\kappa^2-a^2)^{1/2}}\,d\kappa &= \frac{1}{2a^{8/3}}\int_0^\infty x^{-1/2}(x+1)^{-11/6}\,dx \\ &= \frac{\sqrt{\pi}}{2a^{8/3}}\frac{\Gamma\left(\frac{4}{3}\right)}{\Gamma\left(\frac{11}{6}\right)}. \end{aligned} \tag{8.170}$$

In solving this integral, we made the change of variable $\kappa = a\sqrt{x+1}$ and recalled the beta function ($G18$) in Appendix A. The second integral is

$$\begin{aligned} I_2 &= -\frac{k}{L}\int_a^\infty \frac{\kappa^{-14/3}}{(\kappa^2-a^2)^{1/2}}\sin\left(\frac{L\kappa^2}{k}\right)d\kappa \\ &= \frac{k}{L}\,\mathrm{Im}\int_a^\infty \frac{\kappa^{-14/3}}{(\kappa^2-a^2)^{1/2}}\exp\left(-\frac{iL\kappa^2}{k}\right)d\kappa, \end{aligned} \tag{8.171}$$

which, by making the same variable substitution as in Eq. (8.170), leads to

$$\begin{aligned} I_2 &= \frac{k}{2La^{14/3}}\,\mathrm{Im}\left[\exp\left(-\frac{iLa^2}{k}\right)\int_0^\infty \frac{x^{-1/2}}{(x+1)^{17/6}}G_{01}^{10}\left(\frac{iLa^2x}{k}\middle|\, 0\right)dx\right] \\ &= \frac{1}{2\Gamma\left(\frac{17}{6}\right)}\left(\frac{L}{k}\right)^{4/3}\left(\frac{\omega}{\omega_t}\right)^{-14/3}\mathrm{Im}\left\{\exp\left[-i\left(\frac{\omega}{\omega_t}\right)^2\right]G_{12}^{21}\left[i\left(\frac{\omega}{\omega_t}\right)^2\middle|\,\begin{matrix}\frac{1}{2}\\ \frac{7}{3},0\end{matrix}\right]\right\}. \end{aligned} \tag{8.172}$$

Here we have used ($C1$) in Appendix C and ($D11$) in Appendix D. We also replaced $a = \omega/V_\perp$ and introduced the parameter $\omega_t = V_\perp/\sqrt{L/k}$. Now combining Eqs. (8.170) and (8.172) with Eq. (8.169), we get

$$S_{I,pl}(\omega) = \frac{7.12876\sigma_R^2}{\omega_t}\left(\frac{\omega}{\omega_t}\right)^{-8/3} \times \left[\left[1 + 0.34462\left(\frac{\omega_t}{\omega}\right)^2 \mathrm{Im}\left\{\exp\left[-i\left(\frac{\omega}{\omega_t}\right)^2\right] G_{12}^{21}\left[i\left(\frac{\omega}{\omega_t}\right)^2 \middle| \begin{matrix} \frac{1}{2} \\ \frac{7}{3}, 0 \end{matrix}\right]\right\}\right]\right].$$

However, the Meijer G-function above can be replaced by the Whittaker function of the second kind by applying ($C27$) in Appendix C, which in turn can be replaced by ($B54$) in Appendix B. This combination of actions combined with ($CH17$) in Appendix A yields the alternative expression

$$S_{I,pl}(\omega) = \frac{7.129\sigma_R^2}{\omega_t}\left(\frac{\omega}{\omega_t}\right)^{-8/3} \times \left[\left[1 + 1.053\left(\frac{\omega_t}{\omega}\right)^2 \mathrm{Im}\left\{\exp\left[-i\left(\frac{\omega}{\omega_t}\right)^2\right]\left(\frac{i\omega^2}{\omega_t^2}\right)^{-2/3} U\left(\frac{1}{2}; -\frac{4}{3}; \frac{i\omega^2}{\omega_t^2}\right)\right\}\right]\right]. \tag{8.173}$$

Last, by not integrating over the variable ξ first, as done in Eq. (8.166), the term involving $U(a; c; z)$ in Eq. (8.173) is essentially replaced by a Meijer G-function. This leads to the alternative simpler expression

$$S_{I,pl}(\omega) = \frac{7.12876\sigma_R^2}{\omega_t}\left(\frac{\omega}{\omega_t}\right)^{-8/3} \times \mathrm{Re}\left\{1 - 1.0533902\, G_{23}^{21}\left[i\left(\frac{\omega}{\omega_t}\right)^2 \middle| \begin{matrix} 0, \frac{11}{6} \\ \frac{4}{3}, 0, -1 \end{matrix}\right]\right\}. \tag{8.174}$$

To obtain accuracy with the plot in Fig. 8.14, we need to retain a large number of decimal places in the multiplicative constants of Eq. (8.174). We can replace this last expression, however, with the sum of two ${}_2F_2$ functions by using ($MG1$) in Appendix A.

An approximation by Andrews and Phillips (p. 283 of Ref. 1) to the irradiance temporal spectrum was previously developed through replacement of the extended atmospheric turbulence by a single thin phase screen located mid-way between Tx and Rx. A minor adjustment to the multiplicative constants in that approximation leads to

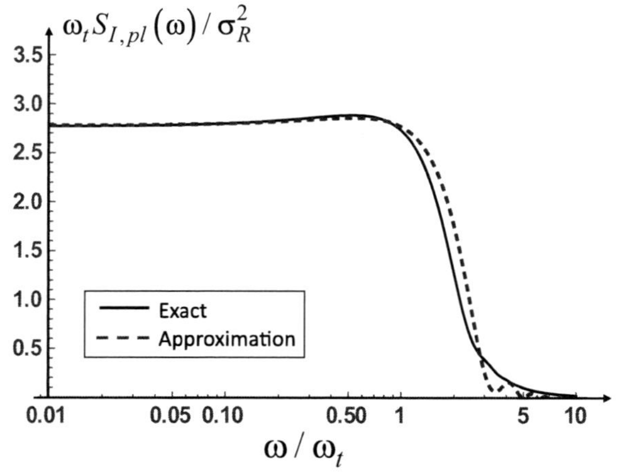

Figure 8.14 Scaled temporal spectrum of irradiance for a plane wave defined by Eq. (8.173) and the approximation of Eq. (8.175). The parameter $\omega_t = V_\perp/(L/k)^{1/2}$.

$$S_{I,pl}(\omega) \approx \frac{7.129\sigma_R^2}{\omega_t}\mathrm{Re}\left\{\left(\frac{\omega}{\omega_t}\right)^{-8/3}\left[1 - {}_1F_1\left(-\frac{5}{6}; -\frac{1}{3}; -\frac{i\omega^2}{2\omega_t^2}\right)\right]\right.$$
$$\left. - 0.779\, i^{4/3}{}_1F_1\left(\frac{1}{2}; \frac{7}{3}; -\frac{i\omega^2}{2\omega_t^2}\right)\right\}. \tag{8.175}$$

The scaled irradiance spectrum model defined by Eq. (8.173) or (8.174) is plotted in Fig. 8.14 (solid line) along with a scaled version of Eq. (8.175) (dashed line). From the plots we observe that the PSD is nearly constant for $\omega < \omega_t$ and drops off as $(\omega/\omega_t)^{-8/3}$ when $\omega > \omega_t$. Thus, the Fresnel parameter ω_t marks the transition frequency at which the spectrum begins to decay.

8.7.2 Plane wave: aperture averaging

The temporal power spectrum of a plane wave under aperture averaging is defined by $(a = \omega/V_\perp)$

$$\begin{aligned}
S_{I,pl}(\omega, D) &= 10.4223 C_n^2 k^2 L \int_0^\infty \kappa^{-8/3} e^{-D^2\kappa^2/16} \int_0^\infty J_0(\kappa V_\perp \tau)\cos(\omega\tau)\, d\tau \\
&\quad \times \left[1 - \frac{k}{L\kappa^2}\sin\left(\frac{L\kappa^2}{k}\right)\right] d\kappa \\
&= \frac{10.4223 C_n^2 k^2 L}{V_\perp}\int_a^\infty \frac{\kappa^{-8/3} e^{-D^2\kappa^2/16}}{(\kappa^2 - a^2)^{1/2}}\left[1 - \frac{k}{L\kappa^2}\sin\left(\frac{L\kappa^2}{k}\right)\right] d\kappa \\
&= \frac{10.4223 C_n^2 k^2 L}{V_\perp}(I_1 + I_2).
\end{aligned} \tag{8.176}$$

The first integral I_1 in Eq. (8.176) after the substitution $\kappa = a\sqrt{x}$ and application of ($C1$) in Appendix C leads to

$$\begin{aligned} I_1 &= \int_a^\infty \frac{\kappa^{-8/3}}{(\kappa^2 - a^2)^{1/2}} \exp\left(-\frac{D^2\kappa^2}{16}\right) d\kappa \\ &= \frac{1}{2a^{8/3}} \int_1^\infty \frac{x^{-11/6}}{(x-1)^{1/2}} G_{01}^{10}\left(\frac{D^2 a^2 x}{16}\middle| 0\right) dx \\ &= \frac{\sqrt{\pi}}{2a^{8/3}} G_{12}^{20}\left(\frac{d^2\omega^2}{4\omega_t^2}\middle| \begin{matrix} \frac{11}{6} \\ \frac{4}{3}, 0 \end{matrix}\right). \end{aligned} \tag{8.177}$$

The integral in the middle line of Eq. (8.177) was evaluated with the help of ($D10$) in Appendix D. We also reset $a = \omega/V_\perp$ in the argument of the Meijer G-function and again defined $d = \sqrt{kD^2/4L}$. The last Meijer G-function in Eq. (8.177) can be expressed in terms of a Whittaker function of the second kind by using ($C25$) in Appendix C. The Whittaker function can be further written in terms of the confluent hypergeometric function of the second kind, but we choose to do neither. Instead, we will leave the result in terms of the Meijer G-function for a simpler formulation. For the integral I_2 in Eq. (8.176), we follow the approach in Section 8.7.1 and write

$$\begin{aligned} I_2 &= -\frac{k}{L}\int_a^\infty \frac{\kappa^{-14/3}}{(\kappa^2 - a^2)^{1/2}} \exp\left(-\frac{D^2\kappa^2}{16}\right) \sin\left(\frac{L\kappa^2}{k}\right) d\kappa \\ &= \frac{k}{L}\mathrm{Im}\int_a^\infty \frac{\kappa^{-14/3}}{(\kappa^2 - a^2)^{1/2}} \exp\left[-\frac{L\kappa^2}{k}\left(\frac{d^2}{4} + i\right)\right] d\kappa. \end{aligned} \tag{8.178}$$

Next, by making the substitution $\kappa = a\sqrt{x+1}$, we get

$$\begin{aligned} I_2 &= \frac{1}{2\Gamma\left(\frac{17}{6}\right)}\left(\frac{L}{k}\right)^{4/3}\left(\frac{\omega}{\omega_t}\right)^{-14/3} \mathrm{Im}\left\{\exp\left[-\frac{\omega^2}{\omega_t^2}\left(\frac{d^2}{4} + i\right)\right]\right. \\ &\quad \times \left.\int_0^\infty \frac{x^{-1/2}}{(x+1)^{17/6}} G_{01}^{10}\left[\frac{\omega^2}{\omega_t^2}\left(\frac{d^2}{4} + i\right)x \middle| 0\right] dx\right\} \\ &= \frac{1}{2\Gamma\left(\frac{17}{6}\right)}\left(\frac{L}{k}\right)^{4/3}\left(\frac{\omega}{\omega_t}\right)^{-14/3} \\ &\quad \times \mathrm{Im}\left\{\exp\left[-\frac{\omega^2}{\omega_t^2}\left(\frac{d^2}{4} + i\right)\right] G_{12}^{21}\left[\frac{\omega^2}{\omega_t^2}\left(\frac{d^2}{4} + i\right)\middle| \begin{matrix} \frac{1}{2} \\ \frac{7}{3}, 0 \end{matrix}\right]\right\}. \end{aligned} \tag{8.179}$$

The Meijer G-function in Eq. (8.179) can also be expressed in terms of a Whittaker function of the second kind by using ($C27$) in Appendix C.

However, we again choose not to do so. Thus, by combining Eqs. (8.176), (8.177), and (8.179) as written, we finally obtain

$$S_{I,pl}(\omega, D) = \frac{7.12876\sigma_R^2}{\omega_t}\left(\frac{\omega}{\omega_t}\right)^{-8/3}\left[\left[1.05339 G_{12}^{20}\left(\frac{d^2\omega^2}{4\omega_t^2}\middle| \begin{matrix} \frac{11}{6} \\ \frac{4}{3}, 0 \end{matrix}\right)\right.\right.$$
$$\left.\left. + 0.34462\left(\frac{\omega_t}{\omega}\right)^2 \mathrm{Im}\left\{\exp\left[-\frac{\omega^2}{\omega_t^2}\left(\frac{d^2}{4}+i\right)\right] G_{12}^{21}\left[\frac{\omega^2}{\omega_t^2}\left(\frac{d^2}{4}+i\right)\middle| \begin{matrix} \frac{1}{2} \\ \frac{7}{3}, 0 \end{matrix}\right]\right\}\right]\right]. \tag{8.180}$$

In Fig. 8.15(a) we plot the temporal spectrum of irradiance scaled by the Rytov variance for Rx aperture diameters corresponding to $d = 0$ and $d = 2$.

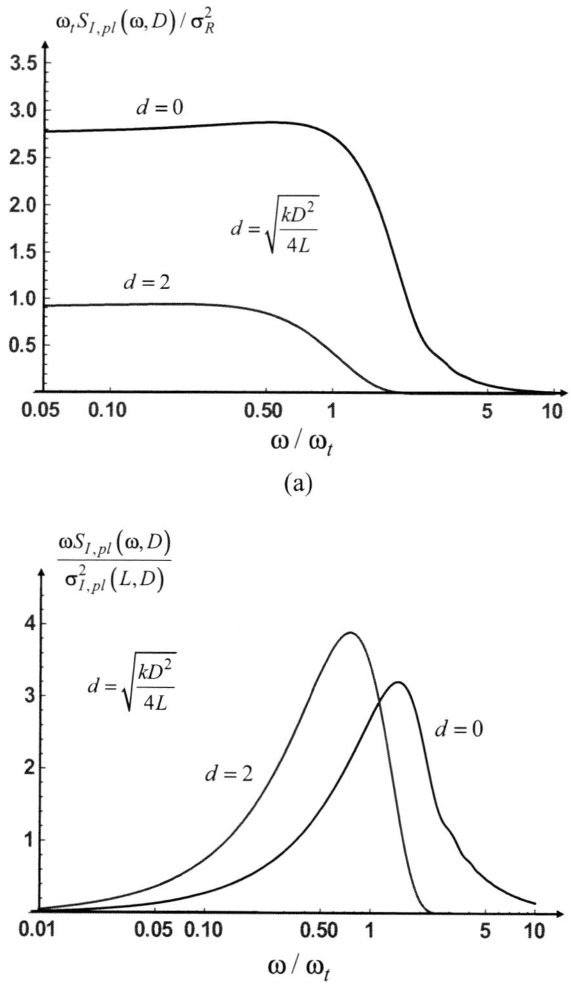

Figure 8.15 Scaled temporal power spectrum of a plane wave in the detector plane under aperture averaging. In both (a) and (b), the aperture diameter is characterized by $d = (kD^2/4L)^{1/2}$ and the parameter $\omega_t = V_\perp/(L/k)^{1/2}$.

This plot shows the transfer to lower frequencies in the temporal power spectrum because the highest frequencies are removed through aperture averaging. We plot the quantity $\omega S_{I,pl}(\omega, D)/\sigma^2_{I,pl}(L, D)$ in Fig. 8.15(b). By plotting the spectrum in this particular fashion, we obtain a clearer idea of which frequencies contribute most to the power of irradiance fluctuations.

8.7.3 Spherical wave

In contrast with the plane wave, we now make the association $\xi\mathbf{p} = \mathbf{V}_\perp\tau$, which accounts for the phase front radius of curvature of a spherical wave. Thus, the temporal covariance function of irradiance for a spherical wave is defined by

$$B_{I,sph}(\tau, L) = 2.606 C_n^2 k^2 L \int_0^1 \int_0^\infty \kappa^{-8/3} J_0(\kappa V_\perp \tau) \left\{1 - \cos\left[\frac{L\kappa^2}{k}\xi(1-\xi)\right]\right\} d\kappa\, d\xi. \tag{8.181}$$

The calculations leading to the temporal spectrum of irradiance are more involved than for the plane wave case. However, following the approach used for the plane wave, the spherical wave temporal power spectrum of irradiance can be written as

$$\begin{aligned} S_{I,sph}(\omega) &= 10.4223 C_n^2 k^2 L \int_0^1 \int_0^\infty \kappa^{-8/3} \int_0^\infty J_0(\kappa V_\perp \tau) \cos(\omega\tau)\, d\tau \\ &\quad \times \left\{1 - \cos\left[\frac{L\kappa^2}{k}\xi(1-\xi)\right]\right\} d\kappa\, d\xi \\ &= \frac{10.4223 C_n^2 k^2 L}{V_\perp} \mathrm{Re} \int_0^1 \int_a^\infty \frac{\kappa^{-8/3}}{(\kappa^2 - a^2)^{1/2}} \left\{1 - \exp\left[-i\frac{L\kappa^2}{k}\xi(1-\xi)\right]\right\} d\kappa\, d\xi \\ &= \frac{10.4223 C_n^2 k^2 L}{V_\perp} \mathrm{Re}(I_1 - I_3), \end{aligned} \tag{8.182}$$

where we recalled Eq. (8.168) and again defined $a = \omega/V_\perp$. The integral I_1 is the same as Eq. (8.170), whereas I_3 takes the form

$$\begin{aligned} I_3 &= \int_0^1 \int_a^\infty \frac{\kappa^{-8/3}}{(\kappa^2 - a^2)^{1/2}} \exp\left[-i\frac{L\kappa^2}{k}\xi(1-\xi)\right] d\kappa\, d\xi \\ &= \frac{1}{2a^{8/3}} \int_0^1 \int_1^\infty \frac{x^{-11/6}}{(x-1)^{1/2}} \exp\left[-i\frac{La^2 x}{k}\xi(1-\xi)\right] dx\, d\xi \\ &= \frac{1}{2a^{8/3}} \int_0^1 \int_1^\infty \frac{x^{-11/6}}{(x-1)^{1/2}} G_{01}^{10}\left[i\frac{\omega^2 x}{\omega_t^2}\xi(1-\xi)\middle|\, 0\right] dx\, d\xi. \end{aligned} \tag{8.183}$$

In arriving at this result, we defined $\kappa = a\sqrt{x}$ and applied property ($C1$) in Appendix C. By use of the integral relation ($D10$) in Appendix D, we find that

$$\begin{aligned} I_3 &= \left(\frac{\omega}{V_\perp}\right)^{-8/3} \frac{1}{2}\Gamma\left(\frac{1}{2}\right) \int_0^1 G_{12}^{20}\left[i\frac{\omega^2}{\omega_t^2}\xi(1-\xi) \middle| \begin{matrix} \frac{11}{6} \\ \frac{4}{3}, 0 \end{matrix} \right] d\xi \\ &= \left(\frac{\omega}{V_\perp}\right)^{-8/3} \left(\frac{\sqrt{\pi}}{2}\right) \frac{1}{2\pi i} \int_C \frac{\Gamma\left(\frac{4}{3}-s\right)\Gamma(-s)}{\Gamma\left(\frac{11}{6}-s\right)} \left(i\frac{\omega^2}{\omega_t^2}\right)^s \int_0^1 \xi^s(1-\xi)^s \, d\xi \, ds, \end{aligned} \tag{8.184}$$

where we have expressed the Meijer G-function in its Barnes integral representation and interchanged the order of integration. With the help of ($G17$) and ($G10$) in Appendix A, the inside integral in Eq. (8.184) leads to

$$\int_0^1 \xi^s(1-\xi)^s \, d\xi = \frac{\Gamma^2(1+s)}{\Gamma(2+2s)} = \frac{\sqrt{\pi}\,\Gamma(1+s)}{2^{2s+1}\Gamma\left(\frac{3}{2}+s\right)}, \tag{8.185}$$

and therefore, we obtain

$$\begin{aligned} I_3 &= \frac{\pi}{4}\left(\frac{\omega}{V_\perp}\right)^{-8/3} \frac{1}{2\pi i} \int_C \frac{\Gamma\left(\frac{4}{3}-s\right)\Gamma(-s)\Gamma(1+s)}{\Gamma\left(\frac{3}{2}+s\right)\Gamma\left(\frac{11}{6}-s\right)} \left(\frac{i\omega^2}{4\omega_t^2}\right)^s ds \\ &= \frac{\pi}{4}\left(\frac{\omega}{V_\perp}\right)^{-8/3} G_{23}^{21}\left(\frac{i\omega^2}{4\omega_t^2} \middle| \begin{matrix} 0, \frac{11}{6} \\ \frac{4}{3}, 0, -\frac{1}{2} \end{matrix} \right). \end{aligned} \tag{8.186}$$

Thus, by combining Eqs. (8.185) and (8.186) with Eq. (8.182), we obtain

$$\begin{aligned} S_{I,sph}(\omega) &= \frac{8.473\sigma_R^2}{\omega_t}\left(\frac{\omega}{\omega_t}\right)^{-8/3} \mathrm{Re}\left[\frac{\sqrt{\pi}}{2}\frac{\Gamma\left(\frac{4}{3}\right)}{\Gamma\left(\frac{11}{6}\right)} - \frac{\pi}{4} G_{23}^{21}\left(\frac{i\omega^2}{4\omega_t^2} \middle| \begin{matrix} 0, \frac{11}{6} \\ \frac{4}{3}, 0, -\frac{1}{2} \end{matrix} \right) \right] \\ &= \frac{7.129\sigma_R^2}{\omega_t}\left(\frac{\omega}{\omega_t}\right)^{-8/3} \mathrm{Re}\left[1 - 0.934\, G_{23}^{21}\left(\frac{i\omega^2}{4\omega_t^2} \middle| \begin{matrix} 0, \frac{11}{6} \\ \frac{4}{3}, 0, -\frac{1}{2} \end{matrix} \right) \right]. \end{aligned} \tag{8.187}$$

Finally, by expressing the Meijer G-function in Eq. (8.187) in the form of ($MG1$) in Appendix A, we are led to the equation derived by Clifford,[17]

$$S_{I,sph}(\omega) = \frac{7.129\sigma_R^2}{\omega_t}\left(\frac{\omega}{\omega_t}\right)^{-8/3}\mathrm{Re}\left[1 - {}_2F_2\left(1, -\frac{5}{6}; \frac{3}{2}, -\frac{1}{3}; -\frac{i\omega^2}{4\omega_t^2}\right)\right.$$
$$\left. - 0.174\left(\frac{i\omega^2}{\omega_t^2}\right)^{4/3} {}_1F_1\left(\frac{1}{2}; \frac{17}{6}; -\frac{i\omega^2}{4\omega_t^2}\right)\right]. \tag{8.188}$$

Similar to the plane wave case, Andrews and Phillips (p. 284 of Ref. 1) derived an approximation to the irradiance temporal spectrum model (8.187), based on replacing the extended atmospheric turbulence by a single phase screen located between Tx and Rx. This approximation is

$$S_{I,sph}(\omega) \approx \frac{5.47\sigma_R^2}{\omega_t}\mathrm{Re}\left\{\left(\frac{\omega}{\omega_t}\right)^{-8/3}\left[1 - {}_1F_1\left(-\frac{5}{6}; -\frac{1}{3}; -\frac{2i\omega^2}{9\omega_t^2}\right)\right]\right.$$
$$\left. - 0.23\, i^{4/3}\, {}_1F_1\left(\frac{1}{2}; \frac{7}{3}; -\frac{2i\omega^2}{9\omega_t^2}\right)\right\}. \tag{8.189}$$

In Fig. 8.16 we plot the temporal spectrum defined by Eq. (8.187) and scaled by the scintillation index for a spherical wave. The general behavior is very similar to that of a plane wave except for the transition frequency where the spectrum begins to decay occurs at a larger value. For comparison, we also plot the approximation given by Eq. (8.189).

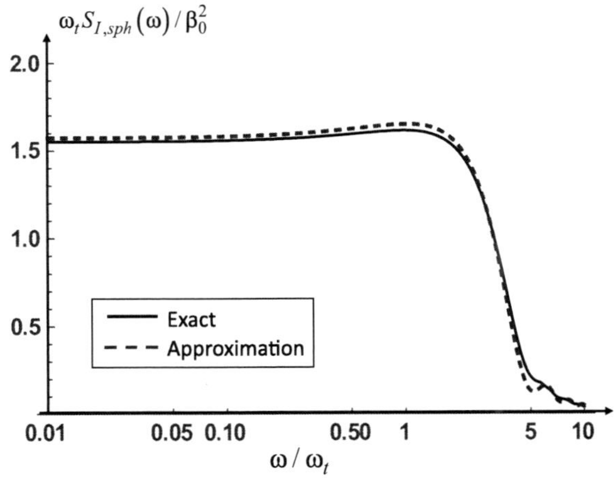

Figure 8.16 Scaled temporal spectrum of irradiance for a spherical wave defined by Eq. (8.187) and the approximation of Eq. (8.189). The parameter $\omega_t = V_\perp/(L/k)^{1/2}$.

8.8 Phase Statistics: Plane Wave

In this section we wish to develop certain statistical models associated with the *phase* of an optical wave. Although the scale sizes that dominate weak irradiance fluctuations are on the order of the Fresnel zone $\sqrt{L/k}$, it is the large-scale structure of the atmosphere that is most important in studying phase fluctuations. For that reason, we will generally ignore inner-scale effects and perform some of the following calculations with the original *von Kármán spectrum*,

$$\Phi_n(\kappa) = \frac{0.033C_n^2}{(\kappa^2 + \kappa_0^2)^{11/6}}. \tag{8.190}$$

8.8.1 Phase structure function

Based on the Rytov approximation for an infinite plane wave, the WSF is defined by

$$D_{pl}(\rho, L) = D_{\chi,pl}(\rho, L) + D_{S,pl}(\rho, L), \tag{8.191}$$

where $D_{\chi,pl}(\rho, L)$ is the *log-amplitude structure function* and $D_{S,pl}(\rho, L)$ is the *phase structure function*. These structure functions together are defined by

$$\left\{ \begin{matrix} D_{\chi,pl}(\rho, L) \\ D_{S,pl}(\rho, L) \end{matrix} \right\} = 4\pi^2k^2L \int_0^1\int_0^\infty \kappa\Phi_n(\kappa)[1 - J_0(\kappa\rho)] \left\{ \begin{matrix} 1 - \cos\left(\dfrac{L\kappa^2\xi}{k}\right) \\ 1 + \cos\left(\dfrac{L\kappa^2\xi}{k}\right) \end{matrix} \right\} d\kappa\, d\xi, \tag{8.192}$$

where the upper (minus) sign corresponds to the log amplitude and the lower (plus) sign to the phase.

Under a geometrical optics approximation, the cosine function in Eq. (8.192) can be replaced by unity; thus, the log-amplitude structure function vanishes, and the phase structure function is the same as the WSF. Based on the spectrum model of Eq. (8.190), this leads to Eq. (8.53), rewritten here in terms of the phase structure function,

$$\begin{aligned} D_{S,pl}(\rho, L) &= 2.206C_n^2k^2L \int_0^\infty \frac{\kappa}{(\kappa^2 + \kappa_0^2)^{11/6}}[1 - J_0(\kappa\rho)]\, d\kappa \\ &= 1.563C_n^2k^2L\kappa_0^{-5/3}[1 - 0.994(\kappa_0\rho)^{5/6} K_{5/6}(\kappa_0\rho)], \quad \rho \gg l_0. \end{aligned} \tag{8.193}$$

To examine diffraction effects, we use Eq. (8.191) to write the phase structure function as

$$D_{S,pl}(\rho, L) = D_{pl}(\rho, L) - D_{\chi,pl}(\rho, L), \tag{8.194}$$

where the WSF is defined by Eq. (8.42) and

$$\begin{aligned} D_{\chi,pl}(\rho, L) &= 2[B_{\chi,pl}(0, L) - B_{\chi,pl}(\rho, L)] \\ &= \frac{1}{2}[B_{I,pl}(0, L) - B_{I,pl}(\rho, L)]. \end{aligned} \tag{8.195}$$

The irradiance covariance function that appears in the last line above is given by Eq. (8.122), so it follows that

$$D_{\chi,pl}(\rho, L) = 1.93\sigma_R^2 \operatorname{Im}\left\{ i^{11/6}\left[1 - {}_1F_1\left(-\frac{11}{6}; 1; \frac{ik\rho^2}{4L}\right)\right]\right\} + 3.76\sigma_R^2\left(\frac{k\rho^2}{4L}\right)^{5/6}, \tag{8.196}$$

which is based on the Kolmogorov spectrum (8.1). Hence, under the Kolmogorov spectrum model, the phase structure function (8.194) takes the form

$$\begin{aligned} D_{S,pl}(\rho, L) &= 3.76\sigma_R^2\left(\frac{k\rho^2}{4L}\right)^{5/6} \\ &\quad + 1.93\sigma_R^2 \operatorname{Im}\left\{ i^{11/6}\left[{}_1F_1\left(-\frac{11}{6}; 1; \frac{ik\rho^2}{4L}\right) - 1\right]\right\}, \\ &\quad l_0 \ll \rho \ll L_0. \end{aligned} \tag{8.197}$$

In Fig. 8.17, the phase structure function (8.197) scaled by the Rytov variance as well as the scaled WSF [recall Eq. (8.42)],

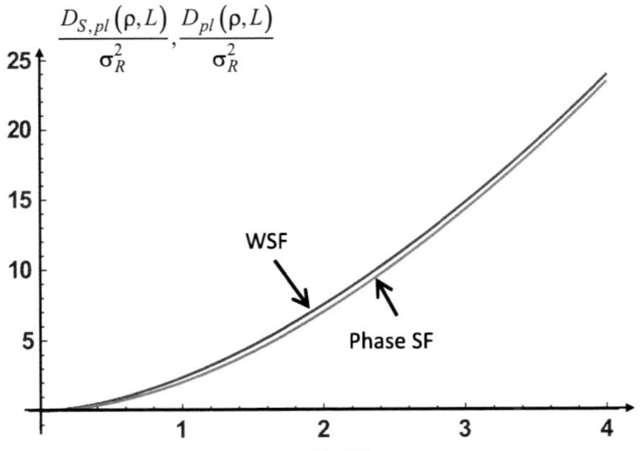

Figure 8.17 Plot of the scaled phase structure function of a plane wave [Eq. (8.197)] and the scaled wave structure function [Eq. (8.198)].

$$D_{pl}(\rho, L)/\sigma_R^2 = 2.914 C_n^2 k^2 L \rho^{5/3}/\sigma_R^2$$
$$= 7.52\left(\frac{k\rho^2}{4L}\right)^{5/6}, \quad l_0 \ll \rho \ll L_0, \tag{8.198}$$

are plotted as a function of $\sqrt{k\rho^2/L}$. Here we see that the phase structure function lies slightly below the WSF for separation distances up to at least four times the size of the Fresnel zone $\sqrt{L/k}$. Although not shown, at separation distances satisfying $\rho \gg \sqrt{L/k}$, the phase structure function (8.197) and WSF are the same [recall the geometrical optics approximation of Eq. (8.193)]. However, for separation distances satisfying $l_0 \ll \rho \ll \sqrt{L/k}$, we can use the small-argument approximation $({}_1F_1 \approx 1)$ for the confluent hypergeometric function in Eq. (8.197), which leads to

$$D_{S,pl}(\rho, L) = 3.76\,\sigma_R^2\left(\frac{k\rho^2}{4L}\right)^{5/6} = \frac{1}{2}D_{pl}(\rho, L), \quad l_0 \ll \rho \ll \sqrt{L/k}. \tag{8.199}$$

In this last case we note that the phase structure function reduces to half of the WSF of a plane wave.

Fried[18] used the above behavior of the phase structure function to define the near field and far field of the Rx aperture. That is, by replacing ρ with the Rx lens diameter D, Fried identified the case $D \gg \sqrt{L/k}$ as the *near field* of the Rx aperture. This represents the case where the WSF and phase structure function are equal. By introducing the Gaussian-beam parameter $\Omega_{Rx} = 2L/kW_{Rx}^2$ that appears in Chapter 9, where W_{Rx} is the Gaussian Rx aperture radius, we can also identify the near field by $\Omega_{Rx} \ll 1$. The case with $D \ll \sqrt{L/k}$, or $\Omega_{Rx} \gg 1$, identifies the *far field* of the Rx aperture where Eq. (8.199) is satisfied.

8.8.2 Angle-of-arrival

Angle-of-arrival (AOA) *fluctuations* of an optical wave in the pupil plane of the Rx are associated with image jitter in the focal plane of an imaging system. These fluctuations can be described in terms of the phase structure function. For example, in the case of an incident plane wave in the near-field (NF) approximation, the AOA variance on a Rx aperture of diameter D is defined by

$$\langle\beta_a^2\rangle_{NF} = \frac{D_{S,pl}(D, L)}{(kD)^2} \approx 0.461\left(\frac{D}{r_0}\right)^{5/3}\left(\frac{\lambda}{D}\right)^2, \tag{8.200}$$

where the Fried parameter is $r_0 = (0.16 C_n^2 k^2 L)^{-3/5}$ and $D = \rho$. Also, we are approximating the phase structure function (8.197) by the plane wave structure function (8.198), assuming that $D \gg \sqrt{L/k}$. In the far field (FF), the phase structure function is only $\frac{1}{2}$ the plane wave structure function as described by Eq. (8.199). Thus, the far-field AOA variance leads to the expression

$$\langle \beta_a^2 \rangle_{FF} = \frac{D_{S,pl}(D, L)}{2(kD)^2} \approx 0.231 \left(\frac{D}{r_0} \right)^{5/3} \left(\frac{\lambda}{D} \right)^2. \tag{8.201}$$

To calculate image jitter in the focal plane, we simply multiply the root mean square (rms) AOA variance by the focal length F_{Rx} of the Rx lens, which leads to

$$\sqrt{\langle \beta_a^2 \rangle} F_{Rx} \approx \begin{cases} 0.679(D/r_0)^{5/6}(\lambda F_{Rx}/D), & \text{(near field)} \\ 0.480(D/r_0)^{5/6}(\lambda F_{Rx}/D), & \text{(far field)}. \end{cases} \tag{8.202}$$

8.8.3 Phase variance

The phase variance is defined by

$$\begin{aligned} \sigma_{S,pl}^2(L) &= \frac{1}{2}[E_2(\mathbf{r}, \mathbf{r}) - E_3(\mathbf{r}, \mathbf{r})] \\ &= 2\pi^2 k^2 L \int_0^1 \int_0^\infty \kappa \Phi_n(\kappa) \left[1 + \cos\left(\frac{L\kappa^2 \xi}{k} \right) \right] d\kappa\, d\xi \\ &= 4\pi^2 k^2 L \int_0^\infty \kappa \Phi_n(\kappa)\, d\kappa, \end{aligned} \tag{8.203}$$

where the last line is the result of a geometrical optics approximation. If we use spectrum model (8.190), the resulting integral leads to

$$\sigma_{S,pl}^2(L) = 0.782 C_n^2 k^2 L \kappa_0^{-5/3}. \tag{8.204}$$

To include diffraction effects, we first write

$$\sigma_{\chi,pl}^2(L) + \sigma_{S,pl}^2(L) = 4\pi^2 k^2 L \int_0^\infty \kappa \Phi_n(\kappa)\, d\kappa, \tag{8.205}$$

where

$$\sigma_{\chi,pl}^2 = \frac{1}{4}\sigma_R^2 = 0.307 C_n^2 k^{7/6} L^{11/6}. \tag{8.206}$$

Hence, it follows that ($Q_0 = L\kappa_0^2/k$)

$$\sigma_{S,pl}^2(L) = 0.782 C_n^2 k^2 L \kappa_0^{-5/3} - 0.307 C_n^2 k^{7/6} L^{11/6} \\ = 0.636 \sigma_R^2 Q_0^{-5/6}(1 - 0.39\, Q_0^{5/6}). \tag{8.207}$$

8.8.4 Phase covariance function

Under the assumption that the turbulence is statistically homogeneous and isotropic, the phase structure function can be expressed as a function of the phase covariance function,

$$D_{S,pl}(\rho, L) = 2[B_{S,pl}(0, L) - B_{S,pl}(\rho, L)]. \tag{8.208}$$

Thus, by comparison with Eq. (8.193), it follows that

$$B_{S,pl}(\rho, L) = 0.777 C_n^2 k^2 L \left(\frac{\rho}{\kappa_0}\right)^{5/6} K_{5/6}(\kappa_0 \rho), \quad \rho \gg l_0. \tag{8.209}$$

Equation (8.209), scaled by the phase variance (8.204), is plotted in Fig. 8.18.

To include diffraction effects, we consider the sum of covariance functions of log amplitude and phase,

$$B_{\chi,pl}(\rho, L) + B_{S,pl}(\rho, L) = 4\pi^2 k^2 L \int_0^\infty \kappa \Phi_n(\kappa) J_0(\kappa\rho) d\kappa. \tag{8.210}$$

From Eq. (8.52), the integral on the right in Eq. (8.210) reduces to

$$\int_0^\infty \frac{\kappa J_0(\kappa\rho)}{(\kappa^2 + \kappa_0^2)^{11/6}} d\kappa = 0.597 \left(\frac{\rho}{\kappa_0}\right)^{5/6} K_{5/6}(\kappa_0 \rho), \tag{8.211}$$

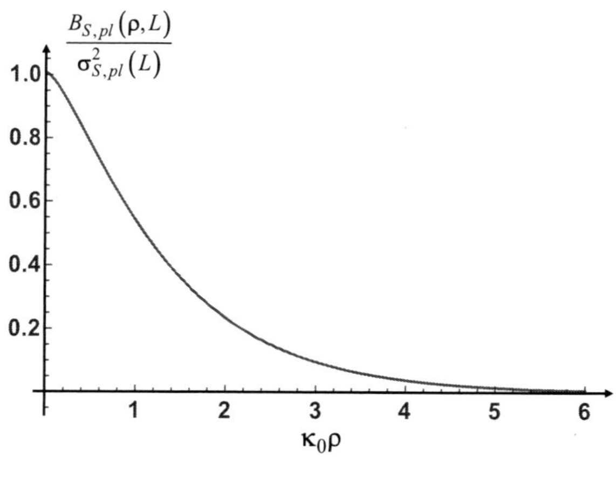

Figure 8.18 Plot of the geometrical optics approximation of the phase covariance function of a plane wave defined by Eq. (8.209) scaled by the phase variance defined by Eq. (8.204).

and hence, combining this last result with Eq. (8.209), we obtain

$$B_{S,pl}(\rho, L) = 0.632\sigma_R^2 Q_0^{-5/6}(\kappa_0\rho)^{5/6}K_{5/6}(\kappa_0\rho) - 0.97\sigma_R^2 \,\mathrm{Im}\left[i^{11/6}\,{}_1F_1\left(-\frac{11}{6}; 1; \frac{ik\rho^2}{4L}\right)\right] + 1.88\sigma_R^2\left(\frac{k\rho^2}{4L}\right)^{5/6}. \tag{8.212}$$

8.8.5 Temporal spectrum of phase

The *phase temporal covariance function* of a plane wave is defined by

$$B_{S,pl}(\tau, L) = 2\pi^2k^2L\int_0^1\int_0^\infty \kappa\Phi_n(\kappa)J_0(\kappa V_\perp\tau)\left[1+\cos\left(\frac{L\kappa^2\xi}{k}\right)\right]d\kappa\,d\xi, \tag{8.213}$$

which differs from the covariance function of the irradiance defined by Eq. (8.117) only by the sign difference between terms in brackets and the multiplicative constant. Under a geometrical optics approximation and the spectrum model of Eq. (8.190), Eq. (8.213) reduces to

$$\begin{aligned}B_{S,pl}(\tau, L) &= 4\pi^2k^2L\int_0^1\int_0^\infty \kappa\Phi_n(\kappa)J_0(\kappa V_\perp\tau)\,d\kappa\,d\xi\\ &= 0.777C_n^2k^2L\left(\frac{V_\perp\tau}{\kappa_0}\right)^{5/6}K_{5/6}(\kappa_0V_\perp\tau),\end{aligned} \tag{8.214}$$

where we are recalling Eq. (8.209) with $\rho = V_\perp\tau$. Consequently, the associated temporal spectrum of the phase is found from the Fourier transform relation

$$\begin{aligned}S_{S,pl}(\omega) &= 3.127C_n^2k^2L\kappa_0^{-5/3}\int_0^\infty(\kappa_0V_\perp\tau)^{5/6}K_{5/6}(\kappa_0V_\perp\tau)\cos(\omega\tau)\,d\tau\\ &= 3.127C_n^2k^2L\kappa_0^{-5/3}\sqrt{\pi}\int_0^\infty(\kappa_0V_\perp\tau)^{5/6}K_{5/6}(\kappa_0V_\perp\tau)G_{02}^{10}\left(\frac{\omega^2\tau^2}{4}\middle|0,\frac{1}{2}\right)d\tau\\ &= 3.127C_n^2k^2L\kappa_0^{-5/3}\frac{2^{5/6}\sqrt{\pi}}{\kappa_0V_\perp}\int_0^\infty x^{-1/12}K_{5/6}(2\sqrt{x})G_{02}^{10}\left(\frac{\omega^2x}{\kappa_0^2V_\perp^2}\middle|0,\frac{1}{2}\right)dx\\ &= 3.127C_n^2k^2L\kappa_0^{-5/3}\frac{\sqrt{\pi}}{2^{1/6}\kappa_0V_\perp}G_{22}^{12}\left(\frac{\omega^2}{\kappa_0^2V_\perp^2}\middle|\begin{matrix}-\frac{1}{3},\frac{1}{2}\\ 0,\frac{1}{2}\end{matrix}\right),\end{aligned} \tag{8.215}$$

where we have used ($C4$) in Appendix C, made the change of variable $\kappa_0V_\perp\tau = 2\sqrt{x}$, and applied ($D13$) in Appendix D. Now, from ($MG3$) in Appendix A, we have

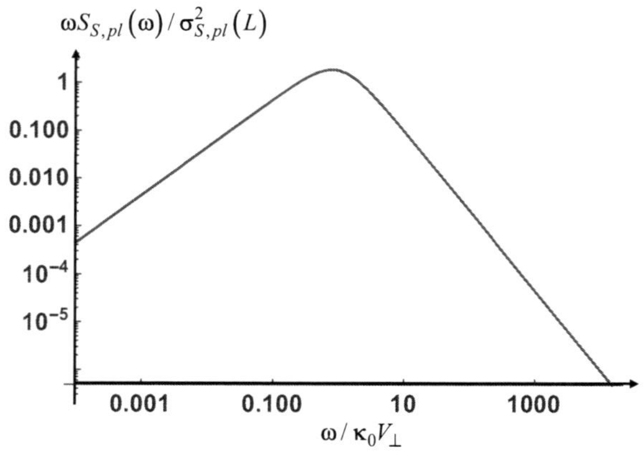

Figure 8.19 Plot of the scaled phase power spectrum of a plane wave defined by Eq. (8.217) under a geometrical optics approximation.

$$\begin{aligned} S_{S,pl}(\omega) &= 3.127 C_n^2 k^2 L \kappa_0^{-5/3} \frac{\sqrt{\pi}}{2^{1/6}\kappa_0 V_\perp} G_{11}^{11}\left(\frac{\omega^2}{\kappa_0^2 V_\perp^2} \middle| \begin{matrix} -\frac{1}{3} \\ 0 \end{matrix} \right) \\ &= 3.127 C_n^2 k^2 L \kappa_0^{-5/3} \frac{\sqrt{\pi}\Gamma\left(\frac{4}{3}\right)}{2^{1/6}\kappa_0 V_\perp} \left(1 + \frac{\omega^2}{\kappa_0^2 V_\perp^2}\right)^{-4/3}. \end{aligned} \tag{8.216}$$

The last step in Eq. (8.216) follows from ($C11$) in Appendix C. Upon simplification of this result, we obtain the temporal phase covariance spectrum given by

$$S_{S,pl}(\omega) = \frac{4.409 C_n^2 k^2 L V_\perp^{5/3}}{(\omega^2 + \kappa_0^2 V_\perp^2)^{4/3}}. \tag{8.217}$$

In Fig. 8.19 we plot the scaled phase power spectrum [Eq. (8.217)] as a function of $\omega/\kappa_0 V_\perp$. Although not shown, the spectrum for a collimated beam wave or spherical wave is nearly the same with small differences occurring only at higher frequencies.[19]

8.9 Phase Statistics: Spherical Wave

Here we follow the analysis presented in Section 8.8 but now for the case of a spherical wave. Once again, we will rely on the von Kármán spectrum given by Eq. (8.190) for some of the calculations.

8.9.1 Phase structure function

For a spherical wave, the phase structure function is defined by

$$D_{S,sph}(\rho, L) = 4\pi^2 k^2 L \int_0^1 \int_0^\infty \kappa \Phi_n(\kappa)[1 - J_0(\kappa\xi\rho)] \times \left\{1 + \cos\left[\frac{L\kappa^2\xi(1-\xi)}{k}\right]\right\} d\kappa\, d\xi. \tag{8.218}$$

Under a geometrical optics approximation, the cosine term in Eq. (8.218) is essentially unity and, thus, the phase structure function is the same as the WSF for a spherical wave (Section 8.3.3). In particular, over the inertial range, we have

$$\begin{aligned} D_{sph}(\rho, L) &= 1.093 C_n^2 k^2 L \rho^{5/3} \\ &= 6.994\, \beta_0^2 \left(\frac{k\rho^2}{4L}\right)^{5/6}, \quad l_0 \ll \rho \ll L_0, \end{aligned} \tag{8.219}$$

where $\beta_0^2 = 0.496 C_n^2 k^{7/6} L^{11/6}$. However, under the von Kármán spectrum model (8.190), we find that the phase structure function is

$$\begin{aligned} D_{S,sph}(\rho, L) &= 2.606 C_n^2 k^2 L \int_0^1 \int_0^\infty \frac{\kappa}{(\kappa^2 + \kappa_0^2)^{11/6}} [1 - J_0(\kappa\xi\rho)]\, d\kappa\, d\xi \\ &= 1.563 C_n^2 k^2 L \kappa_0^{-5/3} \left[1 - 0.994 \int_0^1 (\kappa_0\rho\xi)^{5/6} K_{5/6}(\kappa_0\rho\xi)\, d\xi\right], \end{aligned} \tag{8.220}$$

where we have recalled the plane wave result of Eq. (8.193) with ρ replaced by $\rho\xi$. If we now replace the modified Bessel function in Eq. (8.220) with (*C*23) in Appendix C and set $x = \xi^2$, we are led to

$$D_{S,sph}(\rho, L) = 1.563 C_n^2 k^2 L \kappa_0^{-5/3} \times \left[1 - \frac{0.994}{4} (\kappa_0\rho)^{5/6} \int_0^1 x^{-1/12} G_{02}^{20}\left(\frac{\kappa_0^2\rho^2 x}{4}\middle| \frac{5}{12}, -\frac{5}{12}\right) dx\right], \tag{8.221}$$

which, with the help of (*D*9) in Appendix D, reduces to

$$D_{S,sph}(\rho, L) = 1.563 C_n^2 k^2 L \kappa_0^{-5/3} \left[1 - 0.248 (\kappa_0\rho)^{5/6} G_{13}^{21}\left(\frac{\kappa_0^2\rho^2}{4}\middle| \begin{matrix} \frac{1}{12} \\ \frac{5}{12}, -\frac{5}{12}, -\frac{11}{12} \end{matrix}\right)\right]. \tag{8.222}$$

Although we could use (*MG*1) in Appendix A to express the Meijer *G*-function as the sum of two ${}_1F_2$ functions, we choose not to do so.

Alternatively, we can use the method illustrated in Chapter 6 to find an approximation to the Meijer G-function itself. By using properties ($MG1$) and ($MG2$) in Appendix A, we find the asymptotic relations

$$G_{13}^{21}\left(\frac{\kappa_0^2\rho^2}{4}\middle|\begin{matrix}\frac{1}{12}\\ \frac{5}{12},\ -\frac{5}{12},\ -\frac{11}{12}\end{matrix}\right)\sim\begin{cases}4.023(\kappa_0^2\rho^2)^{-5/12}, & \kappa_0\rho\to 0\\ 0.71(\kappa_0^2\rho^2)^{-11/12}, & \kappa_0\rho\to\infty,\end{cases}$$

and, thus, derive the approximation

$$G_{13}^{21}\left(\frac{\kappa_0^2\rho^2}{4}\middle|\begin{matrix}\frac{1}{12}\\ \frac{5}{12},\ -\frac{5}{12},\ -\frac{11}{12}\end{matrix}\right)\approx\frac{4.023(\kappa_0\rho)^{-5/6}}{\sqrt{1+0.5(\kappa_0\rho)^2}}. \tag{8.223}$$

By using this last result, we can approximate Eq. (8.222) by

$$D_{S,sph}(\rho, L)\approx 1.563C_n^2k^2L\kappa_0^{-5/3}\left(1-\frac{1}{\sqrt{1+0.5(\kappa_0\rho)^2}}\right). \tag{8.224}$$

We plot the scaled phase structure function (8.222) and its approximation (8.224) in Fig. 8.20 along with the scaled phase structure function for a plane wave [Eq. (8.193)], all as a function of $\kappa_0\rho$. In this case the approximation (8.224) is quite good for all values over the range, which shows that the Meijer G-function can also be accurately approximated in some cases by the methods illustrated in Chapter 6. The scaling constant in Fig. 8.20 is twice the phase variance under a geometrical optics approximation. The graphical difference between the scaled plane wave and spherical wave phase structure functions in Fig. 8.20 is caused mostly by the eddy size influence along the path. For

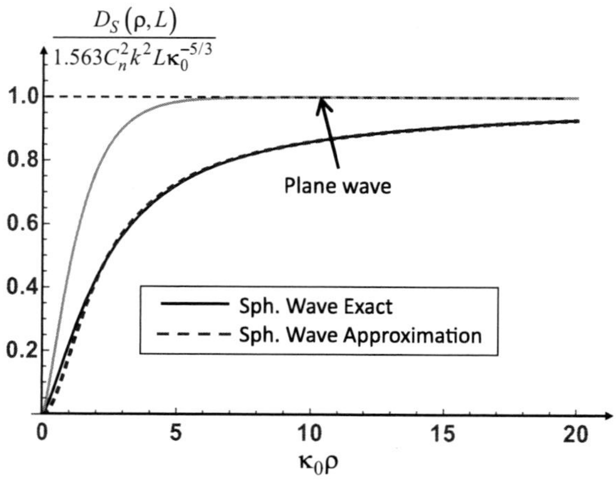

Figure 8.20 Scaled phase structure function of a spherical wave [Eqs. (8.222) and (8.224)] and the scaled phase structure function of a plane wave [Eq. (8.193)], showing the effects of the outer scale under a geometrical optics approximation.

example, in the case of a plane wave, only eddies larger than the separation distance ρ can influence the parallel rays of a plane wave. On the other hand, for a spherical wave, eddies of virtually all sizes along the path have influence on the phase structure function.

To include diffraction effects in the structure function, we will follow the analysis for a plane wave. Let us start by recalling the irradiance covariance function in the more compact form of Eq. (8.132),

$$B_{I,sph}(\rho,L) = -6.99426\beta_0^2\left(\frac{k\rho^2}{4L}\right)^{5/6} - 0.65465\beta_0^2 \mathrm{Re}\left[i^{5/6}G_{23}^{22}\left(-\frac{ik\rho^2}{4L}\middle| \begin{matrix} -\frac{5}{6},\frac{11}{6} \\ \frac{11}{6},0,0 \end{matrix}\right)\right], \tag{8.225}$$

where $B_{I,sph}(0, L) = \beta_0^2$. Therefore, referring back to Eq. (8.194), which is also valid for a spherical wave, we find that

$$D_{\chi,sph} = 0.5\beta_0^2 + 3.497\beta_0^2\left(\frac{k\rho^2}{4L}\right)^{5/6} + 0.327\beta_0^2 \mathrm{Re}\left[i^{5/6}G_{23}^{22}\left(-\frac{ik\rho^2}{4L}\middle| \begin{matrix} -\frac{5}{6},\ \frac{11}{6} \\ \frac{11}{6},\ 0,\ 0 \end{matrix}\right)\right], \tag{8.226}$$

from which we deduce that

$$\begin{aligned} D_{S,sph}(\rho,\ L) &= D_{sph}(\rho,\ L) - D_{\chi,sph}(\rho,\ L) \\ &= 3.497\beta_0^2\left(\frac{k\rho^2}{4L}\right)^{5/6} - 0.5\beta_0^2 - 0.327\beta_0^2 \mathrm{Re}\left[i^{5/6}G_{23}^{22}\left(-\frac{ik\rho^2}{4L}\middle| \begin{matrix} -\frac{5}{6},\ \frac{11}{6} \\ \frac{11}{6},\ 0,\ 0 \end{matrix}\right)\right]. \end{aligned} \tag{8.227}$$

8.9.2 Phase covariance function

Following the analysis for a plane wave phase covariance function, we note that

$$D_{S,sph}(\rho,\ L) = 2[B_{S,sph}(0,\ L) - B_{S,sph}(\rho,\ L)], \tag{8.228}$$

and, thus,

$$B_{S,sph}(\rho,\ L) = B_{S,sph}(0,\ L) - \frac{1}{2}D_{S,sph}(\rho,\ L). \tag{8.229}$$

To include diffraction effects, we consider the sum of covariance functions of log amplitude and phase,

$$B_{\chi,sph}(\rho,\ L) + B_{S,sph}(\rho,\ L) = 4\pi^2k^2L\int_0^1\int_0^\infty \kappa\Phi_n(\kappa)\, J_0(\kappa\rho\xi)d\kappa\, d\xi, \tag{8.230}$$

where

$$B_{\chi,sph}(\rho, L) = \frac{1}{4} B_{I,sph}(\rho, L). \tag{8.231}$$

By the use of ($C9$) in Appendix C, and ($D11$) and ($D9$) in Appendix D, the integral on the right in Eq. (8.230) yields

$$\begin{aligned}
4\pi^2 k^2 L \int_0^1 \int_0^\infty & \kappa \Phi_n(\kappa) J_0(\kappa\rho\xi) d\kappa\, d\xi \\
&= 1.303 C_n^2 k^2 L \int_0^1 \int_0^\infty \frac{\kappa}{(\kappa^2 + \kappa_0^2)^{11/6}} G_{02}^{10}\left(\frac{\kappa^2\rho^2\xi^2}{4}\middle| 0, 0\right) d\kappa\, d\xi \\
&= \frac{1.303}{2\Gamma\left(\frac{11}{6}\right)} C_n^2 k^2 L \kappa_0^{-5/3} \int_0^1 G_{02}^{20}\left(\frac{\kappa_0^2\rho^2\xi^2}{4}\middle| \frac{5}{6}, 0\right) d\xi \\
&= 0.698\, \beta_0^2 Q_0^{-5/6} G_{13}^{21}\left(\frac{\kappa_0^2\rho^2}{4}\middle| \begin{matrix} \frac{1}{2} \\ \frac{5}{6}, 0, -\frac{1}{2} \end{matrix}\right).
\end{aligned} \tag{8.232}$$

The irradiance covariance function in Eq. (8.231) is given by Eq. (8.225), so by combining Eqs. (8.231) and (8.232), we are led to

$$\begin{aligned}
B_{S,sph}(\rho, L) = {} & 0.698\, \beta_0^2 Q_0^{-5/6} G_{13}^{21}\left(\frac{\kappa_0^2\rho^2}{4}\middle| \begin{matrix} \frac{1}{2} \\ \frac{5}{6}, 0, -\frac{1}{2} \end{matrix}\right) \\
& + 0.164\beta_0^2 \mathrm{Re}\left[i^{5/6} G_{23}^{22}\left(-\frac{ik\rho^2}{4L}\middle| \begin{matrix} -\frac{5}{6}, \frac{11}{6} \\ \frac{11}{6}, 0, 0 \end{matrix}\right)\right] + 1.748\beta_0^2 \left(\frac{k\rho^2}{4L}\right)^{5/6}.
\end{aligned} \tag{8.233}$$

By setting $\rho = 0$ in Eq. (8.233), we obtain the spherical wave phase variance

$$\sigma_{S,sph}^2 = 1.575\, \beta_0^2 Q_0^{-5/6}. \tag{8.234}$$

As a final comment here, we should point out that including diffraction effects in the phase statistics for both plane waves and spherical waves produces only a small modification in most cases compared to cases that include only geometrical optics.

8.10 Discussion

In this chapter we used the novel methods illustrated in Chapter 7 to solve a variety of problems in electromagnetic wave propagation. Although some of the mathematical models had been previously derived, the details of the derivations were generally missing in books like Refs. 1 and 2 and in journal articles. Many of the methods we used here are not the original techniques used to derive the math models but are generally more direct and shorter in length than the originals. In addition to presenting details of the derivations, we have also developed several new mathematical models, including those based on the exponential spectrum model. Other new results include the covariance function of irradiance with and without aperture averaging, and the corresponding temporal power spectra of irradiance with and without aperture averaging.

8.11 Historical Remarks

Astronomers were among the first scientists to show interest in optical wave propagation, particularly that associated with the quivering of an astronomical object at the focus of a telescope and temporal fluctuations in received irradiance (star twinkle), the latter commonly called *scintillation*. Sir Isaac Newton (1642–1727), who built the first practical reflecting telescope, was aware of this phenomenon and is credited as the first person to give it a scientific explanation in his book *Opticks*. He attributed the twinkling of stars to the "Tremors of the Atmosphere" and suggested that to reduce the effects of atmospheric turbulence it was best to view the stars from the tops of the highest mountains.

During the 1940s and 1950s, astronomers around the world were busy studying the stars, but the first theoretical studies of scintillation to aid astronomers were done primarily by Russian scientists. This theoretical development was made possible after Andrey Nikolaevich Kolmogorov[20] (1903–1987) introduced his theory of turbulence in 1941. Although based largely on physical insight through a set of hypotheses, the theory of Kolmogorov and his power law spectrum were shown to be correct and eventually led to further theoretical studies soon followed by a series of measurements of optical scintillation. Early theoretical work on scintillation centered around geometrical optics, which imposed severe limitations on the math models. Alexander Mikhailovich Obukhov[21] (1918–1989), who received his Ph.D. under Kolmogorov, was the first to introduce diffraction into the calculations of irradiance fluctuations based on the smooth perturbation theory developed earlier by Sergei Mikhailovich Rytov[22] (1908–1996). The smooth perturbation theory is now called the Rytov approximation in the Western world. In 1956, Obukhov founded the A.M. Obukhov Institute of Atmospheric Physics of the Russian Academy of Sciences in Moscow, one of the leading research centers in the world studying atmospheric science. A competing research center called the V. E. Zuev Institute of Atmospheric Optics of the Siberian Branch of the Russian Academy of Sciences in Tomsk was founded in 1969 by Vladimir Yevseyevich Zuev (1925–2003), who became its first director.

The subject of electromagnetic wave propagation in the atmosphere began to attract interest far beyond the Russian community after the English textbook publications of Lev Aleksandrovich Chernov[23] (1911–1981) in 1960 and Valerian Il'ich Tatarskii[24] (1929–2020) in 1961. It was Tatarskii's Ph.D. thesis that was converted into a book in 1959 and then translated into English in 1961 with the same title. He later developed a more complete version of his earlier text that was based on his Doctor of Science degree and translated into English in 1971.[16] The spectrum model attributed to Tatarskii was first proposed by E. A. Novikov[25] for velocity fluctuations and later adopted by Tatarskii for refractive-index fluctuations. After the laser was introduced on 16 May 1960 by Theodore Maiman (1927–2007), there were myriad scientists all over the world who became occupied with the theoretical and experimental development of electromagnetic wave propagation and its applications that continues to the present day.

References

1. L. C. Andrews and R. L. Phillips, *Laser Beam Propagation through Random Media*, 2nd Edition, SPIE Press, Bellingham, Washington (2005) [doi: 10.1117/3.626196].
2. L. C. Andrews and M. K. Beason, *Laser Beam Propagation in Random Media: New and Advanced Topics*, SPIE Press, Bellingham, Washington (2023) [doi: 10.1117.3.2643989]
3. L. C. Andrews, "An analytical model for the refractive index power spectrum and its application to optical scintillations in the atmosphere," *J. Mod. Opt.* **39**(9), 1849–1853 (1992).
4. F. H. Champagne, C. A. Friehe, J. C. LaRue, and J. C. Wyngaard, "Flux measurements, flux-estimation techniques, and fine-scale turbulence measurements in the unstable surface layer over land," *J. Atmosp. Sci.* **34**(3), 515–530 (1977).
5. R. M. Williams and C. A. Paulson, "Microscale temperature and velocity spectra in the atmospheric boundary layer boundary layer," *J. Fluid Mech.* **83**(3), 547–567 (1977).
6. R. J. Hill, "Models of the scalar spectrum for turbulent advection," *J. Fluid Mech.* **88**(3), 541–662 (1978).
7. L. C. Andrews and W. B. Miller, "Single-pass and double-pass propagation through complex paraxial optical systems," *J. Opt. Soc. Am. A* **12**(1), 137–150 (1995); "Errata," *J. Opt. Soc. Am. A* **12**(1), 2213 (1995).
8. L. C. Andrews, W. B. Miller, and J. C. Ricklin, "Spatial coherence of a Gaussian-beam wave in weak and strong optical turbulence," *J. Opt. Soc. Am. A* **11**(5), 1653–1660 (1994).
9. C. Y. Young and L. C. Andrews, "Effects of a modified spectral model on the spatial coherence of a laser beam," *Waves Random Media* **4**(3), 385–397 (1994).

10. L. C. Andrews, S. Vester, and C. E. Richardson, "Analytic expressions for the wave structure function based on a bump spectral model for refractive index," *J. Mod. Opt.* **40**(5), 931–938 (1993).
11. W. B. Miller, J. C. Ricklin, and L. C. Andrews, "Effects of the refractive index spectral model on the irradiance variance of a Gaussian beam," *J. Opt. Soc. Am. A* **11**(10), 2719–2726 (1994).
12. D. L. Fried and J. D. Cloud, "Propagation of an infinite plane wave in a randomly inhomogeneous medium," *J. Opt. Soc. Am.* **56**(12), 1667–1676 (1966).
13. A. D. Wheelon, *Electromagnetic Scintillation II. Weak Scattering*, Cambridge University Press, Cambridge, New York (2003).
14. D. L. Fried, "Propagation of a spherical wave in a turbulent medium," *J. Opt. Soc. Am.* **57**(2), 175–180 (1967).
15. L. C. Andrews, R. L. Phillips, C. Y. Hopen, and M. A. Al-Habash, "Theory of optical scintillation," *J. Opt. Soc. Am. A* **16**(6), 1417–1429 (1999).
16. V. I. Tatarskii, *The Effects of the Turbulent Atmosphere on Wave Propagation* (trans. from the Russian and issued by the National Technical Information Office, U.S. Dept. of Commerce, Springfield, 1971).
17. S. F. Clifford, "Temporal-frequency spectra for a spherical wave propagating through atmospheric turbulence," *J. Opt. Soc. Am.* **61**(10), 1285–1292 (1971).
18. D. L. Fried, "Optical resolution through a randomly inhomogeneous medium for very long and very short exposures," *J. Opt. Soc. Am.* **56**(10), 1372–1379 (1966).
19. V. P. Lukin and V. V. Pokasov, "Optical wave phase fluctuations," *Appl. Opt.* **20**(1), 121–135 (1981).
20. A. N. Kolmogorov, "The local structure of turbulence in an incompressible viscous fluid for very large Reynolds numbers," *C. R. (Doki) Acad. Sci. U.S.S.R.* **30**, 301–305 (1941).
21. A. M. Obukhov, "Effect of weak inhomogeneities in the atmosphere on sound and light propagation," *Izv. Akad. Nauk SSSR, Ser. Geofiz.* **2**, 155–165 (1953).
22. S. M. Rytov, "Diffraction of light by ultrasonic waves," *Izvestiya Akademii Nauk SSSR, Seriya Fizicheskaya (Bulletin of the Academy of Sciences of the USSR, Physical Series)* **2**, 223–259 (1937).
23. L. A. Chernov, *Wave Propagation in a Random Medium*, McGraw Hill, New York (1960) [trans. from the Russian by R. A. Silverman].
24. V. I. Tatarskii, *Wave Propagation in a Turbulent Medium*, McGraw Hill, New York (1961) [trans. from the Russian by R. A. Silverman].
25. E. A. Novikov, "The energy spectrum of incompressible turbulent flow," *Dokl. Akad. Nauk. SSSR* **139**, 331–333 (1961).

Chapter 9
Imaging Through Atmospheric Turbulence

9.0 Introduction

Small temperature fluctuations in the atmosphere induce index-of-refraction fluctuations that cause *optical scintillation* and adversely affect *imaging systems*. It has long been known (since the time of Newton) that atmospheric turbulence is deleterious to the "seeing" ability of astronomers using a ground-based telescope. That is, the image quality formed in the detector plane may be greatly degraded by absorption, scattering, and atmospheric turbulence. Image degradation effects, such as *image blurring* and *image dancing* (jitter), are associated primarily with phase fluctuations that are dominated by the large-scale turbulent eddies. Image jitter arises from the overall tilt imparted to the optical wavefront by the advection of large eddies. In a short-term exposure, the wavefront tilt displaces the image but does not reduce its sharpness. Long-term image blur is caused by a superposition of small-scale effects and large-scale image jitter.

Many of the adverse effects of atmospheric turbulence on imaging systems were recognized early on by astronomers, but the basic underlying physics wasn't well understood before the 1950s. To overcome some of these adverse effects, various compensation methods were eventually developed. These methods include *postprocessing techniques*, *adaptive optics*, and *hybrid imaging methods*, the latter based on combining the first two methods.[1,2]

- *Postprocessing techniques*: The first methods applied to overcome atmospheric effects on imaging relied on post-detection processing of short-exposure images. The term "short exposure" refers to a single realization of atmospheric turbulence with an exposure time on the order of a few milliseconds. Such techniques are also commonly referred to as speckle imaging in reference to the "speckled" appearance of the images depicting high spatial frequency content. The basic idea is to estimate the modulus and phase of the Fourier transform of the object being imaged.

- *Adaptive optics:* Systems that sense and use mechanical means to correct for atmospherically induced wavefront deformations in real time are called adaptive optics imaging (AOI) systems. The main components are a deformable mirror (DM), wavefront sensor (WFS), and an actuator command computer. Among others, limiting factors that prevent AOI systems from achieving ideal performance are insufficient brightness from the object being imaged and differences between the sensed wavefront and the wavefront of the object being imaged (anisoplanatism).
- *Hybrid imaging techniques*: Both speckle imaging and AOI systems have limitations in overcoming the effects of atmospheric turbulence. Consequently, hybrid methods that combine postprocessing and AOI techniques are sometimes used. Such methods provide a trade-off between pre-detection and post-detection methods, using the best performance features of each type of system.

In this chapter we present various ways in which the image quality of an optical or electro-optical (EO) system can be measured. For some of the analysis, we rely on the notion of a *linear shift-invariant* (LSI) *system*, which is familiar to most engineers in the temporal-frequency study of electric circuits. This linear form of analysis applied to optical and EO imaging systems is similar, except that the time domain is replaced by a two-dimensional spatial domain and the frequency domain is replaced by a two-dimensional spatial frequency domain.

Well-known performance measures that appear in the literature like the *Strehl ratio* and *resolution* are often left in the form of a complicated integral for the practitioner to numerically evaluate. This is also the case when researchers develop mathematical models for describing various types of anisoplanatism for AOI systems. These expressions for anisoplanatism ordinarily involve products of Bessel functions under the integrals arising out of the Zernike filter function analysis. In our treatment here we provide simple algebraic expressions for most of the performance measures that are usually accurate to within a few percent of results obtained through numerical integration. These simplified algebraic models are achieved either using the *ABCD* ray matrix approach of analysis or employing the type of analysis illustrated in Chapter 7 for evaluating integrals, or both. Also, the approximation technique presented in Chapter 6 can often lead to even simpler expressions when generalized hypergeometric functions appear in the final mathematical model.

9.1 Imaging Performance Measures

In vacuum, the far-field diffraction pattern has an angular extent (seeing angle) of roughly λ/D, where D is the aperture diameter of a telescope. However, in atmospheric turbulence, the seeing angle is only λ/r_0, where r_0 is the *Fried parameter* or *atmospheric coherence width* defined by

$$r_0 = \left[0.423k^2 \int_0^L C_n^2(z)\gamma^{5/3} dz\right]^{-3/5}. \tag{9.1}$$

Here, $C_n^2(z)$ is the index-of-refraction structure parameter as a function of propagation distance z. The parameter γ takes on the value $\gamma = 1$ for an infinite plane wave but takes on the value $\gamma = z/L$ for a point source at $z = 0$ and $\gamma = (L - z)/L$ for a point source at $z = L$. The Fried parameter r_0 is related to the spatial coherence radius ρ_0 of the optical wave (Section 8.3) by $r_0 = 2.1\rho_0$.

9.1.1 Point spread function and modulation transfer function

Imaging systems are typically classified as *coherent* or *incoherent systems*. Most active imaging systems are neither of these extremes but generally involve partially coherent light. Nonetheless, it is common practice to treat such a system as an incoherent imaging system. Coherent imaging systems act like a linear system with respect to the optical field, whereas incoherent imaging systems are linear in irradiance. Therefore, the *impulse response function* normally associated with a coherent optical system is replaced by the *point spread function* (PSF) for an incoherent imaging system. The PSF is the mean irradiance of a point source in the optical system image/detector plane—it is sometimes referred to as the "blur spot." The normalized two-dimensional Fourier transform of the PSF is a complex function known as the *optical transfer function* (OTF). The magnitude of the OTF is called the *modulation transfer function* (MTF), which basically describes image quality including wavefront tilt. The phase of the OTF, called the *phase transfer function* (PTF), determines image position and orientation. The OTF, MTF, and PTF all provide a spatial-frequency point of view.

An efficient way to describe an optical wave passing through optical elements is by using 2 × 2 matrices called *ABCD ray matrices*. The use of such matrices allows us to describe the propagation of an optical wave through a train of optical elements by utilizing the *cascade scheme* of multiplying successive matrix representations of each optical element. In particular, if there exists a sequence of N such elements distributed along the propagation path, including those that represent free-space line-of-sight sections between optical elements, the entire propagation path can be represented by a single $ABCD$ ray matrix,

$$\begin{pmatrix} A & B \\ C & \hat{D} \end{pmatrix} = \underbrace{\begin{pmatrix} A_N & B_N \\ C_N & \hat{D}_N \end{pmatrix} \begin{pmatrix} A_{N-1} & B_{N-1} \\ C_{N-1} & \hat{D}_{N-1} \end{pmatrix} \cdots \begin{pmatrix} A_1 & B_1 \\ C_1 & \hat{D}_1 \end{pmatrix}}_{\text{Arranged in reverse order from order of occurrence}}, \tag{9.2}$$

Table 9.1 Ray matrices for various simple optical elements.

Structure	Matrix
Line-of-sight section (length L)	$\begin{pmatrix} 1 & L \\ 0 & 1 \end{pmatrix}$
Thin lens (focal length F_{Rx})	$\begin{pmatrix} 1 & 0 \\ -1/F_{Rx} & 1 \end{pmatrix}$
Gaussian lens (finite radius W_{Rx})	$\begin{pmatrix} 1 & 0 \\ -1/F_{Rx}+2i/kW_{Rx}^2 & 1 \end{pmatrix}$

where we write $\hat{D}$ to avoid confusion with Rx lens diameter D. Here we assume that the optical elements are all properly aligned along the positive z-axis. Some optical elements of importance in imaging systems are listed in Table 9.1. These include free-space line-of-sight sections and matrices for a thin lens and finite aperture stop, the latter known as a *soft aperture* because it doesn't clip the beam. In contrast, a *hard aperture* refers to the physical aperture that clips the beam.

If the input optical field to an $ABCD$ system is given by $U_0(\mathbf{s}, 0)$, the optical field at distance z from the Tx in free space is determined by the *generalized Huygens–Fresnel integral* (pp. 109–112 in Ref. 3):

$$\begin{aligned} U_0(\mathbf{r},z) &= -\frac{ik}{2\pi B}\exp(ikz)\iint_{-\infty}^{\infty} d^2s\, U_0(\mathbf{s},0)\exp\left[\frac{ik}{2B}(As^2-2\mathbf{r}\cdot\mathbf{s}+\hat{D}r^2)\right] \\ &= \iint_{-\infty}^{\infty} d^2s\, U_0(\mathbf{s},0)h(\mathbf{s},\mathbf{r}). \end{aligned} \tag{9.3}$$

Let us assume that the input function to a coherent optical imaging system is a point source described by a delta function (impulse function),

$$U(\mathbf{s}, 0) = \delta(\mathbf{s}). \tag{9.4}$$

In this case, Eq. (9.3) yields

$$U_0(\mathbf{r}, z) = h(0, \mathbf{r}) = -\frac{ik}{2\pi B}\exp\left(ikz + \frac{ik\hat{D}r^2}{2B}\right), \tag{9.5}$$

where $h(0, \mathbf{r})$ is the impulse response function for free-space propagation.

To illustrate the above, let us consider the simple imaging system depicted in Fig. 9.1, which can be described by the $ABCD$ matrix

$$\begin{aligned} \begin{pmatrix} A & B \\ C & \hat{D} \end{pmatrix} &= \begin{pmatrix} 1 & L_f \\ 0 & 1 \end{pmatrix}\begin{pmatrix} 1 & 0 \\ -\dfrac{1}{F_{Rx}}+i\dfrac{2}{kW_{Rx}^2} & 1 \end{pmatrix}\begin{pmatrix} 1 & L \\ 0 & 1 \end{pmatrix} \\ &= \begin{pmatrix} i+i\alpha_{Rx}L_f & L+L_f(1+i\alpha_{Rx}L) \\ i\alpha_{Rx} & 1+i\alpha_{Rx}L \end{pmatrix}, \end{aligned} \tag{9.6}$$

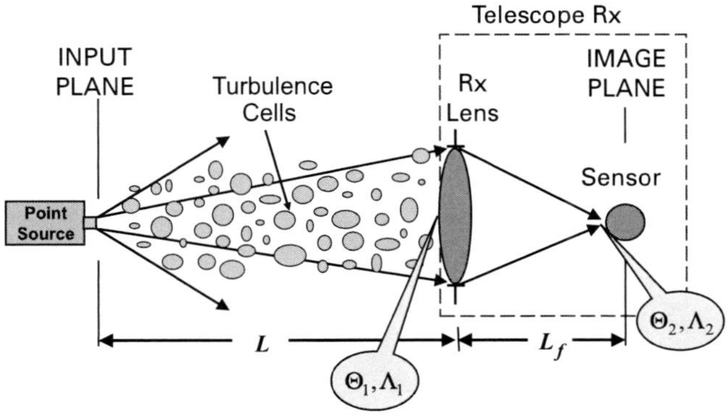

Figure 9.1 Propagation geometry for a point source passing through a lens to the image plane of a simple optical system at distance $z = L + L_f$ (adapted from Ref. 3).

where $i\alpha_{Rx} = -1/F_{Rx} + 2i/kW_{Rx}^2$. Here, W_{Rx} is the radius of the Gaussian lens, and F_{Rx} is its focal length. The two sets of parameters in Fig. 9.1 are free-space nondimensional beam parameters defined for a point source by (p. 407 of Ref. 3)

$$z = L: \begin{cases} \Theta_1 = 0, \\ \Lambda_1 = 0, \end{cases} \qquad z = L + L_f: \begin{cases} \Theta_2 = 0 \\ \Lambda_2 = \dfrac{L}{L_f \Omega_{Rx}} = \dfrac{kW_{Rx}^2}{2L_f}, \end{cases}$$

where $\Omega_{Rx} = 2L/kW_{Rx}^2$.

For an imaging system, the condition $\Theta_2 = 0$ means that the distance L_f behind the collecting lens satisfies a lens law: $1/L + 1/L_f = 1/F_{Rx}$. Therefore, it follows from Eq. (9.6) that $B = iL_f \Omega_{Rx}$ and $\hat{D} = -L/L_f + i\Omega_{Rx}$. If also the propagation distance satisfies $L \gg L_f$ so that $1/L + 1/L_f \approx 1/L_f$, we can set $L_f \approx F_{Rx}$, and the impulse response function [Eq. (9.5)] takes the form

$$h(0, \mathbf{r}) = -\frac{ik}{2\pi B} \exp\left[ik(L + F_{Rx})\right] \exp\left(-\frac{k^2 W_{Rx}^2 r^2}{4F_{Rx}^2} + \frac{ikr^2}{2F_{Rx}}\right). \tag{9.7}$$

The *coherent transfer function* is the normalized two-dimensional Fourier transform of the impulse response function, which, based on Eq. (9.7), leads to

$$H(\nu) = \exp\left[-\left(\frac{\lambda F_{Rx}\nu}{W_{Rx}}\right)^2\right], \tag{9.8}$$

where ν is the magnitude of the two-dimensional *spatial frequency* $\boldsymbol{\nu}$. Both Eqs. (9.7) and (9.8) are considered soft-aperture approximations for a coherent imaging system. The impulse response function and coherent transfer function for a hard-aperture coherent imaging system are given in Ref. 2, pp. 20–25.

Given that $D = 2\sqrt{2}\, W_{Rx}$ is a "rule of thumb" for the equivalent hard-aperture diameter of the Gaussian lens in Fig. 9.1, the normalized PSF for an incoherent optical system in free space is

$$\frac{\mathrm{PSF}_0(\mathbf{r})}{\mathrm{PSF}_0(0)} = \frac{|h(0,\,\mathbf{r})|^2}{|h(0,\,0)|^2} = \exp\left(-\frac{\pi^2 D^2 r^2}{4\lambda^2 F_{Rx}^2}\right), \tag{9.9}$$

whereas that for a hard aperture system is the *Airy disk*:

$$\frac{\mathrm{PSF}_0(\mathbf{r})}{\mathrm{PSF}_0(0)} = \left[\frac{2J_1(\pi Dr/\lambda F_{Rx})}{\pi Dr/\lambda F_{Rx}}\right]^2. \tag{9.10}$$

For both Eqs. (9.9) and (9.10), the corresponding MTFs are the same as the OTFs because the PSF in each case is an even function of r (i.e., the PTF is zero). Therefore, the MTF for the *ABCD* system in Fig. 9.1 is

$$\mathrm{MTF}_0(\nu) = \frac{\iint_{-\infty}^{\infty} e^{-2\pi i\mathbf{r}\cdot\boldsymbol{\nu}}\mathrm{PSF}_0(\mathbf{r})\, d^2r}{\iint_{-\infty}^{\infty} \mathrm{PSF}_0(\mathbf{r})\, d^2r} = \exp\left[-4\left(\frac{\lambda F_{Rx}\nu}{D}\right)^2\right], \tag{9.11}$$

whereas the MTF for the hard-aperture system is

$$\mathrm{MTF}_0(\nu) = \frac{2}{\pi}\left\{\cos^{-1}\left(\frac{\lambda F_{Rx}\nu}{D}\right) - \left(\frac{\lambda F_{Rx}\nu}{D}\right)\left[1-\left(\frac{\lambda F_{Rx}\nu}{D}\right)^2\right]^{1/2}\right\},\ \nu \le \frac{D}{\lambda F_{Rx}}. \tag{9.12}$$

In Example 51 of Section 7.7.6 we presented a derivation of Eq. (9.12) using the two-dimensional Fourier transform.

In Fig. 9.2(a) we plot both Eqs. (9.9) and (9.10), and in Fig. 9.2(b) we plot both Eqs. (9.11) and (9.12). For the soft-aperture PSF, we determine the spatial cutoff at the e^{-4} level, whereas for the hard-aperture Airy disk PSF, it is determined by the first zero of the Bessel function. Hence, the spatial cutoffs lead to $r = 1.27\,\lambda F_{Rx}/D$ and $r = 1.22\,\lambda F_{Rx}/D$, respectively, for Eqs. (9.9) and (9.10). On the other hand, the cutoff spatial frequency for both MTFs given by Eqs. (9.11) and (9.12) is $\nu = D/\lambda F_{Rx}$.

To characterize the atmospheric effects on the performance of an imaging system requires knowledge of the turbulence MTF as well as the free-space MTF. In our analysis we ignore contributions to the MTF from aerosols along the path. Thus, for a point source incident on the imaging system in Fig. 9.1 in the presence of atmospheric turbulence, the PSF is given by (p. 122 of Ref. 4)

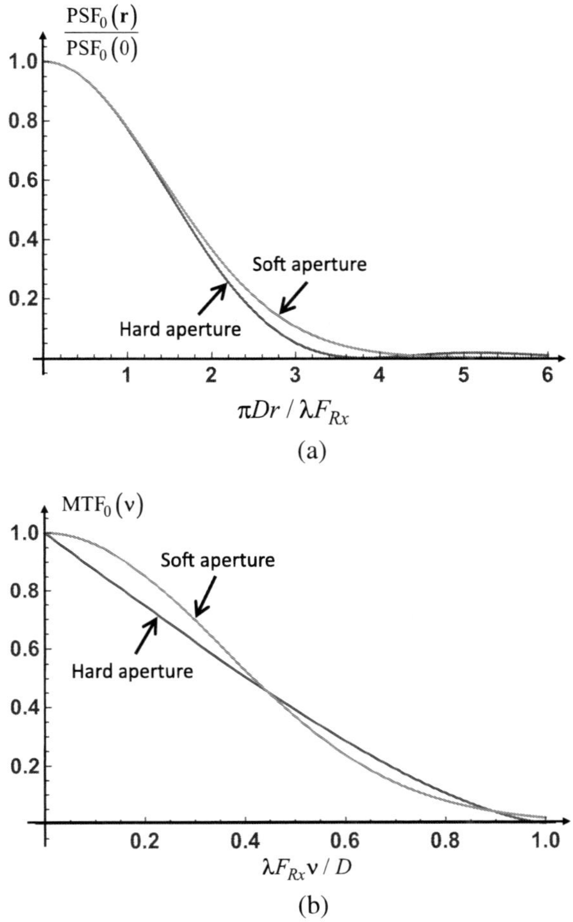

Figure 9.2 (a) PSFs defined by Eqs. (9.9) and (9.10) for an imaging system in free space and (b) MTFs defined by Eqs. (9.11) and (9.12) for the same system.

$$\begin{aligned} \mathrm{PSF}(\mathbf{r}) = \frac{W_{Rx}^2}{W^2} \exp\left(-\frac{2r^2}{W^2}\right) \exp\left\{-4\pi^2 k^2 L \int_0^1 \int_0^\infty \kappa \Phi_n(\kappa)\right. \\ \left. \times \left[1 - \exp\left(-\frac{L\kappa^2\xi^2}{k\Omega_{Rx}}\right) I_0\left(\frac{2\kappa\xi L r}{L_f \Omega_{Rx}}\right)\right] d\kappa\, d\xi \right\}, \end{aligned} \tag{9.13}$$

where $W = 2F_{Rx}/kW_{Rx}$ is the free-space beam radius in the image/detector plane. Upon evaluating the integrals with constant C_n^2, we obtain the approximation

$$\mathrm{PSF}(\mathbf{r}) = \frac{W_{Rx}^2}{W^2[1 + (D/r_0)^{5/3}]^{6/5}} \exp\left\{-\frac{2r^2}{W^2[1 + (D/r_0)^{5/3}]^{6/5}}\right\}, \tag{9.14}$$

where we have identified $1.33\sigma_R^2(L)\Omega_{Rx}^{-5/6} = (D/r_0)^{5/3}$ and the Fried parameter $r_0 = (0.16C_n^2k^2L)^{-3/5}$. It now follows that the normalized two-dimensional Fourier transform of Eq. (9.14) leads to the overall MTF given by

$$\mathrm{MTF}(\nu) = \exp\left\{-4\left(\frac{\lambda F_{Rx}\nu}{D}\right)^2[1 + (D/r_0)^{5/3}]^{6/5}\right\}. \tag{9.15}$$

Both Eqs. (9.14) and (9.15) represent soft-aperture expressions. The hard-aperture overall MTF is

$$\begin{aligned}\mathrm{MTF}(\nu) &= \mathrm{MTF}_0(\nu)\mathrm{MTF}_{atm}(\nu)\\ &= \frac{2}{\pi}\left\{\cos^{-1}\left(\frac{\lambda F_{Rx}\nu}{D}\right) - \left(\frac{\lambda F_{Rx}\nu}{D}\right)\left[1 - \left(\frac{\lambda F_{Rx}\nu}{D}\right)^2\right]^{1/2}\right\}\\ &\times \exp\left[-3.44\left(\frac{\lambda F_{Rx}\nu}{r_0}\right)^{5/3}\right].\end{aligned} \tag{9.16}$$

The first expression in Eq. (9.16) is the MTF for free-space optics [Eq. (9.12)], and the exponential function is the atmospheric MTF obtained by Fried[5] from the mutual coherence function (MCF). It is common practice to simply multiply the two MTFs to obtain the overall MTF. Both Eqs. (9.15) and (9.16) are also valid for astronomical imaging provided we define the Fried parameter r_0 by Eq. (9.1) to account for varying C_n^2 along the path.

9.1.2 Short-exposure imaging

The MTFs defined by Eqs. (9.15) and (9.16) are called *long-exposure* MTFs. In a long-exposure MTF, the optical wave is averaged over many realizations of atmospheric turbulence, whereas a *short-exposure* MTF is that due to one realization of atmospheric turbulence (a few milliseconds). Both long-exposure MTFs are shown in Fig. 9.3(a) for the case where $D/r_0 = 2$. This particular case causes a reduction of roughly 50% in the spatial frequency cutoff. Unfortunately, our statistical analysis for a long-exposure MTF cannot be performed over a short timeframe that essentially “freezes” the atmosphere. Hence, a theoretical model approximation for a short-exposure MTF turns out to be a long-exposure MTF with tilt compensation. That is, wavefront tilt has the effect of shifting the location of the image in the detector plane. It is this random shifting of the image that accounts for most of the degradation in the long-exposure image. However, the shifting caused by tilt does not distort the short-exposure image. For this reason, removing tilt in the long-exposure MTF leads to what is now commonly called a short-exposure MTF.

The long-exposure PSF defined by Eq. (9.14) identifies the square of the long-term image plane spot radius as $W_{LT}^2 = W^2[1 + (D/r_0)^{5/3}]^{6/5}$. To calculate the short-exposure PSF, we simply replace the square of the

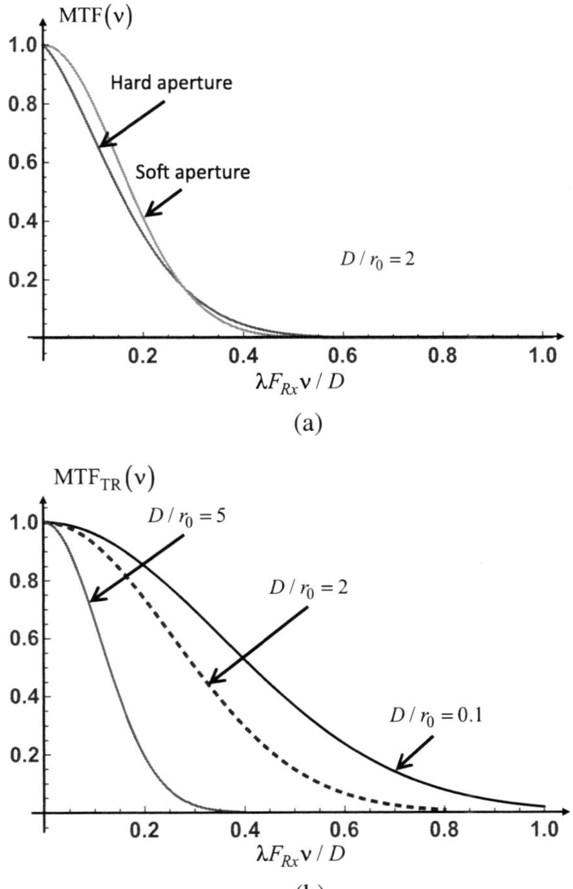

Figure 9.3 Plots of overall (a) long-exposure MTFs defined by Eqs. (9.15) and (9.16) and (b) near-field short-exposure MTF defined by Eq. (9.18) with $b = 1.15$.

long-term spot radius W_{LT} with the square of the image plane short-term spot radius W_{ST}. This is done by subtracting the square of the near-field detector plane image jitter [Eq. (8.202)] from the square of the long-term beam radius (see Section 8.8.1 for definition of near field and far field in the detector/image plane). This action leads to the near-field square of the short-term beam radius, given by

$$\begin{aligned} W_{ST}^2 &= W_{LT}^2 - 0.461(D/r_0)^{5/3}(\lambda F_{Rx}/D)^2 \\ &= W^2\{[1 + (D/r_0)^{5/3}]^{6/5} - 1.15(D/r_0)^{5/3}\}. \end{aligned} \tag{9.17}$$

The final step in Eq. (9.17) is a consequence of the image plane jitter relation $\lambda F_{Rx}/D \approx \pi W/2$. In this case the two-dimensional Fourier transform of the PSF with the short-term spot radius leads to the short-exposure or tilt-removed (TR) MTF given by

$$\text{MTF}_{\text{TR}}(\nu) = \exp\left[\left[-4\left(\frac{\lambda F_{Rx}\nu}{D}\right)^2\left\{[1+(D/r_0)^{5/3}]^{6/5} - b(D/r_0)^{5/3}\right\}\right]\right], \tag{9.18}$$

where $b = 1.15$. For the far-field short-exposure MTF, we again recall Eq. (8.202) and therefore replace b in Eq. (9.18) with $b = \frac{1}{2}(1.15) = 0.57$.

In Fig. 9.3(b) we plot the near-field, short-exposure MTF given by Eq. (9.18) for values $D/r_0 = 0.1$, 2, and 5. The plot with $D/r_0 = 0.1$ is nearly the same as the free-space MTF shown in Fig. 9.2(b).

9.1.3 Strehl ratio

A useful performance measure of an optical or EO imaging system is the *Strehl ratio* (SR). In EO imaging systems, the SR is defined by the ratio of peak irradiance at the output plane of an imaging system to the theoretical maximum peak irradiance of a perfect imaging system (no turbulence).

Given a Rx aperture of diameter D, the SR can also be defined in terms of the normalized volume under the MTF surface of the system. The normalization factor is determined from the volume under the free-space MTF surface described by Eq. (9.11). By using the dummy variable $u = \lambda F_{Rx}\nu/D$, this leads to

$$\int_0^{2\pi}\int_0^{\infty} u \exp(-4u^2)\, du\, d\theta = \frac{\pi}{4}.$$

Consequently, if we use the soft-aperture MTF defined by Eq. (9.15) under atmospheric turbulence conditions, we obtain

$$\begin{aligned} \text{SR} &= \frac{4}{\pi}\int_0^{2\pi}\int_0^{\infty} u\text{MTF}(u)\, du\, d\theta \\ &= \frac{4}{\pi}\int_0^{2\pi}\int_0^{\infty} u \exp\left\{-4u^2[1+(D/r_0)^{5/3}]^{6/5}\right\} du\, d\theta \\ &= \int_0^{\infty} \exp\left\{-t[1+(D/r_0)^{5/3}]^{6/5}\right\} dt, \end{aligned} \tag{9.19}$$

where we have integrated over θ and set $t = 4u^2$. The evaluation of this last integral leads directly to an algebraic expression for SR given by

$$\text{SR} = \frac{1}{[1+(D/r_0)^{5/3}]^{6/5}}. \tag{9.20}$$

On the other hand, for a hard-aperture system, the SR based on the MTF defined by Eq. (9.16) takes the form

$$\mathrm{SR} = \frac{16}{\pi}\int_0^1 u\left(\cos^{-1}u - u\sqrt{1-u^2}\right)\exp\left[-3.44\left(\frac{Du}{r_0}\right)^{5/3}\right]du. \quad (9.21)$$

To obtain an approximation for the integral in Eq. (9.21), we examine asymptotic cases. For $D/r_0 \ll 1$, we find that $\exp[-3.44(Du/r_0)^{5/3}] \sim 1$, so calculating the resulting SR with only the free-space MTF, we have

$$\begin{aligned}\mathrm{SR} &\sim \frac{16}{\pi}\int_0^1 u\left(\cos^{-1}u - u\sqrt{1-u^2}\right)du \\ &\sim 1, \quad D/r_0 \ll 1.\end{aligned} \quad (9.22)$$

For $D/r_0 \gg 1$, we set $t = 3.44(Du/r_0)^{5/3}$ in Eq. (9.21) to get the asymptotic result

$$\begin{aligned}\mathrm{SR} &\sim \frac{24}{5(3.44)^{6/5}}\left(\frac{D}{r_0}\right)^{-2}\int_0^\infty t^{1/5}e^{-t}\,dt \\ &\sim 1.0899\,\Gamma\left(\frac{6}{5}\right)\left(\frac{D}{r_0}\right)^{-2} \\ &\sim \left(\frac{D}{r_0}\right)^{-2}, \quad \frac{D}{r_0} \gg 1.\end{aligned} \quad (9.23)$$

Hence, based on the technique from Chapter 6, the combination of the two asymptotic results [Eqs. (9.22) and (9.23)] can be approximated by the same expression as Eq. (9.20).

In Fig. 9.4 we plot the SR defined by the integral in Eq. (9.21) and the algebraic expression in Eq. (9.20). We have found that the largest difference between these expressions is around 3% and occurs near $D/r_0 = 0.65$.

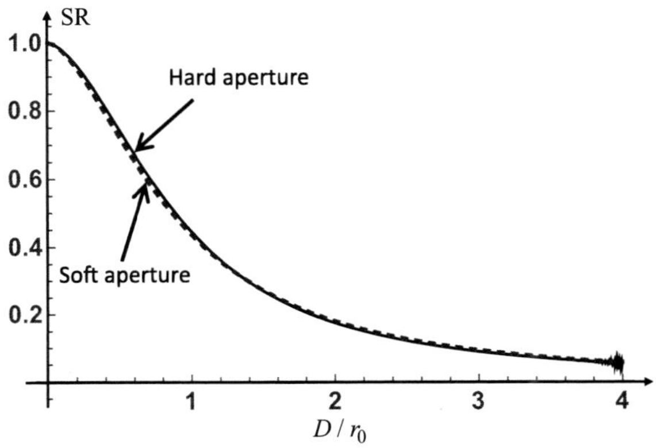

Figure 9.4 Strehl ratios defined by Eqs. (9.20) and (9.21) as a function of D/r_0.

9.1.4 Strehl ratio: outer-scale effect

The SR will increase over that shown in Fig. 9.4 if a finite outer scale is taken into account. The amount of increase will depend on the ratio D/L_0. Based on the SR definition using the mean irradiance $\langle I(0, L+L_f)\rangle / I(0, L+L_f)$ that is commonly used in free-space optical communication (FSOC) systems, we can use the PSF (9.14) to calculate the SR, viz.,

$$\mathrm{SR} = \frac{W^2}{W_{LT}^2} = \frac{1}{(1+T)^{6/5}}. \tag{9.24}$$

The symbol T is defined in the Rx pupil plane by Eq. (8.35), which under the Kolmogorov spectrum (8.1) leads to $T = 1.33\sigma_R^2\Lambda^{5/6}$. In the image plane with spectrum model (8.1) we have already shown that $T = (D/r_0)^{5/3}$, [cf. Eq. (9.14)]. To include the outer scale L_0 in the analysis, we define T in the image plane by

$$T = 1.303C_n^2k^2L\int_0^1\int_0^\infty \kappa^{-8/3}\left(1-e^{-\kappa^2/\kappa_0^2}\right)\left[1-\exp\left(-\frac{D^2\kappa^2\xi^2}{16}\right)\right]d\kappa\, d\xi. \tag{9.25}$$

Note that we introduced the outer-scale effect in Eq. (9.25) with the exponential spectrum (8.4), but here we introduce it with $\kappa_m \to \infty$, i.e.,

$$\Phi_n(\kappa) = 0.033C_n^2\kappa^{-11/3}\left(1-e^{-\kappa^2/\kappa_0^2}\right), \quad \kappa_0 = \frac{2\pi}{L_0}. \tag{9.26}$$

For this analysis, the scaling constant in κ_0 is selected as 2π, but it could be chosen differently. By applying the integral relation of ($D2$) in Appendix D to the inside integral in Eq. (9.25), we find that

$$T = \left(\frac{D}{r_0}\right)^{5/3} - 27.19(r_0\kappa_0)^{-5/3}\int_0^1\left[\left(1+\frac{D^2\kappa_0^2\xi^2}{16}\right)^{5/6}-1\right]d\xi. \tag{9.27}$$

To finish the evaluation of the last integral, we use the Meijer G-function relation ($C11$) in Appendix C combined with ($D9$) in Appendix D. Thus,

$$\begin{aligned} T &= \left(\frac{D}{r_0}\right)^{5/3} - 27.19(r_0\kappa_0)^{-5/3}\left[\frac{1}{2\Gamma\left(-\frac{5}{6}\right)}G_{22}^{12}\left(\frac{D^2\kappa_0^2}{16}\middle|\begin{matrix}\frac{1}{2}, \frac{11}{6}\\ 0, -\frac{1}{2}\end{matrix}\right)-1\right] \\ &= \left(\frac{D}{r_0}\right)^{5/3} - 27.19(r_0\kappa_0)^{-5/3}\left[{}_2F_1\left(\frac{1}{2}, -\frac{5}{6}; \frac{3}{2}; -\frac{D^2\kappa_0^2}{16}\right)-1\right], \end{aligned} \tag{9.28}$$

where the last step follows from ($C19$) in Appendix C. If we now use the small-argument form of the hypergeometric function ($D\kappa_0 < 4$), the SR (9.24) becomes

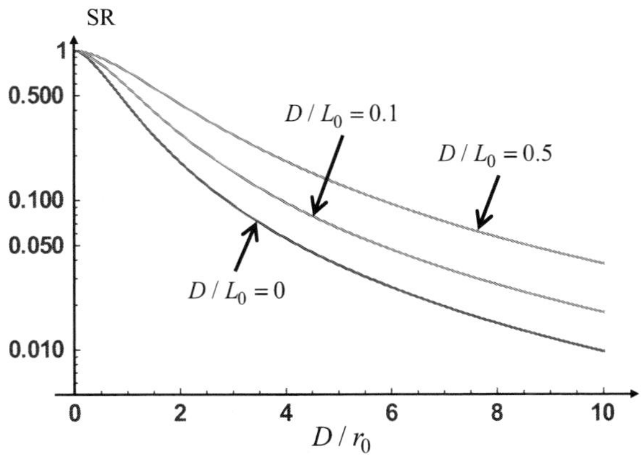

Figure 9.5 Strehl ratio defined by Eq. (9.29) as a function of D/r_0 and D/L_0.

$$\mathrm{SR} \approx \frac{1}{\{1 + (D/r_0)^{5/3}[1 - 0.87(D/L_0)^{1/3}]\}^{6/5}}, \quad \frac{D}{L_0} < \frac{2}{\pi}. \tag{9.29}$$

In Fig. 9.5 we plot the SR given by Eq. (9.29) with outer-scale values leading to $D/L_0 = 0$, 0.1, and 0.5. These results are similar to those generated with the von Kármán spectrum (p. 215 of Ref. 6).

9.1.5 Strehl ratio with tilt removed

In an adaptive optics (AO) system, it is a common practice to use tip-tilt and higher-order AO correction modes to remove the image distortions caused by atmospheric turbulence. For a system that simply removes tilt, we can use the approach introduced for the short-exposure MTF given by Eq. (9.18). Thus, by defining the SR with tilt removed as the normalized volume under the near-field short-exposure MTF surface, we arrive at the expression

$$\mathrm{SR}_{\mathrm{TR}} = \frac{1}{[1 + (D/r_0)^{5/3}]^{6/5} - 1.15(D/r_0)^{5/3}}. \tag{9.30}$$

In Eq. (9.30) we have set the constant $b = 1.15$ for the near-field result.

As an alternative to Eq. (9.30), the SR with tip-tilt can also be calculated by using Eq. (9.24) with a different approach. By applying the principle of reciprocity, we calculate the integral for T in the image plane of the Rx with tilt removed by a Gaussian filter function, which leads to

$$T_{\mathrm{TR}} = 1.303 C_n^2 k^2 L \int_0^1 \int_0^\infty \kappa^{-8/3}[1 - H(\kappa, \xi)]\left[1 - \exp\left(-\frac{D^2\kappa^2\xi^2}{16}\right)\right] d\kappa\, d\xi. \tag{9.31}$$

The term $H(\kappa, \xi)$ is a large-scale filter function for tip-tilt similar to that used in beam wander (see p. 621 of Ref. 3),

$$H(\kappa, \xi) = \exp\left(-\frac{D^2\kappa^2\xi^2}{8}\right). \tag{9.32}$$

Carrying out the integration in Eq. (9.31) by using (*D*2) in Appendix D, we arrive at the SR with tilt removed, given by

$$\begin{aligned} \mathrm{SR_{TR}} &= \frac{1}{(1+T_{\mathrm{TR}})^{6/5}} \\ &= \frac{1}{[1+0.28(D/r_0)^{5/3}]^{6/5}}, \quad \frac{D}{r_0} < 7. \end{aligned} \tag{9.33}$$

We present this alternative approach for the SR with tip-tilt because of its simplicity, and it has compared well in the past with experimental data.[7]

Sasiela provided another approximation for the SR with tilt removed—his approximation is (p. 239 of Ref. 6)

$$\mathrm{SR_{TR}} = \frac{1}{[1+(D/r_0)^{5/3}(1/4.65^{1-D/13.5r_0})^{5/3}]^{6/5}}, \quad \frac{D}{r_0} < 10. \tag{9.34}$$

In Fig. 9.6 we plot the SR with tilt removed given by Eqs. (9.30), (9.33), and (9.34) along with the SR given by Eq. (9.20) without tilt removed. For further comparison, we also show numerical integration results (dots) based on Fried's expression,[5]

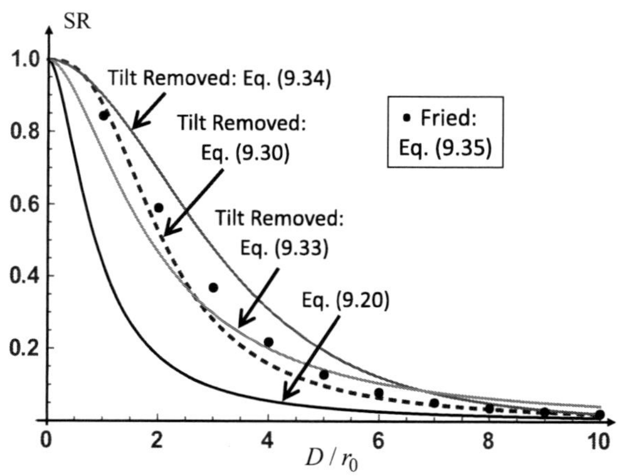

Figure 9.6 Strehl ratio based on Eqs. (9.30), (9.33), and (9.34) with tip-tilt correction compared with Eq. (9.20) as a function of D/r_0. The dots are numerical integration results from Eq. (9.35).

$$\mathrm{SR}_{\mathrm{TR}} = \frac{16}{\pi}\int_0^1 u\left(\cos^{-1}u - u\sqrt{1-u^2}\right)\exp\left[-3.44\left(\frac{Du}{r_0}\right)^{5/3}(1-u^{1/3})\right] du. \tag{9.35}$$

From these plots it is clear that each of the estimates of Strehl ratio with tilt removal shows significant improvement over that with no tilt removal [Eq. (9.20)], but each predicts somewhat different results.

9.1.6 Long-exposure resolution

Resolution is a term that can have multiple definitions. Following the definition introduced by Fried[5] based on a hard aperture, the *long-exposure resolution* is the integral over spatial frequencies of the long-term average MTF, which leads to

$$\begin{aligned} R &= \int_0^{2\pi}\int_0^{D/\lambda F_{Rx}} \nu\,\mathrm{MTF}(\nu)\,d\nu\,d\theta \\ &= \frac{4D^2}{\lambda^2 F_{Rx}^2}\int_0^1 u\left(\cos^{-1}u - u\sqrt{1-u^2}\right)\exp\left[-3.44\left(\frac{Du}{r_0}\right)^{5/3}\right] du, \end{aligned} \tag{9.36}$$

where $u = \lambda F_{Rx}\nu/D$. The maximum resolution $R_{\max} = \pi r_0^2/4\lambda^2 F_{Rx}^2$ is the limiting value of the integral in Eq. (9.36) as the aperture diameter D becomes arbitrarily large [recall Eq. (9.23)]. By dividing the resolution (9.36) by this quantity, the ratio $R/R_{\max}$ becomes

$$\begin{aligned} \frac{R}{R_{\max}} &= \left(\frac{D}{r_0}\right)^2 \underbrace{\frac{16}{\pi}\int_0^1 u\left(\cos^{-1}u - u\sqrt{1-u^2}\right)\exp\left[-3.44\left(\frac{Du}{r_0}\right)^{5/3}\right] du}_{\text{Strehl Ratio}} \\ &= \frac{(D/r_0)^2}{[1+(D/r_0)^{5/3}]^{6/5}}, \end{aligned} \tag{9.37}$$

where we have used the previous expression [Eq. (9.20)] for the SR. The soft-aperture analysis will lead to the same result for $R/R_{\max}$.

9.1.7 Short-exposure resolution

As pointed out in Section 9.1.2, the long exposure of an optical wave is the result of many realizations of atmospheric turbulence. On the other hand, the *short-exposure resolution* is considered to be the long-exposure resolution with tilt removed. Fried[5] provided the first short-exposure analysis for resolution. However, rather than relying on Fried's integral result, we can substitute Eq. (9.30) for the short-exposure SR into Eq. (9.37) to calculate the short-exposure resolution. Thus, we find that the scaled short-exposure resolution becomes

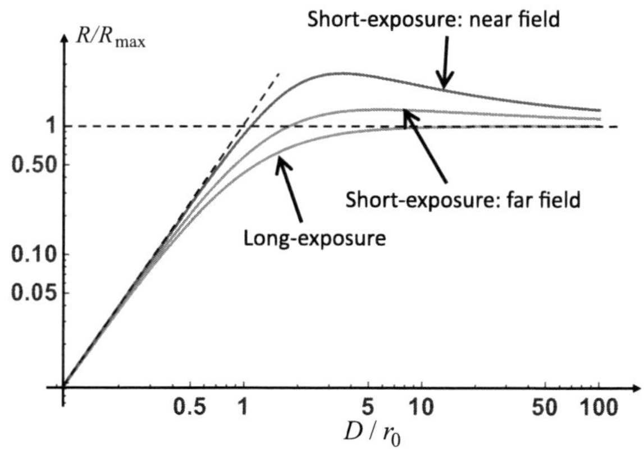

Figure 9.7 Normalized resolution based on the long-exposure resolution (9.37) and on the far-field ($b = 0.57$) and near-field ($b = 1.15$) short-exposure resolution from Eq. (9.38).

$$\frac{R}{R_{max}} = \frac{(D/r_0)^2}{[1 + (D/r_0)^{5/3}]^{6/5} - b(D/r_0)^{5/3}}, \tag{9.38}$$

where $b = 1.15$ for the near-field result. For the far-field short-exposure resolution, we simply set $b = 0.57$ in Eq. (9.38).

In Fig. 9.7 we plot the long-exposure resolution defined by Eq. (9.37) as well as the short-exposure resolution described by Eq. (9.38), the latter for both the far-field and near-field short-exposure resolution.

9.2 Zernike Polynomials and Aperture Filter Functions

Adaptive optics is commonly used in astronomical applications to improve the performance of a ground-based optical system. When light from a star enters the Earth's atmosphere, turbulence distorts and moves the light image in various ways. A telescope equipped with an AOI system first uses light reflected from a DM to a WFS to measure the wavefront distortions generated by variations of the atmospheric index of refraction. Then, the AOI system uses an actuator command computer to apply phase corrections with a DM (e.g., see p. 169 of Ref. 2).

The effective use of AOI relies on the availability of a bright-enough reference source (e.g., a nearby star) within the isoplanatic patch of the object being observed when the object itself is not sufficiently bright. This reference source is then used to retrieve information about the wavefront distortion introduced by atmospheric turbulence. The isoplanatic patch is determined by the *isoplanatic angle* θ_0, which is the largest angular distance from the object being imaged to a reference source over which the atmospheric distortions on both wavefronts are reasonably similar. The isoplanatic angle for path length L through turbulence is defined by

$$\theta_0 = \left[2.91k^2 \int_0^L C_n^2(z) z^{5/3} dz\right]^{-3/5}. \tag{9.39}$$

In analyzing the effects of turbulence on an imaging system, it is advantageous to represent the phase front of the corrupted optical wave in a series of orthogonal polynomials called *Zernike polynomials* (recall Section 2.6). One of the early principal uses of the Zernike polynomials was to represent fixed aberrations in an optical system in the form of a generalized Fourier series of Zernike polynomials (Section 2.6.1). Our interest here in these polynomials, however, is not in fixed aberrations but in diagnosing the random aberration effects of atmospheric turbulence on an optical imaging system.

The Zernike polynomials for our current application are defined by

$$Z_n^m(r, \theta) = \begin{cases} \sqrt{n+1} R_n^0(r), & m = 0, \\ \sqrt{2(n+1)} R_n^m(r) \cos m\theta, & x \text{ component}, \\ \sqrt{2(n+1)} R_n^m(r) \sin m\theta, & y \text{ component}, \end{cases} \tag{9.40}$$

where both m and n are integers, $n \geq 0$, $-n \leq m \leq n$, $n \pm |m|$ is even, and

$$R_n^m(r) = \sum_{k=0}^{(n-|m|)/2} \frac{(-1)^k (n-k)!}{\left(\frac{n+m}{2} - k\right)! \left(\frac{n-m}{2} - k\right)!} \frac{r^{n-2k}}{k!}. \tag{9.41}$$

Lower-order Zernike polynomials are known by names such as piston, tilt, focus, astigmatism, coma, and so forth.

Noll's[8] modification of the notation $Z_n^m(r, \theta)$ that is useful in the analysis here is $Z_j(r, \theta) \equiv Z_j[m, n]$, $j = 1, 2, 3, \ldots$, where the mode-ordering scheme for these polynomials uses the rule that, for a given n, the modes with a smaller m are counted first. For $m > 0$, there are two Zernike functions for each (m,n) pair. For example, $Z_1[0,0]$ is *piston*, $Z_2[1,1]$ and $Z_3[1,1]$ represent *tilt* in the x and y directions, respectively, $Z_4[0,2]$ is *focus*, and $Z_5[2,2]$ and $Z_6[2,2]$ are two components of *astigmatism* (see Table I in Noll[8]).

In an investigation of optical wave propagation through turbulence as presented in Chapter 8, many of the statistical quantities of interest are calculated at a single point in the pupil plane of a Rx. One way to include the finite size of a Rx aperture in the analysis is by the use of *ABCD* ray matrices as done in Section 9.1. An alternative is to use an *aperture filter function.* Basically, this means that we formulate the effect of a finite Rx aperture in terms of wave number instead of physical spatial distance. This can be accomplished by making the substitution $r \to 2r/D$ in the Zernike polynomials followed by the use of the two-dimensional Fourier transform. For example, the two-dimensional Fourier transform of the various Zernike modes, scaled by the area of an aperture of diameter D, are given by the complex aperture filter functions defined by

$$G_j(\kappa, \phi) = \frac{4}{\pi D^2}\int_0^\infty U(D/2 - r)\int_0^{2\pi} e^{ir\kappa\cos(\theta-\phi)} Z_j(r, \theta) r d\theta dr, \qquad j = 1, 2, 3, \ldots, \tag{9.42}$$

where $U(x)$ is the unit step function; i.e., $U(x) = 1,\ x > 0$ and $U(x) = 0,\ x < 0$. In the case of piston, for example, in which $m = n = 0$ and $R_0^0(r) = 1$, the above integral in Eq. (9.42) becomes

$$\begin{aligned} G_1(\kappa, \phi) &= \frac{4}{\pi D^2}\int_0^{D/2} r \int_0^{2\pi} e^{ir\kappa\cos(\theta-\phi)} d\theta\, dr \\ &= \frac{8}{D^2}\int_0^{D/2} r J_0(\kappa r)\, dr. \end{aligned} \tag{9.43}$$

This last integral is well known and leads to

$$G_1(\kappa) = \frac{2J_1(\kappa D/2)}{\kappa D/2}. \tag{9.44}$$

To calculate the remaining aperture filter functions, we substitute the general Zernike polynomials $Z_j(r, \theta)$ with $r \to 2r/D$ into Eq. (9.42). This action leads to the following set of aperture filter functions:

$$\left.\begin{array}{c} G_j(\kappa) \\ G_{j,x}(\kappa, \phi) \\ G_{j,y}(\kappa, \phi) \end{array}\right\} = \sqrt{n+1}\ \frac{2J_{n+1}(\kappa D/2)}{\kappa D/2}\left\{\begin{array}{l} (-1)^{n/2}\ (m = 0) \\ (-1)^{(n-m)/2}\sqrt{2}\, i^m \cos m\phi \\ (-1)^{(n-m)/2}\sqrt{2}\, i^m \sin m\phi. \end{array}\right. \tag{9.45}$$

Because of the presence of the term i^m, the filter functions in Eq. (9.45) are often called *complex filter functions.* These filter functions are used for extracting Zernike modes. However, for calculating variances of statistical quantities, it is the absolute value squared of these complex functions, called simply *filter functions*, that we need. Thus, based on the filter function set described by Eq. (9.45), we obtain the related set of filter functions,

$$\left.\begin{array}{c} F_j(\kappa) \\ F_{j,x}(\kappa, \phi) \\ F_{j,y}(\kappa, \phi) \end{array}\right\} = (n+1)\left[\frac{2J_{n+1}(\kappa D/2)}{\kappa D/2}\right]^2\left\{\begin{array}{l} 1\ (m = 0) \\ 2\cos^2 m\phi \\ 2\sin^2 m\phi. \end{array}\right. \tag{9.46}$$

The filter functions associated with piston and tilt are the most commonly used by researchers in theoretical studies of AOI systems. For piston, the filter function that extracts *phase variance* is

$$F_1(\kappa) = \left[\frac{2J_1(\kappa D/2)}{\kappa D/2}\right]^2. \tag{9.47}$$

The *tilt phase variance* is obtained from Eq. (9.46) by setting $m = n = 1$,

$$\left.\begin{matrix} F_2(\kappa) \\ F_{2,x}(\kappa) \\ F_{3,y}(\kappa) \end{matrix}\right\} = \left[\frac{4J_2(\kappa D/2)}{\kappa D/2}\right]^2 \begin{cases} 1 \\ \cos^2\phi \\ \sin^2\phi. \end{cases} \tag{9.48}$$

To obtain the *tilt angle variance*, we multiply Eq. (9.48) by $(4/kD)^2$.

Instead of defining the filter functions in Eq. (9.42) by integrating over a physical finite circular aperture, let us approximate the finite aperture size in Eq. (9.42) by a Gaussian function. Then, with proper normalization we can replace the Bessel functions with Gaussian function approximations for the first few complex filter functions.[9] That is, we first make the replacement

$$U(D/2 - |r|) \to 2\exp\left(-\frac{8r^2}{D^2}\right),$$

from which the first filter function ($m = n = 0$) in Eq. (9.45), and subsequently that in Eq. (9.47), can be approximated by

$$F_1(\kappa) = |G_1(\kappa)|^2 = \exp\left(-\frac{\kappa^2 D^2}{16}\right). \tag{9.49}$$

Similarly, the filter function for tilt under this approximation becomes

$$F_2(\kappa) \equiv |G_{2,x}(\kappa)|^2 + |G_{3,y}(\kappa)|^2 = \frac{\kappa^2 D^2}{16}\exp\left(-\frac{\kappa^2 D^2}{16}\right). \tag{9.50}$$

It is interesting to note that the exponential filter function in Eq. (9.49) is exactly the same filter function found by the *ABCD* ray matrix method for the aperture-averaged irradiance covariance function of a plane wave [recall Eq. (8.126)].

For an easy illustration of the use of the piston filter function, we consider the case of phase variance over an aperture of diameter D with piston removed. Therefore, based on Eq. (9.47) and the spectrum model of Eq. (8.1), we get

$$\begin{aligned} \sigma_{PR}^2 &= 4\pi^2 k^2 \int_0^L \int_0^\infty \kappa \Phi_n(\kappa, z)\left\{1 - \left[\frac{2J_1(\kappa D/2)}{\kappa D/2}\right]^2\right\} d\kappa\, dz \\ &= 1.033\left(\frac{D}{r_0}\right)^{5/3}. \end{aligned} \tag{9.51}$$

On the other hand, if we use the approximation in Eq. (9.49), we find that

$$\begin{aligned}\sigma_{PR}^2 &\approx 4\pi^2 k^2 \int_0^L \int_0^\infty \kappa\Phi_n(\kappa, z)\left[1 - \exp\left(-\frac{\kappa^2 D^2}{16}\right)\right] d\kappa\, dz \\ &\approx 1.020\left(\frac{D}{r_0}\right)^{5/3}.\end{aligned} \tag{9.52}$$

Consequently, we see that using the approximation in Eq. (9.49) produced only a 1.4% difference in Eq. (9.52) compared with Eq. (9.51) for the piston-removed phase variance. However, Gaussian function approximations for higher-order filter functions are generally not as accurate as that for piston.

9.2.1 Zernike-tilt variance

When viewing an object in space from the ground, tilt jitter caused by atmospheric turbulence becomes an important consideration. In the analysis below we consider the problem of calculating the tilt angle variance, which can involve either a *Zernike-tilt variance* (Z-tilt) or a *gradient-tilt variance* (G-tilt). The correct choice between these two depends on the sensor.

Under isotropic turbulence conditions and a geometrical optics approximation, the x- and y-tilt angle variances are the same and their sum leads to the two-axis Z-tilt angle jitter variance. For the case of an infinite plane wave (or near-field Gaussian-beam wave) on an aperture of diameter D, the two-axis Z-tilt can be described by the integral

$$T_Z^2 = 1.303k^2 \int_0^L dz\, C_n^2(z) \int_0^\infty d\kappa\, \kappa^{-8/3}\left(\frac{4}{kD}\right)^2 \left[\frac{4J_2(\kappa D/2)}{\kappa D/2}\right]^2. \tag{9.53}$$

By making the change of variable $x = \kappa D/2$, this expression becomes

$$T_Z^2 = \frac{105.05}{D^{1/3}} \int_0^L C_n^2(z)\, dz \int_0^\infty x^{-14/3} J_2^2(x)\, dx. \tag{9.54}$$

Now using the integral relation of ($D3$) in Appendix D, it follows that

$$\begin{aligned}T_Z^2 &= \frac{105.05}{D^{1/3}} \frac{\Gamma\left(\frac{1}{6}\right)\Gamma\left(\frac{7}{3}\right)}{2\sqrt{\pi}\,\Gamma\left(\frac{17}{6}\right)\Gamma\left(\frac{29}{6}\right)} \int_0^L C_n^2(z)\, dz \\ &= \frac{6.08}{D^{1/3}} \int_0^L C_n^2(z)\, dz.\end{aligned} \tag{9.55}$$

To use the approximation in Eq. (9.50), the tilt phase variance filter function in Eq. (9.53) is replaced by the expression

$$\left[\frac{4J_2(\kappa D/2)}{\kappa D/2}\right]^2 \approx \frac{\kappa^2 D^2}{16}\exp\left(-\frac{\kappa^2 D^2}{16}\right). \tag{9.56}$$

Making this substitution in Eq. (9.53) leads to the approximation

$$\begin{aligned} T_Z^2 &\approx 1.303 \int_0^L C_n^2(z)\,dz \int_0^\infty \kappa^{-2/3}\exp\left(-\frac{\kappa^2 D^2}{16}\right) d\kappa \\ &\approx \frac{5.756}{D^{1/3}} \int_0^L C_n^2(z)\,dz. \end{aligned} \tag{9.57}$$

Compared with the result in Eq. (9.55), the approximation given by Eq. (9.57) differs by roughly 5%.

9.2.2 Zernike-tilt variance: outer-scale effect

It is known that taking the outer scale into account leads to a significant effect on the SR and also on tilt variance. Although others have used the von Kármán spectrum, we will introduce the outer scale into the analysis by relying once again on the exponential spectrum [Eq. (9.26)]. The inclusion of the outer scale with use of the exponential spectrum leads to an additional term to the integral over variable x in Eq. (9.54), i.e.,

$$\int_0^\infty x^{-14/3} J_2^2(x)\,dx - I.$$

This additional term takes the form

$$\begin{aligned} I &= \int_0^\infty x^{-14/3}\exp\left(-\frac{4x^2}{D^2\kappa_0^2}\right) J_2^2(x)\,dx \\ &= \frac{1}{2\sqrt{\pi}} \int_0^\infty t^{-17/6}\exp\left(-\frac{4t}{D^2\kappa_0^2}\right) G_{13}^{11}\left(t \,\middle|\, \begin{matrix} \frac{1}{2} \\ 2,\,0,\,-2 \end{matrix}\right) dt, \end{aligned} \tag{9.58}$$

where we have used ($C16$) in Appendix C and set $t = x^2$. By applying ($D6$) in Appendix D to Eq. (9.58), we get

$$\begin{aligned} I &= \frac{1}{2\sqrt{\pi}}\left(\frac{D^2\kappa_0^2}{4}\right)^{-11/6} G_{23}^{12}\left(\frac{D^2\kappa_0^2}{4} \,\middle|\, \begin{matrix} \frac{17}{6},\,\frac{1}{2} \\ 2,\,0,\,-2 \end{matrix}\right) \\ &= 0.03452(D\kappa_0)^{1/3}\,{}_2F_2\left(\frac{1}{6}, \frac{5}{2};\; 3, 5;\; -\frac{D^2\kappa_0^2}{4}\right). \end{aligned} \tag{9.59}$$

Use of ($C32$) in Appendix C has led to the last result in Eq. (9.59). Finally, by combining Eq. (9.59) with the result in Eq. (9.55), we have

$$T_Z^2 = \frac{6.08}{D^{1/3}} \int_0^L dz\, C_n^2(z) \left[1 - 0.596(D\kappa_0)^{1/3}\, {}_2F_2\left(\frac{1}{6}, \frac{5}{2}; 3, 5; -\frac{D^2\kappa_0^2}{4}\right)\right]. \tag{9.60}$$

By approximating the generalized hypergeometric function in Eq. (9.60) using the method illustrated in Chapter 6, we have

$${}_2F_2\left(\tfrac{1}{6}, \tfrac{5}{2}; 3, 5; -x\right) \approx (1 + 0.18x)^{-1/6},$$

which allows us to simplify Eq. (9.60) to

$$T_Z^2 \approx \frac{6.08}{D^{1/3}} \int_0^L dz\, C_n^2(z)\ [1 - 0.596(D\kappa_0)^{1/3}(1 + 0.045D^2\kappa_0^2)^{-1/6}]. \tag{9.61}$$

The outer scale may vary as a function of z along a slant path, but a widely accepted model for the outer scale as a function of altitude is not presently available. Therefore, in the graphical analysis of this result, we will tacitly assume that $\kappa_0 = 2\pi/L_0$ is a constant in Eqs. (9.60) and (9.61). Then, in Fig. 9.8 we plot a scaled version of both Eqs. (9.60) and (9.61) as a function of L_0/D. In this plot, the approximation lies directly on top of the exact expression. Although the plots are not entirely correct because κ_0 is generally not a constant except along a horizontal path, these results suggest a substantial reduction in tilt variance caused by the inclusion of the outer scale. Inner-scale effects are not significant unless the inner scale is larger than the telescope diameter (cf. pp. 152–153 of Ref. 6).

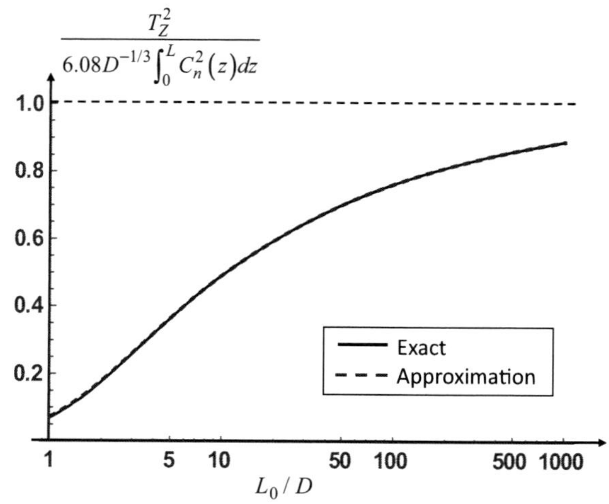

Figure 9.8 The outer-scale effect defined by Eqs. (9.60) and (9.61). These results are based on the exponential spectrum (9.26) for Z-tilt variance as a function of the ratio of outer scale to aperture diameter. The dashed line represents an infinite outer scale.

9.2.3 Gradient-tilt variance

A target position measured with a centroid sensor responds more closely to G-tilt than to Z-tilt. The G-tilt variance can be evaluated in the same fashion as Z-tilt, but now using the filter function $(4/kD)^2 J_1^2(\kappa D/2)$. In this case the tilt angle jitter variance is calculated from the integral

$$\begin{aligned} T_G^2 &= 1.303k^2 \int_0^L dz\, C_n^2(z) \int_0^\infty d\kappa\, \kappa^{-8/3} \left(\frac{4}{kD}\right)^2 J_1^2(\kappa D/2) \\ &= \frac{6.5656}{D^{1/3}} \int_0^L C_n^2(z)\, dz \int_0^\infty x^{-8/3} J_1^2(x)\, dx, \end{aligned} \tag{9.62}$$

where we have made the change of variable $x = \kappa D/2$. Based on the relation ($D3$) in Appendix D, it follows that

$$\begin{aligned} T_G^2 &= \frac{6.5656}{D^{1/3}} \frac{\Gamma\left(\frac{1}{6}\right)\Gamma\left(\frac{4}{3}\right)}{2\sqrt{\pi}\,\Gamma\left(\frac{11}{6}\right)\Gamma\left(\frac{17}{6}\right)} \int_0^L C_n^2(z)\, dz \\ &= \frac{5.675}{D^{1/3}} \int_0^L C_n^2(z)\, dz. \end{aligned} \tag{9.63}$$

This last expression can also be written as

$$T_G^2 = \frac{13.416}{k^2 D^2} \left(\frac{D}{r_0}\right)^{5/3} = 0.340 \left(\frac{D}{r_0}\right)^{5/3} \left(\frac{\lambda}{D}\right)^2. \tag{9.64}$$

Note that Eq. (9.64) is essentially the average of the near-field and far-field AOA variances given by Eqs. (8.200) and (8.201), respectively.

For G-tilt, we can use the approximation in Eq. (9.49) for the filter function in Eq. (9.62), which leads to the same expression as Eq. (9.57) that we got from Z-tilt with the Gaussian filter function approximation. However, the difference here is only around 1.5% compared with that in Eq. (9.63).

9.2.4 Zernike-tilt power spectral density

If we make the association $\rho = V_\perp \tau$ in the Zernike-tilt covariance function for an infinite plane wave, the *temporal Zernike-tilt covariance* is defined by

$$B_{T,Z}(\tau) = 1.303k^2 \int_0^L dz C_n^2(z) \int_0^\infty d\kappa\, \kappa^{-8/3} J_0(\kappa V_\perp \tau) \left(\frac{4}{kD}\right)^2 \left[\frac{4J_2(\kappa D/2)}{\kappa D/2}\right]^2, \tag{9.65}$$

where we are using the frozen turbulence hypothesis. By taking the Fourier transform of Eq. (9.65) using the relation of Eq. (8.164) and employing the result of Eq. (8.168), we obtain the *power spectral density* defined by

$$S_{T,Z}(\omega) = \frac{5336.43}{D^4}\int_0^L dz\, C_n^2(z)\frac{1}{V_\perp}\int_a^\infty \frac{\kappa^{-14/3}}{(\kappa^2 - a^2)^{1/2}} J_2^2(\kappa D/2)\, d\kappa, \qquad (9.66)$$

where $a = \omega/V_\perp$. We now make the change of variable $\kappa = a\sqrt{x}$ and use ($C16$) in Appendix C to obtain

$$S_{T,Z}(\omega) = \frac{105.05 D^{2/3}}{\sqrt{\pi}}\int_0^L dz\, C_n^2(z)\frac{1}{V_\perp}\left(\frac{2V_\perp}{\omega D}\right)^{14/3} \times \int_1^\infty \frac{x^{-17/6}}{(x-1)^{1/2}} G_{13}^{11}\left(\frac{\omega^2 D^2 x}{4V_\perp^2}\middle|\begin{matrix}\frac{1}{2}\\ 2, 0, -2\end{matrix}\right) dx,$$

or

$$S_{T,Z}(\omega) = 105.05 D^{2/3}\int_0^L dz\, C_n^2(z)\frac{1}{V_\perp}\left(\frac{2V_\perp}{\omega D}\right)^{14/3} G_{24}^{21}\left(\frac{\omega^2 D^2}{4V_\perp^2}\middle|\begin{matrix}\frac{1}{2}, \frac{17}{6}\\ \frac{7}{3}, 2, 0, -2\end{matrix}\right). \qquad (9.67)$$

The integral over x was evaluated by using ($D10$) in Appendix D.

At this point we could use ($MG1$) in Appendix A to write the Meijer G-function in Eq. (9.67) as a sum of two ${}_2F_3$ functions. However, for a simpler formulation, we choose to leave Eq. (9.67) in terms of the Meijer G-function. In Fig. 9.9 we plot a scaled version of Eq. (9.67) as a function of $\omega D/2V_\perp$. In this analysis we assume that the wind speed $V_\perp$ is a constant along the path.

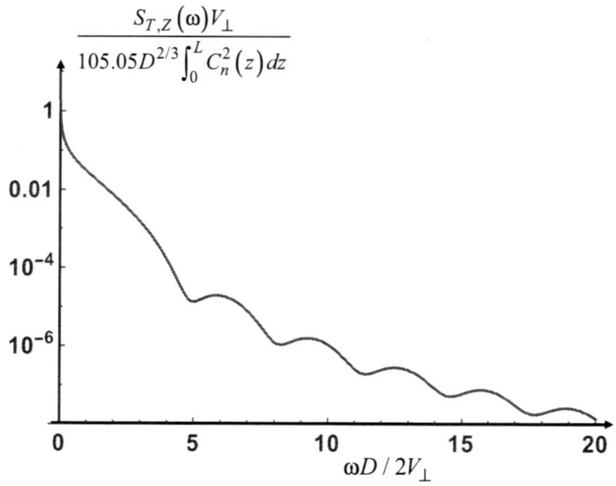

Figure 9.9 Scaled power spectral density of tilt defined by Eq. (9.67). The transverse wind speed $V_\perp$ is assumed constant along the path.

9.3 Anisoplanatism

The principle behind AO systems is to sense the phase distortion of a known source and apply a corrective or conjugate phase to the incoming image or outgoing beam. One of the major problems encountered in astronomical imaging is that the object being imaged (science object) is not sufficiently bright to be a good source used by the WFS for corrections. To overcome this problem, one first looks for a nearby bright star that can be used as a reference beacon called a *natural guide star* (NGS). Unfortunately, such a reference beacon often lies outside the isoplanatic patch which defines the maximum angular separation (around 5–10 μrad for visible light) between the science object and the reference source over which the phase difference is less than 1 rad. This is a limitation that naturally causes a degradation in the performance of an AOI system known as the *anisoplanatic effect*. In such cases the optical paths of the science object and the NGS traverse different regions of atmospheric turbulence along a portion of the path as shown in Fig. 9.10(a), causing distinct wavefront distortions in each optical wave. This particular problem is called *angular anisoplanatism*.

9.3.1 Laser guide star

To overcome the beacon placement problem, a possible solution is to use an artificial *laser guide star* (LGS) that, in theory, can be placed in exactly the same angular direction as the science object. A sodium beacon LGS, for example, can be created by using a laser tuned to 589 nm to energize atoms in the sodium layer of the mesosphere at an altitude of around 90–95 km. However, this causes a different type of problem, as illustrated in Fig. 9.10(b), called *focal anisoplanatism*. That is, the light at a lower altitude emitted from

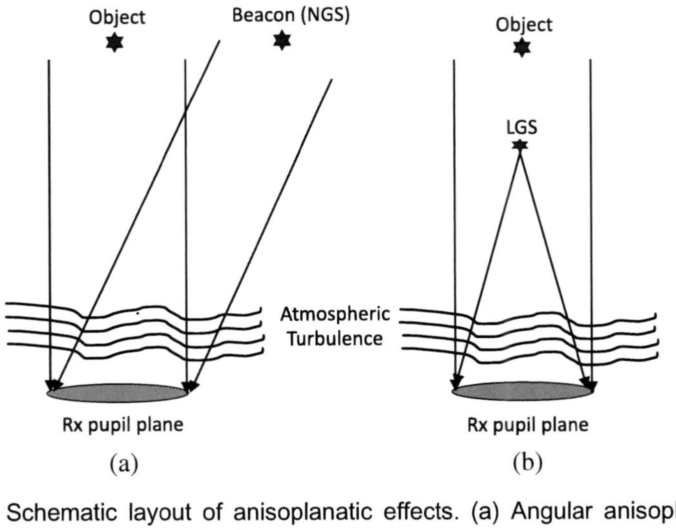

Figure 9.10 Schematic layout of anisoplanatic effects. (a) Angular anisoplanatism and (b) focal anisoplanatism (cone effect).

the LGS acts like a spherical wave, whereas the science object at a far greater distance will act more like an infinite plane wave when entering the atmosphere. This difference in optical wave characteristics is sometimes referred to as the *cone effect*.

In addition to the cone effect, a type of angular anisoplanatism arises as a result of not being able to place a LGS exactly in the direction of the object of interest because of atmospheric turbulence shifting the position of the outgoing upward beam. Hence, the exact location of the focused laser spot is unknown, thus making it impossible to properly measure wavefront tilt caused by the downward propagation of the backscattered LGS. However, if a nearby NGS is used to measure only tilt, then the NGS brightness and angular limit of the isoplanatic patch are not as severe as they are for higher-order corrections.

9.4 Tilt Anisoplanatism

Tilt anisoplanatism is defined by the angular differential tilt jitter between two displaced sources that are received by a single telescope, as illustrated in Fig. 9.11. For example, it accounts for the apparent jitter between two stars that are simultaneously viewed with a single telescope. However, a similar mathematical formulation also applies to the reciprocal situation of a single source and two displaced apertures [e.g., a differential image motion monitor (DIMM)]. We consider the mathematical expressions associated with each configuration in the following sections.

9.4.1 Multi-aperture case

Based on the Kolmogorov spectrum (8.1) and an incident plane wave as the source, we find that the anisoplanatic *tilt angle variance* using Z-tilt in the single-source, multi-aperture case is defined by

$$\begin{pmatrix} \sigma_Z^2 \\ \sigma_{Z\perp}^2 \\ \sigma_{Z\parallel}^2 \end{pmatrix} = \frac{106.16}{D^2}\int_0^L dz\, C_n^2(z)\int_0^\infty d\kappa\,\kappa^{-8/3}\left[\frac{J_2(\kappa D/2)}{\kappa D/2}\right]^2 \times \int_0^{2\pi} d\phi\,[1-\cos(\kappa d\cos\phi)]\begin{pmatrix} 1 \\ \sin^2\phi \\ \cos^2\phi \end{pmatrix}, \tag{9.68}$$

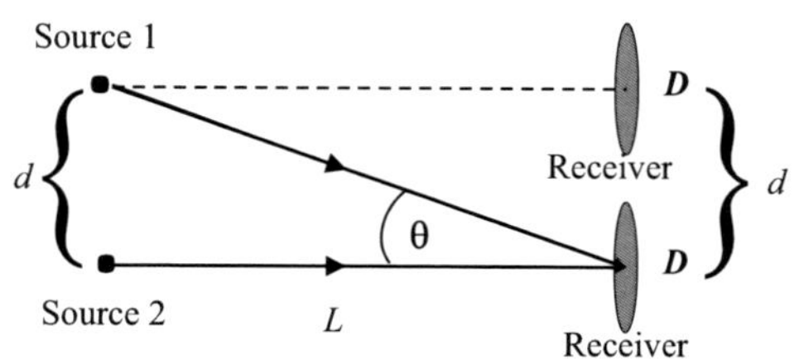

Figure 9.11 Schematic layout of one (or two) source(s) and two (or one) Rx telescope(s).

where $\sigma^2_{Z\perp}$, $\sigma^2_{Z\parallel}$ represent perpendicular and parallel variances, respectively, and whose sum is equal to the total variance σ^2_Z. We initially assume that the separation distance d between two Rx apertures satisfies $d > D$. The case of an incident point source was previously presented in Ref. 4, pp. 357–358.

After integration on the angle variable in Eq. (9.68), we have for the total tilt angle variance

$$\sigma^2_Z = \frac{667.03}{D^2}\int_0^L C_n^2(z)dz\int_0^\infty \kappa^{-8/3}\left[\frac{J_2(\kappa D/2)}{\kappa D/2}\right]^2[1 - J_0(\kappa d)]\,d\kappa. \tag{9.69}$$

With the substitution $x = \kappa D/2$, Eq. (9.69) becomes

$$\begin{aligned}\sigma^2_Z &= \frac{210.1}{D^{1/3}}\int_0^L dz\, C_n^2(z)\int_0^\infty x^{-14/3}J_2^2(x)\left[1 - J_0\left(\frac{2xd}{D}\right)\right]dx\\ &= \frac{210.1}{D^{1/3}}\int_0^L dz\, C_n^2(z)\left[\int_0^\infty x^{-14/3}J_2^2(x)\,dx\right. \\ &\quad \left. - \int_0^\infty x^{-14/3}J_2^2(x)\,J_0\left(\frac{2xd}{D}\right)dx\right].\end{aligned} \tag{9.70}$$

Recalling ($D3$) in Appendix D, the first integral over x in Eq. (9.70) leads to

$$\int_0^\infty x^{-14/3}J_2^2(x)\,dx = \frac{\Gamma\left(\frac{1}{6}\right)\Gamma\left(\frac{7}{3}\right)}{2\sqrt{\pi}\,\Gamma\left(\frac{17}{6}\right)\Gamma\left(\frac{29}{6}\right)} = 0.0579, \tag{9.71}$$

whereas the second integral involves the product of three Bessel functions. Hence, by using ($D4$) in Appendix D and ($MG1$) in Appendix A, we deduce that

$$\begin{aligned}\int_0^\infty x^{-14/3}J_2^2(x)\,J_0\left(\frac{2xd}{D}\right)dx &= \frac{D}{2d\sqrt{\pi}}\,G_{33}^{12}\left(\frac{D^2}{d^2}\left|\begin{matrix}\frac{1}{2},\ -\frac{11}{6},\ \frac{1}{2}\\ -\frac{1}{3},\ -\frac{7}{3},\ -\frac{13}{3}\end{matrix}\right.\right)\\ &= 0.0385\left(\frac{D}{d}\right)^{1/3}{}_3F_2\left(\frac{1}{6},\frac{5}{2},\frac{1}{6};\ 3,\ 5;\ \frac{D^2}{d^2}\right).\end{aligned} \tag{9.72}$$

Now combining the results gives us

$$\sigma^2_Z = \left[\frac{12.16}{D^{1/3}} - \frac{8.09}{d^{1/3}}\,{}_3F_2\left(\frac{1}{6},\frac{5}{2},\frac{1}{6};\ 3,\ 5;\ \frac{D^2}{d^2}\right)\right]\int_0^L C_n^2(z)\,dz, \quad d > D. \tag{9.73}$$

Following the approach above, the perpendicular component in Eq. (9.68) is defined by

$$\begin{aligned}\sigma_{Z\perp}^2 &= \frac{667.03}{D^2}\int_0^L C_n^2(z)dz \int_0^\infty \kappa^{-8/3}\left[\frac{J_2(\kappa D/2)}{\kappa D/2}\right]^2\left[\frac{1}{2}-\frac{J_1(\kappa d)}{\kappa d}\right]d\kappa \\ &= \frac{105.05}{D^{1/3}}\int_0^L C_n^2(z)dz\left[\int_0^\infty x^{-14/3}J_2^2(x)\,dx\right. \\ &\quad \left. -\frac{D}{d}\int_0^\infty x^{-17/3}J_2^2(x)\,J_1\left(\frac{2xd}{D}\right)dx\right].\end{aligned} \tag{9.74}$$

The first integral over x in Eq. (9.74) is the same as that in Eq. (9.71), whereas the second integral is solved by applying ($D4$) in Appendix D and ($MG1$) in Appendix A, from which we find that

$$\begin{aligned}\frac{D}{d}\int_0^\infty x^{-17/3}J_2^2(x)J_1\left(\frac{2xd}{D}\right)dx &= \frac{D^2}{2\sqrt{\pi}d^2}G_{33}^{12}\left(\frac{D^2}{d^2}\middle|\begin{matrix}0,-\frac{7}{3},1\\ -\frac{5}{6},-\frac{17}{6},-\frac{29}{6}\end{matrix}\right) \\ &= 0.0462\left(\frac{D}{d}\right)^{1/3}{}_3F_2\left(\frac{1}{6},\frac{5}{2},-\frac{5}{6};3,5;\frac{D^2}{d^2}\right).\end{aligned} \tag{9.75}$$

Combining the results of Eqs. (9.71) and (9.75) with Eq. (9.74) yields

$$\sigma_{Z\perp}^2 = \left[\frac{6.08}{D^{1/3}}-\frac{4.85}{d^{1/3}}\,{}_3F_2\left(\frac{1}{6},\frac{5}{2},-\frac{5}{6};\,3,\,5;\,\frac{D^2}{d^2}\right)\right]\int_0^L C_n^2(z)\,dz,\quad d>D. \tag{9.76}$$

Finally, to determine the parallel component of Z-tilt, we simply use the relationship

$$\sigma_{Z\parallel}^2 = \sigma_Z^2 - \sigma_{Z\perp}^2. \tag{9.77}$$

For $d > D$, the ${}_3F_2$ functions in both Eqs. (9.73) and (9.76) can be replaced by unity; thus, we have the following approximations:

$$\begin{aligned}\sigma_Z^2 &\approx (12.16D^{-1/3}-8.09d^{-1/3})\int_0^L C_n^2(z)\,dz,\quad d>D,\\ \sigma_{Z\perp}^2 &\approx (6.08D^{-1/3}-4.85\,d^{-1/3})\int_0^L C_n^2(z)\,dz,\quad d>D,\\ \sigma_{Z\parallel}^2 &\approx (6.08D^{-1/3}-3.24\,d^{-1/3})\int_0^L C_n^2(z)\,dz,\quad d>D.\end{aligned} \tag{9.78}$$

It turns out that these expressions in Eq. (9.78) are quite accurate for all $d > D$, yielding an error of roughly 2% or less as compared with the exact results involving the generalized hypergeometric functions. For $d < D$, the expressions with the Meijer G-function can be used.

9.4.2 Multi-aperture: outer-scale effect

The analysis in the last section led to an exact expression for the components of tilt anisoplanatism using Z-tilt in the single-source, multi-aperture case but did not include outer-scale effects. Here we wish to repeat the calculations for a centroid sensor closely approximated by G-tilt but also to include the outer-scale effect by using the exponential spectrum model presented in Eq. (9.26).

The three components of G-tilt are defined by

$$\begin{pmatrix} \sigma_G^2 \\ \sigma_{G\perp}^2 \\ \sigma_{G\parallel}^2 \end{pmatrix} = \frac{6.57}{D^2} \int_0^L dz\, C_n^2(z) \int_0^\infty d\kappa\, \kappa^{-8/3} \left[1 - \exp\left(-\frac{\kappa^2}{\kappa_0^2}\right)\right] J_1^2(\kappa D/2) \times \int_0^{2\pi} d\phi \left[1 - \cos(\kappa d \cos\phi)\right] \begin{pmatrix} 1 \\ \sin^2\phi \\ \cos^2\phi \end{pmatrix}. \tag{9.79}$$

Because of the accuracy in using the exponential function approximation for the filter function in previous calculations, we will perform the calculations here once again with the use of the filter function approximation,

$$J_1^2\left(\frac{\kappa D}{2}\right) \approx \frac{\kappa^2 D^2}{16} \exp\left(-\frac{\kappa^2 D^2}{16}\right). \tag{9.80}$$

After integrating over the angle variable in Eq. (9.79) and making the substitution of Eq. (9.80), we have for total tilt angle variance,

$$\begin{aligned} \sigma_G^2 &\approx 2.58 \int_0^L dz\, C_n^2(z) \int_0^\infty d\kappa\, \kappa^{-2/3} \left[1 - \exp\left(-\frac{\kappa^2}{\kappa_0^2}\right)\right] \\ &\quad \times \exp\left(-\frac{\kappa^2 D^2}{16}\right) [1 - J_0(\kappa d)] \\ &\approx 2.58 \int_0^L dz\, C_n^2(z) (I_1 - I_2), \end{aligned} \tag{9.81}$$

where

$$I_1 = \int_0^\infty \kappa^{-2/3} \left[1 - \exp\left(-\frac{\kappa^2}{\kappa_0^2}\right)\right] \exp\left(-\frac{\kappa^2 D^2}{16}\right) d\kappa, \tag{9.82}$$

$$I_2 = \int_0^\infty \kappa^{-2/3} \left[1 - \exp\left(-\frac{\kappa^2}{\kappa_0^2}\right)\right] \exp\left(-\frac{\kappa^2 D^2}{16}\right) J_0(\kappa d)\, d\kappa. \tag{9.83}$$

Equation (9.82) can be evaluated by recalling ($D2$) in Appendix D, which leads to

$$I_1 = \frac{4.418}{D^{1/3}}\left[1 - \left(1 + \frac{16}{\kappa_0^2 D^2}\right)^{-1/6}\right]. \tag{9.84}$$

For Eq. (9.83), we can refer to ($D1$) in Appendix D, from which we obtain

$$\begin{aligned} I_2 = \frac{4.418}{D^{1/3}}\Bigg\{ {}_1F_1\left(\frac{1}{6};\ 1;\ -\frac{4d^2}{D^2}\right) - \left(1 + \frac{16}{\kappa_0^2 D^2}\right)^{-1/6} \\ \times {}_1F_1\left[\frac{1}{6};\ 1;\ -\frac{4d^2}{D^2(1 + 16/\kappa_0^2 D^2)^2}\right]\Bigg\}. \end{aligned} \tag{9.85}$$

By combining Eq. (9.81) with Eqs. (9.84) and (9.85), we get the result

$$\begin{aligned} \sigma_G^2 \approx \frac{11.40}{D^{1/3}} \int_0^L dz\, C_n^2(z) \Bigg\{1 - \left(1 + \frac{16}{\kappa_0^2 D^2}\right)^{-1/6} - {}_1F_1\left(\frac{1}{6};\ 1;\ -\frac{4d^2}{D^2}\right) \\ + \left(1 + \frac{16}{\kappa_0^2 D^2}\right)^{-1/6} {}_1F_1\left[\frac{1}{6};\ 1;\ -\frac{4d^2}{D^2(1 + 16/\kappa_0^2 D^2)^2}\right]\Bigg\}. \end{aligned} \tag{9.86}$$

For the perpendicular component, we have

$$\begin{aligned} \sigma_{G\perp}^2 &\approx 2.58 \int_0^L dz\, C_n^2(z) \int_0^\infty d\kappa\, \kappa^{-2/3}\left[1 - \exp\left(-\frac{\kappa^2}{\kappa_0^2}\right)\right] \\ &\quad \times \exp\left(-\frac{\kappa^2 D^2}{16}\right)\left[\frac{1}{2} - \frac{J_1(\kappa d)}{\kappa d}\right] \\ &\approx 2.58 \int_0^L dz\, C_n^2(z)(I_3 - I_4), \end{aligned} \tag{9.87}$$

where I_3 is just half of the result in Eq. (9.84), i.e.,

$$I_3 = \frac{1}{2} I_1 = 2.209 D^{-1/3}\left[1 - \left(1 + \frac{16}{\kappa_0^2 D^2}\right)^{-1/6}\right]. \tag{9.88}$$

To evaluate I_4, we again use ($D1$) in Appendix D, which yields

$$\begin{aligned} I_4 &= \frac{1}{d}\int_0^\infty \kappa^{-5/3}\left[1 - \exp\left(-\frac{\kappa^2}{\kappa_0^2}\right)\right] \exp\left(-\frac{\kappa^2 D^2}{16}\right) J_1(\kappa d)\, d\kappa \\ &= \frac{2.209}{D^{1/3}}\Bigg\{ {}_1F_1\left(\frac{1}{6};\ 2;\ -\frac{4d^2}{D^2}\right) \\ &\quad - \left(1 + \frac{16}{\kappa_0^2 D^2}\right)^{-1/6} {}_1F_1\left[\frac{1}{6};\ 2;\ -\frac{4d^2}{D^2(1 + 16/\kappa_0^2 D^2)^2}\right]\Bigg\}. \end{aligned} \tag{9.89}$$

Once more, by combining Eq. (9.87) with Eqs. (9.88) and (9.89), the perpendicular component becomes

$$\sigma_{G\perp}^2 \approx \frac{5.70}{D^{1/3}} \int_0^L dz\, C_n^2(z) \left\{ 1 - \left(1 + \frac{16}{\kappa_0^2 D^2}\right)^{-1/6} - {}_1F_1\left(\frac{1}{6};\ 2;\ -\frac{4d^2}{D^2}\right) \right.$$
$$\left. + \left(1 + \frac{16}{\kappa_0^2 D^2}\right)^{-1/6} {}_1F_1\left[\frac{1}{6};\ 2;\ -\frac{4d^2}{D^2(1+16/\kappa_0^2 D^2)^2}\right] \right\}. \tag{9.90}$$

For an infinitely large outer scale ($\kappa_0 \to 0$), we find that Eqs. (9.86) and (9.90) reduce to

$$\sigma_G^2 \approx \frac{11.40}{D^{1/3}} \left[1 - {}_1F_1\left(\frac{1}{6};\ 1;\ -\frac{4d^2}{D^2}\right)\right] \int_0^L C_n^2(z)\, dz, \tag{9.91}$$

$$\sigma_{G\perp}^2 \approx \frac{5.70}{D^{1/3}} \left[1 - {}_1F_1\left(\frac{1}{6};\ 2;\ -\frac{4d^2}{D^2}\right)\right] \int_0^L C_n^2(z)\, dz. \tag{9.92}$$

Moreover, for $d \gg D$, these expressions reduce further to

$$\sigma_G^2 \approx \left(\frac{11.40}{D^{1/3}} - \frac{8.02}{d^{1/3}}\right) \int_0^L C_n^2(z)\, dz, \quad d \gg D, \tag{9.93}$$

$$\sigma_{G\perp}^2 \approx \left(\frac{5.70}{D^{1/3}} - \frac{4.81}{d^{1/3}}\right) \int_0^L C_n^2(z)\, dz, \quad d \gg D. \tag{9.94}$$

Based on these last results, we find that the parallel component becomes

$$\sigma_{G\parallel}^2 \approx \left(\frac{5.70}{D^{1/3}} - \frac{3.21}{d^{1/3}}\right) \int_0^L C_n^2(z)\, dz, \quad d \gg D. \tag{9.95}$$

Exact expressions for the G-tilt variance with an infinite outer scale ($\kappa_0 = 0$) were previously derived (p. 357 of Ref. 4). Thus, in Fig. 9.12 we compare the scaled exact result (solid black line) for the total G-tilt variance with the scaled result of Eq. (9.91). For graphical purposes with $\kappa_0 \neq 0$, we will treat the outer scale as a constant along the propagation path. In this figure the aperture diameter was selected as $D = 10$ cm and the outer-scale parameter was selected as $\kappa_0 = 5$ and $10\ \text{m}^{-1}$. For $d < 1$ m, there is little to no influence from the outer scale except for the case $\kappa_0 = 10\ \text{m}^{-1}$, which corresponds to a more extreme situation involving an outer scale of less than 1 m.

9.4.3 Multi-source case

When considering objects in the sky, tilt anisoplanatism concerns the atmospheric-induced differential tilt seen by a single telescope observing two light sources separated by a very small angle θ (Fig. 9.11). Also, a simple grid of light-emitting diodes (LEDs) as a source can be used to make a simultaneous measurement of several atmospheric turbulence parameters along a horizontal path.[10,11] In particular, with a LED array, an imaging

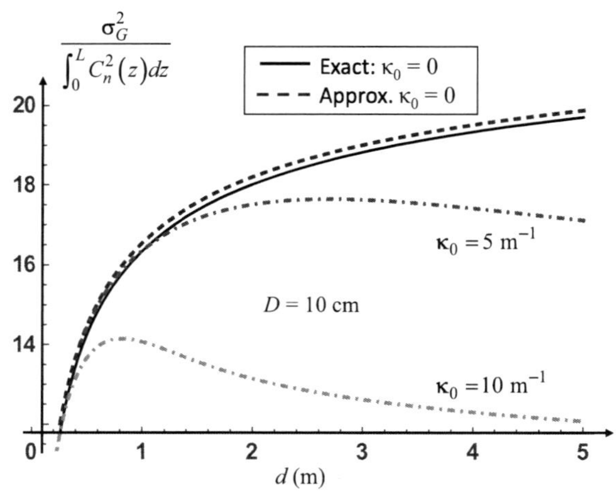

Figure 9.12 Scaled total G-tilt variance defined by Eq. (9.86)] as a function of aperture separation d with three values of outer-scale parameter κ_0 (assumed to be constant). The exact result with $\kappa_0 = 0$ from p. 357 of Ref. 4 is shown by the solid black line. In all cases the aperture diameters were set at $D = 10$ cm.

sensor, and a model of tilt anisoplanatism, one can determine the turbulence parameter C_n^2, outer scale L_0, and mean transverse wind speed. The use of a LED array has the advantage of providing several small light sources at various separation distances.

If D is the diameter of a receiving telescope aperture and $d = L\theta$ is the transverse distance between two point sources as shown in Fig. 9.11, then the centroid-estimation-based tilt anisoplanatism (G-tilt) for multiple point sources is determined from

$$\begin{pmatrix} \sigma_G^2 \\ \sigma_{G\perp}^2 \\ \sigma_{G\parallel}^2 \end{pmatrix} = \frac{41.7}{D^2}\int_0^L dz C_n^2(z)\int_0^\infty d\kappa\kappa^{-8/3}J_1^2(\kappa\gamma D/2) \\ \times \frac{1}{2\pi}\int_0^{2\pi} d\phi\,\{1-\cos[(1-\gamma)\kappa d\cos\phi]\}\begin{pmatrix} 1 \\ \sin^2\phi \\ \cos^2\phi \end{pmatrix}, \tag{9.96}$$

where $\gamma = z/L$ and we are using the Kolmogorov spectrum [Eq. (8.1)]. After integration on ϕ, the total tilt angle variance becomes

$$\sigma_G^2 = \frac{41.7}{D^2}\int_0^L dz C_n^2(z)\int_0^\infty \kappa^{-8/3}J_1^2(\kappa\gamma D/2)\{1-J_0[\kappa d(1-\gamma)]\}\,d\kappa. \tag{9.97}$$

Following the substitution $x = \kappa\gamma D/2$, Eq. (9.97) takes the form

$$\begin{aligned}\sigma_G^2 &= \frac{13.135}{D^{1/3}}\int_0^L dz\, C_n^2(z)\gamma^{5/3}\int_0^\infty x^{-8/3}J_1^2(x)\left\{1 - J_0\left[2x\left(\frac{1}{\gamma}-1\right)\frac{d}{D}\right]\right\}dx\\ &= \frac{13.135}{D^{1/3}}\int_0^L dz\, C_n^2(z)\left(\frac{z}{L}\right)^{5/3}\left\{\int_0^\infty x^{-8/3}J_1^2(x)\,dx\right.\\ &\quad\left. - \int_0^\infty x^{-8/3}J_1^2(x)\,J_0\left[2x\left(\frac{L-z}{z}\right)\frac{d}{D}\right]dx\right\}.\end{aligned} \tag{9.98}$$

With ($D3$) in Appendix D, the first integral over x in this last expression leads to

$$\int_0^\infty x^{-8/3}J_1^2(x)\,dx = \frac{\Gamma\left(\frac{1}{6}\right)\Gamma\left(\frac{4}{3}\right)}{2\sqrt{\pi}\,\Gamma\left(\frac{11}{6}\right)\Gamma\left(\frac{17}{6}\right)} = 0.8644. \tag{9.99}$$

The second integral in Eq. (9.98) can be solved by using ($D4$) in Appendix D followed by ($MG1$) in Appendix A. Doing so, we see that

$$\begin{aligned}\int_0^\infty x^{-8/3}J_1^2(x)\,J_0\left[2x\left(\frac{L-z}{z}\right)\frac{d}{D}\right]dx &= \frac{1}{b\sqrt{\pi}}\,G_{33}^{12}\left(\frac{4}{b^2}\middle|\begin{matrix}\frac{1}{2},\ -\frac{5}{6},\ \frac{1}{2}\\ -\frac{1}{3},\ -\frac{4}{3},\ -\frac{7}{3}\end{matrix}\right)\\ &= \frac{0.777}{b^{1/3}}\,{}_3F_2\left(\frac{1}{6},\frac{3}{2},\frac{1}{6};\ 2, 3;\ \frac{4}{b^2}\right),\end{aligned} \tag{9.100}$$

where $b = 2(L-z)d/zD$. Unfortunately, this last result still leaves a difficult integral involving the ${}_3F_2$ function to evaluate over the propagation distance z. We have found for numerical calculations that it may be best to keep Eq. (9.100) in terms of the Meijer G-function. Then, combining this result and Eq. (9.99) in Eq. (9.98), we arrive at

$$\begin{aligned}\sigma_G^2 &= \frac{11.35}{D^{1/3}}\int_0^L dz\, C_n^2(z)\left(\frac{z}{L}\right)^{5/3}\\ &\quad\times\left\{1 - 0.3264\frac{D}{d}\left(\frac{z}{L-z}\right)G_{33}^{12}\left[\frac{D^2}{d^2}\left(\frac{z}{L-z}\right)^2\middle|\begin{matrix}\frac{1}{2},\ -\frac{5}{6},\ \frac{1}{2}\\ -\frac{1}{3},\ -\frac{4}{3},\ -\frac{7}{3}\end{matrix}\right]\right\}dz.\end{aligned} \tag{9.101}$$

For the perpendicular tilt component in Eq. (9.96), the integration on ϕ yields

$$\sigma_{G\perp}^2 = \frac{41.7}{D^2}\int_0^L dz C_n^2(z)\int_0^\infty \kappa^{-8/3}J_1^2(\kappa\gamma D/2)\left\{\frac{1}{2} - \frac{J_1[\kappa d(1-\gamma)]}{\kappa d(1-\gamma)}\right\}d\kappa. \tag{9.102}$$

Again, we make the substitution $x = \kappa\gamma D/2$, which leads to

$$\sigma^2_{G\perp} = \frac{6.565}{D^{1/3}} \int_0^L dz\, C_n^2(z) \left(\frac{z}{L}\right)^{5/3} \left\{ \int_0^\infty x^{-8/3} J_1^2(x)\, dx \right. \\ \left. - \frac{D}{d}\left(\frac{z}{L-z}\right) \int_0^\infty x^{-11/3} J_1^2(x)\, J_1\left[2x\left(\frac{L-z}{z}\right)\frac{d}{D}\right] dx \right\}. \tag{9.103}$$

The first integral over x is the same as that in Eq. (9.99), whereas the second integral gives us

$$\int_0^\infty x^{-11/3} J_1^2(x)\, J_1\left[2x\left(\frac{L-z}{z}\right)\frac{d}{D}\right] dx = \frac{1}{b\sqrt{\pi}}\, G_{33}^{12}\left(\frac{4}{b^2}\middle|\begin{matrix} 0, -\frac{4}{3}, 1 \\ -\frac{5}{6}, -\frac{11}{6}, -\frac{17}{6}\end{matrix}\right). \tag{9.104}$$

As we did in Eq. (9.101), we choose to leave our result in terms of the Meijer G-function. Therefore, combining the results of Eqs. (9.99) and (9.104) in Eq. (9.103), we get

$$\sigma^2_{G\perp} = \frac{5.677}{D^{1/3}} \int_0^L dz\, C_n^2(z) \left(\frac{z}{L}\right)^{5/3} \\ \times \left\{ 1 - 0.3264\left(\frac{D}{d}\right)^2 \left(\frac{z}{L-z}\right)^2 G_{33}^{12}\left[\frac{D^2}{d^2}\left(\frac{z}{L-z}\right)^2 \middle| \begin{matrix} 0, -\frac{4}{3}, 1 \\ -\frac{5}{6}, -\frac{11}{6}, -\frac{17}{6}\end{matrix}\right]\right\}. \tag{9.105}$$

The corresponding expression for $\sigma^2_{G\|}$ is readily obtained from Eqs. (9.101) and (9.105) by utilizing the identity

$$\sigma^2_{G\|} = \sigma^2_G - \sigma^2_{G\perp}. \tag{9.106}$$

In Fig. 9.13 we plot Eqs. (9.105) and (9.106) scaled by constant C_n^2 as a function of source separation d. The Rx aperture diameter is $D = 10$ cm, and the path length is 1 km.

To avoid use of the Meijer G-function in Eqs. (9.101) and (9.105), fairly accurate approximations for the perpendicular and parallel components were previously calculated (see p. 360 of Ref. 4). These approximations are given by

$$\sigma^2_G \approx \frac{7.15 d^2}{D^{7/3}} \int_0^L C_n^2(z) \left(\frac{z}{L}\right)^{5/3} \left(\frac{L-z}{z}\right)^2 \\ \times \left\{ 1 + \left(\frac{d}{D}\right)^2 \left[1.11 - 0.008\left(\frac{d}{D}\right)\right] \left(\frac{L-z}{z}\right)^2 \right\}^{-1} dz, \tag{9.107}$$

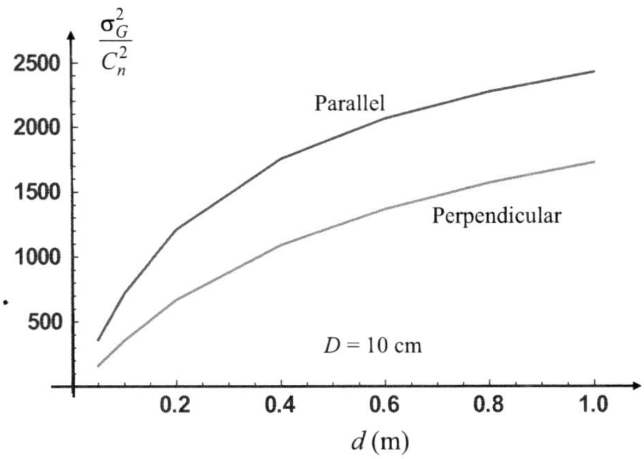

Figure 9.13 Scaled parallel and perpendicular components of tilt anisoplanatism for two point sources separated by distance d and one receiver located at $L = 1$ km. The Rx aperture diameter was set at $D = 10$ cm and C_n^2 is assumed to be constant.

$$\sigma_{G\perp}^2 \approx \frac{1.79d^2}{D^{7/3}} \int_0^L C_n^2(z)\left(\frac{z}{L}\right)^{5/3}\left(\frac{L-z}{z}\right)^2 \times \left\{1+\left(\frac{d}{D}\right)^2\left[0.65-0.007\left(\frac{d}{D}\right)\right]\left(\frac{L-z}{z}\right)^2\right\}^{-1} dz. \tag{9.108}$$

With these results, the parallel component is then calculated from Eq. (9.106).

Exact numerical values scaled by constant C_n^2 are provided in Table 9.2 for the tilt anisoplanatism based on Eqs. (9.101), (9.105), and (9.106) along with approximations based on Eqs. (9.107), (9.108), and (9.106). The exact results for the perpendicular and parallel components are the same as those previously plotted in Fig. 9.13 with $L = 1$ km.

Table 9.2 Exact values and approximations scaled by constant C_n^2 for the total, perpendicular, and parallel components of tilt anisoplanatism with two point sources separated by distance d and one Rx aperture of diameter $D = 10$ cm. The range is set at $L = 1$ km.

d (m)	0.1	0.2	0.4	0.6	0.8	1.0
Total Variance (exact)	1076	1879	2845	3434	3846	4158
Total Variance (approx.)	1133	1958	2907	3462	3845	4139
Perpendicular (exact)	356	670	1091	1368	1571	1728
Perpendicular (approx.)	376	696	1102	1359	1548	1699
Parallel (exact)	720	1209	1754	2065	2275	2430
Parallel (approx.)	757	1262	1805	2102	2297	2440

9.4.4 Multi-source: outer-scale effect

We now wish to repeat the above calculations, but this time with an outer-scale effect and where the filter function is approximated by

$$\left[\frac{J_1(\kappa\gamma D/2)}{\kappa\gamma D}\right]^2 \approx \frac{1}{16}\exp\left(-\frac{\kappa^2\gamma^2 D^2}{16}\right), \quad \gamma = \frac{z}{L}. \tag{9.109}$$

Following the approach in Section 9.4.2, we can repeat that analysis here for the multi-source case. Once again, the calculations will be based on the exponential spectrum model [Eq. (9.26)] along with the exponential function approximation defined by Eq. (9.109) for the Zernike filter function.

To start, we assume two point sources which now lead us to

$$\begin{pmatrix}\sigma_G^2\\ \sigma_{G\perp}^2\\ \sigma_{G\|}^2\end{pmatrix} \approx 2.606\int_0^L dz\gamma^2 C_n^2(z)\int_0^\infty d\kappa\kappa^{-2/3}\left[1-\exp\left(-\frac{\kappa^2}{\kappa_0^2}\right)\right]\exp\left(-\frac{\kappa^2\gamma^2 D^2}{16}\right)$$
$$\times\frac{1}{2\pi}\int_0^{2\pi} d\phi\{1-\cos[(1-\gamma)\kappa d\cos\phi]\}\begin{pmatrix}1\\ \sin^2\phi\\ \cos^2\phi\end{pmatrix}. \tag{9.110}$$

After integration on ϕ, the total tilt variance becomes

$$\begin{aligned}\sigma_G^2 &\approx 2.606\int_0^L dz\,\gamma^2 C_n^2(z)\int_0^\infty d\kappa\kappa^{-2/3}\left[1-\exp\left(-\frac{\kappa^2}{\kappa_0^2}\right)\right]\exp\left(-\frac{\kappa^2\gamma^2 D^2}{16}\right)\\ &\quad\times\{1-J_0[\kappa d(1-\gamma)]\}\\ &\approx 2.606\int_0^L dz\,\gamma^2 C_n^2(z)(I_5-I_6),\end{aligned} \tag{9.111}$$

where

$$I_5 = \int_0^\infty \kappa^{-2/3}\left[1-\exp\left(-\frac{\kappa^2}{\kappa_0^2}\right)\right]\exp\left(-\frac{\kappa^2 D^2\gamma^2}{16}\right)d\kappa \tag{9.112}$$

$$I_6 = \int_0^\infty \kappa^{-2/3}\left[1-\exp\left(-\frac{\kappa^2}{\kappa_0^2}\right)\right]\exp\left(-\frac{\kappa^2 D^2\gamma^2}{16}\right)J_0[\kappa d(1-\gamma)]\,d\kappa. \tag{9.113}$$

Because the integrals in Eqs. (9.112) and (9.113) match the functional forms of I_1 and I_2 in Section 9.4.2, we immediately deduce that

$$I_5 = \frac{4.418}{D^{1/3}\gamma^{1/3}}\left[1-\left(1+\frac{16}{\kappa_0^2 D^2\gamma^2}\right)^{-1/6}\right], \tag{9.114}$$

$$I_6 = \frac{4.418}{D^{1/3}\gamma^{1/3}} \left\{ {}_1F_1\left(\frac{1}{6};\ 1;\ -\frac{4d^2(1-\gamma)^2}{D^2\gamma^2}\right) \right.$$

$$\left. -\left(1+\frac{16}{\kappa_0^2 D^2\gamma^2}\right)^{-1/6} {}_1F_1\left[\frac{1}{6};\ 1;\ -\frac{4d^2(1-\gamma)^2}{D^2\gamma^2\left(1+\frac{16}{\kappa_0^2 D^2\gamma^2}\right)^2}\right]\right\}. \tag{9.115}$$

In this case we find the total G-tilt variance to be

$$\sigma_G^2 \approx \frac{11.51}{D^{1/3}}\int_0^L dz\, C_n^2(z)\gamma^{5/3} \left\{ 1-\left(1+\frac{16}{\kappa_0^2 D^2\gamma^2}\right)^{-1/6} - {}_1F_1\left(\frac{1}{6};1;-\frac{4d^2(1-\gamma)^2}{D^2\gamma^2}\right)\right.$$

$$\left. +\left(1+\frac{16}{\kappa_0^2 D^2\gamma^2}\right)^{-1/6} {}_1F_1\left[\frac{1}{6};1;-\frac{4d^2(1-\gamma)^2}{D^2\gamma^2\left(1+\frac{16}{\kappa_0^2 D^2\gamma^2}\right)^2}\right]\right\}. \tag{9.116}$$

For the perpendicular component, we can also follow the approach used in Section 9.4.2 to obtain

$$\sigma_{G\perp}^2 \approx \frac{5.76}{D^{1/3}}\int_0^L dz\, C_n^2(z)\gamma^{5/3} \left\{ 1-\left(1+\frac{16}{\kappa_0^2 D^2\gamma^2}\right)^{-1/6} - {}_1F_1\left(\frac{1}{6};2;-\frac{4d^2(1-\gamma)^2}{D^2\gamma^2}\right)\right.$$

$$\left. +\left(1+\frac{16}{\kappa_0^2 D^2\gamma^2}\right)^{-1/6} {}_1F_1\left[\frac{1}{6};2;-\frac{4d^2(1-\gamma)^2}{D^2\gamma^2\left(1+\frac{16}{\kappa_0^2 D^2\gamma^2}\right)^2}\right]\right\}. \tag{9.117}$$

Subtracting Eq. (9.117) from Eq. (9.116) gives us the parallel component. Lastly, we note that for the limiting case of an infinite outer scale, Eqs. (9.116), (9.117), and (9.106) reduce to those on pp. 359–360 of Ref. 4:

$$\sigma_G^2 \approx 11.51 D^{-1/3}\int_0^L C_n^2(z)\left(\frac{z}{L}\right)^{5/3}\left\{1 - {}_1F_1\left[\frac{1}{6};\ 1;\ -\frac{4d^2}{D^2}\left(\frac{L-z}{z}\right)^2\right]\right\} dz, \tag{9.118}$$

$$\sigma_{G\perp}^2 \approx 5.76D^{-1/3} \int_0^L C_n^2(z)\left(\frac{z}{L}\right)^{5/3}\left\{1 - {}_1F_1\left[\frac{1}{6};\, 2;\, -\frac{4d^2}{D^2}\left(\frac{L-z}{z}\right)^2\right]\right\} dz, \tag{9.119}$$

$$\sigma_{G\parallel}^2 \approx 5.76D^{-1/3} \int_0^L C_n^2(z)\left(\frac{z}{L}\right)^{5/3} \times\left\{1 - 2\,{}_1F_1\left[\frac{1}{6};\, 1;\, -\frac{4d^2}{D^2}\left(\frac{L-z}{z}\right)^2\right] + {}_1F_1\left[\frac{1}{6};\, 2;\, -\frac{4d^2}{D^2}\left(\frac{L-z}{z}\right)^2\right]\right\} dz. \tag{9.120}$$

9.5 Angular and Focal Anisoplanatism

There are basically three types of anisoplanatism that arise in a LGS AOI system. The first, called *tilt angular anisoplanatism*, arises from the tilt component of the wavefront difference between the LGS and the science object. The second type of anisoplanatism is called *focal anisoplanatism*, which arises from the cone effect illustrated in Fig. 9.10(b). The third type of anisoplanatism arises from the difference in Zernike wavefront terms of higher order than tilt between the LGS and the science object. In this section we examine a combination of focal and angular anisoplanatism.

Degradation in the image quality naturally arises from using a LGS AOI system to make wavefront corrections for a far distant scientific object. That is, the use of a point source from a LGS at an altitude of roughly 90 km to correct for turbulence effects on a plane wave leads to phase variance errors and ultimately lower Strehl ratios, particularly with larger telescopes. Although using a LGS system provides a solution to the lack of an NGS of sufficient brightness, the LGS system cannot be used to measure wavefront tilt, as we previously pointed out. This is called the *tip-tilt determination problem.*[12] Thus, an NGS is still needed to make that measurement. Fortunately, the requirements on the brightness and isoplanatic angle of the NGS are greatly relaxed for making measurements of overall wavefront tilt.

In the treatment here we will use the Zernike filter functions to calculate the aperture-averaged phase error variance difference with piston and tilt components removed between a LGS and a NGS, and then subtract those phase variances from the total phase error variance that has both piston and tilt components present. The expression derived from this operation is called the *effective LGS anisoplanatic error*, which can be used to estimate the Strehl ratio.

The aperture-averaged wavefront phase error variance containing piston and tilt components of a point source due to LGS anisoplanatism is defined under the Kolmogorov spectrum by[13]

$$\sigma_\phi^2 = 2.606k^2 \int_0^L dz\, C_n^2(z) \int_0^\infty d\kappa\kappa^{-8/3}\left[1 - \frac{4L}{\kappa Dz}J_1\left(\frac{\kappa Dz}{2L}\right)J_0(\kappa\theta z)\right], \tag{9.121}$$

where the angle θ represents the NGS offset from the LGS. The phase error variance difference in piston between an LGS and an NGS offset by angle θ is

$$\sigma_P^2 = 5.212k^2 \int_0^L dz\, C_n^2(z) \int_0^\infty d\kappa\kappa^{-8/3} \times \left[\frac{J_1^2(\kappa D/2)}{(\kappa D/2)^2} + \frac{J_1^2(\gamma\kappa D/2)}{(\gamma\kappa D/2)^2} - 2J_0(\kappa\theta z)\frac{J_1(\kappa D/2)J_1(\gamma\kappa D/2)}{\gamma(\kappa D/2)^2}\right], \tag{9.122}$$

where $\gamma = 1$ for the NGS source and $\gamma = 1 - z/L$ for the LGS point source. In likewise fashion, the tilt phase error variance difference is given by

$$\sigma_T^2 = 20.83k^2 \int_0^L dz\, C_n^2(z) \int_0^\infty d\kappa\kappa^{-8/3} \times \left[\frac{J_2^2(\kappa D/2)}{(\kappa D/2)^2} + \frac{J_2^2(\gamma\kappa D/2)}{(\gamma\kappa D/2)^2} - 2J_0(\kappa\theta z)\frac{J_2(\kappa D/2)J_2(\gamma\kappa D/2)}{\gamma(\kappa D/2)^2}\right]. \tag{9.123}$$

Combining Eqs. (9.121), (9.122), and (9.123), the effective LGS anisoplanatic error is defined by

$$\sigma_{eff}^2 = \sigma_\phi^2 - \sigma_P^2 - \sigma_T^2. \tag{9.124}$$

9.5.1 LGS anisoplanatic error: zero NGS offset

Because of the complexity of the above expressions for σ_ϕ^2, σ_P^2, and σ_T^2, we first consider the case where the NGS offset angle is $\theta = 0$. Let us start with Eq. (9.121) for the phase error variance. By making the substitution $x = \kappa Dz/2L$, we find that Eq. (9.121) reduces to

$$\begin{aligned}
\sigma_\phi^2 &= 2.606k^2 \int_0^L dz\, C_n^2(z) \int_0^\infty d\kappa\kappa^{-8/3}\left[1 - \frac{4L}{\kappa Dz}J_1\left(\frac{\kappa Dz}{2L}\right)\right] \\
&= 5.212k^2\left(\frac{D}{2}\right)^{5/3} \int_0^L dz\, C_n^2(z)\left(\frac{z}{L}\right)^{5/3} \int_0^\infty x^{-11/3}\left[\frac{x}{2} - J_1(x)\right] dx \\
&= \frac{5.212k^2}{16}\left(\frac{D}{2}\right)^{5/3} \int_0^L dz\, C_n^2(z)\left(\frac{z}{L}\right)^{5/3} \int_0^\infty x^{-8/3}\,{}_1F_2\left(1;\, 3, 2;\, -\frac{x^2}{4}\right) dx \\
&= \frac{5.212k^2}{8}\left(\frac{D}{2}\right)^{5/3} \int_0^L dz\, C_n^2(z)\left(\frac{z}{L}\right)^{5/3} \int_0^\infty x^{-8/3} G_{13}^{11}\left(\frac{x^2}{4}\middle|\begin{matrix}0\\ 0,\, -2,\, -1\end{matrix}\right) dx \\
&= 0.5007k^2 D^{5/3} \int_0^L dz\, C_n^2(z)\left(\frac{z}{L}\right)^{5/3}.
\end{aligned} \tag{9.125}$$

In arriving at the third line above, we factored out $x/2$ and combined the resulting terms into the generalized hypergeometric function ${}_1F_2$. The last line was obtained from (*D*8) in Appendix D after a variable change.

Next, we write Eq. (9.122) as

$$\sigma_P^2 = 5.212k^2 \int_0^L dz\, C_n^2(z)(P_a + P_b - 2P_{ab}). \tag{9.126}$$

We should point out that there is a singularity at $\kappa = 0$ that exists for the individual terms in Eq. (9.126), but the singularity can be eliminated after summing the individual terms. Thus, we will introduce μ in place of $-14/3$ in all three integrals and then apply the limit $\mu \to -14/3$ to the sum of the individual terms (recall Section 8.2.2). To evaluate P_a, P_b, and P_{ab}, we are first led to

$$\begin{aligned} P_a &= \int_0^\infty \kappa^{2+\mu} \frac{J_1^2(\kappa D/2)}{(\kappa D/2)^2}\, d\kappa \\ &= \left(\frac{D}{2}\right)^{5/3} \int_0^\infty x^\mu J_1^2(x)\, dx, \end{aligned} \tag{9.127}$$

and by using ($D3$) in Appendix D, we obtain

$$P_a = \left(\frac{D}{2}\right)^{5/3} \frac{\Gamma\left(\frac{3+\mu}{2}\right)\Gamma\left(-\frac{\mu}{2}\right)}{2\sqrt{\pi}\,\Gamma\left(\frac{1-\mu}{2}\right)\Gamma\left(\frac{3-\mu}{2}\right)}. \tag{9.128}$$

In a similar fashion, it follows that

$$P_b = \left(\frac{\gamma D}{2}\right)^{5/3} \frac{\Gamma\left(\frac{3+\mu}{2}\right)\Gamma\left(-\frac{\mu}{2}\right)}{2\sqrt{\pi}\,\Gamma\left(\frac{1-\mu}{2}\right)\Gamma\left(\frac{3-\mu}{2}\right)}. \tag{9.129}$$

For the remaining integral in Eq. (9.126), we find that

$$\begin{aligned} P_{ab} &= \int_0^\infty \kappa^{2+\mu} \frac{J_1(\kappa D/2)J_1(\gamma\kappa D/2)}{\gamma(\kappa D/2)^2}\, d\kappa \\ &= \frac{1}{\gamma}\left(\frac{D}{2}\right)^{5/3} \int_0^\infty x^\mu J_1(x)J_1(\gamma x)\, dx. \end{aligned} \tag{9.130}$$

We can evaluate the last integral in Eq. (9.130) with the use of ($C9$) in Appendix C for the second Bessel function, which gives us

$$\begin{aligned} P_{ab} &= \frac{1}{\gamma}\left(\frac{D}{2}\right)^{5/3} \int_0^\infty x^\mu J_1(x) G_{02}^{10}\left(\frac{\gamma^2 x^2}{4}\,\middle|\, \frac{1}{2},\, -\frac{1}{2}\right) dx \\ &= \frac{D^{5/3}}{2^{5/3-\mu}\gamma}\, G_{22}^{11}\left(\gamma^2 \,\middle|\, \begin{matrix} -\frac{\mu}{2},\, 1-\frac{\mu}{2} \\ \frac{1}{2},\, -\frac{1}{2} \end{matrix}\right) \\ &= \frac{D^{5/3}}{2^{5/3-\mu}} \frac{\Gamma\left(\frac{3+\mu}{2}\right)}{\Gamma\left(\frac{1+\mu}{2}\right)}\, {}_2F_1\left(\frac{3+\mu}{2},\, \frac{1+\mu}{2};\, 2;\, \gamma^2\right). \end{aligned} \tag{9.131}$$

In obtaining this last result, we relied on ($D12$) in Appendix D and ($C18$) in Appendix C. By combining Eq. (9.126) with the results of Eqs. (9.128), (9.129), and (9.131), and taking the limit $\mu \to -14/3$, we get the phase error variance difference in piston given by

$$\sigma_P^2 = -0.437089k^2D^{5/3}\int_0^L dz\, C_n^2(z) \times\left\{1+\left(1-\frac{z}{L}\right)^{5/3} - 1.14556\,{}_2F_1\left[-\frac{5}{6},\ -\frac{11}{6};\ 2;\ \left(1-\frac{z}{L}\right)^2\right]\right\}. \tag{9.132}$$

We now repeat the calculations for tilt phase error variance difference given by Eq. (9.123) and written as

$$\sigma_T^2 = 20.83k^2\int_0^L dz\, C_n^2(z)(T_a + T_b - 2T_{ab}). \tag{9.133}$$

The singularity at $\kappa = 0$ does not exist for the individual terms in Eq. (9.133). Therefore, following the calculations above for piston, we arrive at

$$T_a = \int_0^\infty \kappa^{-8/3}\frac{J_2^2(\kappa D/2)}{(\kappa D/2)^2}\, d\kappa = 0.0182309D^{5/3}, \tag{9.134}$$

$$T_b = \int_0^\infty \kappa^{-8/3}\frac{J_2^2(\gamma\kappa D/2)}{(\gamma\kappa D/2)^2}\, d\kappa = 0.0182309(\gamma D)^{5/3}, \tag{9.135}$$

$$\begin{aligned} T_{ab} &= \frac{1}{\gamma}\left(\frac{D}{2}\right)^{5/3}\int_0^\infty x^{-14/3}J_2(x)\,G_{01}^{10}\left(\frac{\gamma^2x^2}{4}\middle|\, 1,\ -1\right)dx \\ &= \frac{D^{5/3}}{2^{19/3}\gamma}\,G_{22}^{11}\left(\gamma^2\middle|\begin{matrix}\frac{11}{6},\ \frac{23}{6}\\ 1,\ -1\end{matrix}\right) \\ &= 0.0200144D^{5/3}\gamma\,{}_2F_1\left(\frac{1}{6},\ -\frac{11}{6};\ 3;\ \gamma^2\right). \end{aligned} \tag{9.136}$$

Equation (9.133) with the results of Eqs. (9.134), (9.135), and (9.136) leads to

$$\sigma_T^2 = 0.37975k^2D^{5/3}\int_0^L dz\, C_n^2(z) \times\left\{1+\left(1-\frac{z}{L}\right)^{5/3} - 2.19566\left(1-\frac{z}{L}\right){}_2F_1\left[\frac{1}{6},\ -\frac{11}{6};\ 3;\ \left(1-\frac{z}{L}\right)^2\right]\right\}. \tag{9.137}$$

The phase error variances given above by Eqs. (9.125), (9.132), and (9.137) are exact. To avoid numerical errors in the calculation of σ_P^2 and σ_T^2, it is necessary to retain a larger number of decimal places than usual, as shown in Eqs. (9.132) and (9.137).

Let us assume that the LGS is at 90-km altitude and use the HAP $C_n^2(z)$ profile model (see Section 9.6) to do the calculations. For illustration, the HAP model parameters are selected as $M = 1$, $p = 4/3$, and $C_n^2(h_0) = 1.7 \times 10^{-14}\ \mathrm{m}^{-2/3}$. The zenith angle is assumed to be zero. The wavelength spread we chose includes the K band, J band, and V band at laser wavelengths of 2.12 μm, 1.25 μm, and 0.55 μm, respectively. For these wavelengths, the summed phase error variances in Eq. (9.124) lead to an effective phase variance at zenith angle zero given by

$$\sigma_{eff}^2 = 0.0122\left(\frac{k_1}{k}\right)^2 D^{5/3} = \begin{cases} 0.0122D^{5/3}\ (\text{K band}) \\ 0.0351D^{5/3}\ (\text{J band}) \\ 0.1811D^{5/3}\ (\text{V band}), \end{cases} \tag{9.138}$$

where we set $k = 2\pi/2.12$ μm (K band). The wave number k_1 is that for a different wavelength of interest. For example, we set $k_1 = 2\pi/1.25$ μm for the J band. Using Eq. (9.138), we plot the Strehl ratio in Fig. 9.14 for all three wavelengths, where we use the Strehl ratio approximation

$$\mathrm{SR} \approx \exp\left(-\sigma_{eff}^2\right). \tag{9.139}$$

The results shown in Fig. 9.14 are based on the assumption of weak irradiance fluctuations. From these results it is clear that larger telescope diameters and shorter wavelengths lead to a stronger cone effect.

9.5.2 LGS anisoplanatic error: nonzero NGS offset

We now consider the case where the NGS offset angle $\theta \neq 0$. This time the calculations will be based on exponential function approximations for the filter functions. By recalling the exponential function approximation to the piston filter given by Eq. (9.49), we have

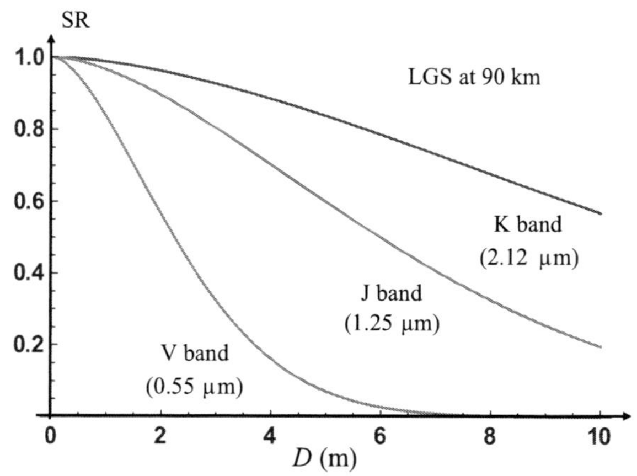

Figure 9.14 Strehl ratio for phase variance defined by Eq. (9.139) with piston and tilt removed as a function of telescope diameter D and offset angle $\theta = 0$.

$$\frac{J_1(\kappa Dz/2L)}{\kappa Dz/2L} \approx \frac{1}{2}\exp\left(-\frac{\kappa^2 D^2 z^2}{32L^2}\right), \tag{9.140}$$

which leads to the phase error variance with piston and tilt present:

$$\begin{aligned}\sigma_\phi^2 &\approx 2.606k^2 \int_0^L dz\, C_n^2(z) \int_0^\infty d\kappa\, \kappa^{-8/3}\left[1 - \exp\left(-\frac{\kappa^2 D^2 z^2}{32L^2}\right) J_0(\kappa\theta z)\right] \\ &\approx 0.48462k^2 D^{5/3}\, {}_1F_1\left(-\frac{5}{6};\ 1;\ -\frac{8\theta^2 L^2}{D^2}\right) \int_0^L C_n^2(z)\left(\frac{z}{L}\right)^{5/3} dz.\end{aligned} \tag{9.141}$$

For the piston term [Eq. (9.122)], the approximation of Eq. (9.140) yields

$$\begin{aligned}\sigma_P^2 \approx 1.303k^2 &\int_0^L dz\, C_n^2(z) \int_0^\infty d\kappa\kappa^{-8/3} \\ &\times \left[\exp\left(-\frac{\kappa^2 D^2}{16}\right) + \exp\left(-\frac{\kappa^2 D^2 \gamma^2}{16}\right)\right. \\ &\left. - 2J_0(\kappa\theta z)\exp\left(-\frac{\kappa^2 D^2}{32}\right)\exp\left(-\frac{\kappa^2 D^2\gamma^2}{32}\right)\right].\end{aligned} \tag{9.142}$$

By recalling ($D1$) and ($D2$) in Appendix D, the evaluation of the inside integral in Eq. (9.142) yields

$$\begin{aligned}\sigma_P^2 \approx &-0.4317k^2 D^{5/3} \int_0^L dz\, C_n^2(z)\left[\left[1 + \left(1 - \frac{z}{L}\right)^{5/3}\right.\right. \\ &\left.\left. -1.1225\left[1 + \left(1 - \frac{z}{L}\right)^2\right]^{5/6} {}_1F_1\left\{-\frac{5}{6};\ 1;\ -\frac{8\theta^2 z^2}{D^2[1 + (1 - z/L)^2]}\right\}\right]\right].\end{aligned} \tag{9.143}$$

Following along the same line for tilt, we start with the exponential function approximation,

$$\frac{J_2(\kappa D/2)}{\kappa D/2} \approx \frac{\kappa D}{16}\exp\left(-\frac{\kappa^2 D^2}{32}\right), \tag{9.144}$$

from which we obtain

$$\begin{aligned}\sigma_T^2 \approx\ &0.08137k^2 D^2 \int_0^L dz\, C_n^2(z) \int_0^\infty d\kappa\kappa^{-2/3}\left[\exp\left(-\frac{\kappa^2 D^2}{16}\right)\right. \\ &\left. + \gamma^2\exp\left(-\frac{\kappa^2 D^2\gamma^2}{16}\right) - 2J_0(\kappa\theta z)\gamma\exp\left(-\frac{\kappa^2 D^2}{32}\right)\exp\left(-\frac{\kappa^2 D^2\gamma^2}{32}\right)\right].\end{aligned} \tag{9.145}$$

Once again, we use ($D1$) and ($D2$) in Appendix D to deduce that

$$\sigma_T^2 \approx 0.3595 k^2 D^{5/3} \int_0^L dz\, C_n^2(z) \left[\left[1 + \left(1 - \frac{z}{L} \right)^{5/3} \right. \right. \\ \left. \left. - 2.2449 \left(1 - \frac{z}{L} \right) \left[1 + \left(1 - \frac{z}{L} \right)^2 \right]^{-1/6} {}_1F_1 \left\{ \frac{1}{6}; 1; -\frac{8\theta^2 z^2}{D^2 [1 + (1 - z/L)^2]} \right\} \right] \right]. \tag{9.146}$$

Shown in Fig. 9.15 are Strehl ratio reductions based on LGS anisoplanatism as a function of telescope aperture diameter D and offset angle θ in arcseconds. To avoid unrealistically low Strehl ratios with nonzero offset angles as defined by Eq. (9.139), we define the Strehl ratio this time by

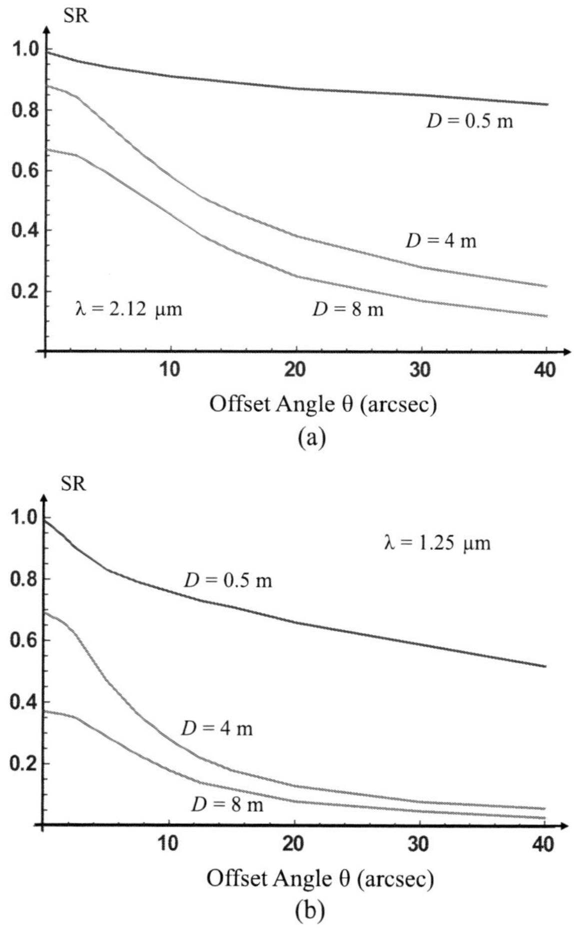

Figure 9.15 (a) Reduction in the K band Strehl ratio due to LGS anisoplanatism vs offset angle. (b) Reduction in the J band Strehl ratio due to LGS anisoplanatism vs offset angle.

$$\mathrm{SR} \approx \exp\left(-\,\sigma_{eff}^2\right) \approx \frac{1}{1+\sigma_{eff}^2}. \tag{9.147}$$

Three different telescope diameters D are represented, corresponding to 0.5, 4, and 8 m. The wavelengths are 2.12 μm (K band) in Fig. 9.15(a) and 1.25 μm (J band) in Fig. 9.15(b). To make the calculations, the zenith angle was chosen as zero and the HAP C_n^2 profile model was used with $M = 1, p = 4/3$, and ground-level refractive-index structure constant $C_n^2(h_0) = 1.7 \times 10^{-14}\ \mathrm{m}^{-2/3}$.

9.5.3 Focus anisoplanatism

In this section we consider a different form of focus anisoplanatism that is used in determining range to an airborne vehicle (pp. 186–188 of Ref. 6). In particular, we will perform an analysis similar to that in Section 9.2.1, but instead of looking at tilt variance, we consider focus variance from the vehicle wavefront return. The *total focus variance* can be found by using the filter function corresponding to the Zernike focus mode $Z_4[0, 2]$, i.e.,

$$F_4(\kappa) = 3\left[\frac{2J_3(\kappa D/2)}{\kappa D/2}\right]^2. \tag{9.148}$$

We will calculate both the total focus variance and the focus correlation function to determine the maximum correlation distance between two displaced apertures. For total focus variance, we have

$$\begin{aligned}\sigma_f^2 &= 3.908k^2 \int_0^L dz\, C_n^2(z) \int_0^\infty d\kappa\kappa^{-8/3}\left[\frac{2J_3(\kappa D/2)}{\kappa D/2}\right]^2\\ &= 4.924k^2 D^{5/3} \int_0^L dz\, C_n^2(z) \int_0^\infty x^{-14/3} J_3^2(x)\, dx,\end{aligned} \tag{9.149}$$

where we have made the change of variable $x = \kappa D/2$. By the use of ($D3$) in Appendix D, this last result reduces to

$$\begin{aligned}\sigma_f^2 &= 0.00983k^2 D^{5/3} \int_0^L C_n^2(z)\, dz\\ &= 0.0232\left(\frac{D}{r_0}\right)^{5/3},\end{aligned} \tag{9.150}$$

where $r_0 = \left[0.423k^2 \int_0^L C_n^2(z)\, dz\right]^{-3/5}$.

The variance difference between two displaced apertures by the distance d leads to the *focus correlation function*,

$$C_f(d) = 1.244k^2 \int_0^L dz\, C_n^2(z) \int_0^\infty d\kappa\kappa^{-8/3} \left[\frac{2J_3(\kappa D/2)}{\kappa D/2}\right]^2 \times \int_0^{2\pi} [1 - \cos(\kappa d \cos\phi)] d\phi$$
$$= 7.816k^2 \int_0^L dz\, C_n^2(z) \int_0^\infty d\kappa\kappa^{-8/3} \left[\frac{2J_3(\kappa D/2)}{\kappa D/2}\right]^2 [1 - J_0(\kappa d)]. \tag{9.151}$$

From the change of variable $x = \kappa D/2$, we find that

$$C_f(d) = 23.28\left(\frac{D}{r_0}\right)^{5/3} \int_0^\infty x^{-14/3} J_3^2(x)\left[1 - J_0\left(\frac{2xd}{D}\right)\right] dx$$
$$= 23.28\left(\frac{D}{r_0}\right)^{5/3} \left[\int_0^\infty x^{-14/3} J_3^2(x)\, dx - \int_0^\infty x^{-14/3} J_3^2(x) J_0\left(\frac{2xd}{D}\right) dx\right]. \tag{9.152}$$

Evaluation of these last two integrals leads to

$$\int_0^\infty x^{-14/3} J_3^2(x)\, dx = \frac{\Gamma\left(\frac{7}{6}\right)\Gamma\left(\frac{7}{3}\right)}{2\sqrt{\pi}\,\Gamma\left(\frac{17}{6}\right)\Gamma\left(\frac{35}{6}\right)} = 0.00199, \tag{9.153}$$

$$\int_0^\infty x^{-14/3} J_3^2(x) J_0\left(\frac{2xd}{D}\right) dx = \frac{D}{2d\sqrt{\pi}} G_{33}^{12}\left(\frac{D^2}{d^2}\middle|\begin{matrix}\frac{1}{2}, & -\frac{11}{6}, & \frac{1}{2}\\ \frac{2}{3}, & -\frac{7}{3}, & -\frac{16}{3}\end{matrix}\right), \tag{9.154}$$

where we have relied on both ($D3$) and ($D4$) in Appendix D. As we have done before, we choose to leave the result in terms of the Meijer G-function in Eq. (9.154). Therefore, Eq. (9.152) takes the form

$$C_f(d) = 23.28\left(\frac{D}{r_0}\right)^{5/3}\left[0.00199 - \frac{D}{2d\sqrt{\pi}} G_{33}^{12}\left(\frac{D^2}{d^2}\middle|\begin{matrix}\frac{1}{2}, & -\frac{11}{6}, & \frac{1}{2}\\ \frac{2}{3}, & -\frac{7}{3}, & -\frac{16}{3}\end{matrix}\right)\right]. \tag{9.155}$$

Finally, by forming the ratio of Eqs. (9.155) and (9.150), we obtain the scaled correlation expression,

$$\frac{C_f(d)}{\sigma_f^2} = 2\left[1 - 141.34\left(\frac{D}{d}\right) G_{33}^{12}\left(\frac{D^2}{d^2}\middle|\begin{matrix}\frac{1}{2}, & -\frac{11}{6}, & \frac{1}{2}\\ \frac{2}{3}, & -\frac{7}{3}, & -\frac{16}{3}\end{matrix}\right)\right]. \tag{9.156}$$

On the other hand, if we use the Gaussian filter function approximation,[9]

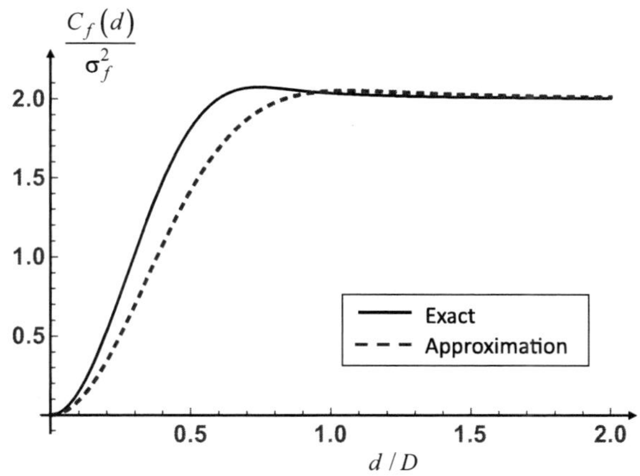

Figure 9.16 Normalized correlation function defined by Eq. (9.156) and the approximation of Eq. (9.158) as a function of scaled separation distance *d* of two apertures, each of diameter *D*.

$$F_4(\kappa) = 3\left[\frac{2J_3(\kappa D/2)}{\kappa D/2}\right]^2 \approx \frac{\kappa^4 D^2}{1024} e^{-\kappa^2 D^2/16}, \tag{9.157}$$

we are led to the simpler expression

$$\frac{C_f(d)}{\sigma_f^2} \approx 2\left[1 - {}_1F_1\left(\frac{7}{6};\ 1;\ -\frac{4d^2}{D^2}\right)\right]. \tag{9.158}$$

We show a plot of the normalized correlation function of focus anisoplanatism [Eq. (9.156)] in Fig. 9.16 as a function of the scaled separation distance. For $d > D$, the term with the Meijer *G*-function in Eq. (9.156) is essentially zero, and therefore the limiting value of Eq. (9.156) is "2." This means that the focus components of the two apertures become uncorrelated at this separation distance. For comparison, we also show the same result [Eq. (9.158)] based on the Gaussian filter function approximation for focus defined by Eq. (9.157). Here it is clear that the Gaussian filter approximation for the focus Zernike mode is not as accurate as it is for piston.

9.6 Discussion

In this chapter we have developed many new mathematical models concerning imaging through turbulence, including new models arising from using the *ABCD* ray matrix approach. The *ABCD* method permits the development of simpler algebraic models for performance measures, such as long-term and short-term exposures of the MTF, Strehl ratio, and resolution. We have also derived several new models concerning anisoplanatism, some with outer-scale effects.

Many of the mathematical models developed in this chapter involve generalized hypergeometric functions. In some cases, we presented simplified algebraic approximations, but not always. However, there were several instances where the math models led to the same confluent hypergeometric functions. Using the method of Chapter 6, two of the most frequently occurring ${}_1F_1$ functions can often be replaced by the approximations

$$\begin{aligned} {}_1F_1\left(\tfrac{1}{6};\ 1;\ -x^2\right) &\approx (1+1.8x^2)^{-1/6}, \\ {}_1F_1\left(\tfrac{1}{6};\ 2;\ -x^2\right) &\approx (1+0.74x^2)^{-1/6}. \end{aligned}$$

Because most of this chapter is concerned with uplink/downlink imaging issues, it becomes necessary to consider the use of a C_n^2 profile model to complete the calculations indicated in the various mathematical models. One of the often used C_n^2 profile models as a function of altitude h is the *Hufnagel–Valley* (HV) *model*:

$$\begin{aligned} C_n^2(h) &= 0.00594\left(\frac{w}{27}\right)^2\left(\frac{h}{10^5}\right)^{10}\exp\left(-\frac{h}{1000}\right) \\ &\quad + 2.7\times 10^{-16}\exp\left(-\frac{h}{1500}\right) + A\exp\left(-\frac{h}{100}\right), \end{aligned} \tag{9.159}$$

where A represents C_n^2 near ground level and w is high-altitude wind speed. By choosing $A = 1.7\times 10^{-14}\ \mathrm{m}^{-2/3}$ and $w = 21$ m/s, the model is referred to in the literature as the HV-5/7 model. A modification of the HV model is the Hufnagel–Andrews–Phillips model, also called the *HAP model* (pp. 12–13 of Ref. 4),

$$\begin{aligned} C_n^2(h) &= M\left[0.00594\left(\frac{w}{27}\right)^2\left(\frac{h+h_g}{10^5}\right)^{10}\exp\left(-\frac{h+h_g}{1000}\right)\right. \\ &\quad \left. + 2.7\times 10^{-16}\exp\left(-\frac{h+h_g}{1500}\right)\right] + C_n^2(h_0)\left(\frac{h_0}{h}\right)^p, \qquad h > h_0, \end{aligned} \tag{9.160}$$

where h_g is the height of the ground above sea level (with a recommended default value of 5–10 m), h_0 represents a reference height above ground for the transmitter/receiver, $C_n^2(h_0)$ is the average refractive-index structure parameter at h_0, and w is the high-altitude average wind speed (typically chosen as 21 m/s). The parameter M (on the order of unity) is an average value of the random background turbulence effect above ground level, and p represents a power law that varies over the course of a day. For the calculations performed in this chapter, we choose $p = 4/3$ and $M = 1$. To be consistent with the parameters of the HV-5/7 model, we also set $C_n^2(h_0) = 1.7 \times 10^{-14}\ \mathrm{m}^{-2/3}$.

A modification of the HAP model has been introduced by Stotts and Andrews[14] that better matches certain atmospheric measurements as a function of altitude. This modification takes the form

$$\begin{aligned} C_n^2(h) = M\bigg[&1.04 \times 10^{-3}\left(\frac{w}{27}\right)^2\left(\frac{h+h_g}{10^5}\right)^{10}\exp\left(-\frac{h+h_g}{1200}\right) \\ &+2.7 \times 10^{-16}\exp\left(-\frac{h+h_g}{1700}\right)\bigg] + C_n^2(h_0)\left(\frac{h_0}{h}\right)^{4/3}, \quad h > h_0. \end{aligned} \tag{9.161}$$

Although the 4/3 power law was utilized in Ref. 14, it can be replaced by the parameter p as in Eq. (9.160) to reflect different times of the day. For nighttime with $p = 2/3$, the last term in (9.160) and (9.161) should be multiplied by $\exp(-h/1000)$.[15]

Many slant-path analyses are performed with a flat-Earth model. As long as the zenith angle is relatively small, the flat-Earth model should work quite well. However, for low elevation angles, it is best to consider the Earth as a round ball having a radius of approximately $R = 6371 \times 10^3$ m. In this case, the height h of the beam above the ground reference height h_0 at zenith angle ζ is described by

$$h(z) = \sqrt{(R+h_0)^2 + 2z(R+h_0)\cos\zeta + z^2} - (R+h_0), \quad h > h_0, \tag{9.162}$$

where z denotes propagation distance in meters along the slant path. To convert from the flat-Earth model to a round-Earth model, one can simply replace h with $h(z)$ defined above and use the relationships

$$\begin{gathered} L - z_0 = \frac{H - h_0}{\cos\zeta}, \quad z_0 = \frac{h_0}{\cos\zeta}, \quad z = \frac{h}{\cos\zeta}, \\ \frac{h - h_0}{H - h_0} = \frac{z - z_0}{L - z_0}. \end{gathered} \tag{9.163}$$

Here, H refers to the altitude of an aircraft or satellite.

References

1. J. E. Pearson, Ed., *Selective Papers on Adaptive Optics for Atmospheric Compensation*, SPIE Milestone Series MS-92, SPIE Press, Bellingham, Washington (1994).
2. M. C. Roggemann and B. Welsh, *Imaging Through Turbulence*, CRC Press, Boca Raton, Florida (1996).

3. L. C. Andrews and R. L. Phillips, *Laser Beam Propagation through Random Media*, 2nd Edition, SPIE Press, Bellingham, Washington (2005) [doi: 10.1117/3.626196].
4. L. C. Andrews and M. K. Beason, *Laser Beam Propagation in Random Media: New and Advanced Topics*, SPIE Press, Bellingham, Washington (2023) [doi: 10.1117/3.2643989].
5. D. L. Fried, "Optical resolution through a randomly inhomogeneous medium for very long and very short exposures," *J. Opt. Soc. Am.* **56**, 1372–1379 (1966).
6. R. J. Sasiela, *Electromagnetic Wave Propagation in Turbulence:Evaluation and Application of Mellin Transforms*, 2nd Edition, SPIE Press, Bellingham, Washington (2007) [doi: 10.1117/3.741372].
7. L. B. Stotts and L. C. Andrews, "Adaptive optics model characterizing turbulence mitigation for free space optical communications link budgets," *Opt. Exp.* **29**(13), 20307–20321 (2021).
8. R. J. Noll, "Zernike polynomials and atmospheric turbulence," *J. Opt. Soc. Am.* **66**(3), 207–211 (1976).
9. M. Assad and L. C. Andrews, "On the use of Gaussian filter functions for adaptive optics," *Proc. SPIE* **6457**, 64570Q (2007) [doi: 10.1117/12.699677].
10. S. Gladysz, M. Segel, C. Eisele, R. Barros, and E. Sucher, "Estimation of turbulence strength, anisotropy, outer scale and spectral slope from an LED array," *Proc. SPIE* **9614**, 961402 (2015) [doi: 10.1117/12.2191287].
11. S. Gladysz, "Nearly complete characterization of optical turbulence with an LED array," in *Imaging and Applied Optics 2017 (3D, AIO, COSI, IS, MATH, pcAOP)*, OSA Technical Digest (online), paper PW1D.1 (2017).
12. F. Rigaut and E. Gendron, "Laser guide star in adaptive optics: The tilt determination problem," *Astron. Astrophys* **261**, 677–684 (1992).
13. M. A. van Dam, R. J. Sasiela, A. H. Bouchez, D. Le Mignant, R. D. Campbell, J. C. Y. Chin, S. K. Hartman, E. M. Johansson, R. E. Lafon, P. J. Stomski, R. E. Lafon, Jr., D. M. Summers, and P. L. Wizinowich "Angular anisoplanatism in laser guide star adaptive optics," *Proc. SPIE* **6272**, 627231 (2006) [doi: 10.1117/12.669737].
14. L. B. Stotts and L. C. Andrews, "Improving the Hufnagel-Andrews-Phillips refractive index structure parameter model using turbulent intensity," *Opt. Express* **31**(9), 14265–14277 (2023).
15. A. Knoedler and F. Moll, "Atmospheric Turbulence Statistics and Profile Modeling. Local to DLR Oberpfaffenhofen," COAT-2019 - workshop (Communications and Observations through Atmospheric Turbulence: characterization and mitigation), ONERA, Châtillon, France 10.34693/COAT2019-S1-001. hal-03206072 (2019).

Chapter 10
Non-Kolmogorov Turbulence

10.0 Introduction

Since the early 1990s, there has been considerable evidence of upper atmospheric *non-Kolmogorov anisotropic turbulence*.[1–3] This has led to numerous theoretical analyses and experimental measurements, including those more recently near ground level that also show evidence of anisotropic turbulence.[4–22]

In the conventional Kolmogorov model of turbulence, the fluctuations of the index of refraction are assumed to be statistically homogeneous and isotropic, and there is a specific mathematical form for the power spectral density of the index-of-refraction fluctuations. Development of the turbulence theory of passive scalar transfer has shown that although the conventional Kolmogorov spectrum model with a $\kappa^{-11/3}$ power-law index is generally correct near the ground (within the inertial subrange), it constitutes only one part of the more general behavior of passive scalar transfer in a turbulent flow—deviations from the conventional Kolmogorov model are possible. A number of studies involving *isotropic* non-Kolmogorov turbulence have emerged in recent years along with those for *anisotropic* turbulence conditions. Some of the published results on isotropic and anisotropic turbulence models are summarized in Chapter 9 of Ref. 23.

In the current chapter, we first introduce both isotropic and anisotropic non-Kolmogorov spatial power spectrum models of atmospheric turbulence that are commonly used by researchers. These spectrum models lead to generalizations of theoretical results presented in Chapters 8 and 9, some of which are new and go beyond what has been previously published.

10.1 Non-Kolmogorov Spectral Models

Conventional Kolmogorov turbulence spectral models were introduced in Chapter 8. By non-Kolmogorov turbulence we mean a spectral index $\kappa^{-\alpha}$ that replaces the $\kappa^{-11/3}$ power law in the conventional Kolmogorov spectrum. Naturally, this change in spectral index has an effect on all statistical quantities involving optical/IR wave propagation through the atmosphere. In addition to a spectral index change, anisotropic parameters might also be introduced into the model. In the following analysis, we will base most of our results on the isotropic non-Kolmogorov spectral model but also include anisotropic conditions in some cases.

10.1.1 Isotropic models

The basic spectrum model that is cited most often in discussing isotropic non-Kolmogorov turbulence is

$$\Phi_n(\kappa) = A(\alpha)\tilde{C}_n^2\kappa^{-\alpha}, \quad 3 < \alpha < 4, \tag{10.1}$$

where $A(\alpha)$ is a dimensionless parameter defined by

$$A(\alpha) = \frac{1}{4\pi^2}\Gamma(\alpha - 1)\cos\left(\frac{\alpha\pi}{2}\right); \quad A\left(\frac{11}{3}\right) = 0.0330054. \tag{10.2}$$

The index-of-refraction parameter $\tilde{C}_n^2$ with units of $m^{3-\alpha}$ is a generalized version of the structure function parameter C_n^2. To relate $\tilde{C}_n^2$ to the conventional structure function parameter, it is often assumed that[23–29] $\tilde{C}_n^2 = \beta\, C_n^2$, where $\beta = \beta(\alpha)$ is a function of α with units $m^{11/3-\alpha}$.

To include the inner scale and outer scale in the model (10.1), we will consider a form of the exponential spectrum [recall Eq. (8.4)],

$$\begin{aligned}\Phi_n(\kappa) &= A(\alpha)\tilde{C}_n^2\kappa^{-\alpha}\exp\left(-\frac{\kappa^2}{\kappa_m^2(\alpha)}\right)\left[1 - \exp\left(-\frac{\kappa^2}{\kappa_0^2}\right)\right] \\ &= A(\alpha)\tilde{C}_n^2\kappa^{-\alpha}\left[\exp\left(-\frac{\kappa^2}{\kappa_m^2(\alpha)}\right) - \exp\left(-\frac{\kappa^2}{\kappa_{m0}^2}\right)\right], \quad 3 < \alpha < 4,\end{aligned} \tag{10.3}$$

where $\kappa_{m0}^2 = \kappa_0^2\kappa_m^2/(\kappa_m^2 + \kappa_0^2)$. However, in the analysis below, we will follow the approach used in Chapter 8 and set $\kappa_{m0} \approx \kappa_0$ under the assumption that $\kappa_m^2 \gg \kappa_0^2$. A suggested choice of scaling constants for $\kappa_m = \kappa_m(\alpha)$ and κ_0 was given in Refs. 9, 28, and in Ref. 23, p. 302.

10.1.2 Anisotropic models

Anisotropy can be modeled by treating the power spectrum as directionally dependent. Following the approach often used in the literature, a general anisotropic spectral model is defined by[13]

$$\Phi_n(\kappa_x, \kappa_y, 0) = \frac{A(\alpha)\tilde{C}_{n,eff}^2\,\mu_x\mu_y}{(\mu_x^2\kappa_x^2 + \mu_y^2\kappa_y^2)^{\alpha/2}}, \quad 3 < \alpha < 4, \tag{10.4}$$

where μ_x, μ_y are the anisotropy coefficients relating inversely to the relative strength of turbulence in the x and y transverse directions. With the assumption of propagation in the positive z-direction, we are emphasizing delta correlation in the direction of beam propagation with $\kappa_z = 0$ in Eq. (10.4).

In general, the spectrum model (10.4) will lead to different statistical values in the horizontal and vertical directions of the derived mathematical models. For that reason, we have designated the "effective" refractive-index structure parameter in Eq. (10.4) as $\tilde{C}^2_{n,eff}$ because it is believed that although the anisotropic structure parameters in the vertical and horizontal directions are not the same, they are coupled. It has been conjectured that $\tilde{C}^2_{n,eff}$ is related to the isotropic value $\tilde{C}^2_n$ by the empirical relationship (pp. 324–325 in Ref. 23)

$$\tilde{C}^2_{n,eff} = \left(\frac{2}{\mu_x^2 + \mu_y^2}\right)^{\eta} \tilde{C}^2_n, \tag{10.5}$$

where η is a parameter to be determined. In estimating the long-term beam size of a Gaussian-beam wave, the authors of Ref. 23 have suggested that $\eta = \frac{1}{2}$ for a collimated beam and $\eta = \frac{1}{4}$ in the case of a focused beam. Although not widely tested, this relationship has led to excellent comparisons with computer simulation data for collimated and focused beam size in the x- and y-directions at a Rx aperture.

To consider equal anisotropy conditions in the transverse directions along the path, we follow the approach introduced by Kon[4] and others.[7,8,12] Namely, we modify the spectrum model (10.4) by setting $\mu_x = \mu_y = \mu$, which leads to

$$\Phi_n(\kappa) = A(\alpha)\tilde{C}^2_{n,eff}\mu^{2-\alpha}\kappa^{-\alpha}, \quad 3 < \alpha < 4. \tag{10.6}$$

Although we don't include it here, inner- and outer-scale modifications have also been introduced for anisotropic spectrum models.[28,30] Last, for $\alpha = 11/3$ and $\mu_x = \mu_y = \mu = 1$, it follows that Eqs. (10.4) and (10.6) reduce to the conventional isotropic Kolmogorov spectrum model defined by Eq. (8.1).

10.1.3 Index-of-refraction structure function

To begin the analysis of non-Kolmogorov turbulence, we first develop the index-of-refraction structure function that was previously defined by Eq. (8.10) for conventional Kolmogorov turbulence. In doing so, we base the analysis here only on isotropic non-Kolmogorov turbulence associated with the exponential spectrum (10.3), taking into account both inner-scale and outer-scale effects.

Based on the spectrum model of Eq. (10.3), the *index-of-refraction structure function* is defined by

$$
\begin{aligned}
D_n(R) &= 8\pi A(\alpha)\tilde{C}_n^2 \int_0^\infty \kappa^{2-\alpha}\left[\exp\left(-\frac{\kappa^2}{\kappa_m^2}\right) - \exp\left(-\frac{\kappa^2}{\kappa_0^2}\right)\right]\left(1 - \frac{\sin\kappa R}{\kappa R}\right) d\kappa \\
&= 8\pi A(\alpha)\tilde{C}_n^2 \sum_{n=1}^{\infty} \frac{(-1)^{n-1}R^{2n}}{(2n+1)!} \int_0^\infty \kappa^{2n+2-\alpha}\left[\exp\left(-\frac{\kappa^2}{\kappa_m^2}\right) - \exp\left(-\frac{\kappa^2}{\kappa_0^2}\right)\right] d\kappa \\
&= 4\pi A(\alpha)\tilde{C}_n^2 \left[\kappa_m^{3-\alpha} \sum_{n=1}^{\infty} \frac{(-1)^{n-1}(\kappa_m R)^{2n}}{(2n+1)!}\Gamma\left(n + \frac{3-\alpha}{2}\right)\right. \\
&\quad \left. - \kappa_0^{3-\alpha} \sum_{n=1}^{\infty} \frac{(-1)^{n-1}(\kappa_0 R)^{2n}}{(2n+1)!}\Gamma\left(n + \frac{3-\alpha}{2}\right)\right],
\end{aligned} \tag{10.7}
$$

where we have written the sine function in a power series and integrated termwise using ($D2$) from Appendix D. Rewriting the last expressions in Eq. (10.7) in terms of Pochhammer symbols from Appendix A, we identify each series plus or minus unity as a confluent hypergeometric function,

$$
\begin{aligned}
D_n(R) &= 4\pi A(\alpha)\Gamma\left(\frac{3-\alpha}{2}\right)\tilde{C}_n^2 \left\{ \kappa_m^{3-\alpha} \sum_{n=1}^{\infty} \frac{(-1)^{n-1}\left(\frac{3-\alpha}{2}\right)_n}{\left(\frac{3}{2}\right)_n n!}\left(\frac{\kappa_m^2 R^2}{4}\right)^n \right. \\
&\quad \left. - \kappa_0^{3-\alpha} \sum_{n=1}^{\infty} \frac{(-1)^{n-1}\left(\frac{3-\alpha}{2}\right)_n}{\left(\frac{3}{2}\right)_n n!}\left(\frac{\kappa_0^2 R^2}{4}\right)^n \right\} \\
&= 4\pi A(\alpha)\Gamma\left(\frac{3-\alpha}{2}\right)\tilde{C}_n^2 \left\{ \kappa_m^{3-\alpha}\left[1 - {}_1F_1\left(\frac{3-\alpha}{2}; \frac{3}{2}; -\frac{\kappa_m^2 R^2}{4}\right)\right]\right. \\
&\quad \left. - \kappa_0^{3-\alpha}\left[1 - {}_1F_1\left(\frac{3-\alpha}{2}; \frac{3}{2}; -\frac{\kappa_0^2 R^2}{4}\right)\right]\right\}, \quad R \geq 0.
\end{aligned} \tag{10.8}
$$

Finally, for vanishing inner scale ($\kappa_m \to \infty$) and infinite outer scale ($\kappa_0 \to 0$), Eq. (10.8) is reduced to the inertial range form

$$
\begin{aligned}
D_n(R) &= -\frac{\Gamma(\alpha-1)}{\sqrt{\pi}\,2^{\alpha-2}\Gamma\left(\frac{\alpha}{2}\right)}\Gamma\left(\frac{3-\alpha}{2}\right)\cos\left(\frac{\alpha\pi}{2}\right)\tilde{C}_n^2 R^{\alpha-3} \\
&= \tilde{C}_n^2 R^{\alpha-3}, \quad l_0 \ll R \ll L_0, \quad 3 < \alpha < 4.
\end{aligned} \tag{10.9}
$$

As a final comment, we note that under anisotropic turbulence conditions described by Eq. (10.4), the inertial range form of the refractive-index structure function becomes[13]

$$D_n(\mathbf{R}) = \tilde{C}^2_{n,eff}\left(\frac{x^2}{\mu_x^2}+\frac{y^2}{\mu_y^2}+z^2\right)^{(\alpha-3)/2}, \quad l_0 \ll R \ll L_0, \quad 3<\alpha<4,$$

where $\mathbf{R} = \left(\frac{x}{\mu_x}, \frac{y}{\mu_y}, z\right)$.

10.2 Generalized Rytov Variance

Because it appears in many of the statistical models to be developed, let us derive the *generalized Rytov variance* with use of the anisotropic spectral model of Eq. (10.4). Assuming a constant $\tilde{C}^2_{n,eff}$, this leads to

$$\begin{aligned}\tilde{\sigma}^2_R(\alpha;\mu_x,\mu_y) &= 4\pi k^2 L\int_0^1\int_{-\infty}^{\infty}\int_{-\infty}^{\infty}\Phi_n(\kappa_x,\kappa_y,0)\left[1-\cos\left(\frac{L\kappa^2\xi}{k}\right)\right]d\kappa_x\,d\kappa_y\,d\xi\\ &= 4\pi k^2 L\int_0^{2\pi}\int_0^1\int_0^{\infty} q\Phi_n(q,0)\\ &\quad\times\left\{1-\cos\left[\frac{Lq^2\xi}{k}\left(\frac{\cos^2\theta}{\mu_x^2}+\frac{\sin^2\theta}{\mu_y^2}\right)\right]\right\}\frac{dq\,d\xi\,d\theta}{\mu_x\mu_y},\end{aligned} \tag{10.10}$$

where $\xi = 1 - z/L$ and $\kappa^2 = \kappa_x^2 + \kappa_y^2$. In the second step, we changed from a stretched coordinate spectral model to one that is isotropic by imposing the following:

$$\begin{aligned}\kappa_x &= \frac{q_x}{\mu_x} = \frac{q\cos\theta}{\mu_x}, \quad \kappa_y = \frac{q_y}{\mu_y} = \frac{q\sin\theta}{\mu_y}, \quad q = \sqrt{q_x^2+q_y^2};\\ d\kappa_x\,d\kappa_y &= \frac{dq_x\,dq_y}{\mu_x\mu_y} = \frac{q\,dq\,d\theta}{\mu_x\mu_y};\\ \Phi_n(q,0) &= \frac{A(\alpha)\tilde{C}^2_{n,eff}\,\mu_x\mu_y}{(\mu_x^2\kappa_x^2+\mu_y^2\kappa_y^2)^{\alpha/2}} = \frac{A(\alpha)\tilde{C}^2_{n,eff}\,\mu_x\mu_y}{q^\alpha}.\end{aligned} \tag{10.11}$$

By introducing the power spectrum model in the last line of Eq. (10.11) into Eq. (10.10), we obtain

$$\begin{aligned}\tilde{\sigma}^2_R(\alpha;\mu_x,\mu_y) &= 4\pi A(\alpha)\tilde{C}^2_{n,eff}\,k^2 L\int_0^{2\pi}\int_0^1\int_0^{\infty} q^{1-\alpha}\\ &\quad\times\left\{1-\cos\left[\frac{Lq^2\xi}{k}\left(\frac{\cos^2\theta}{\mu_x^2}+\frac{\sin^2\theta}{\mu_y^2}\right)\right]\right\}dq\,d\xi\,d\theta.\end{aligned} \tag{10.12}$$

It is useful here to write the cosine function as $\cos x = \mathrm{Re}(e^{-ix})$, from which the evaluation of the inside integral in Eq. (10.12) leads to

$$\tilde{\sigma}_R^2(\alpha;\, \mu_x,\, \mu_y) = -2\pi A(\alpha)\Gamma\left(1-\frac{\alpha}{2}\right)\sin\left(\frac{\alpha\pi}{4}\right)\tilde{C}^2_{n,eff}\, k^2 L\left(\frac{L}{k}\right)^{\alpha/2-1}$$
$$\times \int_0^1 \xi^{\alpha/2-1} d\xi \int_0^{2\pi}\left(\frac{\cos^2\theta}{\mu_x^2}+\frac{\sin^2\theta}{\mu_y^2}\right)^{\alpha/2-1} d\theta. \tag{10.13}$$

We have solved the integral over q by using the same idea illustrated in Eqs. (8.100) through (8.102) and recognized that $\mathrm{Re}(i^{\alpha/2-1}) = \sin(\alpha\pi/4)$. Now completing the last two integrations, we find that

$$\tilde{\sigma}_R^2(\alpha;\, \mu_x,\, \mu_y) = B_1(\alpha)\tilde{C}^2_{n,eff} k^{3-\alpha/2} L^{\alpha/2} \mu_y^{2-\alpha}\, {}_2F_1\left(\frac{1}{2},\, \frac{2-\alpha}{2};\, 1;\, 1-\frac{\mu_y^2}{\mu_x^2}\right), \tag{10.14}$$

where

$$B_1(\alpha) = -\frac{2\Gamma(\alpha-1)}{\alpha}\Gamma\left(1-\frac{\alpha}{2}\right)\sin\left(\frac{\alpha\pi}{4}\right)\cos\left(\frac{\alpha\pi}{2}\right) \tag{10.15}$$

and

$$\int_0^{2\pi}\left(\frac{\cos^2\theta}{\mu_x^2}+\frac{\sin^2\theta}{\mu_y^2}\right)^{\alpha/2-1} d\theta = 2\pi\mu_y^{2-\alpha}\, {}_2F_1\left(\frac{1}{2},\, \frac{2-\alpha}{2};\, 1;\, 1-\frac{\mu_y^2}{\mu_x^2}\right). \tag{10.16}$$

For the special case of isotropic turbulence where $\mu_x = \mu_y = 1$, Eq. (10.14) reduces to

$$\tilde{\sigma}_R^2(\alpha) = B_1(\alpha)\tilde{C}_n^2 k^{3-\alpha/2} L^{\alpha/2}, \tag{10.17}$$

and for $\alpha = 11/3$, Eq. (10.17) further reduces to $\sigma_R^2 = 1.23 C_n^2 k^{7/6} L^{11/6}$.

10.3 Wave Structure Function: Isotropic Turbulence

In the analysis below we will parallel our treatment of the *wave structure function* (WSF) that was presented in Section 8.3.

10.3.1 Plane wave

Starting with the spectrum model (10.3), the WSF of an infinite *plane wave* is defined by

$$D_{pl}(\rho,\alpha) = 8\pi^2 A(\alpha)\tilde{C}_n^2 k^2 L \int_0^\infty \kappa^{1-\alpha}\left[\exp\left(-\frac{\kappa^2}{\kappa_m^2}\right) - \exp\left(-\frac{\kappa^2}{\kappa_0^2}\right)\right][1-J_0(\kappa\rho)]d\kappa. \tag{10.18}$$

If we now replace the Bessel function with its Meijer *G*-function representation (*C*9) in Appendix C, we obtain with the help of (*D*2) and (*D*6) in Appendix D

$$
\begin{aligned}
D_{pl}(\rho,\alpha) &= 8\pi^2 A(\alpha)\tilde{C}_n^2 k^2 L \int_0^\infty \kappa^{1-\alpha}\left[\exp\left(-\frac{\kappa^2}{\kappa_m^2}\right)-\exp\left(-\frac{\kappa^2}{\kappa_0^2}\right)\right] \\
&\quad\times\left[1-G_{02}^{10}\left(\frac{\kappa^2\rho^2}{4}\middle|\, 0,\,0\right)\right]d\kappa \\
&= 4\pi^2 A(\alpha)\tilde{C}_n^2 k^2 L\left\{\kappa_m^{2-\alpha}\left[\Gamma\left(1-\frac{\alpha}{2}\right)-G_{12}^{11}\left(\frac{\kappa_m^2\rho^2}{4}\middle|\begin{matrix}\frac{\alpha}{2}\\ 0,\,0\end{matrix}\right)\right]\right. \\
&\quad\left.-\kappa_0^{2-\alpha}\left[\Gamma\left(1-\frac{\alpha}{2}\right)-G_{12}^{11}\left(\frac{\kappa_0^2\rho^2}{4}\middle|\begin{matrix}\frac{\alpha}{2}\\ 0,\,0\end{matrix}\right)\right]\right\}.
\end{aligned}
\tag{10.19}
$$

Next, by using ($C13$) in Appendix C,we obtain the general expression,

$$
\begin{aligned}
D_{pl}(\rho,\alpha) &= 4\pi^2 A(\alpha)\Gamma\left(1-\frac{\alpha}{2}\right)\tilde{C}_n^2 k^2 L\left\{\kappa_m^{2-\alpha}\left[1-{}_1F_1\left(1-\frac{\alpha}{2};\ 1;\ -\frac{\kappa_m^2\rho^2}{4}\right)\right]\right. \\
&\quad\left.-\kappa_0^{2-\alpha}\left[1-{}_1F_1\left(1-\frac{\alpha}{2};\ 1;\ -\frac{\kappa_0^2\rho^2}{4}\right)\right]\right\}.
\end{aligned}
\tag{10.20}
$$

For $\kappa_0 = 0$ and $l_0 = 0$ (i.e., $\kappa_m \to \infty$), Eq. (10.20) reduces to

$$
D_{pl}(\rho,\alpha) = -\frac{\Gamma(\alpha-1)}{2^{\alpha-2}\Gamma(\alpha/2)}\Gamma\left(1-\frac{\alpha}{2}\right)\cos\left(\frac{\alpha\pi}{2}\right)\tilde{C}_n^2 k^2 L\rho^{\alpha-2}, \quad l_0 \ll \rho \ll L_0.
\tag{10.21}
$$

10.3.2 Spherical wave

In the case of a *spherical wave*, the corresponding WSF is defined by

$$
\begin{aligned}
D_{sph}(\rho,\alpha) &= 8\pi^2 A(\alpha)\tilde{C}_n^2 k^2 L \int_0^1\int_0^\infty \kappa^{1-\alpha}\left[\exp\left(-\frac{\kappa^2}{\kappa_m^2}\right)-\exp\left(-\frac{\kappa^2}{\kappa_0^2}\right)\right] \\
&\quad\times\left[1-J_0(\kappa\xi\rho)\right]d\kappa\, d\xi \\
&= 8\pi^2 A(\alpha)\tilde{C}_n^2 k^2 L \int_0^\infty \kappa^{1-\alpha}\left[\exp\left(-\frac{\kappa^2}{\kappa_m^2}\right)-\exp\left(-\frac{\kappa^2}{\kappa_0^2}\right)\right] \\
&\quad\times\left[1-\frac{1}{2}\,G_{13}^{11}\left(\frac{\kappa^2\rho^2}{4}\middle|\begin{matrix}\frac{1}{2}\\ 0,\,0,\,-\frac{1}{2}\end{matrix}\right)\right]d\kappa,
\end{aligned}
\tag{10.22}
$$

where we have performed the integration on ξ using ($C9$) in Appendix C and ($D9$) in Appendix D. Completing the integration with ($D2$) and ($D6$) in Appendix D gives us

$$D_{sph}(\rho,\alpha) = 4\pi^2 A(\alpha)\tilde{C}_n^2 k^2 L\left\{\left[\Gamma\left(1-\frac{\alpha}{2}\right) - \frac{1}{2}G_{23}^{12}\left(\frac{\kappa_m^2\rho^2}{4}\middle|\begin{matrix}\frac{\alpha}{2},\ \frac{1}{2}\\ 0,\ 0,\ -\frac{1}{2}\end{matrix}\right)\right]\right.$$
$$\left. - \kappa_0^{2-\alpha}\left[\Gamma\left(1-\frac{\alpha}{2}\right) - \frac{1}{2}G_{23}^{12}\left(\frac{\kappa_0^2\rho^2}{4}\middle|\begin{matrix}\frac{\alpha}{2},\ \frac{1}{2}\\ 0,\ 0,\ -\frac{1}{2}\end{matrix}\right)\right]\right\}, \tag{10.23}$$

Now referring to ($C32$) in Appendix C, we get

$$D_{sph}(\rho,\alpha) = 4\pi^2 A(\alpha)\Gamma\left(1-\frac{\alpha}{2}\right)\tilde{C}_n^2 k^2 L\left\{\kappa_m^{2-\alpha}\left[1 - {}_2F_2\left(1-\frac{\alpha}{2},\frac{1}{2};1,\frac{3}{2};-\frac{\kappa_m^2\rho^2}{4}\right)\right]\right.$$
$$\left. - \kappa_0^{2-\alpha}\left[1 - {}_2F_2\left(1-\frac{\alpha}{2},\frac{1}{2};1,\frac{3}{2};-\frac{\kappa_0^2\rho^2}{4}\right)\right]\right\}. \tag{10.24}$$

For a spherical wave with $\kappa_0 = 0$ and $l_0 = 0$, the WSF (10.24) takes the form

$$D_{sph}(\rho,\alpha) = -\frac{\Gamma(\alpha-1)}{2^{\alpha-2}(\alpha-1)\Gamma(\alpha/2)}\Gamma\left(1-\frac{\alpha}{2}\right)\cos\left(\frac{\alpha\pi}{2}\right)\tilde{C}_n^2 k^2 L\rho^{\alpha-2},$$
$$l_0 \ll \rho \ll L_0, \tag{10.25}$$

which is the same as Eq. (10.21) except for the factor of $\alpha - 1$ in the denominator.

In Fig. 10.1 we plot the modulus of the complex *degree of coherence* for a spherical wave, $\mathrm{DOC}(\rho) = \exp\left[-\frac{1}{2}D_{sph}(\rho,\alpha)\right]$, with $\alpha = 3.2, 11/3$, and 3.9.

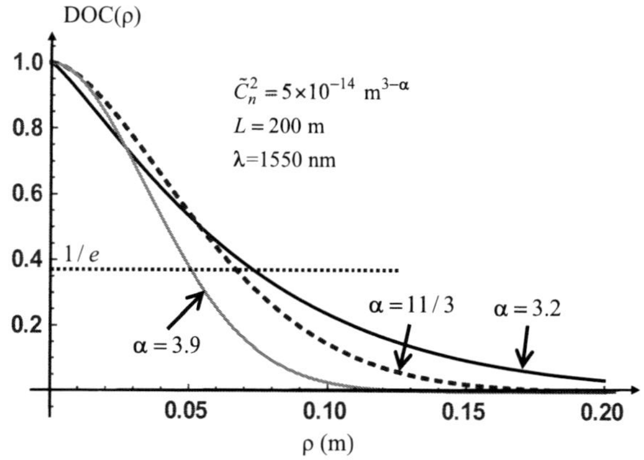

Figure 10.1 The modulus of the complex degree of coherence (DOC) for a spherical wave under non-Kolmogorov turbulence described by Eq. (10.1). The horizontal dotted line is the 1/*e* point that identifies the spatial coherence radii at the points of intersection.

We base these plots on the WSF given in Eq. (10.25). The wavelength is 1.55 μm, path length is 200 m, and $\tilde{C}_n^2 = 5 \times 10^{-14}\,\mathrm{m}^{3-\alpha}$ in all cases. The spatial coherence radius for each value of α is defined by the intersection of DOC(ρ) with the horizontal dotted line. Thus, decreasing values of α generally lead to a larger coherence radius.

Last, by replacing $\tilde{C}_n^2$ with $\tilde{C}_{n,eff}^2$ in Eqs. (10.21) and (10.25), and also multiplying by $\mu^{2-\alpha}$, we obtain the WSF result for both plane waves and spherical waves based on the spectrum model of Eq. (10.6).

10.3.3 Gaussian-beam wave

By using the nondimensional beam parameters defined by Eqs. (8.19) and (8.20) in Section 8.2.1, the WSF for a collimated Gaussian-beam wave with $\kappa_0 = 0$ and $l_0 = 0$ under the non-Kolmogorov spectrum model (10.1) takes the form (assuming that $\mathbf{r}_2 = -\mathbf{r}_1$)

$$
\begin{aligned}
D(\rho,\alpha) &= 8\pi^2 A(\alpha)\tilde{C}_n^2 k^2 L \int_0^1\!\!\int_0^\infty \kappa^{1-\alpha}\exp\left(-\frac{\Lambda L\kappa^2\xi^2}{k}\right)[I_0(\Lambda\rho\xi\kappa)-1]\,d\kappa\,d\xi \\
&+ 8\pi^2 A(\alpha)\tilde{C}_n^2 k^2 L \int_0^1\!\!\int_0^\infty \kappa^{1-\alpha}\exp\left(-\frac{\Lambda L\kappa^2\xi^2}{k}\right)\left\{1 - J_0[(1-\bar{\Theta}\xi)\kappa\rho]\right\}d\kappa\,d\xi,
\end{aligned} \tag{10.26}
$$

which reduces to

$$
\begin{aligned}
D(\rho,L) &= -\frac{4\pi^2 A(\alpha)}{(\alpha-1)}\Gamma\left(1-\frac{\alpha}{2}\right)\tilde{C}_n^2 k^{3-\alpha/2}L^{\alpha/2}\Lambda^{\alpha/2-1}\left[1 - {}_1F_1\left(1-\frac{\alpha}{2};1;\frac{\rho^2}{2W^2}\right)\right] \\
&\quad - 4\pi^2 A(\alpha)\Gamma\left(1-\frac{\alpha}{2}\right)\tilde{C}_n^2 k^{3-\alpha/2}L^{\alpha/2}\Lambda^{\alpha/2-1} \\
&\quad \times \int_0^1 \xi^{\alpha-2}\left\{{}_1F_1\left[1-\frac{\alpha}{2};1;-\frac{(1-\bar{\Theta}\xi)^2 k\rho^2}{4\Lambda L\xi^2}\right]-1\right\}d\xi.
\end{aligned} \tag{10.27}
$$

Next, we use the small argument approximation ($\rho^2/2W \ll 1$) for the first confluent hypergeometric function in Eq. (10.27) and the large-argument approximation for the second confluent hypergeometric function, which leads to

$$
1 - {}_1F_1\left(1-\frac{\alpha}{2};1;\frac{\rho^2}{2W^2}\right) \approx \left(\frac{\alpha-2}{4}\right)\frac{\rho^2}{W^2}, \quad \rho < W, \tag{10.28}
$$

$$
{}_1F_1\left[1-\frac{\alpha}{2};1;-\frac{(1-\bar{\Theta}\xi)^2 k\rho^2}{4\Lambda L\xi^2}\right] - 1 \approx \frac{1}{\Gamma\left(\frac{\alpha}{2}\right)}\left[\frac{(1-\bar{\Theta}\xi)^2 k\rho^2}{4\Lambda L\xi^2}\right]^{\alpha/2-1}. \tag{10.29}
$$

If we now substitute these approximations back into Eq. (10.27) and finish the integration on ξ, we arrive at the expression

$$D(\rho,\alpha) = -\frac{4\pi^2 A(\alpha)\Gamma\left(1-\dfrac{\alpha}{2}\right)}{(\alpha-1)\Gamma\left(\dfrac{\alpha}{2}\right)2^{\alpha-2}}\tilde{C}_n^2 k^{3-\alpha/2}L^{\alpha/2} \times\left[\left(\frac{1-\Theta^{\alpha-1}}{1-\Theta}\right)\left(\frac{k\rho^2}{L}\right)^{\alpha/2-1}+\frac{(\alpha-2)\Gamma\left(\dfrac{\alpha}{2}\right)}{2^{5-\alpha}}\Lambda^{\alpha/2}\left(\frac{k\rho^2}{L}\right)\right], \quad l_0 \ll \rho \ll L_0, \quad 0\le\Theta\le 1. \tag{10.30}$$

10.4 Wave Structure Function: Anisotropic Turbulence

The WSFs developed in Section 10.3 for an infinite plane wave, spherical wave, and Gaussian-beam wave are based on isotropic non-Kolmogorov turbulence. For the anisotropic non-Kolmogorov spectrum defined by Eq. (10.4), we consider only the WSF for a spherical wave, which takes the form[13]

$$\begin{aligned} D_{sph}(\alpha,\mathbf{r}_1,\mathbf{r}_2) &= 4\pi k^2 L\int_0^1\int_{-\infty}^{\infty}\int_{-\infty}^{\infty}\Phi(\kappa_x,\kappa_y,0)[1-\exp(i\xi\mathbf{K}\cdot\mathbf{p})]\,d\kappa_x\,d\kappa_y\,d\xi \\ &= 4\pi A(\alpha)\tilde{C}_n^2 k^2 L\int_0^1\int_0^{\infty}\int_0^{2\pi} q^{1-\alpha} \\ &\quad\times\left\{1-\exp\left[i\xi q\left(\frac{\rho_x}{\mu_x}\cos\theta+\frac{\rho_y}{\mu_y}\sin\theta\right)\right]\right\}d\theta\,dq\,d\xi, \end{aligned} \tag{10.31}$$

where we have introduced

$$\mathbf{p}=\mathbf{r}_1-\mathbf{r}_2=(\rho_x,\rho_y), \qquad \mathbf{K}\cdot\mathbf{p}=q\left(\frac{\rho_x}{\mu_x}\cos\theta+\frac{\rho_y}{\mu_y}\sin\theta\right). \tag{10.32}$$

Equation (10.31) now leads to

$$\begin{aligned} D_{sph}(\alpha,\rho_x,\rho_y) &= 8\pi^2 A(\alpha)\tilde{C}_{n,eff}^2 k^2 L\int_0^1\int_0^{\infty} q^{1-\alpha}[1-J_0(q\xi\tilde{\rho})]\,dq\,d\xi \\ &= \frac{\alpha\,\Gamma(\alpha-1)\,\Gamma(-\alpha/2)}{(\alpha-1)\,2^{\alpha-1}\Gamma(\alpha/2)}\cos\left(\frac{\alpha\pi}{2}\right)\tilde{C}_{n,eff}^2\,k^2 L\,\tilde{\rho}^{\alpha-2}, \end{aligned} \tag{10.33}$$

where we define

$$\tilde{\rho} = \sqrt{\left(\frac{\rho_x}{\mu_x}\right)^2 + \left(\frac{\rho_y}{\mu_y}\right)^2} = \rho\sqrt{\frac{\cos^2\theta}{\mu_x^2} + \frac{\sin^2\theta}{\mu_y^2}}. \tag{10.34}$$

From the results of Eqs. (10.33) and (10.34), we can deduce the spatial coherence radius in the horizontal ($\theta = 0$) and vertical ($\theta = \pi/2$) directions by

$$\rho_{0m,eff} = \mu_m\left[\frac{\alpha\Gamma(\alpha-1)}{(\alpha-1)2^\alpha}\frac{\Gamma(-\alpha/2)}{\Gamma(\alpha/2)}\cos\left(\frac{\alpha\pi}{2}\right)\tilde{C}_{n,eff}^2 k^2 L\right]^{1/(2-\alpha)}, \quad m = x,\, y. \tag{10.35}$$

The WSF and spatial coherence radius for a plane wave differ from those of the spherical wave by the omission of the factor $(\alpha - 1)$ in the denominators of Eqs. (10.33) and (10.35). Because of the conjectured definition of $\tilde{C}_{n,eff}^2$ given by Eq. (10.5), it is clear that the spatial coherence radii defined by Eq. (10.35) of both spherical and plane waves are not independent in the vertical and horizontal directions but are clearly coupled through $\tilde{C}_{n,eff}^2$.

10.5 Scintillation Index: Weak Irradiance Fluctuations

The normalized variance of irradiance, or *scintillation index*, under the influence of conventional Kolmogorov turbulence was provided in Section 8.4. Below we use the same approach to derive the scintillation index in non-Kolmogorov turbulence under weak irradiance fluctuations.[12,30]

10.5.1 Plane wave: isotropic turbulence

The scintillation index of an infinite plane wave is described by

$$\sigma_{I,pl}^2(\alpha, L) = 8\pi^2 A(\alpha)\tilde{C}_n^2 k^2 L \int_0^1\int_0^\infty \kappa^{1-\alpha}\exp\left(-\frac{\kappa^2}{\kappa_m^2}\right)\left[1-\cos\left(\frac{L\kappa^2\xi}{k}\right)\right] d\kappa\, d\xi, \tag{10.36}$$

where the power spectrum we use is Eq. (10.3) with $\kappa_0 = 0$. It is useful here to write the cosine function as $\cos x = \mathrm{Re}(e^{-ix})$, from which Eq. (10.36) becomes

$$\sigma^2_{I,pl}(\alpha, L) = 8\pi^2 A(\alpha)\tilde{C}_n^2 k^2 L \mathrm{Re} \int_0^1 \int_0^\infty \kappa^{1-\alpha} \times \left\{ \exp\left(-\frac{\kappa^2}{\kappa_m^2}\right) - \exp\left[-\frac{\kappa^2(1+iQ_m\xi)}{\kappa_m^2}\right] \right\} d\kappa\, d\xi, \tag{10.37}$$

where $Q_m = L\kappa_m^2/k$. Thus, upon integrating the inside integral in Eq. (10.37) using ($D2$) in Appendix D, we obtain

$$\begin{aligned} \sigma^2_{I,pl}(\alpha, L) &= 4\pi^2 A(\alpha)\Gamma\left(1-\frac{\alpha}{2}\right)\tilde{C}_n^2 k^2 L \kappa_m^{2-\alpha} \\ &\quad \times \mathrm{Re} \int_0^1 [1-(1+iQ_m\xi)^{\alpha/2-1}]\, d\xi \\ &= \frac{2\Gamma(\alpha-1)}{\alpha}\Gamma\left(1-\frac{\alpha}{2}\right)\cos\left(\frac{\alpha\pi}{2}\right)\tilde{C}_n^2 k^2 L \kappa_m^{2-\alpha} \mathrm{Re}\left[\frac{\alpha}{2} - \frac{(1+iQ_m)^{\alpha/2}}{iQ_m}\right]. \end{aligned} \tag{10.38}$$

If we rewrite Eq. (10.38) in terms of its real part and take into account the generalized Rytov variance (10.17), we see that

$$\sigma^2_{I,pl}(\alpha, L) = \frac{\tilde{\sigma}_R^2(\alpha)}{\sin(\alpha\pi/4)} \left\{ (1+1/Q_m^2)^{\alpha/4} \sin\left[\frac{\alpha}{2}\tan^{-1}(Q_m)\right] - \frac{\alpha}{2} Q_m^{1-\alpha/2} \right\}. \tag{10.39}$$

10.5.2 Spherical wave: isotropic turbulence

Again, using the power spectrum model (10.3) with $\kappa_0 = 0$, the scintillation index of a spherical wave is described by

$$\sigma^2_{I,sph}(\alpha, L) = 8\pi^2 A(\alpha)\tilde{C}_n^2 k^2 L \int_0^1 \int_0^\infty \kappa^{1-\alpha} \exp\left(-\frac{\kappa^2}{\kappa_m^2}\right) \times \left\{ 1 - \cos\left[\frac{L\kappa^2 \xi(1-\xi)}{k}\right] \right\} d\kappa\, d\xi. \tag{10.40}$$

By following the approach for the plane wave, Eq. (10.40) can be written as

$$\sigma^2_{I,sph}(\alpha, L) = 4\pi^2 A(\alpha)\Gamma\left(1-\frac{\alpha}{2}\right)\tilde{C}_n^2 k^2 L \kappa_m^{2-\alpha} \times \mathrm{Re} \int_0^1 \{1 - [1 + iQ_m\xi(1-\xi)]^{\alpha/2-1}\}\, d\xi. \tag{10.41}$$

Here we introduce the temporary assumption that $Q_m < 1$ so the binomial function under the integral has a convergent power series representation.

Then termwise integration of the power series based on ($G17$) in Appendix A leads to

$$\int_0^1 [1+iQ_m\xi(1-\xi)]^{\alpha/2-1}d\xi = \sum_{n=0}^{\infty}(-1)^n(1-\alpha/2)_n\frac{(iQ_m)^n}{n!}\int_0^1 \xi^n(1-\xi)^n d\xi$$
$$= \sum_{n=0}^{\infty}(-1)^n\frac{(1-\alpha/2)_n\Gamma^2(n+1)}{\Gamma(2n+2)}\frac{(iQ_m)^n}{n!}, \quad Q_m < 1. \tag{10.42}$$

Upon converting the gamma functions to Pochhammer symbols and summing the series in Eq. (10.42) as a hypergeometric function, we then combine Eqs. (10.41) and (10.42) to arrive at

$$\sigma^2_{I,sph}(\alpha,\, L) = \frac{\alpha\tilde{\sigma}^2_R(\alpha)}{2\sin(\alpha\pi/4)}Q_m^{1-\alpha/2}\mathrm{Re}\left[{}_2F_1\left(1-\frac{\alpha}{2},\, 1;\, \frac{3}{2};\, -\frac{iQ_m}{4}\right)-1\right]. \tag{10.43}$$

In this last expression, we can now remove the assumption that $Q_m < 1$. Thus, in the limit $Q_m \to \infty$, Eq. (10.43) reduces to

$$\sigma^2_{I,sph}(\alpha,\, L) \equiv \tilde{\beta}^2_0(\alpha) = \frac{\Gamma(1+\alpha/2)\,\Gamma(\alpha/2)}{\Gamma(\alpha)}\tilde{\sigma}^2_R(\alpha). \tag{10.44}$$

10.5.3 Gaussian-beam wave: anisotropic turbulence

The longitudinal component of the scintillation index for a Gaussian-beam wave under general anisotropic turbulence [Eq. (10.4)] is defined by[13]

$$\sigma^2_{I,l}(\alpha;\, \mu_x,\, \mu_y) = 4\pi k^2 L\int_0^1\int_{-\infty}^{\infty}\int_{-\infty}^{\infty}\Phi_n(\kappa_x,\, \kappa_y,\, 0)\,\exp\left(-\frac{\Lambda L\kappa^2\xi^2}{k}\right)$$
$$\times\left\{1-\cos\left[\frac{L\kappa^2}{k}\xi(1-\bar{\Theta}\xi)\right]\right\}d\kappa_x\, d\kappa_y\, d\xi. \tag{10.45}$$

By making the substitution of results in Eq. (10.11), we are led to

$$\begin{aligned}\sigma_{I,l}^2(\alpha;\, \mu_x,\, \mu_y) &= -\frac{\Gamma(\alpha-1)}{\pi\alpha}\Gamma\left(1-\frac{\alpha}{2}\right)\cos\left(\frac{\alpha\pi}{2}\right)\tilde{C}_{n,eff}^2 k^{3-\alpha/2}L^{\alpha/2}\\ &\quad\times \mathrm{Re}\left[i^{\alpha/2-1}\,{}_2F_1\left(1-\frac{\alpha}{2},\,\frac{\alpha}{2};\,1+\frac{\alpha}{2};\,\bar{\Theta}+i\Lambda\right)-\frac{\alpha\Lambda^{\alpha/2-1}}{2(\alpha-1)}\right]\\ &\quad\times \int_0^{2\pi}\left(\frac{\cos^2\theta}{\mu_x^2}+\frac{\sin^2\theta}{\mu_y^2}\right)^{\alpha/2-1}d\theta,\end{aligned} \tag{10.46}$$

which, recalling Eq. (10.14), can be expressed as

$$\begin{aligned}\sigma_{I,l}^2(\alpha;\, \mu_x,\, \mu_y) &= \frac{\tilde{\sigma}_R^2(\alpha;\, \mu_x,\, \mu_y)}{\sin(\alpha\pi/4)}\\ &\quad\times \mathrm{Re}\left[i^{\alpha/2-1}\,{}_2F_1\left(1-\frac{\alpha}{2},\,\frac{\alpha}{2};\,1+\frac{\alpha}{2};\,\bar{\Theta}+i\Lambda\right)-\frac{\alpha\Lambda^{\alpha/2-1}}{2(\alpha-1)}\right].\end{aligned} \tag{10.47}$$

Comparable expressions for the limiting cases of a plane wave $(\Theta = 1,\ \Lambda = 0)$ and spherical wave $(\Theta = \Lambda = 0)$ are easily found as special cases of Eq. (10.47). For the isotropic non-Kolmogorov case, we simply set $\mu_x = \mu_y = 1$ and replace $\tilde{\sigma}_R^2(\alpha;\, \mu_x,\, \mu_y)$ with $\tilde{\sigma}_R^2(\alpha)$.

10.6 Scintillation Index: Strong Irradiance Fluctuations

We now wish to develop the scintillation index in weak-to-strong irradiance fluctuations under non-Kolmogorov turbulence.[11,12] However, we will provide full detail only for the case of an infinite plane wave in isotropic turbulence. The method we use is based on the *extended Rytov theory*, which uses the notion that small-scale irradiance fluctuations of the optical wave are modulated by large-scale irradiance fluctuations (recall Section 8.6). Hence, the spectrum model is defined by

$$\Phi_{n,e}(\kappa) = \Phi_n(\kappa)[G_X(\kappa) + G_Y(\kappa)], \tag{10.48}$$

where $G_X(\kappa)$ and $G_Y(\kappa)$ are the large-scale and small-scale filter functions, respectively, and $\Phi_n(\kappa)$ is the non-Kolmogorov power spectrum (10.1). Although the approach we use here is based on the extended Rytov theory, these filter functions and the asymptotic behavior based on the conventional Kolmogorov theory must all be recalculated for non-Kolmogorov turbulence. The impact of the inner and outer scales will be ignored in the following analysis.

10.6.1 Plane wave: isotropic turbulence

We start by introducing the shorthand notation

$$\begin{aligned} B_1(\alpha) &= -2\frac{\Gamma(\alpha-1)}{\alpha}\Gamma\left(1-\frac{\alpha}{2}\right)\sin\left(\frac{\alpha\pi}{4}\right)\cos\left(\frac{\alpha\pi}{2}\right), \\ B_2(\alpha) &= \frac{\alpha\Gamma(\alpha-1)}{2^{\alpha-1}}\,\frac{\Gamma\left(-\frac{\alpha}{2}\right)}{\Gamma\left(\frac{\alpha}{2}\right)}\cos\left(\frac{\alpha\pi}{2}\right). \end{aligned} \tag{10.49}$$

Hence, the generalized Rytov variance becomes $\tilde{\sigma}_R^2(\alpha) = B_1(\alpha)\tilde{C}_n^2 k^{3-\alpha/2}L^{\alpha/2}$ and the WSF for a plane wave is $D_{pl}(\rho, \alpha) = B_2(\alpha)\tilde{C}_n^2 k^2 L\,\rho^{\alpha-2}$, where we are using the inertial range form of the WSF.

In the saturation regime $[\tilde{\sigma}_R^2(\alpha) \to \infty]$, the scintillation index for a plane wave takes the asymptotic form

$$\begin{aligned} \sigma_{I,pl}^2(\alpha, L) \sim 1 + 32\pi^2 A(\alpha)\tilde{C}_n^2 k^2 L \int_0^1\int_0^\infty \kappa^{1-\alpha}\sin^2\left[\frac{L\kappa^2}{2k}w(\xi, \xi)\right] \\ \times \exp\left\{-\int_0^1 D_{pl}\left[\frac{L\kappa}{k}w(\tau, \xi)\right]d\tau\right\}d\kappa\, d\xi, \end{aligned} \tag{10.50}$$

where

$$w(\tau, \xi) = \begin{cases} \tau, & \tau < \xi \\ \xi, & \tau > \xi. \end{cases} \tag{10.51}$$

Upon a partial evaluation of Eq. (10.50), following the analysis given in Ref. 31, pp. 330–331, we find that

$$\begin{aligned} \sigma_{I,pl}^2(\alpha, L) \sim 1 + \frac{8\pi^2 A(\alpha)}{\alpha-2}\Gamma\left(\frac{4}{\alpha-2}-1\right)[B_2(\alpha)]^{\frac{\alpha-6}{\alpha-2}}\left[\frac{\tilde{\sigma}_R^2(\alpha)}{B_1(\alpha)}\right]^{\frac{2(\alpha-4)}{\alpha-2}} \\ \times \int_0^1 \xi^{\alpha-4}\left[1-\left(\frac{\alpha-2}{\alpha-1}\right)\xi\right]^{\frac{\alpha-6}{\alpha-2}}d\xi. \end{aligned} \tag{10.52}$$

The remaining integral in Eq. (10.52) yields

$$\begin{aligned} \int_0^1 \xi^{\alpha-4}\left[1-\left(\frac{\alpha-2}{\alpha-1}\right)\xi\right]^{\frac{\alpha-6}{\alpha-2}}d\xi &= \sum_{n=0}^{\infty}\frac{\left(\frac{6-\alpha}{\alpha-2}\right)_n}{n!}\left(\frac{\alpha-2}{\alpha-1}\right)^n\int_0^1 \xi^{n+\alpha-4}\,d\xi \\ &= \frac{\Gamma(\alpha-3)}{\Gamma(\alpha-2)}\,{}_2F_1\left(\frac{6-\alpha}{\alpha-2}, \alpha-3; \alpha-2; \frac{\alpha-2}{\alpha-1}\right), \end{aligned} \tag{10.53}$$

and thus, in the saturation regime, we deduce that

$$\sigma_{I,pl}^2(\alpha, L) \sim 1 + \frac{B_{3,pl}(\alpha)}{[\tilde{\sigma}_R^2(\alpha)]^{\frac{2(4-\alpha)}{\alpha-2}}}, \quad \tilde{\sigma}_R^2(\alpha) \to \infty, \tag{10.54}$$

where

$$\begin{aligned} B_{3,pl}(\alpha) &= 2\Gamma(\alpha-3)\Gamma\left(\frac{6-\alpha}{\alpha-2}\right)\cos\left(\frac{\alpha\pi}{2}\right)[B_2(\alpha)]^{\frac{\alpha-6}{\alpha-2}}[B_1(\alpha)]^{\frac{2(4-\alpha)}{\alpha-2}} \\ &\quad \times {}_2F_1\left(\frac{6-\alpha}{\alpha-2}, \alpha-3; \alpha-2; \frac{\alpha-2}{\alpha-1}\right). \end{aligned} \tag{10.55}$$

Next, based on the large-scale filter function, $G_X(\kappa) = \exp(-\kappa^2/\kappa_X^2)$, the large-scale log-irradiance variance is defined by

$$\begin{aligned} \sigma_{\ln X}^2 &= 8\pi^2 A(\alpha)\tilde{C}_n^2 k^2 L \int_0^1\int_0^\infty \kappa^{1-\alpha}\exp\left(-\frac{\kappa^2}{\kappa_X^2}\right)\left[1-\cos\left(\frac{L\kappa^2}{k}\xi\right)\right]d\kappa\, d\xi \\ &\approx 2\pi^2 A(\alpha)\frac{\tilde{\sigma}_R^2(\alpha)}{B_1(\alpha)}\int_0^1 \xi^2 d\xi \int_0^\infty \eta^{2-\alpha/2}\exp\left(-\frac{\eta}{\eta_X}\right)d\eta, \end{aligned} \tag{10.56}$$

where we used $1-\cos x \approx x^2/2$ for small arguments (i.e., geometrical optics approximation) and set $\eta = L\kappa^2/k$. Completing the integration in Eq. (10.56) yields

$$\sigma_{\ln X}^2 = \frac{2}{3}\pi^2 A(\alpha)\Gamma\left(3-\frac{\alpha}{2}\right)\frac{\tilde{\sigma}_R^2(\alpha)}{B_1(\alpha)}\eta_X^{3-\alpha/2}, \tag{10.57}$$

where

$$\eta_X \approx \frac{1}{c_1 + c_2\left(\dfrac{L}{k\rho_0^2}\right)} = \frac{1}{c_1 + c_2\left[\dfrac{B_2(\alpha)}{2B_1(\alpha)}\tilde{\sigma}_R^2(\alpha)\right]^{\frac{2}{\alpha-2}}}. \tag{10.58}$$

The constants c_1 and c_2 are determined by matching asymptotic relations for weak and strong irradiance fluctuations. For example, under weak irradiance fluctuations, $\sigma_{\ln X}^2(\alpha, L) \approx 0.49\tilde{\sigma}_R^2(\alpha)$, $\tilde{\sigma}_R^2(\alpha) \ll 1$, from which we find that

$$c_1 = \left[\frac{\frac{2}{3}\pi^2 A(\alpha)\Gamma\left(3-\frac{\alpha}{2}\right)}{0.49B_1(\alpha)}\right]^{\frac{2}{6-\alpha}}, \tag{10.59}$$

and under strong irradiance fluctuations suggested by Eq. (10.54),

$$c_2 = \left[\frac{4\pi^2 A(\alpha)}{3B_{3,pl}(\alpha)}\Gamma\left(3-\frac{\alpha}{2}\right)\right]^{\frac{2}{6-\alpha}}\left[\frac{2}{B_2(\alpha)}\right]^{\frac{2}{\alpha-2}}. \tag{10.60}$$

Last, by combining the above results, Eq. (10.57) now takes the final form

$$\sigma^2_{\ln X} \approx \frac{0.49\,\tilde{\sigma}^2_R(\alpha)}{\left\{1+\left[\frac{0.98}{B_{3,pl}(\alpha)}\right]^{\frac{2}{6-\alpha}}[\tilde{\sigma}^2_R(\alpha)]^{\frac{2}{\alpha-2}}\right\}^{\frac{6-\alpha}{2}}}. \tag{10.61}$$

With the filter function $G_Y(\kappa) = \kappa^\alpha/(\kappa^2+\kappa_Y^2)^{\alpha/2}$, the small-scale log-irradiance variance is

$$\begin{aligned}\sigma^2_{\ln Y} &= 8\pi^2 A(\alpha)\tilde{C}_n^2 k^2 L\int_0^1\int_0^\infty \frac{\kappa}{(\kappa^2+\kappa_Y^2)^{\alpha/2}}\left[1-\cos\left(\frac{L\kappa^2}{k}\xi\right)\right]d\kappa\, d\xi\\ &\approx 4\pi^2 A(\alpha)\frac{\tilde{\sigma}^2_R(\alpha)}{B_1(\alpha)}\int_0^\infty \frac{1}{(\eta+\eta_Y)^{\alpha/2}}\,d\eta,\end{aligned} \tag{10.62}$$

which simplifies to

$$\sigma^2_{\ln Y} \approx 8\pi^2\frac{A(\alpha)}{(\alpha-2)}\frac{\tilde{\sigma}^2_R(\alpha)}{B_1(\alpha)}\eta_Y^{1-\alpha/2}. \tag{10.63}$$

Here,

$$\eta_Y \approx c_3 + c_4\left(\frac{L}{k\rho_0^2}\right) = c_3 + c_4\left[\frac{B_2(\alpha)}{2B_1(\alpha)}\tilde{\sigma}^2_R(\alpha)\right]^{\frac{2}{\alpha-2}}, \tag{10.64}$$

where c_3 and c_4 are determined by matching weak and strong irradiance fluctuation terms. For example, $\sigma^2_{\ln Y}(\alpha, L) \approx 0.51\tilde{\sigma}^2_R(\alpha)$, $\tilde{\sigma}^2_R(\alpha) \ll 1$, and $\sigma^2_{\ln Y}(\alpha, L) \approx \ln 2$, $\tilde{\sigma}^2_R(\alpha) \gg 1$. This time we find that

$$c_3 = \left[\frac{8\pi^2 A(\alpha)}{0.51(\alpha-2)B_1(\alpha)}\right]^{\frac{2}{\alpha-2}} \tag{10.65}$$

and

$$c_4 = \left[\frac{16\pi^2 A(\alpha)}{(\alpha-2)B_2(\alpha)\ln 2}\right]^{\frac{2}{\alpha-2}}. \tag{10.66}$$

Putting the above results together, Eq. (10.63) yields

$$\sigma^2_{\ln Y} \approx \frac{0.51\tilde{\sigma}^2_R(\alpha)}{\left\{1 + [0.736\,\tilde{\sigma}^2_R(\alpha)]^{\frac{2}{\alpha-2}}\right\}^{\alpha/2-1}}. \tag{10.67}$$

Consequently, based on the results of Eqs. (10.61) and (10.67), the total scintillation index under moderate-to-strong irradiance fluctuations becomes

$$\sigma^2_{I,pl}(\alpha, L) = \exp\left(\sigma^2_{\ln X} + \sigma^2_{\ln Y}\right) - 1 \sim \begin{cases} \sigma^2_{\ln X} + \sigma^2_{\ln Y}, & \tilde{\sigma}^2_R(\alpha) \ll 1 \\ 1 + 2\sigma^2_{\ln X}, & \tilde{\sigma}^2_R(\alpha) \gg 1. \end{cases} \tag{10.68}$$

In Fig. 10.2 we plot the plane wave scintillation index (10.68) as a function of range up to 3000 m. The structure constant is fixed at the same value, $\tilde{C}_n^2 = 2 \times 10^{-13}\,\mathrm{m}^{3-\alpha}$, for all three cases: $\alpha = 3.2$, $11/3$, and 3.9. The wavelength chosen is 1.55 μm.

10.6.2 Plane wave: anisotropic turbulence

Before introducing the spherical wave and Gaussian-beam wave cases in isotropic turbulence, we now develop the scintillation index for a plane wave in anisotropic turbulence. Most of the analysis will parallel that presented for the isotropic case. However, we will need the transformation definitions provided in Eq. (10.11) and the generalized Rytov variance defined by Eq. (10.14). Also, we need the WSF in anisotropic turbulence given by

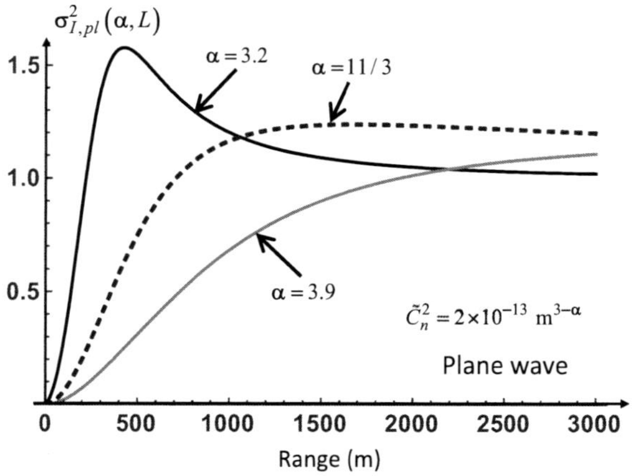

Figure 10.2 Scintillation index [Eq. (10.68)] for a plane wave with $\lambda = 1.55$ μm as a function of range and three different values of the spectral index.

$$
\begin{aligned}
D_{pl}(\alpha, \rho_x, \rho_y) &= B_2(\alpha)\tilde{C}^2_{n,eff}k^2L\tilde{\rho}^{\alpha-2} \\
&= B_2(\alpha)\tilde{C}^2_{n,eff}k^2L\rho^{\alpha-2}\left(\frac{\cos^2\theta}{\mu_x^2}+\frac{\sin^2\theta}{\mu_y^2}\right)^{(\alpha-2)/2},
\end{aligned} \tag{10.69}
$$

where $\tilde{\rho} = \sqrt{(\rho_x/\mu_x)^2 + (\rho_y/\mu_y)^2}$ and $\rho_x = \rho\cos\theta,\ \ \rho_y = \rho\sin\theta$.

To start, we follow the analysis in the previous section, which ultimately leads to a solution for the asymptotic result defined by Eq. (10.50),

$$
\sigma^2_{I,pl}(\alpha;\, \mu_x, \mu_y) \sim 1 + \frac{B_{3,pl}(\alpha;\, \mu_x, \mu_y)}{[\tilde{\sigma}^2_R(\alpha;\, \mu_x, \mu_y)]^{2(4-\alpha)/(\alpha-2)}}, \quad \tilde{\sigma}^2_R(\alpha;\, \mu_x, \mu_y) \to \infty, \tag{10.70}
$$

where

$$
\begin{aligned}
B_{3,pl}(\alpha;\, \mu_x, \mu_y) = {} & 2\Gamma(\alpha-3)\Gamma\left(\frac{6-\alpha}{\alpha-2}\right)\cos\left(\frac{\alpha\pi}{2}\right)[B_2(\alpha)]^{(\alpha-6)/(\alpha-2)} \\
& \times {}_2F_1\left(\frac{1}{2}, 4-\alpha; 1; 1-\frac{\mu_y^2}{\mu_x^2}\right) {}_2F_1\left(\frac{6-\alpha}{\alpha-2}, \alpha-3; \alpha-2; \frac{\alpha-2}{\alpha-1}\right) \\
& \times \left[B_1(\alpha)\,{}_2F_1\left(\frac{1}{2}, \frac{1-\alpha}{2}; 1; 1-\frac{\mu_y^2}{\mu_x^2}\right)\right]^{2(4-\alpha)/(\alpha-2)}.
\end{aligned} \tag{10.71}
$$

Note that when $\mu_x = \mu_y = 1$, Eq. (10.71) reduces to Eq. (10.55).

The large-scale log-irradiance takes the form

$$
\begin{aligned}
\sigma^2_{\ln X} = {} & 4\pi^2 A(\alpha)\tilde{C}^2_{n,eff}k^2L\left(\frac{L}{k}\right)^2\left(\frac{1}{2\pi}\right)\int_0^{2\pi}\left(\frac{\cos^2\theta}{\mu_x^2}+\frac{\sin^2\theta}{\mu_y^2}\right)^2 \\
& \times \int_0^1 \xi^2 d\xi \int_0^\infty q^{5-\alpha}\exp\left[-\frac{q^2}{\kappa_X^2}\left(\frac{\cos^2\theta}{\mu_x^2}+\frac{\sin^2\theta}{\mu_y^2}\right)\right] dq\, d\theta,
\end{aligned} \tag{10.72}
$$

where we have again imposed the geometrical optics approximation. From the integral relations

$$
\begin{aligned}
& \int_0^1 \xi^2 d\xi \int_0^\infty q^{5-\alpha}\exp\left[-\frac{q^2}{\kappa_X^2}\left(\frac{\cos^2\theta}{\mu_x^2}+\frac{\sin^2\theta}{\mu_y^2}\right)\right] dq \\
& \qquad = \frac{1}{6}\kappa_X^{6-\alpha}\Gamma\left(\frac{6-\alpha}{2}\right)\left(\frac{\cos^2\theta}{\mu_x^2}+\frac{\sin^2\theta}{\mu_y^2}\right)^{(\alpha-6)/2}
\end{aligned} \tag{10.73}
$$

and Eq. (10.16), we find that Eq. (10.72) becomes

$$
\sigma^2_{\ln X} = \frac{2\pi^2}{3}A(\alpha)\Gamma\left(\frac{6-\alpha}{2}\right)\frac{\tilde{\sigma}^2_R(\alpha;\, \mu_x, \mu_y)}{B_1(\alpha)}\eta_X^{3-\alpha/2}. \tag{10.74}
$$

The parameter η_X is found in the same manner as in Section 10.6.1, which leads to

$$\sigma^2_{\ln X} = \frac{0.49\tilde{\sigma}^2_R(\alpha;\, \mu_x,\, \mu_y)}{\left\{1 + \left[\frac{0.98}{B_{3,pl}(\alpha;\, \mu_x,\, \mu_y)}\right]^{\frac{2}{6-\alpha}} [\tilde{\sigma}^2_R(\alpha;\, \mu_x,\, \mu_y)]^{\frac{2}{\alpha-2}}\right\}^{\frac{6-\alpha}{2}}}. \tag{10.75}$$

The small-scale log-irradiance variance is given by

$$\begin{aligned}\sigma^2_{\ln Y} &= 8\pi^2 A(\alpha)\tilde{C}^2_{n,eff} k^2 L\left(\frac{1}{2\pi}\right)\int_0^{2\pi}\left(\frac{\cos^2\theta}{\mu_x^2} + \frac{\sin^2\theta}{\mu_y^2}\right)^{\alpha/2} \\ &\times \int_0^\infty q\left[\left(\frac{\cos^2\theta}{\mu_x^2} + \frac{\sin^2\theta}{\mu_y^2}\right)q^2 + \kappa_Y^2\right]^{-\alpha/2} dq\, d\theta.\end{aligned} \tag{10.76}$$

By applying the integral relation

$$\int_0^\infty q\left[\left(\frac{\cos^2\theta}{\mu_x^2} + \frac{\sin^2\theta}{\mu_y^2}\right)q^2 + \kappa_Y^2\right]^{-\alpha/2} dq = \frac{\kappa_Y^{2-\alpha}}{(\alpha-2)\left(\frac{\cos^2\theta}{\mu_x^2} + \frac{\sin^2\theta}{\mu_y^2}\right)} \tag{10.77}$$

along with Eq. (10.16), we get

$$\begin{aligned}\sigma^2_{\ln Y} &= \frac{8\pi^2 A(\alpha)\tilde{\sigma}^2_R(\alpha;\, \mu_x,\, \mu_y)}{(\alpha-2)B_1(\alpha)}\eta_Y^{1-\alpha/2} \\ &= \frac{0.51\tilde{\sigma}^2_R(\alpha;\, \mu_x,\, \mu_y)}{\left\{1 + [0.736\,\tilde{\sigma}^2_R(\alpha;\, \mu_x,\, \mu_y)]^{2/(\alpha-2)}\right\}^{(\alpha-2)/2}}.\end{aligned} \tag{10.78}$$

Last, by combining Eqs. (10.75) and (10.78), which look remarkedly similar to the isotropic case in Section 10.6.1, we arrive at

$$\sigma^2_{I,pl}(\alpha;\, \mu_x,\, \mu_y) = \exp(\sigma^2_{\ln X} + \sigma^2_{\ln Y}) - 1, \quad 0 \le \tilde{\sigma}^2_R(\alpha;\, \mu_x,\, \mu_y) < \infty. \tag{10.79}$$

10.6.3 Spherical wave: isotropic turbulence

In the case of a spherical wave in isotropic turbulence, the calculations are very similar to those of a plane wave. For example, in the saturation regime,

$$\sigma^2_{I,sph}(\alpha,\, L) \sim 1 + \frac{B_{3,sph}(\alpha)}{\left[\tilde{\sigma}^2_R(\alpha)\right]^{\frac{2(4-\alpha)}{\alpha-2}}}, \quad \tilde{\sigma}^2_R(\alpha) \to \infty, \tag{10.80}$$

where

$$B_{3,sph}(\alpha) = 2\frac{\Gamma^2(\alpha-3)}{\Gamma(2\alpha-6)}\Gamma\left(\frac{6-\alpha}{\alpha-2}\right)\Gamma(\alpha-2)\cos\left(\frac{\alpha\pi}{2}\right)\left[\frac{B_2(\alpha)}{\alpha-1}\right]^{\frac{\alpha-6}{\alpha-2}} \times [B_1(\alpha)]^{\frac{2(4-\alpha)}{\alpha-2}}. \tag{10.81}$$

By comparing the weak irradiance fluctuations and strong irradiance fluctuations with the large-scale log-irradiance variance, it can be shown that

$$\sigma^2_{\ln X} \approx \frac{0.49\tilde{\beta}_0^2(\alpha)}{\left\{1+\left[\frac{1.39}{B_{3,sph}(\alpha)}\right]^{\frac{2}{6-\alpha}}[\tilde{\beta}_0^2(\alpha)]^{\frac{2}{\alpha-2}}\right\}^{\frac{6-\alpha}{2}}}, \tag{10.82}$$

where $\tilde{\beta}_0^2(\alpha)$ is defined by Eq. (10.44). Similarly, the small-scale log-irradiance variance is

$$\sigma^2_{\ln Y} \approx \frac{0.51\tilde{\beta}_0^2(\alpha)}{\left\{1+[0.736\,\tilde{\beta}_0^2(\alpha)]^{\frac{2}{\alpha-2}}\right\}^{\frac{\alpha-2}{2}}}. \tag{10.83}$$

Consequently, the scintillation index for a spherical wave becomes

$$\sigma^2_{I,sph}(\alpha, L) = \exp(\sigma^2_{\ln X} + \sigma^2_{\ln Y}) - 1, \quad 0 \le \tilde{\beta}_0^2(\alpha) < \infty. \tag{10.84}$$

In Fig. 10.3 we show plots of the scintillation index for a spherical wave under the same conditions as those for the plane wave in Fig. 10.2. In this case we show results out to 5000 m for all three values of α. The wavelength

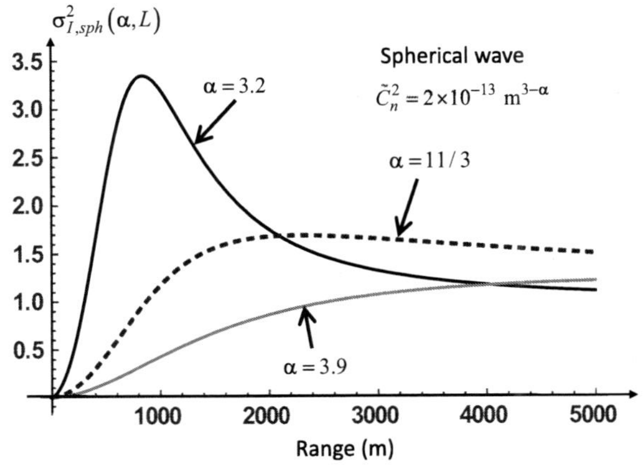

Figure 10.3 Scintillation index [Eq. (10.84)] for a spherical wave with $\lambda = 1.55\ \mu$m as a function of range and three different values of the spectral index.

is 1.55 μm and $\tilde{C}_n^2 = 2 \times 10^{-13}\,\mathrm{m}^{-2/3}$. These plots are similar to those in Fig. 10.2, but the maximum values of the scintillation index are higher than for the plane wave case.

10.6.4 Gaussian-beam wave: isotropic turbulence

Once again, following the analysis presented for the infinite plane wave, we now take up the case for the on-axis component of a Gaussian-beam wave. Here the saturation expression for the on-axis component is given by

$$\sigma_I^2(\alpha, 0, L) \sim 1 + \frac{B_3(\alpha)}{[\tilde{\sigma}_R^2(\alpha)]^{\frac{2(4-\alpha)}{\alpha-2}}}, \quad \tilde{\sigma}_R^2(\alpha) \to \infty, \tag{10.85}$$

where we make the approximation

$$B_3(\alpha) \approx \Theta B_{3,pl}(\alpha) + \bar{\Theta} B_{3,sph}(\alpha). \tag{10.86}$$

For the large-scale log-irradiance variance, we follow the weak fluctuation analysis of the plane wave and spherical wave, which leads to $\sigma_{\ln X}^2 \approx 0.49\tilde{\sigma}_B^2(\alpha, L)$, where

$$\tilde{\sigma}_B^2(\alpha, L) = \frac{\tilde{\sigma}_R^2(\alpha)}{\sin(\alpha\pi/4)} \operatorname{Re}\left[i^{\alpha/2-1}\, {}_2F_1\left(1 - \frac{\alpha}{2}, \frac{\alpha}{2}; 1 + \frac{\alpha}{2}; \bar{\Theta} + i\Lambda\right) - \frac{\alpha\Lambda^{\alpha/2-1}}{2(\alpha-1)}\right]. \tag{10.87}$$

In strong irradiance fluctuations, $\sigma_{\ln X}^2 \sim \frac{1}{2} B_3(\alpha)/[\tilde{\sigma}_R^2(\alpha)]^{2(4-\alpha)/(\alpha-2)}$. Thus, by imposing the weak and strong asymptotic results, and making adjustments so that the plane wave and spherical-wave expressions are specializations that lead to Eqs. (10.61) and (10.75), the combination gives us

$$\sigma_{\ln X}^2 \approx \frac{0.49\tilde{\sigma}_B^2(\alpha, L)}{\left\{1 + \left[\frac{1.39 - 0.41\Theta}{B_3(\alpha)}\right]^{\frac{2}{6-\alpha}} [\tilde{\sigma}_B^2(\alpha, L)]^{\frac{2}{\alpha-2}}\right\}^{\frac{6-\alpha}{2}}}. \tag{10.88}$$

Similarly, the small-scale log-irradiance variance is represented by

$$\sigma_{\ln Y}^2 \approx \frac{0.51\tilde{\sigma}_B^2(\alpha, L)}{\left\{1 + [0.736\,\tilde{\sigma}_B^2(\alpha, L)]^{\frac{2}{\alpha-2}}\right\}^{\frac{\alpha-2}{2}}}. \tag{10.89}$$

Combining the results of Eqs. (10.88) and (10.89) leads to

$$\sigma_I^2(\alpha, L) = \exp\left(\sigma_{\ln X}^2 + \sigma_{\ln Y}^2\right) - 1, \quad 0 \le \tilde{\sigma}_R^2(\alpha) < \infty. \tag{10.90}$$

10.7 Aperture Averaging: Weak Irradiance Fluctuations

Figure 10.4 is basically a simplified diagram for a free-space optical communication (FSOC) system with a laser beam as the source. However, the special cases we consider are a large-aperture collimated beam (essentially a plane wave) and a small aperture divergent beam (essentially a spherical wave). By treating the lens as part of the Rx system, atmospheric effects are taken into account only before the lens. Accordingly, we assume that the distance from the Tx to the lens is L and the distance from the lens to the output plane is L_f. The total path length is therefore the sum, $L + L_f$.

The distance L_f is chosen in Fig. 10.4 to place the photo-detector in the image plane. Our objective is to find the reduced scintillation index behind a Rx lens due to aperture averaging.[10]

10.7.1 Flux variance: plane wave

After passing through a Rx lens of diameter D, the reduced scintillation index of an optical wave is referred to as the *flux variance*. In the case of a plane wave, the flux variance under weak irradiance fluctuations and isotropic spectrum model (10.1) is defined by

$$\sigma^2_{I,pl}(\alpha,D)=8\pi^2A(\alpha)\tilde{C}_n^2k^2L\int_0^1\int_0^\infty \kappa^{1-\alpha}\exp\left(-\frac{D^2\kappa^2}{16}\right)\left[1-\cos\left(\frac{L\kappa^2\xi}{k}\right)\right]d\kappa d\xi, \tag{10.91}$$

which reduces to[7,10]

$$\begin{aligned}\sigma^2_{I,pl}(\alpha,D)&=\frac{8\pi^2A(\alpha)}{\alpha}\Gamma\left(1-\frac{\alpha}{2}\right)\tilde{C}_n^2k^2L\left(\frac{16}{D^2}\right)^{2-\alpha}\mathrm{Re}\left[\frac{\alpha}{2}-\frac{\left(1+\frac{16iL}{D^2}\right)^{\alpha/2}}{\frac{16iL}{D^2}}\right]\\&=\frac{\tilde{\sigma}_R^2(\alpha)}{\sin(\alpha\pi/4)}\left\{\left[1+\left(\frac{kD^2}{16L}\right)^2\right]^{\alpha/4}\sin\left[\frac{\alpha}{2}\tan^{-1}\left(\frac{16L}{kD^2}\right)\right]-\frac{\alpha}{2}\left(\frac{16L}{kD^2}\right)^{1-\alpha/2}\right\}.\end{aligned} \tag{10.92}$$

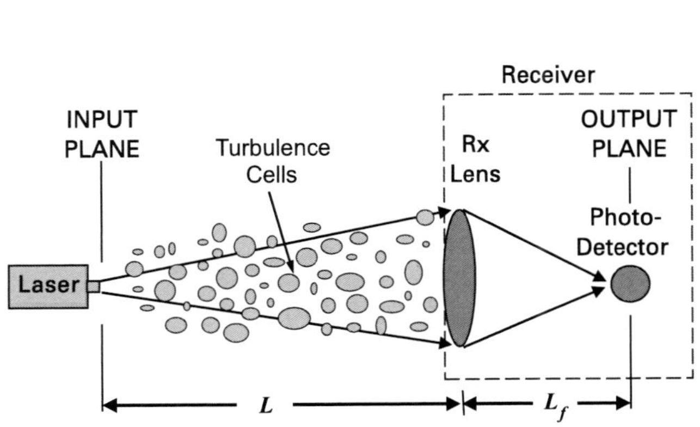

Figure 10.4 Propagation geometry for an optical wave propagating through a thin Gaussian lens at the Rx. Total path length is $L + L_f$ (figure adapted from Ref. 23).

This last result follows directly from Eqs. (10.38) and (10.39) by making the replacement $Q_m \to 16L/kD^2$.

10.7.2 Flux variance: spherical wave

In an analogous fashion, the flux variance of a spherical wave incident on an aperture of diameter D under weak irradiance fluctuations and spectrum model (10.1) is

$$\begin{aligned}\sigma^2_{I,sph}(\alpha, L) &= 8\pi^2 A(\alpha)\tilde{C}_n^2 k^2 L \int_0^1\int_0^\infty \kappa^{1-\alpha}\exp\left(-\frac{D^2\kappa^2\xi^2}{16}\right)\\&\quad\times\left\{1-\cos\left[\frac{L\kappa^2\xi(1-\xi)}{k}\right]\right\}d\kappa\, d\xi\\&= 8\pi^2 A(\alpha)\tilde{C}_n^2 k^2 L\,\mathrm{Re}\left\{\frac{\Gamma\left(1-\frac{\alpha}{2}\right)}{2(\alpha-1)}\left(\frac{D^2}{16}\right)\right.\\&\quad\left.-\frac{\Gamma\left(1-\frac{\alpha}{2}\right)}{\alpha}\left(\frac{iL}{k}\right)^{\alpha/2-1}{}_2F_1\left(1-\frac{\alpha}{2},\frac{\alpha}{2};1+\frac{\alpha}{2};1+\frac{ikD^2}{16L}\right)\right\},\end{aligned} \tag{10.93}$$

from which we deduce that

$$\begin{aligned}\sigma^2_{I,sph}(\alpha, D) &= \frac{\tilde{\sigma}^2_R(\alpha)}{\sin(\alpha\pi/4)}\times\left\{\mathrm{Re}\left[i^{\alpha/2-1}{}_2F_1\left(1-\frac{\alpha}{2},\frac{\alpha}{2};1+\frac{\alpha}{2};1+\frac{ikD^2}{16L}\right)\right]\right.\\&\quad\left.-\frac{\alpha}{2(\alpha-1)}\left(\frac{kD^2}{16L}\right)^{\alpha/2-1}\right\}.\end{aligned} \tag{10.94}$$

10.8 Irradiance Covariance Function

As related examples featuring isotropic non-Kolmogorov turbulence, we now consider the irradiance covariance function of a plane wave and spherical wave. Our treatment here will follow that for the case of conventional Kolmogorov turbulence in Section 8.5. However, because the inner scale seemed to have only a modest effect on the normalized covariance function under conventional Kolmogorov turbulence, we assume that the same is true for non-Kolmogorov turbulence.

10.8.1 Plane wave

If ρ denotes the distance between two points on the wavefront in the Rx pupil plane, the plane wave irradiance covariance function is described by

$$\begin{aligned} B_{I,pl}(\rho,\alpha,L) &= 8\pi^2 A(\alpha)\tilde{C}_n^2 k^2 L \int_0^1\int_0^\infty \kappa\Phi_n(\kappa)J_0(\kappa\rho)\left[1-\cos\left(\frac{L\kappa^2\xi}{k}\right)\right]d\kappa\, d\xi \\ &= 8\pi^2 A(\alpha)\tilde{C}_n^2 k^2 L \int_0^\infty \kappa^{1-\alpha}J_0(\kappa\rho)\left[1-\frac{k}{L\kappa^2}\sin\left(\frac{L\kappa^2}{k}\right)\right]d\kappa, \end{aligned} \tag{10.95}$$

where we have performed the integration on ξ. To evaluate the remaining integral in Eq. (10.95), we use ($C9$) in Appendix C for the Bessel function and split the integral into a sum, which yields [temporarily assuming that $\frac{1}{2} < \alpha < 2$ up until we reach Eq. (10.98) to avoid a singularity at $\kappa = 0$]

$$\begin{aligned} \int_0^\infty \kappa^{1-\alpha}\, G_{02}^{10}\left(\frac{\kappa^2\rho^2}{4}\middle|\, 0,\ 0\right)d\kappa &= \frac{1}{2}\int_0^\infty t^{-\alpha/2}\, G_{02}^{10}\left(\frac{t\rho^2}{4}\middle|\, 0,\ 0\right)dt \\ &= \frac{\Gamma\left(1-\frac{\alpha}{2}\right)}{2\Gamma\left(\frac{\alpha}{2}\right)}\left(\frac{\rho^2}{4}\right)^{\alpha/2-1}, \end{aligned} \tag{10.96}$$

$$\begin{aligned} &\frac{k}{L}\int_0^\infty \kappa^{-1-\alpha}\, e^{-iL\kappa^2/k}G_{02}^{10}\left(\frac{\kappa^2\rho^2}{4}\middle|\, 0,\ 0\right)d\kappa \\ &= \frac{k}{2L}\int_0^\infty t^{-1-\alpha/2}\, e^{-iLt/k}G_{02}^{10}\left(\frac{t\rho^2}{4}\middle|\, 0,\ 0\right)dt \\ &= \frac{1}{2}\left(\frac{L}{k}\right)^{\alpha/2-1} i^{\alpha/2}G_{12}^{11}\left(-\frac{ik\rho^2}{4L}\middle|\, \begin{matrix} 1+\frac{\alpha}{2} \\ 0,\ 0 \end{matrix}\right) \\ &= \frac{1}{2}\left(\frac{L}{k}\right)^{\alpha/2-1}\Gamma\left(-\frac{\alpha}{2}\right) i^{\alpha/2}\, {}_1F_1\left(-\frac{\alpha}{2};\ 1;\ \frac{ik\rho^2}{4L}\right), \end{aligned} \tag{10.97}$$

where we have used ($D8$) and ($D6$) in Appendix D. Combining Eqs. (10.96) and (10.97) with Eq. (10.95), we get

$$\begin{aligned} B_{I,pl}(\rho,\alpha,L) &= \frac{\tilde{\sigma}_R^2(\alpha)}{\sin\left(\frac{\alpha\pi}{4}\right)}\mathrm{Im}\left[i^{\alpha/2}\, {}_1F_1\left(-\frac{\alpha}{2};\ 1;\ \frac{ik\rho^2}{4L}\right)\right] \\ &\quad - \frac{\alpha\tilde{\sigma}_R^2(\alpha)}{2\Gamma\left(\frac{\alpha}{2}\right)\sin\left(\frac{\alpha\pi}{4}\right)}\left(\frac{k\rho^2}{4L}\right)^{\alpha/2-1}. \end{aligned} \tag{10.98}$$

In Fig. 10.5 we plot the scaled covariance function based on Eq. (10.98) with $\alpha = 3.2$ and $\alpha = 3.9$. For comparison, we also show the normalized covariance function for conventional Kolmogorov turbulence with $\alpha = 11/3$.

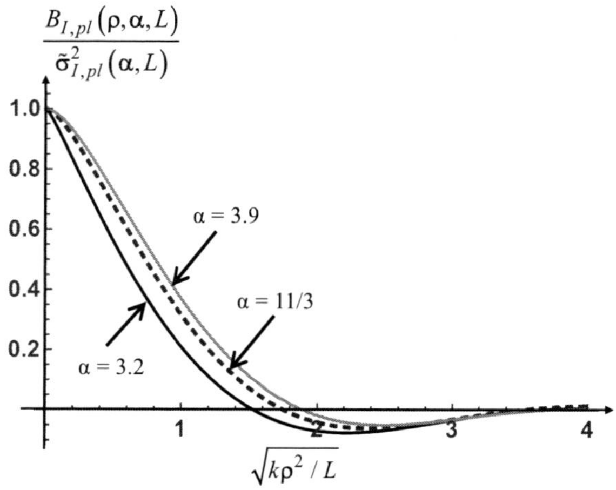

Figure 10.5 Scaled irradiance covariance function for a plane wave in non-Kolmogorov turbulence [Eq. (10.98)] with $\alpha = 3.2$ and $\alpha = 3.9$. Also shown is the scaled covariance function under conventional Kolmogorov turbulence with $\alpha = 11/3$.

10.8.2 Spherical wave

For the spherical wave case, we will follow the analysis presented above for the irradiance covariance function of a plane wave. Thus, we start with ($C9$) in Appendix C, which leads to

$$\begin{aligned} B_{I,sph}(\rho,\alpha,L) &= 8\pi^2 A(\alpha)\tilde{C}_n^2 k^2 L \mathrm{Re} \int_0^1\int_0^\infty \kappa^{1-\alpha} J_0(\kappa\xi\rho) \\ &\quad \times \left\{1-\exp\left[-\frac{iL\kappa^2}{k}\xi(1-\xi)\right]\right\} d\kappa\, d\xi \\ &= 8\pi^2 A(\alpha)\tilde{C}_n^2 k^2 L \left\{ \int_0^1\int_0^\infty \kappa^{1-\alpha} G_{02}^{10}\left(\frac{\rho^2\xi^2\kappa^2}{4}\middle| 0,0\right) d\kappa\, d\xi \right. \\ &\quad \left. - \mathrm{Re}\int_0^1\int_0^\infty \kappa^{1-\alpha} \exp\left[-\frac{iL\kappa^2}{k}\xi(1-\xi)\right] G_{02}^{10}\left(\frac{\rho^2\xi^2\kappa^2}{4}\middle| 0,0\right) d\kappa\, d\xi \right\}. \end{aligned} \tag{10.99}$$

Based on ($D8$) and ($D6$) in Appendix D, we obtain

$$\begin{aligned} B_{I,sph}(\rho,\alpha,L) &= 4\pi^2 A(\alpha)\tilde{C}_n^2 k^2 L \left[\left[\frac{\Gamma\left(1-\frac{\alpha}{2}\right)}{(\alpha-1)\Gamma\left(\frac{\alpha}{2}\right)} \left(\frac{\rho^2}{4}\right)^{\alpha/2-1} \right.\right. \\ &\quad \left.\left. - \mathrm{Re}\left\{ \left(\frac{iL}{k}\right)^{\alpha/2-1} \int_0^1 [\xi(1-\xi)]^{\alpha/2-1} G_{12}^{11}\left[\frac{k\rho^2\xi}{4iL(1-\xi)}\middle| \begin{matrix} \frac{\alpha}{2} \\ 0,\ 0 \end{matrix}\right] d\xi \right\} \right]\right]. \end{aligned} \tag{10.100}$$

To complete the integration in Eq. (10.100), we let $x = \xi/(1-\xi)$ followed by use of ($D11$) in Appendix D,

$$\int_0^1 [\xi(1-\xi)]^{\alpha/2-1} G_{12}^{11}\left[\frac{k\rho^2\xi}{4iL(1-\xi)}\middle|\begin{matrix}\frac{\alpha}{2}\\0,0\end{matrix}\right]d\xi = \int_0^\infty \frac{x^{\alpha/2-1}}{(x+1)^\alpha}G_{12}^{11}\left(-\frac{ik\rho^2 x}{4L}\middle|\begin{matrix}\frac{\alpha}{2}\\0,0\end{matrix}\right)dx$$
$$= \frac{1}{\Gamma(\alpha)}G_{23}^{22}\left(-\frac{ik\rho^2}{4L}\middle|\begin{matrix}1-\frac{\alpha}{2},\frac{\alpha}{2}\\\frac{\alpha}{2},0,0\end{matrix}\right). \tag{10.101}$$

Note that we can write Eq. (10.101) in terms of generalized hypergeometric functions by using ($MG1$) in Appendix A to obtain the sum

$$G_{23}^{22}\left(-\frac{ik\rho^2}{4L}\middle|\begin{matrix}1-\frac{\alpha}{2},\ \frac{\alpha}{2}\\\frac{\alpha}{2},\ 0,\ 0\end{matrix}\right) = \Gamma^2\left(\frac{\alpha}{2}\right)\Gamma\left(1-\frac{\alpha}{2}\right){}_1F_1\left(\frac{\alpha}{2};\ 1;\ -\frac{ik\rho^2}{4L}\right)$$
$$+\frac{\Gamma\left(-\frac{\alpha}{2}\right)\Gamma[\alpha]}{\Gamma\left(1-\frac{\alpha}{2}\right)}\left(-\frac{ik\rho^2}{4L}\right)^{\alpha/2}{}_2F_2\left(\alpha,\ 1;\ 1+\frac{\alpha}{2},\ 1+\frac{\alpha}{2};\ -\frac{ik\rho^2}{4L}\right). \tag{10.102}$$

However, for the sake of notational simplicity, we choose to leave the covariance function in terms of the Meijer G-function, which leads to

$$B_{I,sph}(\rho,\alpha,L) = 4\pi^2 A(\alpha)\tilde{C}_n^2 k^{3-\alpha/2}L^{\alpha/2}\left\{\frac{\Gamma\left(1-\frac{\alpha}{2}\right)}{(\alpha-1)\Gamma\left(\frac{\alpha}{2}\right)}\left(\frac{k\rho^2}{4L}\right)^{\alpha/2-1}\right.$$
$$\left.-\frac{1}{\Gamma(\alpha)}\mathrm{Re}\left[i^{\alpha/2-1}G_{23}^{22}\left(-\frac{ik\rho^2}{4L}\middle|\begin{matrix}1-\frac{\alpha}{2},\ \frac{\alpha}{2}\\\frac{\alpha}{2},\ 0,\ 0\end{matrix}\right)\right]\right\}. \tag{10.103}$$

In Fig. 10.6 we plot Eq. (10.103) for the scaled irradiance covariance function with $\alpha = 3.2$ and $\alpha = 3.9$. Recall that the spherical-wave scintillation index is given by Eq. (10.44). For comparison, we also show the normalized covariance function for conventional Kolmogorov turbulence with $\alpha = 11/3$. Note that the plot in Fig. 10.6 is similar to that in Fig. 10.5 where we featured the plane wave; however, the zero crossings occur at a larger value of $\sqrt{k\rho^2/L}$.

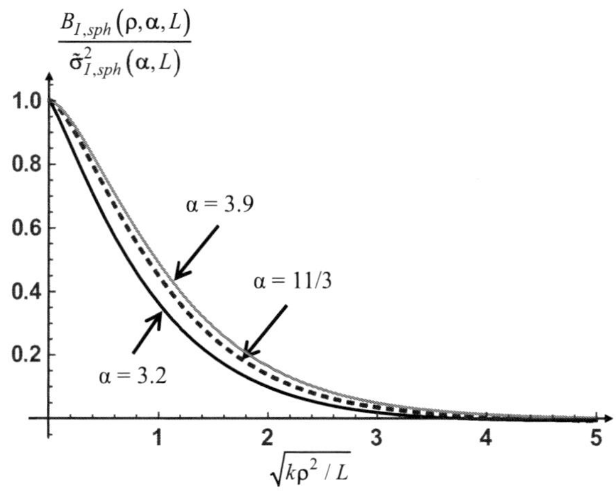

Figure 10.6 Scaled irradiance covariance function for a spherical wave in non-Kolmogorov turbulence [Eq. (10.103)] with $\alpha = 3.2$ and $\alpha = 3.9$. Also shown is the scaled covariance function under conventional Kolmogorov turbulence with $\alpha = 11/3$.

10.9 Temporal Spectrum of Irradiance

As noted previously, the relation between spatial and temporal statistics is generally determined by applying *Taylor's frozen turbulence hypothesis*. This requires an estimate of the mean wind velocity transverse to the direction of propagation. We will follow the derivation given in Section 8.7 but here with the development for the case of isotropic non-Kolmogorov turbulence.

10.9.1 Plane wave

Based on Eq. (10.95), the temporal covariance function for an infinite plane wave with $\mathbf{p} = \mathbf{V}_\perp \tau$ takes the form

$$B_{I,pl}(\alpha,\, \tau,\, L) = 8\pi^2 A(\alpha)\tilde{C}_n^2 k^2 L \int_0^\infty \kappa^{1-\alpha} J_0(\kappa V_\perp \tau)\left[1 - \frac{k}{L\kappa^2}\sin\left(\frac{L\kappa^2}{k}\right)\right] d\kappa, \tag{10.104}$$

where we have performed the integration on ξ. By taking the Fourier transform of Eq. (10.104) through making use of Eq. (8.153), we obtain

$$\begin{aligned} S_{I,pl}(\omega) &= \frac{315.827 A(\alpha)\tilde{C}_n^2 k^2 L}{V_\perp} \int_a^\infty \frac{\kappa^{1-\alpha}}{(\kappa^2 - a^2)^{1/2}}\left[1 - \frac{k}{L\kappa^2}\sin\left(\frac{L\kappa^2}{k}\right)\right] d\kappa \\ &= \frac{315.827 A(\alpha)\tilde{C}_n^2 k^2 L}{V_\perp}(I_1 + I_2), \end{aligned} \tag{10.105}$$

where we have defined $a = \omega/V_\perp$. To solve for the integral I_1, we make the change of variable $\kappa = a\sqrt{x+1}$ and recall the beta function ($G18$) in Appendix A; thus,

$$\begin{aligned} I_1 &= \int_a^\infty \frac{\kappa^{1-\alpha}}{(\kappa^2 - a^2)^{1/2}}\,d\kappa \\ &= \frac{1}{2a^{\alpha-1}} \int_0^\infty x^{-1/2}(x+1)^{-\alpha/2}\,dx \\ &= \frac{\sqrt{\pi}}{2a^{\alpha-1}} \frac{\Gamma\left(\frac{\alpha-1}{2}\right)}{\Gamma\left(\frac{\alpha}{2}\right)}. \end{aligned} \tag{10.106}$$

For the second integral I_2, we start with $-\sin x = \mathrm{Im}(e^{-ix})$ and find that

$$I_2 = \frac{k}{L} \mathrm{Im} \int_a^\infty \frac{\kappa^{-1-\alpha}}{(\kappa^2 - a^2)^{1/2}} \exp\left(-\frac{iL\kappa^2}{k}\right) d\kappa. \tag{10.107}$$

By using ($C1$) in Appendix C and ($D11$) in Appendix D, we obtain

$$\begin{aligned} I_2 &= \frac{k}{2La^{\alpha+1}} \mathrm{Im}\left[\exp\left(-\frac{iLa^2}{k}\right) \int_0^\infty \frac{x^{-1/2}}{(x+1)^{1+\alpha/2}} G_{01}^{10}\left(\frac{iLa^2 x}{k}\,\middle|\, 0\right) dx\right] \\ &= \frac{1}{2\Gamma\left(1+\frac{\alpha}{2}\right)} \left(\frac{L}{k}\right)^{(\alpha-1)/2} \left(\frac{\omega}{\omega_t}\right)^{-(\alpha+1)} \\ &\quad \times \mathrm{Im}\left\{\exp\left[-i\left(\frac{\omega}{\omega_t}\right)^2\right] G_{12}^{21}\left[i\left(\frac{\omega}{\omega_t}\right)^2 \middle|\, \begin{matrix} \frac{1}{2} \\ \frac{1+\alpha}{2},\ 0 \end{matrix}\right]\right\}. \end{aligned} \tag{10.108}$$

Last, by combining Eqs. (10.106) and (10.108) with Eq. (10.105), we arrive at

$$\begin{aligned} S_{I,pl}(\omega) &= \frac{157.91\sqrt{\pi}A(\alpha)\tilde{C}_n^2 k^{3-\alpha/2} L^{\alpha/2}}{\omega_t} \frac{\Gamma\left(\frac{\alpha-1}{2}\right)}{\Gamma\left(\frac{\alpha}{2}\right)} \left(\frac{\omega}{\omega_t}\right)^{1-\alpha} \\ &\quad \times \left[\left[1 + \frac{\Gamma\left(\frac{\alpha}{2}\right)}{\sqrt{\pi}\,\Gamma\left(\frac{\alpha-1}{2}\right)\Gamma\left(1+\frac{\alpha}{2}\right)} \left(\frac{\omega_t}{\omega}\right)^2 \right.\right. \\ &\quad \left.\left. \times \mathrm{Im}\left\{\exp\left[-i\left(\frac{\omega}{\omega_t}\right)^2\right] G_{12}^{21}\left[i\left(\frac{\omega}{\omega_t}\right)^2 \middle|\, \begin{matrix} \frac{1}{2} \\ \frac{1+\alpha}{2},\ 0 \end{matrix}\right]\right\}\right]\right]. \end{aligned} \tag{10.109}$$

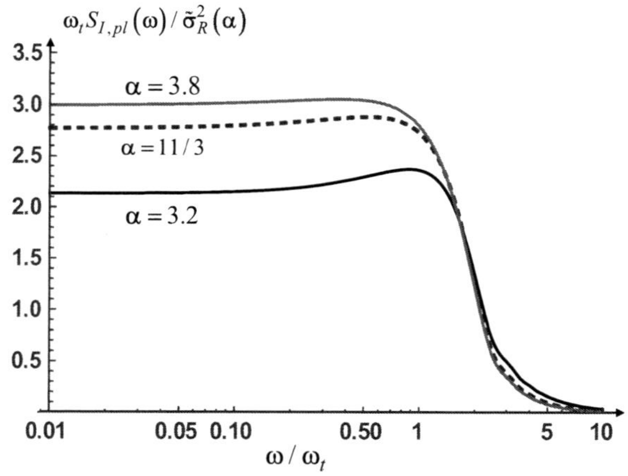

Figure 10.7 Scaled temporal power spectrum of irradiance for a plane wave in non-Kolmogorov turbulence [Eq. (10.109)] with $\alpha = 3.2$ and $\alpha = 3.8$. Also shown is the scaled temporal spectrum of irradiance under conventional Kolmogorov turbulence with $\alpha = 11/3$. The parameter $\omega_t = V_\perp/(L/k)^{1/2}$.

Shown in Fig. 10.7 are plots of the scaled temporal power spectrum derived from Eq. (10.109) for $\alpha = 3.2$, $11/3$, and 3.8. The behavior in all cases is nearly the same for $\omega > \omega_t$, but each curve approaches a different limiting value for frequencies $\omega \ll \omega_t$.

10.9.2 Spherical wave

By making the association $\xi\mathbf{p} = \mathbf{V}_\perp\tau$, the temporal power spectrum for a spherical wave takes the form

$$\begin{aligned} B_{I,sph}(\alpha, \tau, L) = 8\pi^2 A(\alpha)\tilde{C}_n^2 k^2 L \int_0^1 \int_0^\infty \kappa^{1-\alpha} J_0(\kappa V_\perp \tau) \\ \times \left\{1 - \cos\left[\frac{L\kappa^2}{k}\xi(1-\xi)\right]\right\} d\kappa\, d\xi. \end{aligned} \tag{10.110}$$

At this point we basically follow the derivation given in Section 8.7.3. First we apply the Fourier transform to Eq. (10.110) and make use of Eq. (8.153) to obtain

$$\begin{aligned} S_{I,sph}(\omega) &= \frac{315.827\tilde{C}_n^2 k^2 L}{V_\perp} \mathrm{Re} \int_0^1 \int_a^\infty \frac{\kappa^{1-\alpha}}{(\kappa^2 - a^2)^{1/2}} \\ &\quad \times \left\{1 - \exp\left[-i\frac{L\kappa^2}{k}\xi(1-\xi)\right]\right\} d\kappa\, d\xi \\ &= \frac{315.827\tilde{C}_n^2 k^2 L}{V_\perp} \mathrm{Re}(I_1 - I_2). \end{aligned} \tag{10.111}$$

The integral I_1 is recognized to be the same as Eq. (10.106), whereas

$$\begin{aligned} I_2 &= \int_0^1\int_a^\infty \frac{\kappa^{1-\alpha}}{(\kappa^2-a^2)^{1/2}}\exp\left[-i\frac{L\kappa^2}{k}\xi(1-\xi)\right]d\kappa\, d\xi \\ &= \frac{1}{2a^{\alpha-1}}\int_0^1\int_1^\infty \frac{x^{-\alpha/2}}{(x-1)^{1/2}}\exp\left[-i\frac{La^2x}{k}\xi(1-\xi)\right]dx\, d\xi \\ &= \frac{1}{2a^{\alpha-1}}\int_0^1\int_1^\infty \frac{x^{-\alpha/2}}{(x-1)^{1/2}}G_{01}^{10}\left[-i\frac{\omega^2x}{\omega_t^2}\xi(1-\xi)\middle|\, 0\right]dx\, d\xi. \end{aligned} \tag{10.112}$$

Here we defined $\kappa = a\sqrt{x}$ and applied property (C1) in Appendix C. Next, we use the integral relation (D10) in Appendix D and express the G-function in its Barnes integral form to find that

$$\begin{aligned} I_2 &= \left(\frac{\omega}{V_\perp}\right)^{1-\alpha}\frac{1}{2}\Gamma\left(\frac{1}{2}\right)\int_0^1 G_{12}^{20}\left[-i\frac{\omega^2}{\omega_t^2}\xi(1-\xi)\middle|\begin{matrix}\frac{\alpha}{2}\\ \frac{\alpha-1}{2},\ 0\end{matrix}\right]d\xi \\ &= \left(\frac{\omega}{V_\perp}\right)^{1-\alpha}\left(\frac{\sqrt{\pi}}{2}\right)\frac{1}{2\pi i}\int_C \frac{\Gamma\left(\frac{\alpha-1}{2}-s\right)\Gamma(-s)}{\Gamma\left(\frac{\alpha}{2}-s\right)}\left(-i\frac{\omega^2}{\omega_t^2}\right)^s\int_0^1 \xi^s(1-\xi)^s\,d\xi\, ds. \end{aligned} \tag{10.113}$$

Continuing to follow the analysis in Section 8.7.3, we have

$$\begin{aligned} I_2 &= \frac{\pi}{4}\left(\frac{\omega}{V_\perp}\right)^{1-\alpha}\frac{1}{2\pi i}\int_C \frac{\Gamma\left(\frac{\alpha-1}{2}-s\right)\Gamma(-s)\Gamma(1+s)}{\Gamma\left(\frac{3}{2}+s\right)\Gamma\left(\frac{\alpha}{2}-s\right)}\left(-\frac{i\omega^2}{4\omega_t^2}\right)^s ds \\ &= \frac{\pi}{4}\left(\frac{\omega}{V_\perp}\right)^{1-\alpha}G_{23}^{21}\left(-\frac{i\omega^2}{4\omega_t^2}\middle|\begin{matrix}0,\ \frac{\alpha}{2}\\ \frac{\alpha-1}{2},\ 0,\ -\frac{1}{2}\end{matrix}\right), \end{aligned} \tag{10.114}$$

and by combining the results of Eqs. (10.106) and (10.114) with Eq. (10.111), we finally obtain

$$\begin{aligned} S_{I,sph}(\omega) &= \frac{157.91\sqrt{\pi}A(\alpha)\tilde{C}_n^2k^{3-\alpha/2}L^{\alpha/2}}{\omega_t}\frac{\Gamma\left(\frac{\alpha-1}{2}\right)}{\Gamma\left(\frac{\alpha}{2}\right)}\left(\frac{\omega}{\omega_t}\right)^{1-\alpha} \\ &\quad\times \mathrm{Re}\left[1-\frac{\sqrt{\pi}\Gamma\left(\frac{\alpha}{2}\right)}{2\Gamma\left(\frac{\alpha-1}{2}\right)}G_{23}^{21}\left(-\frac{i\omega^2}{4\omega_t^2}\middle|\begin{matrix}0,\ \frac{\alpha}{2}\\ \frac{\alpha-1}{2},\ 0,\ -\frac{1}{2}\end{matrix}\right)\right]. \end{aligned} \tag{10.115}$$

We plot the scaled temporal power spectrum of irradiance for a spherical wave in Fig. 10.8. Analogous to the plane wave case, we set $\alpha = 3.2,\ 11/3,$

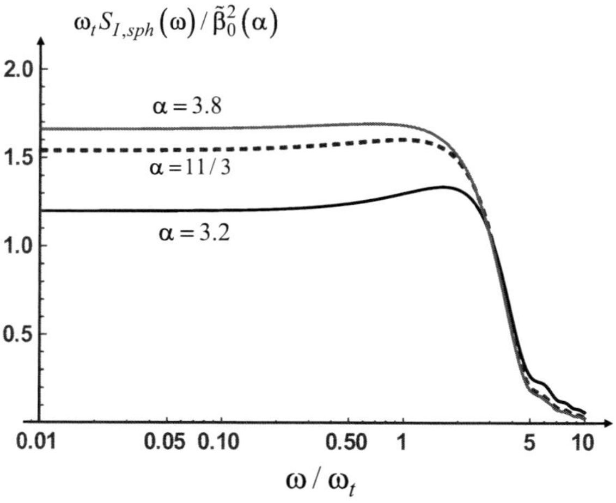

Figure 10.8 Scaled temporal power spectrum of irradiance for a spherical wave in non-Kolmogorov turbulence [Eq. (10.115)] with $\alpha = 3.2$ and $\alpha = 3.8$. Also shown is the scaled temporal spectrum of irradiance in conventional Kolmogorov turbulence.

and 3.8. Here again, we note that the plots are all basically the same for $\omega > 2\omega_t$ and all approach a different limiting value for $\omega \ll \omega_t$.

10.10 Phase Statistics: Plane Wave

We now consider two statistical quantities associated with the phase of an infinite plane wave. Following the treatment in Chapter 8, we will perform the following calculations with a variation of the *von Kármán spectrum*:

$$\Phi_n(\kappa) = \frac{A(\alpha)\tilde{C}_n^2}{(\kappa^2 + \kappa_0^2)^{\alpha/2}}, \quad 3 < \alpha < 4. \tag{10.116}$$

10.10.1 Phase structure function

For a plane wave, the isotropic phase structure function is defined by

$$\begin{aligned} D_{S,pl}(\rho, L) &= 4\pi^2\tilde{C}_n^2 k^2 L A(\alpha)\int_0^\infty \frac{\kappa}{(\kappa^2+\kappa_0^2)^{\alpha/2}}[1 - J_0(\kappa\rho)]\,d\kappa \\ &= \frac{4\pi^2\tilde{C}_n^2 k^2 L}{\alpha - 2}A(\alpha)\kappa_0^{2-\alpha}\left[1 - \frac{(\alpha-2)}{\Gamma\left(\frac{\alpha}{2}\right)}\left(\frac{\kappa_0\rho}{2}\right)^{\alpha/2-1}K_{\alpha/2-1}(\kappa_0\,\rho)\right], \quad \rho \gg l_0, \end{aligned} \tag{10.117}$$

which is based on a geometrical optics approximation. In this derivation we have relied on ($C9$) in Appendix C and ($D11$) in Appendix D.

10.10.2 Phase variance

Under a geometrical optics approximation, the phase variance is defined by

$$\tilde{\sigma}_S^2(\alpha) = 4\pi^2 \tilde{C}_n^2 k^2 L A(\alpha) \int_0^\infty \frac{\kappa}{(\kappa^2+\kappa_0^2)^{\alpha/2}}\, d\kappa = \frac{4\pi^2 \tilde{C}_n^2 k^2 L}{\alpha-2} A(\alpha)\kappa_0^{2-\alpha}. \tag{10.118}$$

To include diffraction effects, we first write

$$\sigma_{\chi,pl}^2(\alpha, L) + \sigma_{S,pl}^2(\alpha, L) = 4\pi^2 k^2 L \int_0^\infty \kappa \Phi_n(\kappa)\, d\kappa, \tag{10.119}$$

where

$$\sigma_{\chi,pl}^2(\alpha, L) = \frac{1}{4}\sigma_{I,pl}^2(\alpha, L). \tag{10.120}$$

Hence, it follows that

$$\sigma_{S,pl}^2(\alpha, L) = \frac{4\pi^2 \tilde{C}_n^2 k^2 L}{\alpha-2} A(\alpha)\kappa_0^{2-\alpha} - \frac{1}{4}\tilde{\sigma}_R^2(\alpha), \tag{10.121}$$

where $\sigma_{I,pl}^2(\alpha, L) \equiv \tilde{\sigma}_R^2(\alpha)$ is defined by Eq. (10.17).

10.11 Imaging Performance Measures

In this section we examine performance measures in an imaging system under isotropic non-Kolmogorov turbulence conditions. Again, we assume that the optical wave is a point source. To begin, we note that we can derive the spatial coherence radius of a spherical wave by referring to Eq. (10.25) and setting $D_{sph}(\rho_0, \alpha) = 2$. Then solving for ρ_0, we see that

$$\rho_{0,sph}(\alpha) = \left[\frac{\alpha\Gamma(\alpha-1)}{(\alpha-1)2^\alpha}\frac{\Gamma(-\alpha/2)}{\Gamma(\alpha/2)}\cos\left(\frac{\alpha\pi}{2}\right)\right]^{1/(2-\alpha)} (\tilde{C}_n^2 k^2 L)^{1/(2-\alpha)}. \tag{10.122}$$

This same result follows from Eq. (10.35) by setting $\mu_x = \mu_y = 1$ and $\tilde{C}_{n,eff}^2 = \tilde{C}_n^2$. The corresponding Fried parameter is related by $r_0 = 2.1\, \rho_{0,sph}(\alpha)$.

10.11.1 Long-term beam radius: detector plane

The mean on-axis irradiance of a spherical wave passing through a Rx lens of diameter D is given by

$$\langle I(0, L+L_f)\rangle \approx \frac{W_{Rx}^2}{W_{LT}^2}, \tag{10.123}$$

where $W_{Rx} = D/2\sqrt{2}$ is the Gaussian beam radius of the Rx lens and W_{LT} is the long-term spot radius in the image (detector) plane. By taking the non-

Kolmogorov spectrum (10.1), we find that the *long-term spot radius* in the image plane of an incident spherical wave is

$$W_{LT} \approx W_2[1+T(\alpha)]^m, \quad m = \begin{cases} \frac{1}{2} \text{ (near field)} \\ \frac{3}{5} \text{ (far field)}, \end{cases} \tag{10.124}$$

where W_2 is the free-space spot radius and

$$T(\alpha) = 4\pi^2 A(\alpha) k^2 L \tilde{C}_n^2 \int_0^1 \int_0^\infty \kappa^{1-\alpha} \left[1 - \exp\left(-\frac{\kappa^2 D^2 \xi^2}{16}\right)\right] d\kappa\, d\xi. \tag{10.125}$$

Evaluation of the integrals in Eq. (10.125) gives us

$$T(\alpha) = \frac{\alpha\, \tilde{\sigma}_R^2(\alpha)}{4(\alpha-1)\sin(\alpha\pi/4)} \left(\frac{kD^2}{16L}\right)^{\alpha/2-1} = \left(\frac{D}{r_0}\right)^{\alpha-2}. \tag{10.126}$$

10.11.2 Point spread function and modulation transfer function

For a point source incident on an imaging system in the presence of atmospheric turbulence, the PSF in the output detector plane is given by

$$\begin{aligned} \mathrm{PSF}(\mathbf{r}) = {} & \frac{W_{Rx}^2}{W_2^2} \exp\left(-\frac{2r^2}{W_2^2}\right) \exp\left\{-4\pi^2 k^2 L \int_0^1 \int_0^\infty \kappa \Phi_n(\kappa) \right. \\ & \left. \times \left[1 - \exp\left(-\frac{L\kappa^2\xi^2}{k\Omega_{Rx}}\right) I_0\left(\frac{2\kappa\xi L r}{L_f \Omega_{Rx}}\right)\right] d\kappa\, d\xi \right\}, \end{aligned} \tag{10.127}$$

where $\Omega_{Rx} = 2L/kW_{Rx}^2 = 16L/kD^2$ and we are recalling Eq. (9.13). Upon evaluating the integrals in Eq. (10.127) with the spectrum model of Eq. (10.1), we obtain the approximation

$$\mathrm{PSF}(\mathbf{r}) = \frac{W_{Rx}^2}{W_2^2[1+(D/r_0)^{\alpha-2}]^{6/5}} \exp\left\{-\frac{2r^2}{W_2^2[1+(D/r_0)^{\alpha-2}]^{6/5}}\right\}. \tag{10.128}$$

The corresponding total MTF (optics plus atmosphere) is obtained by taking the normalized two-dimensional Fourier transform of Eq. (10.128), from which we calculate

$$\mathrm{MTF}(\nu) = \exp\left\{-4\left(\frac{\lambda F_{Rx}\nu}{D}\right)^2 [1+(D/r_0)^{\alpha-2}]^{6/5}\right\}. \tag{10.129}$$

10.11.3 Strehl ratio

In imaging applications, the Strehl ratio (SR) is one of the most important performance measures. Based on Eqs. (9.19) and (10.129), the SR is

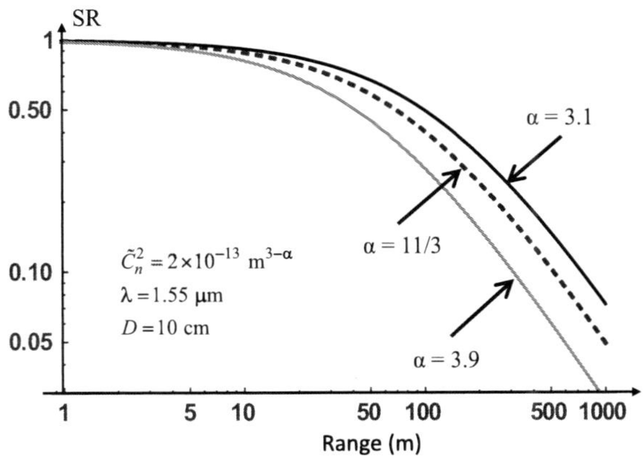

Figure 10.9 A comparison of Strehl ratio (10.130) for non-Kolmogorov turbulence and Eq. (10.131) for conventional Kolmogorov turbulence. The Rx aperture diameter is 10 cm, and the non-Kolmogorov spectral index is $\alpha = 3.1$ and $\alpha = 3.9$.

$$\mathrm{SR} = \frac{1}{[1 + (D/r_0)^{\alpha-2}]^{6/5}}, \quad 3 < \alpha < 4. \tag{10.130}$$

For $\alpha = 11/3$, the Strehl ratio (10.130) reduces to Eq. (9.20), i.e.,

$$\mathrm{SR} = \frac{1}{[1 + (D/r_0)^{5/3}]^{6/5}}. \tag{10.131}$$

In Fig. 10.9, the SR defined by Eq. (10.130) is plotted as a function of range out to 1 km with Rx aperture diameter $D = 10\,\mathrm{cm}$, $\tilde{C}_n^2 = 2\times10^{-13}\,\mathrm{m}^{3-\alpha}$, and $\lambda = 1.55\,\mu\mathrm{m}$. The featured non-Kolmogorov spectral indices are $\alpha = 3.1$ and $\alpha = 3.9$. Also shown is the conventional Kolmogorov SR [Eq. (10.131)].

10.12 Tilt Variances

Below we develop tilt variances comparable to those in Chapter 9 but now under isotropic non-Kolmogorov turbulence.

10.12.1 Zernike-tilt variance

The two-axis Z-tilt variance for an infinite plane wave on an aperture of diameter D under a geometrical optics approximation involves the product of two Bessel functions of the same order and argument. Under non-Kolmogorov turbulence defined by Eq. (10.1), this leads to the tilt angle variance

$$
\begin{aligned}
T_Z^2 &= 4\pi^2 A(\alpha)k^2 \int_0^L dz\tilde{C}_n^2(z) \int_0^\infty d\kappa\, \kappa^{1-\alpha}\left(\frac{16}{kD}\right)^2 \left[\frac{J_2(\kappa D/2)}{\kappa D/2}\right]^2 \\
&= \frac{10{,}106.47A(\alpha)}{2^{\alpha-2}D^{4-\beta}} \int_0^L \tilde{C}_n^2(z)\, dz \int_0^\infty x^{-\alpha-1} J_2^2(x)\, dx \\
&= \frac{2851A(\alpha)}{2^{\alpha-2}D^{4-\alpha}} \frac{\Gamma\left(2-\frac{\alpha}{2}\right)}{\Gamma\left(\frac{1+\alpha}{2}\right)\Gamma\left(1+\frac{\alpha}{2}\right)\Gamma\left(3+\frac{\alpha}{2}\right)} \int_0^L \tilde{C}_n^2(z)\, dz,
\end{aligned}
\tag{10.132}
$$

where we have used ($D3$) in Appendix D to evaluate the inside integral. For $\alpha = 11/3$, this last result reduces to Eq. (9.55).

The scaled Zernike-tilt variance is shown in Fig. 10.10 as a function of spectral index α. The dashed vertical line locates the $\alpha = 11/3$ value associated with the Kolmogorov spectrum.

10.12.2 Gradient-tilt variance

The G-tilt variance can be evaluated in the same manner as done in Section 10.12.1 but now with the filter function $(4/kD)^2 J_1^2(\kappa D/2)$. In this case we again use the non-Kolmogorov spectrum model (10.1), from which the G-tilt angle variance is

$$
\begin{aligned}
T_G^2 &= 4\pi^2 A(\alpha)k^2 \int_0^L dz\tilde{C}_n^2(z) \int_0^\infty d\kappa\, \kappa^{1-\alpha}\left(\frac{4}{kD}\right)^2 J_1^2(\kappa D/2) \\
&= \frac{631.65A(\alpha)}{2^{\alpha-2}D^{4-\alpha}} \int_0^\infty x^{1-\alpha} J_1^2(x)\, dx \int_0^L \tilde{C}_n^2(z)\, dz,
\end{aligned}
\tag{10.133}
$$

where we have made the change of variable $x = \kappa D/2$. It now follows from ($D3$) in Appendix D that

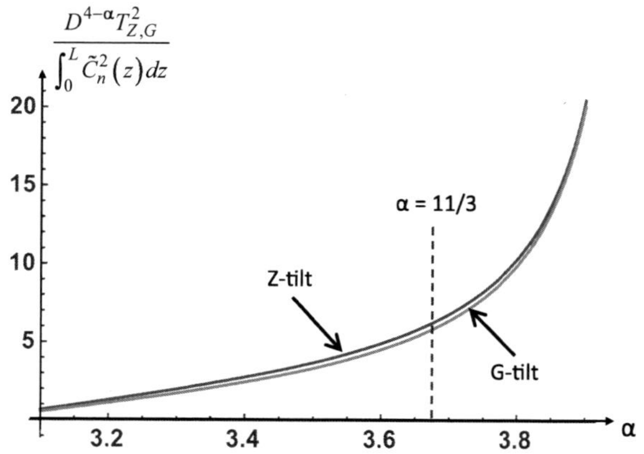

Figure 10.10 Scaled tilt variance based on Z-tilt [Eq. (10.132)] and G-tilt [Eq. (10.134)] as a function of spectral index.

$$T_G^2 = \frac{178.19A(\alpha)}{2^{\alpha-2}D^{4-\alpha}} \frac{\Gamma\left(2-\frac{\alpha}{2}\right)\Gamma\left(\frac{\alpha-1}{2}\right)}{\Gamma\left(\frac{\alpha}{2}\right)\Gamma\left(1+\frac{\alpha}{2}\right)} \int_0^L \tilde{C}_n^2(z)\,dz. \qquad (10.134)$$

The scaled gradient-tilt variance is plotted in Fig. 10.10 along with the scaled Zernike-tilt variance. Clearly, the difference between Z-tilt and G-tilt is small for all α in the interval $3 < \alpha < 4$.

10.12.3 Zernike-tilt power spectral density

In this section we follow the analysis in Section 9.2.4 to calculate the power spectral density of Zernike tilt. Starting with the temporal covariance function,

$$\begin{aligned} B_{T,Z}(\alpha, \tau) = {} & 39.48A(\alpha)k^2 \int_0^L dz\tilde{C}_n^2(z) \int_0^\infty d\kappa\, \kappa^{1-\alpha} J_0(\kappa V_\perp \tau) \\ & \times \left(\frac{4}{kD}\right)^2 \left[\frac{4J_2(\kappa D/2)}{\kappa D/2}\right]^2, \end{aligned} \qquad (10.135)$$

where we have set $\rho = V_\perp \tau$, the Fourier transform leads to

$$S_{T,Z}(\alpha, \omega) = \frac{161{,}710.08}{D^4 V_\perp} A(\alpha) \int_0^L dz\tilde{C}_n^2(z) \int_a^\infty \frac{\kappa^{-\alpha-1}}{(\kappa^2 - a^2)^{1/2}} J_2^2(\kappa D/2)\,d\kappa. \qquad (10.136)$$

Here, $a = \omega/V_\perp$ and we are treating $V_\perp$ as a constant. Next, we make the change of variable $\kappa = a\sqrt{x}$ and use ($C16$) in Appendix C to find that

$$\begin{aligned} S_{T,Z}(\alpha, \omega) = {} & \frac{40{,}425.80D^{\alpha-3}}{V_\perp 2^\alpha \sqrt{\pi}} A(\alpha) \left(\frac{2V_\perp}{\omega D}\right)^{\alpha+1} \int_0^L dz\tilde{C}_n^2(z) \\ & \times \int_1^\infty \frac{x^{-(1+\alpha/2)}}{(x-1)^{1/2}} G_{13}^{11}\left(\frac{\omega^2 D^2 x}{4V_\perp^2}\middle|\begin{matrix} \frac{1}{2} \\ 2,\ 0,\ -2 \end{matrix}\right) dx. \end{aligned}$$

Then, evaluating the integral over x in this last equation with ($D10$) in Appendix D, we are led to

$$\begin{aligned} S_{T,Z}(\alpha, \omega) = {} & \frac{40{,}425.80D^{\alpha-3}}{2^\alpha V_\perp} A(\alpha) \left(\frac{2V_\perp}{\omega D}\right)^{\alpha+1} G_{24}^{21}\left(\frac{\omega^2 D^2}{4V_\perp^2}\middle|\begin{matrix} \frac{1}{2},\ 1+\frac{\alpha}{2} \\ \frac{\alpha+1}{2},\ 2,\ 0,\ -2 \end{matrix}\right) \\ & \times \int_0^L \tilde{C}_n^2(z)\,dz. \end{aligned} \qquad (10.137)$$

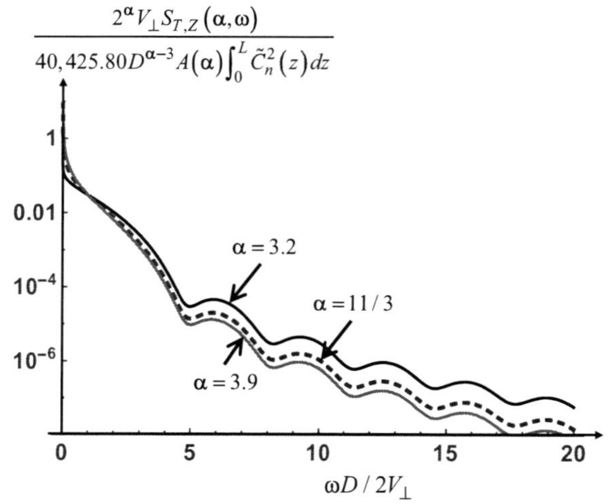

Figure 10.11 Scaled power spectral density of tilt. The transverse velocity $V_{\perp}$ is assumed constant.

We plot scaled versions of the Zernike-tilt power spectral density defined by Eq. (10.137) in Fig. 10.11 as a function of $\omega D/2V_{\perp}$ for spectral index values $\alpha = 3.2$, $11/3$, and 3.9. Although the transverse velocity generally varies along the path, we will consider it a constant in Fig. 10.11.

10.13 Tilt Anisoplanatism

The use of a NGS or LGS beacon leads to angular and focal forms of anisoplanatism. To begin, we examine *tilt anisoplanatism* under the assumption of non-Kolmogorov turbulence. As before, we separately consider cases of a single source and two Rx apertures and, conversely, the case of two point sources and one Rx aperture.

10.13.1 Multi-apertures

For a reference NGS, the received waveform is essentially that of an incident plane wave. Based on the non-Kolmogorov spectrum (10.1), we find that tilt anisoplanatism using Z-tilt in the single-source, multi-aperture case leads to the expressions for total, perpendicular, and parallel tilt variance, defined by

$$\begin{pmatrix} \sigma_Z^2(\alpha) \\ \sigma_{Z\perp}^2(\alpha) \\ \sigma_{Z\parallel}^2(\alpha) \end{pmatrix} = \frac{3217A(\alpha)}{D^2}\int_0^L dz\tilde{C}_n^2(z)\int_0^\infty d\kappa\,\kappa^{1-\alpha}\left[\frac{J_2(\kappa D/2)}{\kappa D/2}\right]^2 \times \int_0^{2\pi} d\phi\,[1-\cos(\kappa d\cos\phi)]\begin{pmatrix} 1 \\ \sin^2\phi \\ \cos^2\phi \end{pmatrix}, \tag{10.138}$$

where the separation distance d between apertures satisfies $d > D$.

For the total variance, we have

$$\sigma_Z^2(\alpha) = \frac{20{,}213}{D^2} A(\alpha) \int_0^L dz \tilde{C}_n^2(z) \int_0^\infty \kappa^{1-\alpha} \left[\frac{J_2(\kappa D/2)}{\kappa D/2}\right]^2 [1 - J_0(\kappa d)]\, d\kappa, \tag{10.139}$$

and with the substitution $x = \kappa D/2$, Eq. (10.139) becomes

$$\begin{aligned}\sigma_Z^2(\alpha) &= \frac{20{,}213A(\alpha)}{2^{\alpha-2}D^{4-\alpha}} \int_0^L dz \tilde{C}_n^2(z) \int_0^\infty x^{-\alpha-1} J_2^2(x) \left[1 - J_0\left(\frac{2xd}{D}\right)\right] dx \\ &= \frac{20{,}213A(\alpha)}{2^{\alpha-2}D^{4-\alpha}} \int_0^L dz \tilde{C}_n^2(z) \left[\int_0^\infty x^{-\alpha-1} J_2^2(x)\, dx \right. \\ &\quad \left. - \int_0^\infty x^{-\alpha-1} J_2^2(x)\, J_0\left(\frac{2xd}{D}\right) dx\right].\end{aligned} \tag{10.140}$$

By using the result of ($D3$) in Appendix D, we find that

$$\int_0^\infty x^{-\alpha-1} J_2^2(x)\, dx = \frac{\Gamma\left(2 - \frac{\alpha}{2}\right)\Gamma\left(\frac{1+\alpha}{2}\right)}{2\sqrt{\pi}\Gamma\left(1 + \frac{\alpha}{2}\right)\Gamma\left(3 + \frac{\alpha}{2}\right)}, \tag{10.141}$$

and for the second integral in Eq. (10.140), we use ($D4$) in Appendix D and ($MG1$) in Appendix A to get

$$\begin{aligned}\int_0^\infty x^{-\alpha-1} J_2^2(x)\, J_0\left(\frac{2xd}{D}\right) dx &= \frac{D}{2d\sqrt{\pi}}\, G_{33}^{12}\left(\frac{D^2}{d^2}\middle|\begin{matrix}\frac{1}{2},\ -\frac{\alpha}{2},\ \frac{1}{2} \\ 2 - \frac{\alpha+1}{2},\ -\frac{\alpha+1}{2},\ -\frac{\alpha+1}{2} - 2\end{matrix}\right) \\ &= 0.0078 \frac{\Gamma\left(2 - \frac{\alpha}{2}\right)}{\Gamma\left(\frac{\alpha}{2} - 1\right)} \left(\frac{D}{d}\right)^{4-\alpha} {}_3F_2\left(2 - \frac{\alpha}{2},\ \frac{5}{2},\ 2 - \frac{\alpha}{2};\ 3,\ 5;\ \frac{D^2}{d^2}\right).\end{aligned} \tag{10.142}$$

However, over the range of values $3 < \alpha < 4$, the ${}_3F_2$ function is essentially unity for $d > D$, so in that case the total tilt variance can be closely approximated by

$$\sigma_Z^2(\alpha) \approx A(\alpha)\left[\frac{5702}{2^{\alpha-2}D^{4-\alpha}}\frac{\Gamma\left(2-\dfrac{\alpha}{2}\right)\Gamma\left(\dfrac{1+\alpha}{2}\right)}{\Gamma\left(1+\dfrac{\alpha}{2}\right)\Gamma\left(3+\dfrac{\alpha}{2}\right)} - \frac{44.48}{2^{\alpha-2}d^{4-\alpha}}\frac{\Gamma\left(2-\dfrac{\alpha}{2}\right)}{\Gamma\left(\dfrac{\alpha}{2}-1\right)}\right]$$
$$\times \int_0^L \tilde{C}_n^2(z)\,dz, \quad d > D, \tag{10.143}$$

with an error on the order of 1–2% or less.

The perpendicular component similarly leads to

$$\begin{aligned}\sigma_{Z\perp}^2(\alpha) &= \frac{20{,}213}{D^2}A(\alpha)\int_0^L dz\tilde{C}_n^2(z)\int_0^\infty \kappa^{1-\alpha}\left[\frac{J_2(\kappa D/2)}{\kappa D/2}\right]^2\left[\frac{1}{2} - \frac{J_1(\kappa d)}{\kappa d}\right]d\kappa \\ &= \frac{10{,}106.5A(\alpha)}{2^{\alpha-2}D^{4-\alpha}}\int_0^L dz\tilde{C}_n^2(z)\left[\int_0^\infty x^{-\alpha-1}J_2^2(x)\,dx\right. \\ &\quad \left. - \frac{D}{d}\int_0^\infty x^{-\alpha-2}J_2^2(x)\,J_1\left(\frac{2xd}{D}\right)dx\right].\end{aligned} \tag{10.144}$$

The first integral in Eq. (10.144) is the same as Eq. (10.141), whereas the second integral yields

$$\begin{aligned}\frac{D}{d}\int_0^\infty x^{-\alpha-2}J_2^2(x)\,J_1\left(\frac{2xd}{D}\right)dx &= \frac{D}{d\sqrt{\pi}}\int_0^\infty J_1\left(\frac{2xd}{D}\right) \\ &\quad \times G_{13}^{11}\left(x^2\left|\begin{matrix} -\frac{\alpha+1}{2} \\ 1-\frac{\alpha}{2},\ -1-\frac{\alpha}{2},\ -3-\frac{\alpha}{2}\end{matrix}\right.\right)dx \\ &= \frac{D^2}{2\sqrt{\pi}\,d^2}G_{33}^{12}\left(\frac{D^2}{d^2}\left|\begin{matrix} 0,\ -\frac{\alpha+1}{2},\ 1 \\ 1-\frac{\alpha}{2},\ -1-\frac{\alpha}{2},\ -3-\frac{\alpha}{2}\end{matrix}\right.\right).\end{aligned} \tag{10.145}$$

With the help of ($MG1$) in Appendix A, this last expression becomes

$$\frac{D}{d}\int_0^\infty x^{-\alpha-2}J_2^2(x)\,J_1\left(\frac{2xd}{D}\right)dx$$
$$= 0.0078\frac{\Gamma\left(2-\frac{\alpha}{2}\right)}{\Gamma\left(\frac{\alpha}{2}\right)}\left(\frac{D}{d}\right)^{4-\alpha}{}_3F_2\left(2-\frac{\alpha}{2},\ \frac{5}{2},\ 1-\frac{\alpha}{2};\ 3,\ 5;\ \frac{D^2}{d^2}\right), \tag{10.146}$$

so by combining the results of Eqs. (10.141) and (10.146), we obtain

$$\sigma_{Z\perp}^2(\alpha) \approx A(\alpha)\left[\frac{2851}{2^{\alpha-2}D^{4-\alpha}}\frac{\Gamma\left(2-\frac{\alpha}{2}\right)\Gamma\left(\frac{1+\alpha}{2}\right)}{\Gamma\left(1+\frac{\alpha}{2}\right)\Gamma\left(3+\frac{\alpha}{2}\right)} - \frac{22.24}{2^{\alpha-2}d^{4-\alpha}}\frac{\Gamma\left(2-\frac{\alpha}{2}\right)}{\Gamma\left(\frac{\alpha}{2}\right)}\right]$$
$$\times\int_0^L \tilde{C}_n^2(z)\,dz, \quad d > D, \tag{10.147}$$

where we have replaced the ${}_3F_2$ function in Eq. (10.146) with unity for all $d > D$. The parallel component of Z-tilt is now found from the relation

$$\sigma_{Z\parallel}^2(\alpha) = \sigma_Z^2(\alpha) - \sigma_{Z\perp}^2(\alpha). \tag{10.148}$$

To include the case where $d < D$, it may be best to use the Meijer G-function expressions in Eqs. (10.142) and (10.145).

10.13.2 Multi-source

For a single aperture of diameter D and two point sources separated by distance d, a centroid sensor responds more closely to G-tilt than to Zernike tilt. Thus, tilt anisoplanatism using G-tilt for multiple sources with a single Rx aperture is given by

$$\begin{pmatrix}\sigma_{LED}^2(\alpha)\\ \sigma_{LED\perp}^2(\alpha)\\ \sigma_{LED\parallel}^2(\alpha)\end{pmatrix} = \frac{1263.3}{D^2}A(\alpha)\int_0^L dz\tilde{C}_n^2(z)\int_0^\infty d\kappa\kappa^{1-\alpha}J_1^2(\kappa\gamma D/2)$$
$$\times\frac{1}{2\pi}\int_0^{2\pi}d\phi\left\{1-\cos\left[(1-\gamma)\kappa d\cos\phi\right]\right\}\begin{pmatrix}1\\ \sin^2\phi\\ \cos^2\phi\end{pmatrix}, \tag{10.149}$$

where $\gamma = z/L$ and we are using the non-Kolmogorov spectrum (10.1). Here we are deriving the multi-source case for an LED array that may include a slant path. After integration on ϕ, the total tilt variance becomes

$$\sigma^2_{LED}(\alpha) = \frac{1263.3}{D^2} A(\alpha) \int_0^L dz\, \tilde{C}_n^2(z) \int_0^\infty \kappa^{1-\alpha} J_1^2(\kappa\gamma D/2)\{1 - J_0[\kappa d(1-\gamma)]\}\, d\kappa. \tag{10.150}$$

If we make the substitution $x = \kappa\gamma D/2$, then Eq. (10.150) takes the form

$$\begin{aligned}\sigma^2_{LED}(\alpha) &= \frac{1263.3 A(\alpha)}{2^{\alpha-2} D^{4-\alpha}} \int_0^L dz\, \tilde{C}_n^2(z)\gamma^{\alpha-2} \int_0^\infty x^{1-\alpha} J_1^2(x)\left\{1 - J_0\left[2x\left(\frac{1}{\gamma} - 1\right)\frac{d}{D}\right]\right\} dx \\ &= \frac{1263.3 A(\alpha)}{2^{\alpha-2} D^{4-\alpha}} \int_0^L dz\, \tilde{C}_n^2(z)\left(\frac{z}{L}\right)^{\alpha-2}\left\{\int_0^\infty x^{1-\alpha} J_1^2(x)\, dx \right. \\ &\quad \left. - \int_0^\infty x^{1-\alpha} J_1^2(x) J_0\left[2x\left(\frac{L-z}{z}\right)\frac{d}{D}\right] dx\right\}.\end{aligned} \tag{10.151}$$

From ($D3$) in Appendix D, the first integral above yields

$$\int_0^\infty x^{1-\alpha} J_1^2(x)\, dx = \frac{\Gamma\left(2 - \frac{\alpha}{2}\right)\Gamma\left(\frac{\alpha-1}{2}\right)}{2\sqrt{\pi}\,\Gamma\left(\frac{\alpha}{2}\right)\Gamma\left(1 + \frac{\alpha}{2}\right)}, \tag{10.152}$$

whereas the second integral can be solved directly from use of ($D4$) in Appendix D; that is,

$$\begin{aligned}&\int_0^\infty x^{1-\alpha} J_1^2(x) J_0\left[2x\left(\frac{L-z}{z}\right)\frac{d}{D}\right] dx \\ &= \int_0^\infty J_0\left[2x\left(\frac{L-z}{z}\right)\frac{d}{D}\right] G_{13}^{11}\left(x^2 \middle| \begin{matrix} 1-\frac{\alpha}{2} \\ \frac{3-\alpha}{2},\ \frac{1-\alpha}{2},\ -\frac{1+\alpha}{2}\end{matrix}\right) dx \\ &= \frac{D}{2\sqrt{\pi}\, d}\left(\frac{z}{L-z}\right) G_{33}^{12}\left[\frac{D^2}{d^2}\left(\frac{z}{L-z}\right)^2 \middle| \begin{matrix} \frac{1}{2},\ 1-\frac{\alpha}{2},\ \frac{1}{2} \\ \frac{3-\alpha}{2},\ \frac{1-\alpha}{2},\ -\frac{1+\alpha}{2}\end{matrix}\right].\end{aligned} \tag{10.153}$$

Now converting the Meijer G-function to a generalized hypergeometric function using ($MG1$) in Appendix A, we have

$$\int_0^\infty x^{1-\alpha} J_1^2(x)\, J_0\left[2x\left(\frac{L-z}{z}\right)\frac{d}{D}\right]dx = \frac{\Gamma\left(2-\frac{\alpha}{2}\right)}{4\Gamma\left(\frac{\alpha}{2}-1\right)}\left[\frac{D}{d}\left(\frac{z}{L-z}\right)\right]^{4-\alpha}$$
$$\times\, {}_3F_2\left[2-\frac{\alpha}{2},\ \frac{3}{2},\ 2-\frac{\alpha}{2};\ 2,\ 3;\ \frac{D^2}{d^2}\left(\frac{z}{L-z}\right)^2\right]. \tag{10.154}$$

However, as explained in Section 9.4.3, we believe that for numerical calculations it is best to use Eq. (10.153) with the Meijer *G*-function. Then, by combining Eqs. (10.152) and (10.153) with Eq. (10.151), we arrive at

$$\sigma^2_{LED}(\alpha) = \frac{1263.3A(\alpha)}{2^{\alpha-2}D^{4-\alpha}}\int_0^L dz\,\tilde{C}_n^2(z)\left(\frac{z}{L}\right)^{\alpha-2}\left\{\frac{\Gamma\left(2-\frac{\alpha}{2}\right)\Gamma\left(\frac{\alpha-1}{2}\right)}{2\sqrt{\pi}\,\Gamma\left(\frac{\alpha}{2}\right)\Gamma\left(1+\frac{\alpha}{2}\right)}\right.$$
$$\left. - \frac{D}{2\sqrt{\pi}\,d}\left(\frac{z}{L-z}\right)G_{33}^{12}\left[\frac{D^2}{d^2}\left(\frac{z}{L-z}\right)^2\middle|\begin{matrix}\frac{1}{2},\ 1-\frac{\alpha}{2},\ \frac{1}{2}\\ \frac{3-\alpha}{2},\ \frac{1-\alpha}{2},\ -\frac{1+\alpha}{2}\end{matrix}\right]\right\}. \tag{10.155}$$

Following the integration on ϕ, the perpendicular tilt component in Eq. (10.149) leads to

$$\sigma^2_{LED\perp}(\alpha) = \frac{1263.3}{D^2}A(\alpha)\int_0^L dz\tilde{C}_n^2(z)\int_0^\infty \kappa^{1-\alpha}J_1^2(\kappa\gamma D/2)\left\{\frac{1}{2}-\frac{J_1[\kappa d(1-\gamma)]}{\kappa d(1-\gamma)}\right\}d\kappa. \tag{10.156}$$

Again, the substitution $x = \kappa\gamma D/2$ yields

$$\sigma^2_{LED\perp}(\alpha) = \frac{631.65A(\alpha)}{2^{\alpha-2}D^{4-\alpha}}\int_0^L dz\tilde{C}_n^2(z)\left(\frac{z}{L}\right)^{\alpha-2}\left\{\int_0^\infty x^{1-\alpha}J_1^2(x)\,dx\right.$$
$$\left. - \left(\frac{z}{L-z}\right)\frac{D}{d}\int_0^\infty x^{-\alpha}J_1^2(x)\,J_1\left[2x\left(\frac{L-z}{z}\right)\frac{d}{D}\right]dx\right\}. \tag{10.157}$$

The first integral in Eq. (10.157) is the same as Eq. (10.152), and the second integral based on ($D4$) in Appendix D leads to

$$\int_0^\infty x^{-\alpha}J_1^2(x)\,J_1\left[2x\left(\frac{L-z}{z}\right)\frac{d}{D}\right]dx$$
$$=\frac{D}{2\sqrt{\pi}\,d}\left(\frac{z}{L-z}\right)G_{33}^{12}\left[\frac{D^2}{d^2}\left(\frac{z}{L-z}\right)^2\middle|\begin{matrix}0,\ \frac{1-\alpha}{2},\ 1\\ 1-\frac{\alpha}{2},\ -\frac{\alpha}{2},\ -\frac{\alpha+2}{2}\end{matrix}\right],\tag{10.158}$$

and by combining Eqs. (10.152) and (10.158) with Eq. (10.157), the result is

$$\sigma_{LED\perp}^2(\alpha)=\frac{631.65A(\alpha)}{2^{\alpha-2}D^{4-\alpha}}\int_0^L dz\tilde{C}_n^2(z)\left(\frac{z}{L}\right)^{\alpha-2}\left\{\frac{\Gamma\left(2-\frac{\alpha}{2}\right)\Gamma\left(\frac{\alpha-1}{2}\right)}{2\sqrt{\pi}\,\Gamma\left(\frac{\alpha}{2}\right)\Gamma\left(1+\frac{\alpha}{2}\right)}\right.$$
$$\left.-\frac{D}{2\sqrt{\pi}\,d}\left(\frac{z}{L-z}\right)G_{33}^{12}\left[\frac{D^2}{d^2}\left(\frac{z}{L-z}\right)^2\middle|\begin{matrix}0,\ \frac{1-\alpha}{2},\ 1\\ 1-\frac{\alpha}{2},\ -\frac{\alpha}{2},\ -\frac{\alpha+2}{2}\end{matrix}\right]\right\}.\tag{10.159}$$

Last, the parallel component of tilt anisoplanatism is calculated from

$$\sigma_{LED\parallel}^2(\alpha)=\sigma_{LED}^2(\alpha)-\sigma_{LED\perp}^2(\alpha).\tag{10.160}$$

10.14 Focus Anisoplanatism

To examine the concept of focus anisoplanatism, we start with the total focus variance

$$\sigma_f^2=118.435A(\alpha)k^2\int_0^L dz\tilde{C}_n^2(z)\int_0^\infty d\kappa\,\kappa^{1-\alpha}\left[\frac{2J_3(\kappa D/2)}{\kappa D/2}\right]^2$$
$$=\frac{1894.96}{2^\alpha}A(\alpha)k^2D^{\alpha-2}\int_0^L dz\tilde{C}_n^2(z)\int_0^\infty x^{-\alpha-1}J_3^2(x)\,dx,\tag{10.161}$$

where we have set $x=\kappa D/2$. By the use of ($D3$) in Appendix D, this last result reduces to

$$\sigma_f^2 = \frac{1894.96}{2^{\alpha+1}\sqrt{\pi}} \frac{\Gamma\left(3-\frac{\alpha}{2}\right)\Gamma\left(\frac{1+\alpha}{2}\right)}{\Gamma\left(1+\frac{\alpha}{2}\right)\Gamma\left(4+\frac{\alpha}{2}\right)} A(\alpha)k^2 D^{\alpha-2}\int_0^L \tilde{C}_n^2(z)\,dz$$
$$= \frac{534.56}{2^{\alpha}} \frac{\Gamma\left(3-\frac{\alpha}{2}\right)\Gamma\left(\frac{1+\alpha}{2}\right)}{\Gamma\left(1+\frac{\alpha}{2}\right)\Gamma\left(4+\frac{\alpha}{2}\right)} A(\alpha)k^2 D^{\alpha-2}\int_0^L \tilde{C}_n^2(z)\,dz. \tag{10.162}$$

The variance difference between two displaced apertures by the distance d leads to the focus correlation function

$$C_f(d) = 236.86A(\alpha)k^2\int_0^L dz\tilde{C}_n^2(z)\int_0^\infty d\kappa\,\kappa^{-\alpha+1}\left[\frac{2J_3(\kappa D/2)}{\kappa D/2}\right]^2[1-J_0(\kappa d)]$$
$$= \frac{3789.92}{2^{\alpha}}A(\alpha)k^2D^{\alpha-2}\int_0^L dz\tilde{C}_n^2(z)\left\{\int_0^\infty x^{-\alpha-1}J_3^2(x)\,dx - \int_0^\infty x^{-\alpha-1}J_3^2(x)\,J_0\left(\frac{2xd}{D}\right)dx\right\}, \tag{10.163}$$

where again we made the change of variable $x = \kappa D/2$. The first integral over x is evaluated by the use of ($D3$) in Appendix D. The second integral is of the form given by ($D4$) in Appendix D, which leads to

$$\int_0^\infty x^{-\alpha-1}J_3^2(x)J_0\left(\frac{2xd}{D}\right)dx = \frac{D}{2d\sqrt{\pi}}G_{33}^{12}\left(\frac{D^2}{d^2}\middle|\begin{matrix}\frac{1}{2},\ -\frac{\alpha}{2},\ \frac{1}{2}\\ \frac{5-\alpha}{2},\ -\frac{\alpha+1}{2},\ -\frac{7+\alpha}{2}\end{matrix}\right). \tag{10.164}$$

Thus, Eq. (10.163) becomes

$$C_f(d) = \frac{534.56}{2^{\alpha}} \frac{\Gamma\left(3-\frac{\alpha}{2}\right)\Gamma\left(\frac{1+\alpha}{2}\right)}{\Gamma\left(1+\frac{\alpha}{2}\right)\Gamma\left(4+\frac{\alpha}{2}\right)} A(\alpha)k^2 D^{\alpha-2}\int_0^L \tilde{C}_n^2(z)\,dz$$
$$\times 2\left[1-\frac{\Gamma\left(1+\frac{\alpha}{2}\right)\Gamma\left(4+\frac{\alpha}{2}\right)}{\Gamma\left(3-\frac{\alpha}{2}\right)\Gamma\left(\frac{1+\alpha}{2}\right)}\left(\frac{D}{d}\right)G_{33}^{12}\left(\frac{D^2}{d^2}\middle|\begin{matrix}\frac{1}{2},\ -\frac{\alpha}{2},\ \frac{1}{2}\\ \frac{5-\alpha}{2},\ -\frac{\alpha+1}{2},\ -\frac{7+\alpha}{2}\end{matrix}\right)\right]. \tag{10.165}$$

If we now divide Eq. (10.165) by Eq. (10.162), we obtain the scaled correlation expression

$$\frac{C_f(d)}{\sigma_f^2} = 2\left[1 - \frac{\Gamma\left(1+\frac{\alpha}{2}\right)\Gamma\left(4+\frac{\alpha}{2}\right)}{\Gamma\left(3-\frac{\alpha}{2}\right)\Gamma\left(\frac{1+\alpha}{2}\right)}\left(\frac{D}{d}\right)G_{33}^{12}\left(\frac{D^2}{d^2}\middle|\begin{matrix}\frac{1}{2}, & -\frac{\alpha}{2}, & \frac{1}{2}\\ \frac{5-\alpha}{2}, & -\frac{\alpha+1}{2}, & -\frac{7+\alpha}{2}\end{matrix}\right)\right]. \quad (10.166)$$

As in the case of conventional Kolmogorov turbulence (Section 9.5.3), the limiting value of Eq. (10.166) for all $d > D$ is "2."

10.15 Discussion

Throughout the chapter we have presented results based on isotropic non-Kolmogorov turbulence and, in some cases, anisotropic turbulence. Experimental measurements have shown that anisotropic turbulence conditions are prevalent in the free atmosphere above 1–2 km, but also near the ground under proper conditions.[15,20] Toselli[32] introduced the concept of anisotropy for different scale sizes of turbulence where anisotropy is the case $\mu_x = \mu_y = \mu$. In his model, turbulence cells are geometrically considered to be ellipsoids with circular cross sections perpendicular to the direction of propagation. It was also suggested that anisotropy is higher for larger eddies and nearly isotropic for eddies close to the inner-scale size. If so, anisotropic conditions with varying large eddy size with altitude may play a major role in future uplink/downlink propagation studies.

References

1. G. M. Grechko, A. S. Gurvich, V. Kan, S. V. Kireev, and S. A. Savchenko, "Anisotropy of spatial structures in the middle atmosphere," *Adv. Space Res.* **12**(10), 169–175 (1992).
2. D. T. Kyrazis, J. Wissler, D. D. B. Keating, A. J. Preble, and K. P. Bishop, "Measurement of optical turbulence in the upper troposphere and lower stratosphere," *Proc. SPIE* **2120**, 43–55 (1994) [doi: 10.1117/12.177698].
3. M. S. Belen'kii, J. D. Barchers, S. J. Karis, C. L. Osmon, J. M. Brown II, and R. Q. Fugate, "Preliminary experimental evidence of anisotropy of turbulence and the effect of non-Kolmogorov turbulence on wavefront tilt statistics," *Proc. SPIE* **3762**, 396–406 (1999) [doi: 10.1117/12.363597].
4. A. I. Kon, "Qualitative theory of amplitude and phase fluctuations in a medium with anisotropic turbulent irregularities," *Waves Random Media* **4**(3), 297–305 (1994).

5. B. E. Stribling, B. M. Welsh, and M. C. Roggemann, "Optical propagation in non-Kolmogorov atmospheric turbulence," *Proc. SPIE* **2471**, 181–196 (1995) [doi: 10.1117/12.211927].
6. R. R. Beland, "Some aspects of propagation through weak isotropic non-Kolmogorov turbulence," *Proc. SPIE* **2375**, 6–16 (1995) [doi: 10.1117/12.206979].
7. I. Toselli, L. C. Andrews, R. L. Phillips, and V. Ferrero, "Free-space optical system performance for laser beam propagation through non-Kolmogorov turbulence," *Opt. Eng.* **47**(2), 026003 (2008) [doi: 10.1117/1.2870113].
8. I. Toselli, L. C. Andrews, R. L. Phillips, and V. Ferrero, "Free space optical system performance for a Gaussian beam propagating through non-Kolmogorov turbulence," *IEEE Trans. Antennas Propag.* **57**(6), 1783–1788 (2009).
9. L. Y. Cui, B. D. Xue, X. G. Cao, J. K. Dong, and J. N. Wang, "Generalized atmospheric turbulence MTF for wave propagating through non-Kolmogorov turbulence," *Opt. Exp.* **18**(20), 21269–21283 (2010).
10. I. Toselli, B. Agrawal, and S. Restaino, "Light propagation through anisotropic turbulence," *J. Opt. Soc. Am. A* **28**(3), 493–488 (2011).
11. L. Cui, B. Xue, S. Zheng, W. Xue, X. Bai, X. Cao, and F. Zhou, "Atmospheric spectral model and theoretical expressions of irradiance scintillation index for optical wave propagating through moderate-to-strong non-Kolmogorov turbulence," *J. Opt. Soc. Am. A* **29**(6), 1091–1098 (2012).
12. L. C. Andrews, R. L. Phillips, R. Crabbs, and T. Leclerc, "Deep turbulence propagation of a Gaussian-beam wave in anisotropic non-Kolmogorov turbulence," *Proc. SPIE* **8874**, 887402 (2013) [doi: 10.1117/12.2026405].
13. L. C. Andrews, R. L. Phillips, and R. Crabbs, "Propagation of a Gaussian-beam wave in general anisotropic turbulence," *Proc. SPIE* **9224**, 922402 (2014) [doi: 10.1117/12.2061892].
14. V. P. Lukin, E. V. Nosov, V. V. Nosov, and A. V. Torgaev, "Causes of non-Kolmogorov turbulence in the atmosphere," *Appl. Opt.* **55**(12), B163–B168 (2016).
15. X. Xiao, D. G. Voelz, I. Toselli, and O. Korotkova, "Gaussian beam propagation in anisotropic turbulence along horizontal links: theory, simulation, and laboratory implementation," *Appl. Opt.* **55**(15), 4079–4083 (2016).
16. D. A. Paulson, C. Wu, and C. C. Davis, "A detailed comparison of non-Kolmogorov and anisotropic optical turbulence theories using wave optics simulations," *Proc. SPIE* **10770**, 107700X (2018) [doi: 10.117/12.2323342].
17. O. Korotkova and I. Toselli, "Non-classic optical turbulence: review," *Appl. Sci.* **11**(18), 8487 (2021).

18. M. Beason, L. Andrews, and R. Phillips, "Study on the effect of anisotropy on a propagating beam," *Proc. SPIE* **10408**, 104080B (2017) [doi: 10.1117/12.2274197].
19. F. Wang, I. Toselli, J. Li, and O. Korotkova, "Measuring anisotropy ellipse of atmospheric turbulence by intensity correlations of laser light," *Opt. Lett.* **42**(6), 1129–1132 (2017).
20. M. Beason, F. Sanzone, J. Coffaro, C. Smith, J. Spychalsky, B. Berry, F. Titus, R. Crabbs, L. Andrews, and R. Phillips, "Near ground measurements of beam shaping and anisotropic turbulence over concrete runway and grass range," *Proc. SPIE* **10770**, 107700L (2018) [doi: 10.1117/12.2323317].
21. M. Beason, F. Sanzone, B. Berry, J. Coffaro, J. Spychalski, F. Titus, R. Crabbs, L. Andrews, and R. Phillips, "Considering power law, optical refractivity, and anisotropy using Gaussian beam statistics," in *Imaging and Applied Optics 2019 (COSI, IS, MATH, pcAOP)*, OSA Technical *Digest* (Optical Society of America, 2019), paper PTh2D.2.
22. A. A. Consortini, L. Ronchi, and L. Stefanutti, "Investigation of atmospheric turbulence by narrow laser beams," *Appl. Opt.* **9**(11), 2543–2547 (1970).
23. L. C. Andrews and M. K. Beason, *Laser Beam Propagation in Random Media: New and Advanced Topics*, SPIE Press, Bellingham, Washington (2023) [doi: 10.1117/3.2643989].
24. Y. Baykal and H. Gerçekcioğlu, "Equivalence of structure constants in non-Kolmogorov and Kolmogorov spectra," *Opt. Lett.* **36**(23), 4554–4556 (2011).
25. M. Charnotskii, "Intensity fluctuations of flat-topped beam in non-Kolmogorov weak turbulence: comment," *J. Opt. Soc. Am. A* **29**(9), 1838–1840 (2012).
26. Y. Li, W. Zhu, X. Wu, and R. Rao, "Equivalent refractive-index structure constant of non-Kolmogorov turbulence," *Opt. Exp.* **23**(18), 23004–23012 (2015).
27. I. Toselli and S. Gladysz, "Equivalent refractive-index structure constant on non-Kolmogorov turbulence: comment," *Opt. Exp.* **30**(22), 40965–40967 (2022).
28. M. Beason, L. Andrews, and I. Toselli, "Calculating structure function constant from measured C_n^2 in non-Kolmogorov and anisotropic turbulence including inner scale effects," *Appl. Opt.* **58**(25), 6813–6819 (2019).
29. C. Zhai, "A novel idea: Calculating anisotropic turbulence only by Kolmogorov structure constant C_n^2 and power law α," *Results in Physics* **19**, 03483 (2020).
30. L. Cui, B. Xue, L. Cao, S. Zheng, W. Xue, X. Bai, X. Cao, and F. Zhou, "Irradiance scintillation for Gaussian-beam wave propagating through

weak non-Kolmogorov turbulence," *Opt. Exp.* **19**(18), 16872–16884 (2011).

31. L. C. Andrews and R. L. Phillips, *Laser Beam Propagation through Random Media*, 2nd Edition, SPIE Press, Bellingham, Washington (2005) [doi: 10.1117/3.626196].
32. I. Toselli, "Introducing the concept of anisotropy at different scales for modeling optical turbulence," *J. Opt. Soc. Am. A* **31**(8), 1868–1875 (2014).

Chapter 11
Miscellaneous Applications

11.0 Introduction

In this final chapter we briefly discuss some additional applications that also naturally lead to special functions. For that reason, the emphasis is primarily on the mathematics leading to special functions, not the subject itself. We first examine the basic theory of signal-to-noise ratio for certain electrical or acoustic communication systems and then turn to some examples taken from the field of fluid mechanics.

11.1 Statistical Communication Theory

Statistical communication theory is basically the application of probability theory to problems in the communication process.[1,2] Inherent in communication systems is the presence of random noise in the system. Let us assume that random noise exists at the output of a linear filter with center frequency $f_0 = \omega_0/2\pi$ and bandwidth $B \ll f_0$. In such cases the noise is considered *narrowband* and can be modeled by the expression

$$\mathbf{n}(t) = \mathbf{x}(t)\cos\omega_0 t - \mathbf{y}(t)\sin\omega_0 t, \tag{11.1}$$

where $\mathbf{x}(t)$ and $\mathbf{y}(t)$ are independent *Gaussian* (or *normal*) *random processes* with zero means and equal mean-squared values (variances) given by

$$\langle \mathbf{x}^2(t) \rangle = \langle \mathbf{y}^2(t) \rangle = \langle \mathbf{n}^2(t) \rangle = N. \tag{11.2}$$

The brackets $\langle\rangle$ indicate an ensemble average. Under the stated assumptions, the joint Gaussian probability distribution for $\mathbf{x}(t)$ and $\mathbf{y}(t)$ is

$$p_{\mathbf{xy}}(x, y) = \frac{1}{2\pi N} e^{-(x^2+y^2)/2N}. \tag{11.3}$$

In addition to noise, there is generally a signal that we will consider to be an unmodulated carrier of amplitude A and frequency $f_0 = \omega_0/2\pi$. Because the phase of the signal has little effect on the following analysis concerning specific components of a communication system, we set it to zero. In this case the composite signal plus narrowband noise $\mathbf{u}(t)$ has the representation

$$\begin{aligned}\mathbf{u}(t) &= A\cos\omega_0 t + \mathbf{x}(t)\cos\omega_0 t - \mathbf{y}(t)\sin\omega_0 t\\ &= [\mathbf{x}(t) + A]\cos\omega_0 t - \mathbf{y}(t)\sin\omega_0 t\\ &= \mathbf{r}(t)\cos[\omega_0 t + \boldsymbol{\theta}(t)],\end{aligned} \tag{11.4}$$

where $\mathbf{r}(t)$ and $\boldsymbol{\theta}(t)$ represent the random envelope and random phase, respectively, of the composite signal and noise. These random quantities are defined by

$$\mathbf{r}(t) = \sqrt{[A + \mathbf{x}(t)]^2 + \mathbf{y}^2(t)},$$
$$\boldsymbol{\theta}(t) = \tan^{-1}\left[\frac{\mathbf{y}(t)}{A + \mathbf{x}(t)}\right].$$

In terms of $\mathbf{r}(t)$ and $\boldsymbol{\theta}(t)$, the joint density function described by Eq. (11.3), but now including the signal, becomes

$$\begin{aligned}p_{\mathbf{r}\boldsymbol{\theta}}(r,\theta) &= \frac{r}{2\pi N}\exp\left[-\frac{(r\cos\theta - A)^2 + (r\sin\theta)^2}{2N}\right]\\ &= \frac{r}{2\pi N}\exp\left[-\frac{(r^2 + A^2 - 2Ar\cos\theta)}{2N}\right].\end{aligned} \tag{11.5}$$

The marginal distribution of the envelope $\mathbf{r}(t)$ is found from

$$p_{\mathbf{r}}(r) = \int_0^{2\pi} p_{\mathbf{r}\boldsymbol{\theta}}(r,\theta)\,d\theta, \tag{11.6}$$

which, using (*BI*5) in Appendix A, reduces to

$$p_{\mathbf{r}}(r) = \frac{r}{N}e^{-(r^2+A^2)/2N}I_0\left(\frac{Ar}{N}\right), \quad r > 0. \tag{11.7}$$

This distribution is known as the *Rician* or *Rice–Nakagami distribution*, where $I_0(x)$ is the modified Bessel function of the first kind.

11.1.1 Nonlinear devices

Statistical communication systems are broadly classified in terms of *linear operations* such as amplification and filtering, and *nonlinear operations* such as modulation and detection. Let us assume that the output $\mathbf{u}(t)$ of the narrowband filter in Fig. 11.1 is passed through a nonlinear device classified as a rectifier, detector, or nonlinear amplifier. In detection devices, it is then passed through a lowpass filter for further processing. In the case of a

Figure 11.1 Block diagram for a nonlinear device system.

nonlinear amplifier or limiter, the output is passed through a zonal filter at the same frequency as the input signal. Nonlinear devices commonly fall under the category of a half-wave, full-wave (even), or full-wave (odd) *νth-law device*. The full-wave (even) and half-wave devices are used with detection circuits, whereas the full-wave (odd) device is sometimes used with nonlinear amplifiers. The output of a νth-law device can be represented by the Fourier series (pp. 280–288 of Ref. 2)

$$\mathbf{w}(t) = \mathbf{r}^{\nu}(t)\left\{\frac{1}{2}c_0(\nu) + \sum_{n=1}^{\infty} c_n(\nu)\cos\left[n\,\omega_0 t + n\,\boldsymbol{\theta}(t)\right]\right\}, \tag{11.8}$$

where

- *Half-wave device*: $\mathbf{w}(t) = \begin{cases} \mathbf{u}^{\nu}(t), & \mathbf{u}(t) > 0 \\ 0, & \mathbf{u}(t) < 0 \end{cases}$

$$c_n(\nu) = \frac{a\,\Gamma(1+\nu)}{2^{\nu}\Gamma\left(1+\dfrac{\nu+n}{2}\right)\Gamma\left(1+\dfrac{\nu-n}{2}\right)}, \qquad n = 0,\ 1,\ 2,\ldots$$

- *Full-wave (even) device*: $\mathbf{w}(t) = |\mathbf{u}(t)|^{\nu}$

$$c_n(\nu) = \frac{a\,\Gamma(1+\nu)}{2^{\nu-1}\Gamma\left(1+\dfrac{\nu+n}{2}\right)\Gamma\left(1+\dfrac{\nu-n}{2}\right)}, \qquad n = 0,\ 2,\ 4,\ldots$$

- *Full-wave (odd) device*: $\mathbf{w}(t) = \begin{cases} \mathbf{u}^{\nu}(t), & \mathbf{u}(t) > 0 \\ -|\mathbf{u}(t)|^{\nu}, & \mathbf{u}(t) < 0 \end{cases}$

$$c_n(\nu) = 0, \qquad n = 0,\ 2,\ 4,\ldots$$

$$c_n(\nu) = \frac{a\,\Gamma(1+\nu)}{2^{\nu-1}\Gamma\left(1+\dfrac{\nu+n}{2}\right)\Gamma\left(1+\dfrac{\nu-n}{2}\right)}, \qquad n = 1,\ 3,\ 5,\ldots.$$

The parameter a is a scaling constant and $\mathbf{u}(t)$ is defined by Eq. (11.4).

A full-wave (even) device followed by a lowpass filter leads to the output

$$\mathbf{z}(t) = \frac{1}{2}c_0(\nu)\mathbf{r}^\nu(t) = \frac{2\Gamma[(1+\nu)/2]}{\sqrt{\pi}\Gamma(1+\nu/2)}\,\mathbf{r}^\nu(t), \tag{11.9}$$

where we have applied ($G10$) in Appendix A and set the scaling constant $a = 2$. One of the key performance measures of the output $\mathbf{z}(t)$ is the mean *signal-to-noise ratio* (SNR). Based on Eq. (11.9), the *output signal power* S_0 and *output noise power* N_0 are defined, respectively, by

$$\begin{aligned} S_0 &= \Big[\langle\mathbf{z}\rangle - \langle\mathbf{z}|A=0\rangle\Big]^2, \\ N_0 &= \langle\mathbf{z}^2\rangle - \langle\mathbf{z}\rangle^2. \end{aligned} \tag{11.10}$$

Equations (11.9) and (11.10) make it clear that we need to calculate the moments of $\mathbf{r}^\nu$ using the probability density function of Eq. (11.7). This was previously done in Section 7.1.2, which led to

$$\begin{aligned} \langle\mathbf{r}^\nu\rangle &= \int_0^\infty r^\nu p_{\mathbf{r}}(r)\,dr \\ &= (2N)^{\nu/2}\Gamma\left(1+\frac{\nu}{2}\right){}_1F_1\left(-\frac{\nu}{2};\ 1;\ -\frac{A^2}{2N}\right). \end{aligned} \tag{11.11}$$

Now referring back to Eqs. (11.9) and (11.10), we find that the output signal power is

$$S_0 = \frac{1}{4}c_0^2(\nu)(2N)^\nu\Gamma^2\left(1+\frac{\nu}{2}\right)\left[{}_1F_1\left(-\frac{\nu}{2};\ 1;\ -s\right) - 1\right]^2. \tag{11.12}$$

The function ${}_1F_1$ is the confluent hypergeometric function of the first kind, and we have introduced $s = A^2/2N$, which is the SNR at the output of the bandpass filter.

In a similar manner, we use Eqs. (11.10) and (11.11) to find the output noise power

$$\begin{aligned} N_0 = \frac{1}{4}&c_0^2(\nu)(2N)^\nu \\ &\times\left[\Gamma(1+\nu)\,{}_1F_1(-\nu;\ 1;\ -s) - \Gamma^2(1+\nu/2)\,{}_1F_1^2\left(-\frac{\nu}{2};\ 1;\ -s\right)\right]. \end{aligned} \tag{11.13}$$

Hence, the output SNR denoted by SNR_0 is obtained from the ratio S_0/N_0, which for the output defined by Eq. (11.9) leads to

$$\mathrm{SNR}_0 \equiv \frac{S_0}{N_0} = \frac{\Gamma^2\left(1+\frac{\nu}{2}\right)\left[{}_1F_1\left(-\frac{\nu}{2};\ 1;\ -s\right)-1\right]^2}{\Gamma(1+\nu)\,{}_1F_1(-\nu;\ 1;\ -s)-\Gamma^2(1+\nu/2)\,{}_1F_1^2\left(-\frac{\nu}{2};\ 1;\ -s\right)}. \tag{11.14}$$

For small and large values of input SNR, Eq. (11.14) reduces to

$$\mathrm{SNR}_0 \sim \begin{cases} \dfrac{\Gamma^2(1+\nu/2)\nu^2 s^2}{4[\Gamma(1+\nu)-\Gamma^2(1+\nu/2)]}, & s \ll 1 \\ \dfrac{2s}{\nu^2}, \quad s \gg 1. \end{cases} \tag{11.15}$$

In arriving at Eq. (11.15), we used the small-argument and large-argument approximations of the confluent hypergeometric functions from (*CH*6) in Appendix A. Thus, similar to a quadratic detector, we see that $\mathrm{SNR}_0 \sim s^2$ for all full-wave (even) and half-wave νth-law detectors when the input SNR is weak. Also, for strong input signals, we have $\mathrm{SNR}_0 \sim s$, which is essentially like that of a linear system.

The cases $\nu = 1$ and $\nu = 2$ in Eq. (11.14) represent a *linear rectifier* and a *square-law detector*, respectively. Let us first consider the simpler case $\nu = 2$. Here we find from the power series definition of ${}_1F_1$ that

$$\begin{aligned} {}_1F_1(-1;\ 1;\ -s) &= 1+s, \\ {}_1F_1(-2;\ 1;\ -s) &= \frac{1}{2}(s^2+4s+2), \end{aligned} \tag{11.16}$$

which are exact results because the power series for both confluent hypergeometric functions terminate. Also, based on (*B*22) in Appendix B, these confluent hypergeometric functions are equal to the Laguerre polynomials, i.e., ${}_1F_1(-n;\ 1;\ -s) = L_n(-s)$, $n = 1, 2, \ldots$. Thus, from Eqs. (11.14) and (11.16), the output SNR for a square-law detector reduces to

$$\mathrm{SNR}_0 = \frac{s^2}{1+2s}. \tag{11.17}$$

On the other hand, for the linear rectifier with $\nu = 1$, we first follow the approximation technique illustrated in Chapter 6 to find that

$${}_1F_1\left(-\frac{1}{2};\ 1;\ -s\right) \approx (1+1.25s)^{1/2}. \tag{11.18}$$

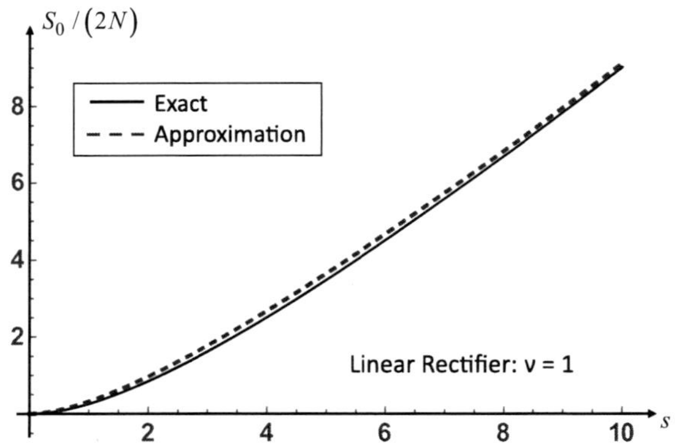

Figure 11.2 Mean scaled output signal power for a linear rectifier ($\nu = 1$) based on Eq. (11.19) as a function of mean input SNR.

Hence, the output signal power becomes

$$\begin{aligned} S_0 &= \frac{4}{\pi}(2N)\left[{}_1F_1\left(-\frac{1}{2};\ 1;\ -s\right) - 1\right]^2. \\ &\approx \frac{4}{\pi}(2N)\left[(1 + 1.25s)^{1/2} - 1\right]^2. \end{aligned} \tag{11.19}$$

In Fig. 11.2 we plot the scaled signal power $S_0/(2N)$ with $\nu = 1$ for the exact expression and that with the approximation in Eq. (11.19).

For the case $\nu = 1$, we find that although the approximation in Eq. (11.19) works quite well for output signal power, it doesn't work well for output SNR because of the behavior of the difference in confluent hypergeometric functions in the denominator of Eq. (11.14). However, there is another approximation that can be used for the output SNR. That is, let us use the technique from Chapter 6 and the asymptotic relations given in Eq. (11.15) for SNR_0 with $\nu = 1$. In this case we arrive at the asymptotic relations

$$\text{SNR}_0 \sim \begin{cases} 0.915s^2, & s \ll 1 \\ 2s, & s \gg 1, \end{cases}$$

from which we construct

$$\text{SNR}_0 \approx \frac{0.9s^2}{(1 + 0.61\sqrt{s})^2}. \tag{11.20}$$

In Fig. 11.3 we compare the actual output SNR_0 taken from Eq. (11.14) with the approximation of Eq. (11.20). The constants that appear in Eq. (11.20) represent a compromise between small and large arguments.

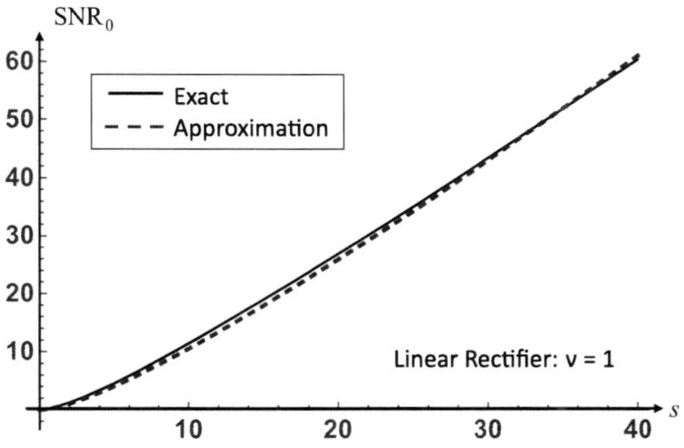

Figure 11.3 Mean output SNR_0 for a linear rectifier ($\nu = 1$) based on Eqs. (11.14) and (11.20) as a function of mean input SNR.

Last, another way to avoid the confluent hypergeometric function in the output signal-to-noise ratio with $\nu = 1$ is to relate the confluent hypergeometric function to the modified Bessel function through use of the relationship ($B12$) in Appendix B,

$$ {}_1F_1\left(-\frac{1}{2};\ 1;\ -s\right) = e^{-s/2}\left[(1+s)I_0\left(\frac{s}{2}\right) + sI_1\left(\frac{s}{2}\right)\right]. \tag{11.21} $$

Hence, using Eq. (11.21), we find that the exact result of Eq. (11.14) for the output SNR in the case of a linear rectifier can be expressed as

$$ \mathrm{SNR}_0 = \frac{\frac{\pi}{4}\left\{e^{-s/2}\left[(1+s)I_0\left(\frac{s}{2}\right) + sI_1\left(\frac{s}{2}\right)\right] - 1\right\}^2}{1 + s - \frac{\pi}{4}e^{-s}\left[(1+s)I_0\left(\frac{s}{2}\right) + sI_1\left(\frac{s}{2}\right)\right]^2}. \tag{11.22} $$

11.1.2 Ideal bandpass limiter

For a nonlinear amplifier, we replace the lowpass filter in Fig. 11.1 with a zonal filter at the first harmonic of the signal. In this case the nonlinear amplifier is a full-wave (odd) νth-law device with an output of the zonal filter described by ($a = 1$)

$$ \mathbf{z}(t) = \frac{\Gamma(1+\nu)}{2^{\nu-1}\Gamma\left(\frac{3+\nu}{2}\right)\Gamma\left(\frac{1+\nu}{2}\right)}\mathbf{r}^{\nu}(t)\cos\left[\omega_0 t + \boldsymbol{\theta}(t)\right]. \tag{11.23} $$

For illustrative purposes, we will restrict the analysis to the special case $\nu = 0$, which is called an *ideal bandpass limiter*. To calculate the mean SNR, this time we need both an ensemble average and a time average. Also, we need the joint density function in Eq. (11.5) to calculate the statistical moments of Eq. (11.23). Thus, the first moment leads to

$$\langle \mathbf{z} \rangle = \frac{2}{\pi^2 N} \int_0^\infty r e^{-(r^2+A^2)/2N} \int_0^{2\pi} \cos(\omega_0 t + \theta) \exp\left(\frac{Ar}{N}\cos\theta\right) d\theta\, dr. \quad (11.24)$$

From the integral relation (cf. Section 7.1.2),

$$\int_0^{2\pi} \cos(\omega_0 t + \theta) \exp\left(\frac{Ar}{N}\cos\theta\right) d\theta = 2\pi I_1\left(\frac{Ar}{N}\right) \cos\omega_0 t,$$

we calculate with the help of $I_1(z) = i^{-1} J_1(iz)$ and ($C9$) in Appendix C,

$$\begin{aligned} \langle \mathbf{z} \rangle &= \frac{4}{\pi N} e^{-A^2/2N} \cos\omega_0 t \int_0^\infty r e^{-r^2/2N} I_1\left(\frac{Ar}{N}\right) dr \\ &= \frac{2}{i\pi N} e^{-A^2/2N} \cos\omega_0 t \int_0^\infty e^{-x/2N} G_{02}^{10}\left(-\frac{A^2 x}{4\, N^2}\middle|\, \frac{1}{2},\ -\frac{1}{2}\right) dx \\ &= 2\sqrt{\frac{s}{\pi}}\, {}_1F_1\left(\frac{1}{2};\ 2;\ -s\right) \cos\omega_0 t. \end{aligned} \quad (11.25)$$

The last result in Eq. (11.25) follows from ($D6$) in Appendix D and ($C14$) in Appendix C. We also used ($CH4$) in Appendix A. The time-averaged mean signal power is therefore (the overbar means "time-average")

$$\begin{aligned} \overline{S_0} &= \overline{[\langle \mathbf{z} \rangle - \langle \mathbf{z} | s = 0 \rangle]^2} = \frac{4s}{\pi}\left[{}_1F_1\left(\frac{1}{2};\ 2;\ -s\right)\right]^2 \overline{\cos^2\omega_0 t} \\ &= \frac{2s}{\pi}\left[{}_1F_1\left(\frac{1}{2};\ 2;\ -s\right)\right]^2. \end{aligned} \quad (11.26)$$

By employing the technique used in Eqs. (6.10) through (6.13), we first approximate the output signal power (11.26) through the relationship

$$\begin{aligned} {}_1F_1\left(\frac{1}{2};\ 2;\ -s\right) &= 1 - \frac{s}{4}\, {}_2F_2\left(\frac{3}{2},\ 1;\ 2{,}3;\ -s\right) \\ &\approx 1 - \frac{1}{4}\left(\frac{s}{1 + 0.283s}\right). \end{aligned} \quad (11.27)$$

Based on Eqs. (11.26) and (11.27), we obtain the approximation for the mean output signal power:

$$\overline{S_0} \approx \frac{2s}{\pi}\left[1 - \frac{1}{4}\left(\frac{s}{1 + 0.283s}\right)\right]^2. \quad (11.28)$$

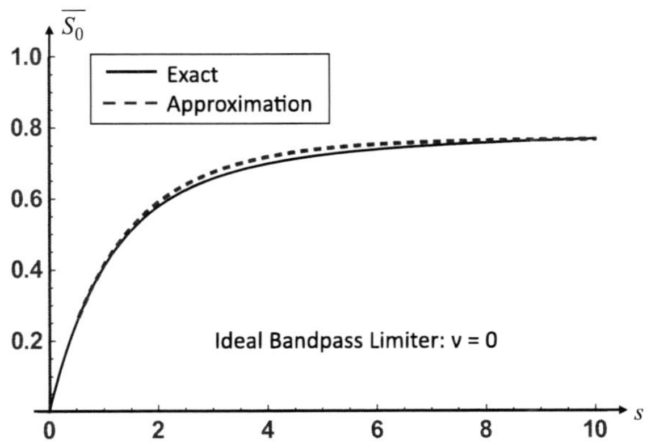

Figure 11.4 Time-average mean output signal power of an ideal bandpass limiter based on Eqs. (11.26) and (11.28) as a function of mean input SNR.

In Fig. 11.4 we plot the exact output signal power (11.26) and the approximation (11.28) as a function of input SNR.

The time-averaged mean second moment leads to

$$\overline{\langle \mathbf{z}^2 \rangle} = \frac{8}{\pi^2}\left[1 + \frac{1}{s}(s - 1 + e^{-s})\overline{\cos 2\omega_0 t}\right] = \frac{8}{\pi^2}. \tag{11.29}$$

The approximation in Eq. (11.27) works quite well in predicting the output signal power in Eq. (11.28) but, again, works not so well for the output SNR. However, similar to Eq. (11.21), the confluent hypergeometric function in Eq. (11.26) is directly tied to the modified Bessel functions of the first kind through the relation (*B*11) in Appendix B,

$${}_1F_1\left(\frac{1}{2}; 2, -s\right) = e^{-s/2}\left[I_0\left(\frac{s}{2}\right) + I_1\left(\frac{s}{2}\right)\right]. \tag{11.30}$$

In this case, based on Eqs. (11.26), (11.29), and (11.30), we find the time-average mean output SNR for the ideal limiter to be

$$\overline{\mathrm{SNR}_0} = \frac{se^{-s}\left[I_0\left(\frac{s}{2}\right) + I_1\left(\frac{s}{2}\right)\right]^2}{\frac{4}{\pi} - se^{-s}\left[I_0\left(\frac{s}{2}\right) + I_1\left(\frac{s}{2}\right)\right]^2} \sim \begin{cases} \frac{\pi s}{4}, & s \ll 1 \\ 2s, & s \gg 1. \end{cases} \tag{11.31}$$

Last, by using the asymptotic results in Eq. (11.31) and the approximation techniques from Chapter 6, we can approximate the mean output SNR in Eq. (11.31) by the expression

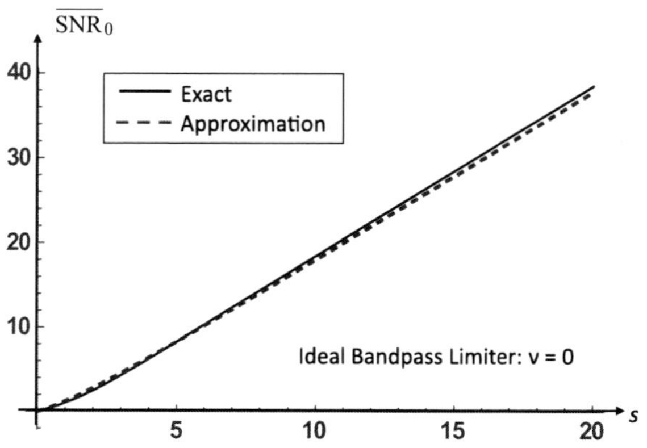

Figure 11.5 Time-average mean output SNR for an ideal bandpass limiter based on Eqs. (11.31) and (11.32).

$$\overline{\text{SNR}}_0 \approx \frac{s\left(s+\frac{\pi}{4}\right)}{1+0.5s}. \tag{11.32}$$

We plot the mean output SNR for the ideal bandpass limiter given by Eqs. (11.31) and (11.32) in Fig. 11.5. Here we see that the simple expression in Eq. (11.32) provides a good approximation to Eq. (11.31).

11.1.3 Joint moments of the envelopes

In some communication systems, there might be two narrowband noise inputs (see Section 11.2). In this case, the joint distribution of the envelopes and phases at the output of a pair of bandpass filters with the same center frequency is given by

$$p_2(r_1, r_2; \theta_1, \theta_2) = \frac{r_1 r_2}{(2\pi N)^2(1-\rho^2)} \exp\left\{-\frac{\left[r_1^2 + r_2^2 - 2\rho r_1 r_2 \cos(\theta_1 - \theta_2)\right]}{2N(1-\rho^2)}\right\}, \tag{11.33}$$

where ρ is the correlation coefficient between the noise inputs. For some cases, we may only be concerned with the moments of the envelopes $\mathbf{r}_1$ and $\mathbf{r}_2$. Thus, based on Eq. (11.33) and property (*BI*5) in Appendix A, the joint density function for the envelopes alone is

$$\begin{aligned} p_2(r_1, r_2) &= \int_0^{2\pi}\int_0^{2\pi} p_2(r_1, r_2; \theta_1, \theta_2)\, d\theta_1 d\theta_2 \\ &= \frac{r_1 r_2}{N^2(1-\rho^2)} \exp\left[-\frac{(r_1^2+r_2^2)}{2N(1-\rho^2)}\right] I_0\left[\frac{\rho r_1 r_2}{N(1-\rho^2)}\right]. \end{aligned} \tag{11.34}$$

Now, by use of Eq. (11.34), we can calculate the joint moments

$$\begin{aligned} \langle \mathbf{r}_1^{\mu}\mathbf{r}_2^{\nu}\rangle &= \int_0^{\infty}\int_0^{\infty} r_1^{\mu} r_2^{\nu} p(r_1, r_2)\, dr_1\, dr_2 \\ &= \frac{1}{N^2(1-\rho^2)} \int_0^{\infty} r_1^{\mu+1} \exp\left[-\frac{r_1^2}{2N(1-\rho^2)}\right] \\ &\quad \times \int_0^{\infty} r_2^{\nu+1} \exp\left[-\frac{r_2^2}{2N(1-\rho^2)}\right] I_0\left[\frac{\rho r_1 r_2}{N(1-\rho^2)}\right] dr_2\, dr_1. \end{aligned} \tag{11.35}$$

The inside integral in Eq. (11.35) is roughly the same as that in Eq. (7.34), which gives us

$$\begin{aligned} \langle \mathbf{r}_1^{\mu}\mathbf{r}_2^{\nu}\rangle &= \frac{[2N(1-\rho^2)]^{\nu/2}}{N} \int_0^{\infty} r_1^{\mu+1} \exp\left[-\frac{r_1^2}{2N(1-\rho^2)}\right] G_{12}^{11}\left[-\frac{\rho^2 r_1^2}{2N(1-\rho^2)}\middle|\begin{matrix} -\frac{\nu}{2} \\ 0,0 \end{matrix}\right] dr_1 \\ &= \frac{1}{2N}[2N(1-\rho^2)]^{(\mu+\nu)/2+1} G_{22}^{12}\left(-\rho^2\middle|\begin{matrix} -\frac{\mu}{2}, -\frac{\nu}{2} \\ 0,0 \end{matrix}\right). \end{aligned} \tag{11.36}$$

In the last step, we used ($D6$) from Appendix D. Also, the last Meijer G-function in Eq. (11.36) based on ($C19$) in Appendix C leads to

$$G_{22}^{12}\left(-\rho^2\middle|\begin{matrix} -\frac{\mu}{2}, & -\frac{\nu}{2} \\ 0, & 0 \end{matrix}\right) = \Gamma\left(1+\frac{\mu}{2}\right)\Gamma\left(1+\frac{\nu}{2}\right) {}_2F_1\left(1+\frac{\mu}{2},\ 1+\frac{\nu}{2};\ 1;\ \rho^2\right), \tag{11.37}$$

and, thus, from ($HY8$) in Appendix A, we obtain the result

$$\langle \mathbf{r}_1^{\mu}\mathbf{r}_2^{\nu}\rangle = (2N)^{(\mu+\nu)/2}\Gamma\left(1+\frac{\mu}{2}\right)\Gamma\left(1+\frac{\nu}{2}\right) {}_2F_1\left(-\frac{\mu}{2},\ -\frac{\nu}{2};\ 1;\ \rho^2\right). \tag{11.38}$$

The cases of greatest interest in applications involving Eq. (11.38) are those for which $\mu = \nu = 1, 2$. The first case, $\mu = \nu = 1$, corresponds to a *half-wave linear detector*, which yields

$$\langle \mathbf{r}_1\mathbf{r}_2 \rangle = \frac{N\pi}{2}\, {}_2F_1\left(-\frac{1}{2},\ -\frac{1}{2};\ 1;\ \rho^2\right) \approx \frac{N\pi}{2}(1 + 0.273\rho^2), \qquad -1 \le \rho \le 1, \tag{11.39}$$

where we have introduced an approximation for the hypergeometric function in the last expression. For the special cases of no correlation ($\rho = 0$) or perfect correlation ($\rho = 1$), we find that

$$\langle \mathbf{r}_1\mathbf{r}_2 \rangle \sim \begin{cases} N\pi/2, & \rho = 0 \\ 2N, & \rho = 1. \end{cases} \tag{11.40}$$

In the *square-law* case defined by $\mu = \nu = 2$, we obtain

$$\langle \mathbf{r}_1^2\mathbf{r}_2^2 \rangle = (2N)^2\, {}_2F_1(-1,\ -1;\ 1;\ \rho^2) = (2N)^2(1 + \rho^2). \tag{11.41}$$

11.2 Cross Correlators

A true estimate over a finite observation time of the normalized correlation function between two random signals can be obtained by the operation of a *cross correlator*. Cross correlation techniques for processing signals embedded in additive random noise have been extensively employed in communication systems such as radar and sonar devices.[3–9] Many of the early research studies on cross correlators were concerned with the output probability density function—others directed their attention to the output SNR as a performance measure.

A block diagram of an analog cross correlator that consists of a multiplier and lowpass filter is featured in Fig. 11.6. Based on such a system, we first consider the case of two noise inputs to identical narrowband filters where the noise terms are correlated with *normalized correlation coefficient* ρ. If each

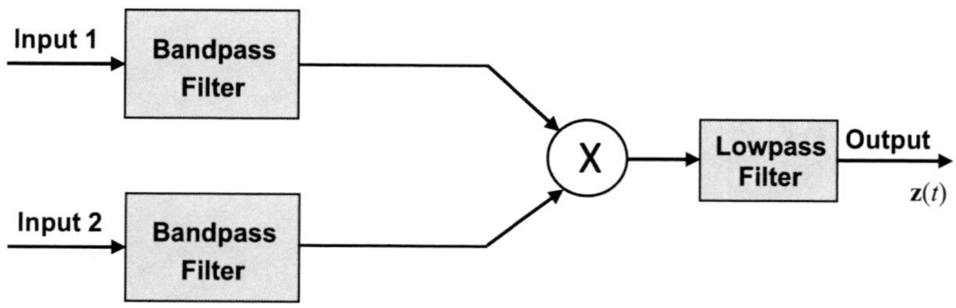

Figure 11.6 Block diagram for a cross correlator.

noise term at the output of the narrowband filters is governed by a Gaussian random process with zero mean and the same variance N, then they each can be represented by an expression like Eq. (11.1). In this case, the joint distribution of the envelopes and phases at the output of the bandpass filters is that given by Eq. (11.33).

After passing the product of two narrowband noise processes through the lowpass filter in Fig. 11.6, the output is

$$\begin{aligned}\mathbf{z} &= \frac{1}{2}(\mathbf{x}_1\mathbf{x}_2 + \mathbf{y}_1\mathbf{y}_2)\\ &= \frac{1}{2}\mathbf{r}_1\mathbf{r}_2\cos(\boldsymbol{\theta}_1 - \boldsymbol{\theta}_2).\end{aligned} \tag{11.42}$$

Thus, from the second result in Eq. (11.42), the mean value of the output is

$$\begin{aligned}\langle \mathbf{z}\rangle &= \frac{1}{2(2\pi N)^2(1-\rho^2)}\int_0^\infty\int_0^\infty\int_0^{2\pi}\int_0^{2\pi} r_1^2 r_2^2 \cos(\theta_1-\theta_2)\\ &\quad\times \exp\left\{-\frac{\left[r_1^2+r_2^2-2\rho r_1 r_2\cos(\theta_1-\theta_2)\right]}{2N(1-\rho^2)}\right\}d\theta_1 d\theta_2 dr_1 dr_2.\end{aligned} \tag{11.43}$$

By starting with the two phase integrals, we have from Section 7.1.2,

$$\int_0^{2\pi}\int_0^{2\pi}\cos(\theta_1-\theta_2)\exp\left[\frac{r_1 r_2\rho}{N(1-\rho^2)}\cos(\theta_1-\theta_2)\right]d\theta_1 d\theta_2 = 4\pi^2 I_1\left[\frac{r_1 r_2 \rho}{N(1-\rho^2)}\right], \tag{11.44}$$

and the remaining integrations on the envelope integrals in Eq. (11.43) lead to

$$\begin{aligned}\langle \mathbf{z}\rangle &= \frac{1}{2\,N^2(1-\rho^2)}\int_0^\infty r_1^2 \exp\left[-\frac{r_1^2}{2N(1-\rho^2)}\right]\\ &\quad\times\int_0^\infty r_2^2\exp\left[-\frac{r_2^2}{2N(1-\rho^2)}\right] I_1\left[\frac{r_1 r_2\rho}{N(1-\rho^2)}\right]dr_2\,dr_1.\end{aligned} \tag{11.45}$$

The integral over r_2 yields (see Section 7.1.2)

$$\int_0^\infty r_2^2\exp\left[-\frac{r_2^2}{2N(1-\rho^2)}\right] I_1\left[\frac{r_1 r_2\rho}{N(1-\rho^2)}\right]dr_2 = N\rho(1-\rho^2)r_1\exp\left[\frac{\rho^2 r_1^2}{2N(1-\rho^2)}\right], \tag{11.46}$$

and, thus, Eq. (11.45) reduces to a multiple of the correlation coefficient ρ, i.e.,

$$\langle \mathbf{z} \rangle = \frac{\rho}{2N} \int_0^\infty r_1^3 \exp\left(-\frac{r_1^2}{2N}\right) dr_1 = N\rho. \tag{11.47}$$

Equation (11.47) also follows directly from the upper expression in Eq. (11.42) with $\langle \mathbf{x}_1 \mathbf{x}_2 \rangle = \langle \mathbf{y}_1 \mathbf{y}_2 \rangle = N\rho$. In the presence of signal plus noise at each narrowband filter output, the mean output is

$$\begin{aligned} \langle \mathbf{z} \rangle &= \frac{1}{2}\left[\langle (A_1 + \mathbf{x}_1)(A_2 + \mathbf{x}_2) \rangle + \langle \mathbf{y}_1 \mathbf{y}_2 \rangle\right] \\ &= N(\sqrt{s_1 s_2} + \rho), \end{aligned} \tag{11.48}$$

where $s_1 = A_1^2/2N$ and $s_2 = A_2^2/2N$. Thus, the mean output signal power is

$$S_0 = \left[\langle \mathbf{z} \rangle - \langle \mathbf{z} | s_1, s_2 = 0 \rangle\right]^2 = N^2 s_1 s_2. \tag{11.49}$$

To calculate the output noise power alone without signals, we start with the second moment,

$$\begin{aligned} \langle \mathbf{z}^2 \rangle = {} & \frac{1}{4(2\pi N)^2(1-\rho^2)} \int_0^\infty \int_0^\infty \int_0^{2\pi} \int_0^{2\pi} r_1^3 r_2^3 \cos^2(\theta_1 - \theta_2) \\ & \times \exp\left\{-\frac{\left[r_1^2 + r_2^2 - 2\rho r_1 r_2 \cos(\theta_1 - \theta_2)\right]}{2N(1-\rho^2)}\right\} d\theta_1 d\theta_2 dr_1\, dr_2. \end{aligned} \tag{11.50}$$

From the phase integrals, this time we have

$$\begin{aligned} & \int_0^{2\pi} \int_0^{2\pi} \cos^2(\theta_1 - \theta_2) \exp\left[\frac{r_1 r_2 \rho}{N(1-\rho^2)} \cos(\theta_1 - \theta_2)\right] d\theta_1 d\theta_2 \\ & \quad = 4\pi^2 \frac{N(1-\rho^2)}{\rho r_1 r_2} I_1\left[\frac{r_1 r_2 \rho}{N(1-\rho^2)}\right] + 4\pi^2 I_2\left[\frac{r_1 r_2 \rho}{N(1-\rho^2)}\right]. \end{aligned} \tag{11.51}$$

The remaining integrations on the envelope integrals take the form

$$\begin{aligned} \langle \mathbf{z}^2 \rangle = {} & \frac{1}{4N^2(1-\rho^2)} \int_0^\infty r_1^3 \exp\left[-\frac{r_1^2}{2N(1-\rho^2)}\right] \int_0^\infty \exp\left[-\frac{r_2^2}{2N(1-\rho^2)}\right] \\ & \times \left\{\frac{N(1-\rho^2)}{\rho r_1} r_2^2 I_1\left[\frac{r_1 r_2 \rho}{N(1-\rho^2)}\right] + r_2^3 I_2\left[\frac{r_1 r_2 \rho}{N(1-\rho^2)}\right]\right\} dr_2\, dr_1. \end{aligned} \tag{11.52}$$

The integral over r_2 based on the first modified Bessel function in Eq. (11.52) is the same as Eq. (11.46). Thus, we are left with

$$\langle \mathbf{z}^2 \rangle = \frac{1}{4N^2(1-\rho^2)} \int_0^\infty r_1^3 \exp\left[-\frac{r_1^2}{2N(1-\rho^2)}\right] \left\{ N^2(1-\rho^2)^2 \exp\left[\frac{\rho^2 r_1^2}{2N(1-\rho^2)}\right] \right.$$
$$\left. + \int_0^\infty r_2^3 \exp\left[-\frac{r_2^2}{2N(1-\rho^2)}\right] I_2\left[\frac{r_1 r_2 \rho}{N(1-\rho^2)}\right] \right\} dr_2\, dr_1. \tag{11.53}$$

Completing the inside integration in Eq. (11.53) in a manner similar to deriving Eq. (7.34) by first recognizing that $I_2(ar_2) = -\,G_{02}^{10}\left(-\frac{a^2 r_2^2}{4}\middle|\ 1,\ -1\right)$ yields

$$\langle \mathbf{z}^2 \rangle = \frac{1}{4N} \int_0^\infty r_1^3 \exp\left(-\frac{r_1^2}{2N}\right) \left[N(1-\rho^2) + \rho^2 r_1^2\right] dr_1, \tag{11.54}$$

which finally reduces to

$$\langle \mathbf{z}^2 \rangle = \frac{1}{2} N^2 (1 + 3\rho^2). \tag{11.55}$$

Hence, the output noise power in the presence of noise alone at the output of the bandpass filters becomes

$$N_0 = \langle \mathbf{z}^2 \rangle - \langle \mathbf{z} \rangle^2 = \frac{1}{2} N^2 (1 + \rho^2). \tag{11.56}$$

We calculated the output moments with $\mathbf{r}$ and $\boldsymbol{\theta}$ for the noise-only case mainly for illustrative purposes. However, the more interesting case is when both signal and noise are present at the output of the bandpass filters. Thus, following calculation of the mean $\langle \mathbf{z} \rangle$ with Gaussian random variables $\mathbf{x}_1$, $\mathbf{x}_2$ and $\mathbf{y}_1$, $\mathbf{y}_2$ at the output of the bandpass filters, we can also use them to compute the variance

$$N_0 = \frac{1}{4} \langle [(A_1 + \mathbf{x}_1)(A_2 + \mathbf{x}_2) + \mathbf{y}_1 \mathbf{y}_2]^2 \rangle - N^2(\sqrt{s_1 s_2} + \rho)^2$$
$$= \frac{1}{2} N^2 (1 + s_1 + s_2 + 2\rho\sqrt{s_1 s_2} + \rho^2),$$

leading to SNR_0 defined by

$$\mathrm{SNR}_0 \equiv \frac{S_0}{N_0} = \frac{2 s_1 s_2}{1 + s_1 + s_2 + 2\rho\sqrt{s_1 s_2} + \rho^2}. \tag{11.57}$$

11.2.1 Cross correlator with limiter in one channel

The analog correlator presented in the last section is the one studied most often. Although the essential feature in all cross correlators is the multiplication of two waveforms, the hardware used to calculate this product varies from one system to another. For example, some correlators "clip" the narrowband filter output of each channel before the multiplier to reduce the complexity of the hardware—only the polarity of the incoming waveform is retained.[6–8] In the case we first look at here, only one of the signals is clipped with a hard limiter, as shown in Fig. 11.7. This still reduces some of the dynamic range of the multiplier without losing all amplitude information.

The output of the lowpass filter in Fig. 11.7 is

$$\mathbf{z} = \frac{2\mathbf{r}_1}{\pi}\cos(\boldsymbol{\theta}_1 - \boldsymbol{\theta}_2). \tag{11.58}$$

In the analysis below, we will take the case where signals are present in each channel but the noise terms are not correlated. To calculate the moments of Eq. (11.58) leading to the output SNR, we use the moment-generating property of the characteristic function. It has been shown that the characteristic function of the output $\mathbf{z}$ is given by the expected value[7]

$$\begin{aligned} C_{\mathbf{z}}(u) &= E[e^{iu\mathbf{z}}] \\ &= e^{-iNu^2/\pi^2}\sum_{n=0}^{\infty}\varepsilon_n f(n;\, s_2) I_n\left(\frac{2iu}{\pi}\sqrt{2Ns_1}\right), \end{aligned} \tag{11.59}$$

where ε_n is the Neumann factor equal to one for $n = 0$ and two for all other values of n. For algebraic simplicity, we have introduced the notation

$$\begin{aligned} f(n;\, s) &= \frac{s^{n/2}}{n!}\Gamma\left(1+\frac{n}{2}\right) {}_1F_1\left(\frac{n}{2};\, 1+n;\, -s\right) \\ &= \frac{1}{2}\sqrt{\pi s}\, e^{-s/2}\left[I_{(n-1)/2}\left(\frac{s}{2}\right) + I_{(n+1)/2}\left(\frac{s}{2}\right)\right]. \end{aligned} \tag{11.60}$$

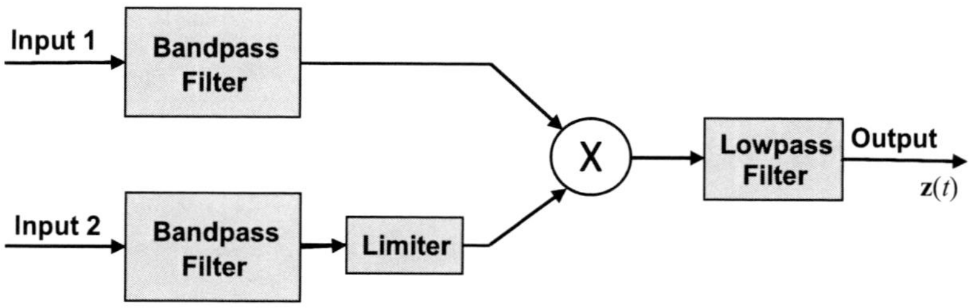

Figure 11.7 Block diagram for a cross correlator with one limiter.

In arriving at the second line in Eq. (11.60), we have used ($CH8$) in Appendix A and ($B19$) in Appendix B. Upon using the characteristic function in Eq. (11.59) to calculate the first two moments of the output $\mathbf{z}$ in Eq. (11.58), we have

$$\begin{aligned}\langle \mathbf{z} \rangle = -iC'(0) &= N\sqrt{\frac{2s_1s_2}{\pi}}\,{}_1F_1\left(\frac{1}{2};\,2,\,-s_2\right)\\ &= N\sqrt{\frac{2s_1s_2}{\pi}}\,e^{-s_2/2}\left[I_0\left(\frac{s_2}{2}\right)+I_1\left(\frac{s_2}{2}\right)\right],\end{aligned} \tag{11.61}$$

$$\langle \mathbf{z}^2 \rangle = -C''(0) = \frac{4N^2}{\pi^2}\left[1+s_1+\frac{1}{2}s_1s_2\,{}_1F_1(1;\,3,\,-s_2)\right]. \tag{11.62}$$

Based on ($B14$) in Appendix B, the confluent hypergeometric function in Eq. (11.62) leads to

$${}_1F_1(1;\,3,\,-s_2) = \frac{2}{s_2^2}(s_2 - 1 + e^{-s_2}). \tag{11.63}$$

Hence, the output SNR is given by

$$\mathrm{SNR}_0 = \frac{\frac{\pi}{2}s_1s_2e^{-s_2}\left[I_0\left(\frac{s_2}{2}\right)+I_1\left(\frac{s_2}{2}\right)\right]^2}{1+s_1+\frac{s_1}{s_2}(s_2-1+e^{-s_2})-\frac{\pi}{2}s_1s_2e^{-s_2}\left[I_0\left(\frac{s_2}{2}\right)+I_1\left(\frac{s_2}{2}\right)\right]^2}. \tag{11.64}$$

Of special interest may be the limiting cases of small-input SNR and large-input SNR (in one channel). These special cases correspond to the asymptotic expressions

$$\mathrm{SNR}_0 \sim \begin{cases} \dfrac{1}{2}\pi s_1s_2, & s_1,\, s_2 \ll 1 \\ 2s_2, & s_2 \ll 1,\; s_1 \gg 1. \end{cases} \tag{11.65}$$

11.2.2 Polarity coincidence correlator

A cross correlator with hard limiters, or clippers, in both channels is known as a *polarity coincidence correlator* (PCC).[6,8] A block diagram of a PCC is shown in Fig. 11.8. As in the case of a cross correlator with only one limiter, this model with two limiters offers hardware advantages over the analog correlator by greatly reducing the dynamic range of the multiplier.

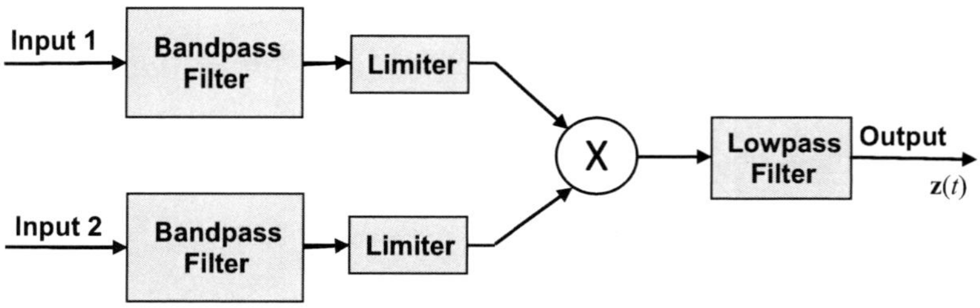

Figure 11.8 Block diagram for a polarity coincidence correlator.

For the SNR analysis of the PCC, we will again take the case of uncorrelated noise terms. The output of the lowpass filter in Fig. 11.8 is

$$\begin{aligned}\mathbf{z} &= \frac{8}{\pi^2}\sum_{n=0}^{\infty}\frac{\cos[(2n+1)(\boldsymbol{\theta}_1-\boldsymbol{\theta}_2)]}{(2n+1)^2}\\ &= 1-\frac{2}{\pi}|\boldsymbol{\theta}_1-\boldsymbol{\theta}_2|, \quad |\boldsymbol{\theta}_1-\boldsymbol{\theta}_2|<\pi,\end{aligned} \tag{11.66}$$

where the second line is the sum of the series (p. 99 of Ref. 10). It has been shown that the output PDF for $\mathbf{z}$ is given by

$$\begin{aligned}p_{\mathbf{z}}(z) = \frac{1}{2} + \sum_{n=1}^{\infty}(-1)^n\Big\{&f(2n;\, s_1)f(2n;\, s_2)\cos(n\pi z)\\ &+f(2n-1;\, s_1)f(2n-1;\, s_2)\sin\left[\frac{1}{2}(2n-1)\pi z\right]\Big\}.\end{aligned} \tag{11.67}$$

Therefore, the moments required for calculating the output SNR can be determined from

$$\langle \mathbf{z}^k\rangle = \int_{-1}^{1} z^k p_{\mathbf{z}}(z)\,dz, \quad k=1,2. \tag{11.68}$$

In this case, the mean output leads to the infinite series

$$\langle \mathbf{z}\rangle = \frac{8N}{\pi^2}\sum_{n=1}^{\infty}\frac{f(2n-1;\, s_1)f(2n-1;\, s_2)}{(2n-1)^2}. \tag{11.69}$$

However, the series in Eq. (11.69) converges fast enough that only the first term ($n=1$) is significant for most values of input SNR. Thus, by recalling Eq. (11.60), we have

$$\langle \mathbf{z}\rangle \approx \frac{2N}{\pi}\sqrt{s_1 s_2}\, e^{-(s_1+s_2)/2}\left[I_0\left(\frac{s_1}{2}\right)+I_1\left(\frac{s_1}{2}\right)\right]\left[I_0\left(\frac{s_2}{2}\right)+I_1\left(\frac{s_2}{2}\right)\right]. \tag{11.70}$$

Similarly, the second moment yields

$$\begin{aligned}\langle \mathbf{z}^2 \rangle &= \frac{N^2}{3} + \frac{4N^2}{\pi^2} \sum_{n=1}^{\infty} \frac{f(2n;\, s_1) f(2n;\, s_2)}{n^2} \\ &\approx \frac{N^2}{3} + \frac{4N^2}{\pi^2 s_1 s_2} (s_1 - 1 + e^{-s_1})(s_2 - 1 + e^{-s_2}),\end{aligned} \tag{11.71}$$

where we have used Eq. (11.63). The output SNR is then determined from S_0/N_0, based on relations in Eq. (11.10). Asymptotic formulas derived for the output SNR under small- and large-input SNRs are given by

$$\mathrm{SNR}_0 \sim \begin{cases} \dfrac{12}{\pi^2} s_1 s_2, & s_1,\, s_2 \ll 1 \\ \dfrac{48}{\pi^3} s_2, & s_2 \ll 1,\ s_1 \gg 1. \end{cases} \tag{11.72}$$

11.3 Free-Space Optical Communication

Up to this point we have considered only electrical or acoustic signals and noise. We now consider the case of a free-space optical communication (FSOC) system. FSOC technology started in the 1960s, but deleterious atmospheric effects on optical waves together with the invention of optical fibers in the early 1970s caused a decline in its immediate use. Now, with the advanced signal coding and modulation techniques, breakthroughs in fiber telecommunication components, and the loss of the available RF spectrum to the cellular industry, there is a strong revival in FSOC system development for 100 Gbps (gigabits per sec) or higher transient/front-haul/back-haul networks.[11]

A FSOC link budget permits an engineer to predict system performance under a broad range of conditions and enables effective operation planning. It is a valuable tool but requires underlying models that accurately capture system and component performance over a wide range of environmental conditions. In many FSOC links, it may be necessary to use an adaptive optics (AO) system and/or an optical automatic gain controller (OAGC) system to improve performance. The AO structure may consist of only tip-tilt corrections or, for more sophisticated systems, may include a number of AO higher-order Zernike correction modes. The OAGC module is designed to boost a degraded optical signal to a level that the communications modem can decipher with reduced bit-error-rate (BER). Other mitigating techniques such as a forward-error-control (FEC) coding scheme or spatial diversity in terminal architectures can be used or added to the system layout to possibly improve network/link performance.

A *direct detection system* is one that responds only to the instantaneous power of the collected waveform (see Fig. 11.9). Unfortunately, the optical

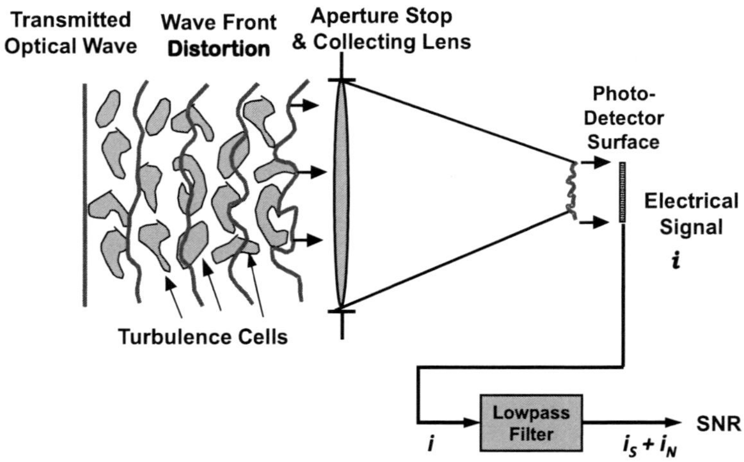

Figure 11.9 Block diagram of a direct detection system (adapted from Ref. 12).

field is always detected in the presence of noise throughout the receiver (Rx), called detector noise or shot noise. Also, circuit and electronic thermal noise are generated in the processing electronics following photo-detection. All electrical noise sources are treated as an additive corrupting signal, although we will only consider shot noise in the following analysis. The lowpass filter shown in the diagram in Fig. 11.9 eliminates much of the random noise.

11.3.1 Threshold detection

At the output of the lowpass filter in Fig. 11.9, the electrical signal current is given by the sum of signal current i_S and noise current i_N, i.e.,

$$i = i_S + i_N. \tag{11.73}$$

The signal current in the absence of atmospheric turbulence is defined by

$$i_S = \frac{\eta e P_{S,0}}{h\nu}, \tag{11.74}$$

where $P_{S,0}$ is signal power in watts collected by the lens, η is detector quantum efficiency in electrons/photon, e is electric charge in coulombs, h is Planck's constant in joule-seconds, and ν is optical frequency in hertz. In many cases, the noise current is assumed to be Gaussian with $\langle i_N \rangle = 0$.

The most basic form of pulse modulation is *on-off keying* (OOK). When the output of a detector exceeds a set *threshold* i_T, we say that a signal is present, called *detection.* A *false alarm* arises when noise alone exceeds the threshold and is interpreted as signal. When the combination of signal plus noise does not exceed the threshold, we say that a *missed detection* or *fade* has

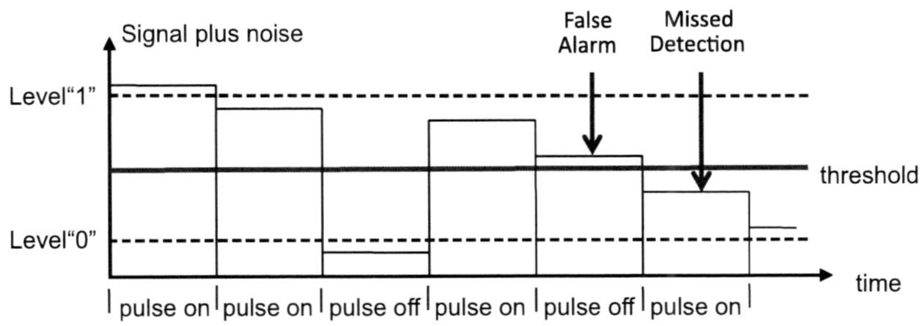

Figure 11.10 Threshold setting for a digital OOK signal (adapted from Ref. 12).

occurred. Threshold concepts involving false alarm and missed detection for an OOK system are illustrated in Fig. 11.10.

If the noise alone is a zero-mean Gaussian process, then its probability density function (PDF) is described by

$$p_{\mathbf{n}}(i) = \frac{1}{\sqrt{2\pi}\sigma_N} \exp\left(-\frac{i^2}{2\sigma_N^2}\right), \tag{11.75}$$

where $\sigma_N^2 = \langle i_N^2 \rangle = 2eBi_S$ represents noise power and B is the bandwidth (positive frequencies only) of the lowpass filter. In this case, the total signal current [Eq. (11.73)] is governed by the non-zero mean Gaussian PDF,

$$p_{s+\mathbf{n}}(i) = \frac{1}{\sqrt{2\pi}\sigma_N} \exp\left[-\frac{(i-i_S)^2}{2\sigma_N^2}\right]. \tag{11.76}$$

Consequently, the *probability of false alarm* and *probability of detection* (*fade*) are given, respectively, by expressions involving the complementary error function,

$$\begin{aligned} \mathrm{Pr}_{fa} &= \int_{i_T}^{\infty} p_{\mathbf{n}}(i)\,di = \frac{1}{2}\mathrm{erfc}\left(\frac{i_T}{\sqrt{2}\sigma_N}\right), \\ \mathrm{Pr}_d &= \int_{i_T}^{\infty} p_{s+\mathbf{n}}(i)\,di = \frac{1}{2}\mathrm{erfc}\left(\frac{i_T - i_S}{\sqrt{2}\sigma_N}\right), \\ \mathrm{Pr}_{fade} &= 1 - \mathrm{Pr}_d\,. \end{aligned} \tag{11.77}$$

Another quantity of interest is the *false alarm rate* (FAR) or the expected number of fades in the absence of signal,

$$\mathrm{FAR} = \frac{B}{2\sqrt{3}} \exp\left(-\frac{1}{2}\mathrm{TNR}^2\right), \tag{11.78}$$

where TNR is the *threshold-to-noise ratio* defined by $\mathrm{TNR} = i_T/\sigma_N$.

11.3.2 Performance measures

In the presence of atmospheric turbulence, we find it useful to introduce the notion of performance measures. That is, the reliability of a FSOC system can be deduced from the analysis of several statistical performance measures: Strehl ratio, optical power, SNR, and BER.

- *Strehl Ratio*: We have already discussed the Strehl ratio in Section 9.1.3 for imaging systems. However, the Strehl ratio in a FSOC system is defined at the Rx pupil plane (RP) as well as at the detector plane (DP) located beyond the Rx aperture. At the RP, we have

$$\mathrm{SR_{RP}} \approx \frac{1}{\left[1+m(D_{Tx}/r_{0\mathrm{T}})^{5/3}\right]^{6/5}}, \quad m = \begin{cases} 1, \text{ no AO} \\ 0.28, \text{ with Tx TT,} \end{cases} \tag{11.79}$$

where TT means tip-tilt. At the DP,

$$\mathrm{SR_{DP}} \approx \frac{1}{\left[1+m(D_{Rx}/r_{0\mathrm{R}})^{5/3}\right]^{6/5}}, \quad m = \begin{cases} 1, \text{ no AO} \\ 0.28, \text{ with Rx TT.} \end{cases} \tag{11.80}$$

The quantities D_{Tx} and D_{Rx} represent aperture diameters at the Tx and Rx, respectively. The two Fried parameters appearing in the Strehl ratios are illustrated in Fig. 11.11. They are defined by

$$\begin{aligned} r_{0\mathrm{T}} &= \left[0.423k^2 \int_0^L C_n^2(z)\left(1-\frac{z}{L}\right)^{5/3} dz\right]^{-3/5}, \\ r_{0\mathrm{R}} &= \left[0.423k^2 \int_0^L C_n^2(z)\left(\frac{z}{L}\right)^{5/3} dz\right]^{-3/5}. \end{aligned} \tag{11.81}$$

- *Optical Power* (power in the bucket, PIB): In FSOC systems, an important measure of system performance is the amount of optical power that falls on the Rx collecting lens. At the laser Tx, the optical power is denoted by P_{Tx} [watts]. The average power at the detector plane, called the *power in the bucket* or average PIB, is defined by

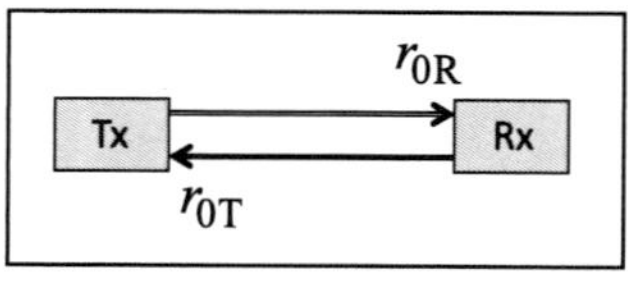

Figure 11.11 Fried parameters at the Tx and Rx (adapted from Ref. 12).

$$\langle \mathrm{PIB} \rangle = P_{Tx} \cdot \frac{D_{Rx}^2}{8W^2} \cdot \tau_{\mathrm{atm}} \cdot \tau_{\mathrm{opt}} \cdot \mathrm{SR}_{\mathrm{RP}}; \quad D_{Rx} \le 2\sqrt{2}W, \tag{11.82}$$

where W is the free-space Gaussian beam radius at the Rx lens, τ_{atm} is the atmospheric transmission loss caused by aerosols, etc., and τ_{opt} is the Tx optical element transmission loss.

- *Optical Power* (power in the fiber, PIF): In some FSOC systems the received waveform is launched into an optical fiber before being converted to an electrical signal for processing. The average amount of power that gets into the fiber is called the *power in the fiber* or average PIF. The average PIB given by Eq. (11.82) is reduced by this process according to

$$\langle \mathrm{PIF} \rangle = \langle \mathrm{PIB} \rangle \cdot \eta \cdot \tau_{\mathrm{opt}} \cdot \tau_{\mathrm{circ}} \cdot \mathrm{SR}_{DP}, \tag{11.83}$$

where η is the fiber coupling efficiency loss, τ_{circ} is the circulator loss (for duplex systems), and τ_{opt} accounts for any Rx optical element loss.

- *Signal-to-Noise Ratio* (SNR): In the absence of atmospheric turbulence, the shot-noise optical SNR is

$$\mathrm{SNR}_0 = \frac{i_S}{\sigma_N} = \sqrt{\frac{\eta P_{S,0}}{2h\nu B}}, \tag{11.84}$$

where $P_{S,0} = $ PIB in the absence of turbulence. The expression (11.84) is the ratio of optical signal to rms noise power, not electrical signal to noise power (which is the square of optical SNR). In the presence of atmospheric turbulence, the mean signal current and noise power become

$$\begin{aligned} \langle i_S \rangle &= \frac{\eta e \langle P_S \rangle}{h\nu}, \\ \langle i_S^2 \rangle &= \left(\frac{\eta e}{h\nu}\right)^2 \langle P_S^2 \rangle, \end{aligned} \tag{11.85}$$

$$\langle i_N^2 \rangle = 2eB\langle i_S \rangle = \frac{2\eta e^2 B \langle P_S \rangle}{h\nu}. \tag{11.86}$$

The variance of signal plus noise [Eq. (11.73)] is now

$$\begin{aligned} \sigma_{SN}^2 &= \langle i_S^2 \rangle - \langle i_S \rangle^2 + \langle i_N^2 \rangle \\ &= \left(\frac{\eta e}{h\nu}\right)^2 \left[\langle P_S^2 \rangle - \langle P_S \rangle^2\right] + \frac{2\eta e^2 B \langle P_S \rangle}{h\nu}, \end{aligned} \tag{11.87}$$

and, consequently, the optical output mean SNR becomes

$$\langle \mathrm{SNR} \rangle = \frac{\langle i_S \rangle}{\sigma_{SN}} = \frac{\langle P_S \rangle}{\sqrt{\left[\langle P_S^2 \rangle - \langle P_S \rangle^2\right] + \dfrac{2h\nu B \langle P_S \rangle}{\eta}}}. \tag{11.88}$$

Upon multiplying both numerator and denominator by SNR_0 and dividing by $\langle P_S \rangle = \langle \mathrm{PIB} \rangle$, the mean SNR becomes

$$\langle \mathrm{SNR} \rangle = \frac{\mathrm{SNR}_0}{\sqrt{\dfrac{P_{S,0}}{\langle \mathrm{PIB} \rangle} + \sigma_I^2(D_{Rx})\mathrm{SNR}_0^2}}, \tag{11.89}$$

where $\sigma_I^2(D_{Rx}) = [\langle P_S^2 \rangle - \langle P_S \rangle^2]/\langle P_S \rangle^2$ is the flux variance.

- *Bit Error Rate* (BER): In digital communications it is common to use the BER as a measure of performance. For an OOK system, the BER in free space with equal probability of a "0" or "1" is defined by

$$\mathrm{BER} = \mathrm{erfc}\left(\frac{\mathrm{SNR}_0}{2\sqrt{2}}\right), \tag{11.90}$$

where the threshold is set at $i_T = 0.5 i_S$. The function $\mathrm{erfc}(x)$ is the complementary error function (Section 3.10.1). In atmospheric turbulence, the mean BER becomes

$$\langle \mathrm{BER} \rangle = \frac{1}{2}\int_0^\infty p_I(u)\,\mathrm{erfc}\left(\frac{\langle \mathrm{SNR} \rangle u}{2\sqrt{2}}\right) du. \tag{11.91}$$

The function $p_I(u)$ is the PDF describing irradiance fluctuations at the Rx caused by atmospheric turbulence.

By representing both the PDF and the complementary error function as Meijer *G*-functions, we can get a closed-form solution of Eq. (11.91). For example, if we assume that the PDF model is the gamma-gamma distribution, it can be represented in terms of a Meijer *G*-function as

$$\begin{aligned} p_I(u) &= \frac{2(\alpha\beta)^{(\alpha+\beta)/2}}{\Gamma(\alpha)\Gamma(\beta)} u^{(\alpha+\beta)/2-1} K_{\alpha-\beta}(2\sqrt{\alpha\beta u}) \\ &= \frac{1}{\Gamma(\alpha)\Gamma(\beta)u} G_{02}^{20}(\alpha\beta u|\alpha,\, \beta), \quad u > 0, \end{aligned} \tag{11.92}$$

where we are using (*C*23) in Appendix C. By assuming a perfectly tracked beam (no beam wander), the PDF parameters are defined by

$$\frac{1}{\alpha} = \exp\left(\sigma^2_{\ln X}\right) - 1, \quad \frac{1}{\beta} = \exp\left(\sigma^2_{\ln Y}\right) - 1,$$

where $\sigma^2_{\ln X}$ and $\sigma^2_{\ln Y}$ are the large-scale and small-scale log-irradiance variances. If we apply (*C*25) in Appendix C, the complementary error function with $\eta = \langle \mathrm{SNR} \rangle / 2\sqrt{2}$ can likewise be expressed as

$$\mathrm{erfc}(\eta u) = \frac{1}{\sqrt{\pi}} G^{20}_{12}\left(\eta^2 u^2 \middle| \begin{matrix} 1 \\ 0, \frac{1}{2} \end{matrix} \right). \tag{11.93}$$

By inserting both Eqs. (11.92) and (11.93) into Eq. (11.91) and expressing Eq. (11.93) in its Barnes integral representation, we obtain

$$\begin{aligned} \langle \mathrm{BER} \rangle &= \frac{1}{2\sqrt{\pi}\,\Gamma(\alpha)\Gamma(\beta)} \int_0^\infty u^{-1} G^{20}_{02}(\alpha\beta u|\alpha, \beta)\, G^{20}_{12}\left(\eta^2 u^2 \middle| \begin{matrix} 1 \\ 0, \frac{1}{2} \end{matrix} \right) du \\ &= \frac{1}{2\sqrt{\pi}\,\Gamma(\alpha)\Gamma(\beta)} \frac{1}{2\pi i} \int_C \frac{\Gamma(-s)\Gamma\left(\frac{1}{2} - s\right)}{\Gamma(1-s)} \\ &\quad \times \eta^{2s} \int_0^\infty u^{2s-1} G^{20}_{02}(\alpha\beta u|\alpha, \beta)\, du\, ds, \end{aligned} \tag{11.94}$$

where we have interchanged the order of integration. Based on (*D*8) in Appendix D, this last expression reduces to

$$\begin{aligned} \langle \mathrm{BER} \rangle = {} & \frac{1}{2\sqrt{\pi}\,\Gamma(\alpha)\Gamma(\beta)} \frac{1}{2\pi i} \int_C \frac{\Gamma(-s)\Gamma\left(\frac{1}{2} - s\right)\Gamma(\alpha + 2s)\Gamma(\beta + 2s)}{\Gamma(1-s)} \\ & \times \left(\frac{\eta}{\alpha\beta}\right)^{2s} ds. \end{aligned} \tag{11.95}$$

Finally, by applying (*G*10) in Appendix A to $\Gamma(\alpha + 2s)$ and $\Gamma(\beta + 2s)$, we find that

$$\begin{aligned} \langle \mathrm{BER} \rangle = {} & \frac{2^{\alpha+\beta-3}}{\pi^{3/2}\Gamma(\alpha)\Gamma(\beta)} \times \frac{1}{2\pi i} \int_C \\ & \times \frac{\Gamma(-s)\Gamma\left(\frac{1}{2} - s\right)\Gamma\left(\frac{\alpha}{2} + s\right)\Gamma\left(\frac{\beta}{2} + s\right)\Gamma\left(\frac{\alpha+1}{2} + s\right)\Gamma\left(\frac{\beta+1}{2} + s\right)}{\Gamma(1-s)} \\ & \times \left(\frac{4\eta}{\alpha\beta}\right)^{2s} ds \end{aligned}$$

or, upon recognizing the Meijer *G*-function in this last expression, we arrive at (now replacing $\eta = \langle \mathrm{SNR} \rangle / 2\sqrt{2}$)

$$\langle \mathrm{BER} \rangle = \frac{2^{\alpha+\beta-3}}{\pi^{3/2}\Gamma(\alpha)\Gamma(\beta)} \times G_{52}^{24}\left(\frac{2\langle \mathrm{SNR} \rangle^2}{\alpha^2\beta^2}\,\middle|\, \begin{matrix} 1-\frac{\alpha}{2},\ 1-\frac{\beta}{2},\ \frac{1-\alpha}{2},\ \frac{1-\beta}{2},\ 1 \\ 0,\ \frac{1}{2} \end{matrix}\right). \tag{11.96}$$

11.4 Fluid Mechanics

In Sections 11.1 through 11.3 we investigated some *random processes* associated with various communication systems. In this section we wish to briefly discuss a totally different application area involving the flow of fluids. *Fluid mechanics* is the theory of the motions of liquids and gases. Although actually three dimensional, a fluid flow problem is considered two-dimensional if the velocity vector **V** is always parallel to a fixed plane (xy-plane) and if the velocity components u and v parallel to this plane along with pressure p and fluid density ρ are all constant along any normal to this plane. If the stress on an element of area is wholly normal and independent of the orientation of the area, it is called an *ideal fluid* (or *inviscid fluid*). When the stress is no longer normal to the small element of area it is called a *viscous fluid.*

11.4.1 Unsteady hydrodynamic flow past an infinite plate

A basic problem in hydrodynamics concerns the flow of a viscous fluid past a continuously moving flat plate. Let x denote the distance along a porous plate and y denote the distance in the normal direction. We choose the parameters u and v to correspond, respectively, to the velocity components in the x and y directions. Let us assume that the decelerated fluid particles are removed from the boundary layer through the porous plate by suction. Let us also assume that the velocity component u equals zero at the surface of the plate and approaches $f(t)$ as $y \to \infty$ for all time $t > 0$. Thus $u(y, 0) = f(0)$ at time $t = 0$. The problem we wish to solve is a boundary-value problem for $u(y, t)$, described by

$$\begin{gathered} \frac{\partial u}{\partial t} - \upsilon_s \frac{\partial u}{\partial y} = \frac{\partial f}{\partial t} + \mu \frac{\partial^2 u}{\partial y^2}, \quad 0 < y < \infty, \quad t > 0 \\ u(y, 0) = f(0), \quad 0 < y < \infty \\ u(0, t) = 0, \quad u(y, t) \to f(t) \text{ as } y \to \infty, \quad t > 0, \end{gathered} \tag{11.97}$$

where υ_s is a constant suction velocity and μ is the kinematic viscosity of the fluid.

To solve this problem, we use the Laplace transform which, in the transform domain, gives us

$$\mu \frac{\partial^2 U}{\partial y^2} + \upsilon_s \frac{\partial U}{\partial y} - sU = -sF(s), \qquad (11.98)$$
$$U(0, s) = 0, \quad U(y, s) \to F(s) \text{ as } y \to \infty.$$

Here we have introduced

$$\mathscr{L}\{u(y, t)\} = U(y, s), \quad \mathscr{L}\left\{\frac{df(t)}{dt}\right\} = sF(s) - f(0),$$

along with

$$\mathscr{L}\left\{\frac{\partial^2 u}{\partial y^2}\right\} = \frac{\partial^2 U}{\partial y^2}, \quad \mathscr{L}\left\{\frac{\partial u}{\partial y}\right\} = \frac{\partial U}{\partial y}, \quad \mathscr{L}\left\{\frac{\partial u}{\partial t}\right\} = sU(y, s) - f(0).$$

For a fixed value of s, we can treat the partial DE in Eq. (11.98) as an ordinary DE in the variable y subject to the prescribed boundary conditions. In this case, the solution is found to be

$$U(y, s) = F(s)\left\{1 - \exp\left[-\left(a + \frac{1}{2}\sqrt{b + cs}\right)y\right]\right\}, \qquad (11.99)$$

where the constants a, b, and c are defined by

$$a = \frac{\upsilon_s}{2\mu}, \quad b = \left(\frac{\upsilon_s}{\mu}\right)^2, \quad c = \frac{4}{\mu}.$$

Without providing the details, the inversion of Eq. (11.99) based on Laplace transform methods when $f(t) = u_0$ (constant) is given by[13]

$$u(y, t) = u_0\left[1 - \frac{y}{2\sqrt{\mu\pi}} e^{-\upsilon_s y/2\mu} \int_0^t \tau^{-3/2} e^{-\upsilon_s^2 \tau/4\mu} e^{-y^2/4\mu\tau}\, d\tau\right], \quad t > 0. \qquad (11.100)$$

Suppose that we first consider the steady-state solution $(t \to \infty)$. The integral in Eq. (11.100) then takes the following form [recall Eq. (7.9) and (*BK*9) in Appendix A]:

$$\int_0^\infty \tau^{-3/2} e^{-\upsilon_s^2 \tau/4\mu} e^{-y^2/4\mu\tau}\, d\tau = \frac{2\sqrt{\pi\mu}}{y} \exp\left(-\frac{\upsilon_s y}{2\mu}\right), \qquad (11.101)$$

from which we deduce the steady-state solution,

$$\lim_{t\to\infty} u(y, t) = u_0\left[1 - \exp\left(-\frac{\upsilon_s y}{\mu}\right)\right]. \qquad (11.102)$$

If we next take the limiting case $e^{-\upsilon_s^2\tau/4\mu} \approx 1$ with the change of variable $\tau = t/(s+1)$, the resulting integral in Eq. (11.100) becomes

$$\begin{aligned}\frac{e^{-y^2/4\mu t}}{\sqrt{t}}\int_0^\infty e^{-sy^2/4\mu t}(s+1)^{-1/2}\,ds &= \frac{e^{-y^2/4\mu t}}{\sqrt{t}}\,U\left(1;\,\frac{3}{2};\,\frac{y^2}{4\mu t}\right)\\ &= \frac{2\sqrt{\mu}}{y}\,e^{-y^2/4\mu t}\,U\left(\frac{1}{2};\,\frac{1}{2};\,\frac{y^2}{4\mu t}\right),\end{aligned} \tag{11.103}$$

where we have used ($CH16$) and ($CH17$) in Appendix A. In this case, the solution described by Eq. (11.100) takes the form

$$u(y,\,t) \sim u_0\left[1 - e^{-\upsilon_s y/2\mu}\,\text{erfc}\left(\frac{y}{2\sqrt{\mu t}}\right)\right], \quad \upsilon_s t \ll 4\mu. \tag{11.104}$$

In deriving this final answer, we have used the relation ($B42$) in Appendix B.

11.4.2 Irrotational flow of an ideal fluid

Consider the two-dimensional steady flow of an ideal fluid through a slit in a plane rigid boundary (x-axis). If the fluid is introduced normal to the region with prescribed velocity $v = f(x)$, the resulting velocity components u and v in the upper half-plane $y \geq 0$ can be determined by solving the boundary value problem

$$\begin{aligned}&\frac{\partial^2\phi}{\partial x^2} + \frac{\partial^2\phi}{\partial y^2} = 0, \qquad -\infty < x < \infty,\ y > 0\\ &\frac{\partial\phi}{\partial y} = -f(x)U(a - |x|), \qquad -\infty < x < \infty\\ &\frac{\partial\phi}{\partial x},\ \frac{\partial\phi}{\partial y} \to 0 \text{ as } y \to \infty.\end{aligned} \tag{11.105}$$

The function $U(a - |x|)$ is the unit step function and $\phi(x,\,y)$ is called the *velocity potential function.* The potential function is related to the velocity components u and v in the x and y directions, respectively, according to

$$u = -\frac{\partial\phi}{\partial x}, \qquad v = -\frac{\partial\phi}{\partial y}. \tag{11.106}$$

By use of the Fourier transform, it has been shown that the solution of the boundary value problem (11.105) can be expressed as (p. 271 of Ref. 14)

$$\phi(x,\,y) = \frac{1}{\sqrt{2\pi}}\int_{-\infty}^{\infty}\frac{F(\xi)}{|\xi|}e^{i\xi x - |\xi| y}\,d\xi, \tag{11.107}$$

where $F(\xi)$ is the Fourier transform of $f(x)U(a-|x|)$. If we take the special case where

$$f(x)=(a^2-x^2)^{-1/2}, \tag{11.108}$$

then

$$F(\xi)=\frac{1}{\sqrt{2\pi}}\int_{-a}^{a}\frac{e^{-i\xi x}}{(a^2-x^2)^{1/2}}=\sqrt{\frac{2}{\pi}}\int_0^a\frac{\cos(\xi x)}{(a^2-x^2)^{1/2}}\,dx. \tag{11.109}$$

By replacing the cosine function with the Meijer G-function ($C4$) in Appendix C and making the change of variable $x^2=a^2t$, we find that

$$\begin{aligned}F(\xi)&=\frac{1}{2}\sqrt{\frac{2}{\pi}}\int_0^1 t^{-1/2}(1-t)^{-1/2}\,G_{02}^{10}\left(\frac{a^2\xi^2t}{4}\,\middle|\,0,\ \frac{1}{2}\right)dt\\&=\sqrt{\frac{\pi}{2}}\,G_{02}^{10}\left(\frac{a^2\xi^2}{4}\,\middle|\,0,\ 0\right)\\&=\sqrt{\frac{\pi}{2}}\,J_0(a\xi),\end{aligned} \tag{11.110}$$

where we have used ($D9$) in Appendix D and ($C9$) in Appendix C.

It follows now from Eq. (11.107) that the velocity potential in this case is

$$\phi(x,y)=\frac{1}{2}\int_{-\infty}^{\infty}\frac{J_0(a\xi)}{|\xi|}e^{i\xi x-|\xi|y}\,d\xi. \tag{11.111}$$

Hence, the corresponding velocity components in Eq. (11.106), with the help of ($C6$) and ($C7$) in Appendix C, can then be expressed as

$$\begin{aligned}u(x,y)&=\int_0^\infty e^{-y\xi}\sin(x\xi)J_0(a\xi)\,d\xi\\&=\mathrm{Im}\int_0^\infty e^{-(y-ix)\xi}J_0(a\xi)\,d\xi\\&=\sqrt{\pi}\,\mathrm{Im}\int_0^\infty J_0(a\xi)\,G_{02}^{10}\left[\frac{(x+iy)^2\xi^2}{4}\,\middle|\,\frac{1}{2},\ 0\right]d\xi,\end{aligned} \tag{11.112}$$

$$\begin{aligned}v(x,y)&=\int_0^\infty e^{-y\xi}\cos(x\xi)J_0(a\xi)\,d\xi\\&=\mathrm{Re}\int_0^\infty e^{-(y-ix)\xi}J_0(a\xi)\,d\xi\\&=\sqrt{\pi}\,\mathrm{Re}\int_0^\infty J_0(a\xi)\,G_{02}^{10}\left[\frac{(x+iy)^2\xi^2}{4}\,\middle|\,0,\ \frac{1}{2}\right]d\xi,\end{aligned} \tag{11.113}$$

where we are imposing $\cosh z = \cos iz$ and $\sinh z = -i \sin iz$. Upon using ($D12$) in Appendix D and ($MG3$) in Appendix A to evaluate these integrals and simplify the results, the velocity components in Eqs. (11.112) and (11.113) reduce to

$$u(x, y) = \frac{\sqrt{\pi}}{a} \operatorname{Im} G_{11}^{10}\left[\left(\frac{x+iy}{a}\right)^2 \middle| \begin{matrix} \frac{1}{2} \\ 0 \end{matrix}\right] = \frac{1}{a} \operatorname{Im}\left[1 - \left(\frac{x+iy}{a}\right)^2\right]^{-1/2}, \quad (11.114)$$

$$v(x, y) = \frac{\sqrt{\pi}}{a} \operatorname{Re} G_{11}^{10}\left[\left(\frac{x+iy}{a}\right)^2 \middle| \begin{matrix} \frac{1}{2} \\ 0 \end{matrix}\right] = \frac{1}{a} \operatorname{Re}\left[1 - \left(\frac{x+iy}{a}\right)^2\right]^{-1/2}. \quad (11.115)$$

References

1. D. Middleton, *An Introduction to Statistical Communication Theory*, McGraw Hill, New York (1960).
2. W. B. Davenport, Jr. and W. L. Root, *An Introduction to the Theory of Random Signals and Noise*, IEEE Press, New York (1987).
3. D. G. Lampard, "The probability distribution for the filtered output of a multiplier whose inputs are correlated, stationary, Gaussian time-series," *IEEE Trans. Inform. Theory* **IT-2**, 4–11 (1956).
4. L. C. Andrews, "The probability density function for the output of a cross-correlator with bandpass inputs," *IEEE Trans. Inform. Theory* **IT-19**, 13–19 (1973).
5. C. Brice and L. C. Andrews, "New expressions for the pdf and cdf of the filtered output of an analog cross correlator," *IEEE Trans. Inform. Theory* **IT-28**, 668–676 (1982).
6. L. C. Andrews, "The output pdf of a polarity coincidence correlation detector," *IEEE Trans. Aerosp. Electron. Syst.* **AES-10**, 712–715 (1974).
7. L. C. Andrews, "Analysis of a cross correlator with a clipper in one channel," *IEEE Trans. Inform. Theory* **IT-26**, 743–746 (1980).
8. S. G. Gopani, R. L. Phillips, and L. C. Andrews, "A theoretical and experimental comparison of output signal-to-noise ratio of three types of cross-correlators," *IEEE Trans. Acoust., Speech, Signal Processing*, **ASSP-31**(1), 202–205 (1983).
9. A. V. Forman, Jr. and L. C. Andrews, "Approximate expressions for the output signal-to-noise ratios for two types of cross correlators with correlated bandpass inputs," *IEEE Trans. Acoust., Speech, Signal Processing* **ASSP-31**(3), 740–742 (1983).
10. L. B. W. Jolley, *Summation of Series*, 2nd Edition, Dover Publications, New York (1961).
11. L. B. Stotts and L. C. Andrews, "Optical communications in turbulence: a tutorial," *Opt. Eng.* **63**(4), 041207 (2024) [doi: 10.1117/1.OE.63.4.041207].

12. L. C. Andrews, *Field Guide to Atmospheric Optics*, 2nd Edition, SPIE Press, Bellingham, Washington (2019) [doi: 10.1117/3.2318080].
13. K. Vajravelu, L. C. Andrews, and R. N. Mohapatra, "Exact solutions of the unsteady hydrodynamic and hydromagnetic flows past an infinite plate," *Acta Mech.* **74**, 185–193 (1988).
14. I. N. Sneddon, *Fourier Transforms*, McGraw Hill, New York (1951).

Appendix A
Special Function Identities

Listed below is a summary of important identities associated with most of the special functions appearing throughout this text.

(*G*) Gamma Function

$$\Gamma(z) = \int_0^\infty e^{-t} t^{z-1} dt, \quad \text{Re}(z) > 0$$

$(G1)$: $\Gamma(1) = 1$

$(G2)$: $\Gamma\left(\frac{1}{2}\right) = \sqrt{\pi}$

$(G3)$: $\Gamma(z+1) = z\Gamma(z)$ (Recurrence formula)

$(G4)$: $\Gamma(n+1) = n!, \quad n = 0, 1, 2, \ldots$

$(G5)$: $\Gamma\left(n + \frac{1}{2}\right) = \frac{(2n)!}{2^{2n} n!}\sqrt{\pi}, \quad n = 0, 1, 2, \ldots$

$(G6)$: $\frac{1}{\Gamma(-n)} = 0, \quad n = 0, 1, 2, \ldots$

$(G7)$: $\Gamma(z)\Gamma(1-z) = \frac{\pi}{\sin \pi z}$

$(G8)$: $\Gamma\left(\frac{1}{2} + z\right)\Gamma\left(\frac{1}{2} - z\right) = \frac{\pi}{\cos \pi z}$

$(G9)$: $\Gamma(z)\Gamma(-z) = -\frac{\pi}{z \sin \pi z}$

$(G10)$: $\sqrt{\pi}\Gamma(2z) = 2^{2z-1}\Gamma(z)\Gamma\left(z + \frac{1}{2}\right)$ (Duplication formula)

$$(G11):\ \binom{a}{n} = \frac{\Gamma(a+1)}{n!\Gamma(a-n+1)}, \quad n = 0, 1, 2, \ldots \quad \text{(Binomial coefficient)}$$

$$(G12):\ \frac{\Gamma(k-n)}{\Gamma(-n)} = \begin{cases} \dfrac{(-1)^k n!}{(n-k)!}, & 0 \le k \le n \quad (k,\ n \text{ non-negative integers}) \\ 0, & k > n \end{cases}$$

$$(G13):\ \frac{1}{\Gamma(z)} = ze^{\gamma z}\Pi_{n=1}^{\infty}\left(1+\frac{z}{n}\right)e^{-z/n}$$

$$(G14):\ \Gamma(z) = \frac{\Gamma(z+k)}{z(z+1)(z+2)\ldots(z+k-1)}, \quad k = 1, 2, 3, \ldots$$

$$(G15):\ \Gamma(z+1) \sim \sqrt{2\pi z}\, z^z e^{-z}, \quad |z| \to \infty \quad \text{(Stirling's formula)}$$

$$(G16):\ \int_0^{\pi/2} \cos^{2x-1}\theta \sin^{2y-1}\theta\, d\theta = \frac{\Gamma(x)\Gamma(y)}{2\Gamma(x+y)}, \quad x > 0,\ y > 0$$

$$(G17):\ B(x, y) = \int_0^1 t^{x-1}(1-t)^{y-1} dt = \frac{\Gamma(x)\Gamma(y)}{\Gamma(x+y)}, \ x > 0,\ y > 0$$

$$(G18):\ B(x, y) = \int_0^{\infty} \frac{t^{x-1}}{(1+t)^{x+y}}\, dt = \frac{\Gamma(x)\Gamma(y)}{\Gamma(x+y)}, \ x > 0,\ y > 0$$

$$(G19):\ \Gamma(z) = s^z \int_0^{\infty} e^{-st} t^{z-1}\, dt, \quad \text{Re}(z) > 0$$

$$(G20):\ \Gamma'(1) = -\gamma = \int_0^{\infty} e^{-t} \ln t\, dt = -0.577216\ldots$$

(*PS*) Pochhammer Symbol

$$(a)_n = \frac{\Gamma(a+n)}{\Gamma(a)}, \quad n = 0, 1, 2, \ldots$$

$$(PS1):\ (0)_0 = 1, \quad (1)_n = n!$$

$$(PS2):\ (a)_{n+k} = (a)_k (a+k)_n$$

$$(PS3):\ (a)_{n-k} = \frac{(-1)^k (a)_n}{(1-a-n)_k}$$

$$(PS4):\ (a)_{-n} = \frac{(-1)^n}{(1-a)_n}$$

$$(PS5):\ (-k)_n = \begin{cases} \dfrac{(-1)^n k!}{(k-n)!}, & 0 \le n \le k \\ 0, \quad n > k \end{cases}$$

$$(PS6):\ (n+k)! = n!(n+1)_k = k!(k+1)_n$$

$$(PS7):\ \frac{1}{(n-k)!} = \frac{(-1)^k(-n)_k}{n!}$$

$$(PS8):\ \Gamma(a+1-n) = \frac{(-1)^n\Gamma(a+1)}{(-a)_n}, \qquad n = 0, 1, 2, \ldots$$

$$(PS9):\ \binom{-a}{n} = \frac{(-1)^n}{n!}(a)_n$$

$$(PS10):\ (2n)! = 2^{2n}\left(\frac{1}{2}\right)_n n!$$

$$(PS11):\ (2n+1)! = 2^{2n}\left(\frac{3}{2}\right)_n n!$$

$$(PS12):\ (a)_{2n} = 2^{2n}\left(\frac{a}{2}\right)_n \left(\frac{a+1}{2}\right)_n$$

$$(PS13):\ \frac{1}{n+a} = \frac{(a)_n}{a(a+1)_n}$$

$$(PS14):\ \binom{a+k-1}{n} = \frac{(-1)^n(1-a)_n(a)_k}{n!(a-n)_k}$$

(*HY*) Hypergeometric Function

$${}_2F_1(a, b; c; z) = \sum_{n=0}^{\infty} \frac{(a)_n(b)_n}{(c)_n}\frac{z^n}{n!}, \qquad |z| < 1$$

$$(HY1):\ {}_2F_1(a, b; c; z) = {}_2F_1(b, a; c; z)$$

$$(HY2):\ \frac{d^k}{dz^k}\,{}_2F_1(a,\,b;\,c;\,z)=\frac{(a)_k(b)_k}{(c)_k}\,{}_2F_1(a+k,\,b+k;\,c+k;\,z),\quad k=1,\,2,\,3,\,\ldots$$

$$(HY3):\ {}_2F_1(a,b;c;z)=\frac{\Gamma(c)}{\Gamma(b)\Gamma(c-b)}\int_0^1 t^{b-1}(1-t)^{c-b-1}(1-zt)^{-a}\,dt,\quad c>b>0$$

$$(HY4):\ {}_2F_1(a,\,b;\,c;\,1)=\frac{\Gamma(c)\Gamma(c-a-b)}{\Gamma(c-a)\Gamma(c-b)},\qquad c-a-b>0$$

$$(HY5):\ {}_2F_1(a,b;1+a-b;-1)=\frac{\Gamma(1+a-b)\Gamma\left(1+\frac{a}{2}\right)}{\Gamma\left(1-b+\frac{a}{2}\right)\Gamma(1+a)},\quad 1-b+\frac{a}{2}>0$$

$$(HY6):\ {}_2F_1\left(2a,\,2b;\,a+b+\frac{1}{2};\,\frac{1}{2}\right)=\frac{\sqrt{\pi}\,\Gamma\left(a+b+\frac{1}{2}\right)}{\Gamma\left(a+\frac{1}{2}\right)\Gamma\left(b+\frac{1}{2}\right)},\quad a+b+\frac{1}{2}\neq 0,\,-1,\,-2,\,\ldots$$

$$(HY7):\ {}_2F_1(a,\,b;\,c;\,-z)=(1+z)^{-a}{}_2F_1\left(a,\,c-b;\,c;\,\frac{z}{1+z}\right)$$

$$(HY8):\ {}_2F_1(c-a,\,c-b;\,c;\,z)=(1-z)^{a+b-c}\,{}_2F_1(a,\,b;\,c;\,z)$$

$$(HY9):\ \left[{}_2F_1(a,\,b;\,a+b+\frac{1}{2};\,z)\right]^2={}_3F_2\left(2a,\,a+b,\,2b;\,a+b+\frac{1}{2},\,a+2b;\,z\right)$$

$$(HY10):\ {}_2F_1(a,b;c;-z)\sim\begin{cases}1-\dfrac{ab}{c}z, & z\to 0\\[2mm] \dfrac{\Gamma(c)\Gamma(b-a)z^{-a}}{\Gamma(b)\Gamma(c-a)}+\dfrac{\Gamma(c)\Gamma(a-b)z^{-b}}{\Gamma(a)\Gamma(c-b)}, & z\to\infty\end{cases}$$

$$\begin{aligned}(HY11):\ (b-a)(1-z)\,{}_2F_1(a,\,b;\,c;\,z)&=(c-a)\,{}_2F_1(a-1,\,b;\,c;\,z)\\&\quad-(c-b)\,{}_2F_1(a,\,b-1;\,c;\,z)\end{aligned}$$

$$\begin{aligned}(HY12):\ (1-z)\,{}_2F_1(a,\,b;\,c;\,z)&={}_2F_1(a-1,\,b;\,c;\,z)\\&\quad-c^{-1}(c-b)z\,{}_2F_1(a,\,b;\,c;\,z)\end{aligned}$$

$$(HY13):\ (1-z)\,{}_2F_1(a,b;c;z) = {}_2F_1(a,b-1;c;z) - c^{-1}(c-a)z\,{}_2F_1(a,b;c+1;z)$$

$$(HY14):\ (a-b)\,{}_2F_1(a,b;c;z) = a\,{}_2F_1(a+1,b;c;z) - b\,{}_2F_1(a,b+1;c;z)$$

$$(HY15):\ (c-a-b)\,{}_2F_1(a,b;c;z) + b(1-z)\,{}_2F_1(a,b+1;c;z) = (c-a)\,{}_2F_1(a-1,b;c;z)$$

$$(HY16):\ (c-b-1)\,{}_2F_1(a,b;c;z) + b\,{}_2F_1(a,b+1;c;z) = (c-1)\,{}_2F_1(a,b;c-1;z)$$

$$(HY17):\ [2b-c+(a-b)z]\,{}_2F_1(a,b;c;z) = b(1-z)\,{}_2F_1(a,b+1;c;z) - (c-b)\,{}_2F_1(a,b-1;c;z)$$

$$(HY18):\ [a-1-(c-b-1)z]\,{}_2F_1(a,b;c;z) + (c-a)\,{}_2F_1(a-1,b;c;z) = (c-1)(1-z)\,{}_2F_1(a,b;c-1;z)$$

(*CH*) Confluent Hypergeometric Functions

$${}_1F_1(a;c;z) = \sum_{n=0}^{\infty} \frac{(a)_n}{(c)_n}\frac{z^n}{n!}, \quad |z| < \infty$$

$$(CH1):\ \frac{d^k}{dz^k}\,{}_1F_1(a;c;z) = \frac{(a)_k}{(c)_k}\,{}_1F_1(a+k;c+k;z), \quad k = 1, 2, 3, \ldots$$

$$(CH2):\ {}_1F_1(a;c;z) = \frac{\Gamma(c)}{\Gamma(a)\Gamma(c-a)}\int_0^1 e^{zt}t^{a-1}(1-t)^{c-a-1}\,dt,\ c>a>0$$

$$(CH3):\ {}_1F_1(a;c;-z) = \frac{\Gamma(c)2^{1-c}}{\Gamma(a)\Gamma(c-a)}e^{-z/2}\int_{-1}^1 e^{zt/2}(1-t)^{a-1}(1+t)^{c-a-1}dt$$

$$(CH4):\ {}_1F_1(a;c;z) = e^z\,{}_1F_1(c-a;c;-z)$$

$$(CH5):\ {}_1F_1(a;2a;2z) = e^z\,{}_0F_1\left(-;a+\frac{1}{2};\frac{z^2}{4}\right)$$

$$(CH6):\ {}_1F_1(a;c;-z) \sim \begin{cases} 1-\dfrac{a}{c}z, & z \to 0 \\ \dfrac{\Gamma(c)}{\Gamma(c-a)}z^{-a}, & z \to \infty \end{cases}$$

$$(CH7):\ {}_1F_1(a;\ c;\ z)\,{}_1F_1(a;\ c;\ -z) = {}_2F_3\left(a,\ c-a;\ c,\ \frac{c}{2},\ \frac{c+1}{2};\ \frac{z^2}{4}\right)$$

$$(CH8):\ c(c-1)\,{}_1F_1(a;\ c-1;\ z) = c(z+c-1)\,{}_1F_1(a;\ c;\ z) + z(a-c)\,{}_1F_1(a;\ c+1;\ z)$$

$$(CH9):\ (c-a-1)\,{}_1F_1(a;c;z) + a\,{}_1F_1(a+1;c;z) = (c-1)\,{}_1F_1(a;c-1;z)$$

$$(CH10):\ c\,{}_1F_1(a;\ c;\ z) - c\,{}_1F_1(a-1;\ c;\ z) = z\,{}_1F_1(a;\ c+1;\ z)$$

$$(CH11):\ c(a-1+z)\,{}_1F_1(a;\ c;\ z) = c(c-1)\,{}_1F_1(a;\ c-1;\ z) + (c-a)z\,{}_1F_1(a;\ c+1;\ z)$$

$$(CH12):\ c(a+z)\,{}_1F_1(a;c;z) - ac\,{}_1F_1(a+1;c;z) = (c-a)z\,{}_1F_1(a;c+1;z)$$

$$(CH13):\ (2a-c+z)\,{}_1F_1(a;c;z) + (c-a)\,{}_1F_1(a-1;c;z) = a\,{}_1F_1(a+1;c;z)$$

$$(CH14):\ c(1-a+z)\,{}_1F_1(a;\ c;\ z) + c(c-1)\,{}_1F_1(a-1;\ c-1;\ z) = az\,{}_1F_1(a+1;\ c+1;\ z)$$

$$(CH15):\ \frac{d}{dz}\,U(a;\ c;\ z) = -aU(a+1;\ c+1;\ z)$$

$$(CH16):\ U(a;c;z) = \frac{1}{\Gamma(a)}\int_0^\infty e^{-zt}t^{a-1}(1+t)^{c-a-1}\,dt,\quad a>0,\quad \mathrm{Re}(z)>0$$

$$(CH17):\ U(a;\ c;\ z) = z^{1-c}U(1+a-c;\ 2-c;\ z)$$

$$(CH18):\ U(a;\ c;\ z) = \frac{\Gamma(1-c)}{\Gamma(1+a-c)}\,{}_1F_1(a;\ c;\ z) + \frac{\Gamma(c-1)}{\Gamma(a)}z^{1-c}\,{}_1F_1(1+a-c;\ 2-c;\ z)$$

$$(CH19):\ U(a;\ c;\ z) \sim \begin{cases} \dfrac{\Gamma(1-c)}{\Gamma(1+a-c)} + \dfrac{\Gamma(c-1)}{\Gamma(a)}z^{1-c}, & z\to 0 \\ z^{-a}, \quad z\to\infty \end{cases}$$

$$(CH20):\ U(-n;\ c;\ z) = (-1)^n(c)_n\,{}_1F_1(-n;\ c;\ z),\quad n = 0,\ 1,\ 2,\ \ldots$$

(*BJ*) Bessel Function of the First Kind

$$J_p(z) = \left(\frac{z}{2}\right)^p \sum_{k=0}^{\infty} \frac{(-1)^k (z/2)^{2k}}{\Gamma(1+p+k)k!}$$

$(BJ1)$: $J_{-n}(z) = (-1)^n J_n(z), \ n = 0,\ 1,\ 2,\ \ldots$

$(BJ2)$: $\exp\left[\frac{z}{2}\left(t - \frac{1}{t}\right)\right] = \sum_{n=-\infty}^{\infty} J_n(z) t^n, \quad t \neq 0$

$(BJ3)$: $e^{\pm iz\cos\theta} = J_0(z) + 2\sum_{n=0}^{\infty} (\pm i)^n J_n(z) \cos n\theta, \quad i = \sqrt{-1}$

$(BJ4)$: $J_n(x+y) = \sum_{k=-\infty}^{\infty} J_k(x) J_{n-k}(y), \quad n = 0, 1, 2, \ldots$ (Addition formula)

$(BJ5)$: $J_0(\sqrt{a^2 + b^2 - 2ab\cos\theta}) = \sum_{k=-\infty}^{\infty} J_k(a) J_k(b)\, e^{ik\theta}$

$(BJ6)$: $J_0(z) = \frac{1}{2\pi} \int_0^{2\pi} e^{\pm iz\cos\theta}\, d\theta$

$(BJ7)$: $J_n(z) = \frac{1}{\pi} \int_0^{\pi} \cos(n\theta - z\sin\theta)\, d\theta, \quad n = 0,\ 1,\ 2,\ \ldots$

$(BJ8)$: $\frac{d}{dz}[z^p J_p(z)] = z^p J_{p-1}(z)$

$(BJ9)$: $\frac{d}{dz}[z^{-p} J_p(z)] = -z^{-p} J_{p+1}(z)$

$(BJ10)$: $J'_p(z) + \frac{p}{z} J_p(z) = J_{p-1}(z)$

$(BJ11)$: $J'_p(z) - \frac{p}{z} J_p(z) = -J_{p+1}(z)$

$(BJ12)$: $J_{p-1}(z) - J_{p+1}(z) = 2J'_p(z)$

$(BJ13)$: $J_{p-1}(z) + J_{p+1}(z) = \frac{2p}{z} J_p(z)$ (Recurrence formula)

$(BJ14)$: $\int z^p J_{p-1}(z)\, dz = z^p J_p(z) + C$

$$(BJ15):\quad \int z^{-p} J_{p+1}(z)\,dz = -z^{-p} J_p(z) + C$$

$$(BJ16):\quad \int_0^b x J_p(k_m z) J_p(k_n z)\,dz = 0, \quad J_p(k_n b) = 0 \quad (m \neq n)$$

(Orthogonality)

$$(BJ17):\ J_p(z) = \frac{(z/2)^p}{\sqrt{\pi}\,\Gamma\left(p+\frac{1}{2}\right)} \int_0^{\pi} \cos(z\cos\theta)\sin^{2p}\theta\,d\theta, \quad p > -\frac{1}{2}, \ |z| > 0$$

$$(BJ18):\quad J_p(z) = \frac{(z/2)^p}{\sqrt{\pi}\,\Gamma\left(p+\frac{1}{2}\right)} \int_{-1}^{1} e^{izt}(1-t^2)^{p-\frac{1}{2}}dt, \quad p > -\frac{1}{2}, \ |z| > 0$$

$$(BJ19):\ J_\mu(z)J_v(z) = \frac{(z/2)^{\mu+v}}{\Gamma(1+\mu)\Gamma(1+v)} \times {}_2F_3\left(\frac{\mu+v+1}{2}, \frac{\mu+v+2}{2}; \mu+1, v+1, \mu+v+1; -z^2\right)$$

$$(BJ20):\quad J_{1/2}(z) = \sqrt{\frac{2}{\pi z}}\sin z, \quad J_{-1/2}(z) = \sqrt{\frac{2}{\pi z}}\cos z, \quad \mathrm{Re}(z) > 0$$

(*BY*) Bessel Function of the Second Kind

$$Y_p(z) = \frac{(\cos p\pi)J_p(z) - J_{-p}(z)}{\sin p\pi}, \quad p > 0$$

$$(BY1):\quad Y_{-n}(z) = (-1)^n Y_n(z), \ n = 0, 1, 2, \ldots$$

$$(BY2):\quad \frac{d}{dz}[z^p Y_p(z)] = z^p Y_{p-1}(z)$$

$$(BY3):\quad \frac{d}{dz}[z^{-p} Y_p(z)] = -z^{-p} Y_{p+1}(z)$$

$$(BY4):\quad Y'_p(z) + \frac{p}{z} Y_p(z) = Y_{p-1}(z)$$

$$(BY5):\quad Y'_p(z) - \frac{p}{z} Y_p(z) = -Y_{p+1}(z)$$

$$(BY6):\ Y_{p-1}(z) - Y_{p+1}(z) = 2Y'_p(z)$$

$$(BY7):\ Y_{p-1}(z) + Y_{p+1}(z) = \frac{2p}{z} Y_p(z) \quad \text{(Recurrence formula)}$$

$$(BY8):\ Y_p(z) = \frac{2(z/2)^p}{\sqrt{\pi}\,\Gamma\left(p+\frac{1}{2}\right)}\left[\int_0^1 \sin(z\tau)(1-\tau^2)^{p-\frac{1}{2}}d\tau - \int_0^\infty e^{-z\tau}(1+\tau^2)^{p-\frac{1}{2}}d\tau\right]$$

$$(BY9):\ Y_0(z) = \frac{2}{\pi} J_0(z)\left(\gamma + \ln\frac{z}{2}\right) - \frac{2}{\pi}\sum_{k=1}^{\infty}\frac{(-1)^k(z/2)^{2k}}{(k!)^2}\left(1+\frac{1}{2}+\cdots+\frac{1}{k}\right), \quad \mathrm{Re}(z) > 0$$

(*BI*) Modified Bessel Function of the First Kind

$$I_p(z) = i^{-p}J_p(iz) = \left(\frac{z}{2}\right)^p \sum_{k=0}^{\infty}\frac{(z/2)^{2k}}{\Gamma(1+p+k)k!}$$

$$(BI1):\ I_{-n}(z) = I_n(z),\ n = 0, 1, 2, \ldots$$

$$(BI2):\ \exp\left[\frac{z}{2}\left(t+\frac{1}{t}\right)\right] = \sum_{n=-\infty}^{\infty} I_n(z)t^n, \quad t \neq 0$$

$$(BI3):\ e^{z\cos\theta} = I_0(z) + 2\sum_{n=1}^{\infty} I_n(z)\cos n\theta$$

$$(BI4):\ I_n(x+y) = \sum_{k=-\infty}^{\infty} I_k(x)I_{n-k}(y), \quad n = 0, 1, 2, \ldots$$

$$(BI5):\ I_0(z) = \frac{1}{2\pi}\int_0^{2\pi} e^{\pm z\cos\theta}\,d\theta$$

$$(BI6):\ I_p(z) = \frac{(z/2)^p}{\sqrt{\pi}\Gamma\left(p+\frac{1}{2}\right)}\int_0^{\pi} e^{\pm z\cos\theta}\sin^{2p}\theta\,d\theta, \quad p > -\frac{1}{2}$$

$$(BI7):\ I_p(z) = \frac{(z/2)^p}{\sqrt{\pi}\Gamma\left(p+\frac{1}{2}\right)}\int_{-1}^{1} e^{-zt}(1-t^2)^{p-1/2}dt, \quad p > -\frac{1}{2},\ |z| > 0$$

$(BI8)$: $\dfrac{d}{dz}[z^p I_p(z)] = z^p I_{p-1}(z)$

$(BI9)$: $\dfrac{d}{dz}[z^{-p} I_p(z)] = z^{-p} I_{p+1}(z)$

$(BI10)$: $I'_p(z) + \dfrac{p}{z} I_p(z) = I_{p-1}(z)$

$(BI11)$: $I'_p(z) - \dfrac{p}{z} I_p(z) = I_{p+1}(z)$

$(BI12)$: $I_{p-1}(z) + I_{p+1}(z) = 2I'_p(z)$

$(BI13)$: $I_{p-1}(z) - I_{p+1}(z) = \dfrac{2p}{z} I_p(z)$ (Recurrence formula)

$(BI14)$: $\displaystyle\int z^p I_{p-1}(z)\,dz = z^p I_p(z) + C$

$(BI15)$: $\displaystyle\int z^{-p} I_{p+1}(z)\,dz = z^{-p} I_p(z) + C$

$(BI16)$: $I_{1/2}(z) = \sqrt{\dfrac{2}{\pi z}} \sinh z, \quad I_{-1/2}(z) = \sqrt{\dfrac{2}{\pi z}} \cosh z, \quad \mathrm{Re}(z) > 0$

(*BK*) Modified Bessel Function of the Second Kind

$$K_p(z) = \frac{\pi}{2} \frac{I_{-p}(z) - I_p(z)}{\sin p\pi}, \quad p \neq 0, \quad |\arg(z)| < \pi$$

$(BK1)$: $K_{-p}(z) = K_p(z)$

$(BK2)$: $K_p(z) = \dfrac{\sqrt{\pi}(z/2)^p}{\Gamma\left(p + \dfrac{1}{2}\right)} \displaystyle\int_1^\infty e^{-zt}(t^2 - 1)^{p-1/2} dt, \quad p > -\dfrac{1}{2}, \ |z| > 0$

$(BK3)$: $\dfrac{d}{dz}[z^p K_p(z)] = -z^p K_{p-1}(z)$

$(BK4)$: $\dfrac{d}{dz}[z^{-p} K_p(z)] = -z^{-p} K_{p+1}(z)$

$(BK5)$: $K'_p(z) + \dfrac{p}{z} K_p(z) = -K_{p-1}(z)$

$(BK6)$: $K'_p(z) - \dfrac{p}{z} K_p(z) = -K_{p+1}(z)$

$$(BK7):\ K_{p-1}(z) + K_{p+1}(z) = -2K'_p(z)$$

$$(BK8):\ K_{p-1}(z) - K_{p+1}(z) = -\frac{2p}{z}K_p(z) \quad \text{(Recurrence formula)}$$

$$(BK9):\ K_{1/2}(z) = \sqrt{\frac{\pi}{2z}}\,e^{-z}$$

$$(BK10):\ K_0(z) = -I_0(z)\left(\gamma + \ln\frac{z}{2}\right) + \sum_{k=1}^{\infty}\frac{(z/2)^{2k}}{(k!)^2}\left(1 + \frac{1}{2} + \cdots + \frac{1}{k}\right),\ \mathrm{Re}(z) > 0$$

(*SBj*) Spherical Bessel Function of the First Kind

$$j_n(z) = \sqrt{\frac{\pi}{2z}}\,J_{n+1/2}(z), \quad n = 0, 1, 2, \ldots$$

$$(SBj1):\ j_0(0) = 1; \quad j_n(0) = 0, \quad n = 1, 2, 3, \ldots$$

$$(SBj2):\ \frac{d}{dz}[z^{n+1}j_n(z)] = z^{n+1}j_{n-1}(z), \quad n = 1, 2, 3, \ldots$$

$$(SBj3):\ \frac{d}{dz}[z^{-n}j_n(z)] = -z^{-n}j_{n+1}(z), \quad n = 1, 2, 3, \ldots$$

$$(SBj4):\ j'_n(z) = j_{n-1}(z) - \frac{n+1}{x}j_n(z), \quad n = 1, 2, 3, \ldots$$

$$(SBj5):\ j'_n(z) = \frac{n}{z}j_n(z) - j_{n+1}(z), \quad n = 1, 2, 3, \ldots$$

$$(SBj6):\ (2n+1)j'_n(z) = nj_{n-1}(z) - (n+1)j_{n+1}(z), \quad n = 1, 2, 3, \ldots$$

$$(SBj7):\ j_{n-1}(z) + j_{n+1}(z) = \frac{2n+1}{z}j_n(z), \quad n = 1, 2, 3, \ldots$$

(Recurrence formula)

$$(SBj8):\ j_n(z) \sim \frac{n!(2z)^n}{(2n+1)!}, \quad n = 0, 1, 2, \ldots, \quad z \to 0$$

$$(SBj9):\ j_n(z) \sim \frac{1}{z}\sin\left(z - \frac{n\pi}{2}\right), \quad n = 0, 1, 2, \ldots, \quad z \to \infty$$

(*SBy*) Spherical Bessel Function of the Second Kind

$$y_n(z) = \sqrt{\frac{\pi}{2z}}\, Y_{n+1/2}(z), \quad n = 0, 1, 2, \ldots$$

$$(SBy1):\ y_n(0) = -\infty, \quad n = 0, 1, 2, \ldots$$

$$(SBy2):\ \frac{d}{dz}[z^{n+1} y_n(z)] = z^{n+1} y_{n-1}(z), \quad n = 1, 2, 3, \ldots$$

$$(SBy3):\ \frac{d}{dz}[z^{-n} y_n(z)] = -z^{-n} y_{n+1}(z), \quad n = 1, 2, 3, \ldots$$

$$(SBy4):\ y'_n(z) = y_{n-1}(z) - \frac{n+1}{x} y_n(z), \quad n = 1, 2, 3, \ldots$$

$$(SBy5):\ y'_n(z) = \frac{n}{z} y_n(z) - y_{n+1}(z), \quad n = 1, 2, 3, \ldots$$

$$(SBy6):\ (2n+1)y'_n(z) = ny_{n-1}(z) - (n+1)y_{n+1}(z), \quad n = 1, 2, 3, \ldots$$

$$(SBy7):\ y_{n-1}(z) + y_{n+1}(z) = \frac{2n+1}{z} y_n(z), \quad n = 1, 2, 3, \ldots$$

(Recurrence formula)

$$(SBy8):\ y_n(z) \sim -\frac{2(2n)!}{n!(2z)^{n+1}}, \quad n = 0, 1, 2, \ldots, \quad z \to 0$$

$$(SBy9):\ y_n(z) \sim -\frac{1}{z}\cos\left(z - \frac{n\pi}{2}\right), \quad n = 0, 1, 2, \ldots, \quad z \to \infty$$

(*MG*) Meijer *G*-Function

$$G_{p,q}^{m,n}\left(z \,\middle|\, \begin{matrix} a_1, \ldots, a_p \\ c_1, \ldots, c_q \end{matrix}\right) = \frac{1}{2\pi i}\int_C \frac{\Pi_{j=1}^{m}\Gamma(c_j - s)\Pi_{j=1}^{n}\Gamma(1 - a_j + s)}{\Pi_{j=m+1}^{q}\Gamma(1 - c_j + s)\Pi_{j=n+1}^{p}\Gamma(a_j - s)} z^s ds,$$
$$0 \le n \le p, \quad 0 \le m \le q$$

$$(MG1):\ G_{p,q}^{m,n}\left(z \middle| \begin{matrix} a_1, \ldots a_p \\ c_1, \ldots, c_q \end{matrix}\right) = \sum_{k=1}^{m} \frac{\Pi'^{m}_{j=1}\Gamma(c_j - c_k)\Pi_{j=1}^{n}\Gamma(1 + c_k - a_j)}{\Pi_{j=m+1}^{q}\Gamma(1 + c_k - c_j)\Pi_{j=n+1}^{p}\Gamma(a_j - c_k)} z^{c_k}$$
$$\times\, {}_pF_{q-1}[1 + c_k - a_1, \ldots, 1 + c_k - a_p; 1 + c_k - c_1, \ldots, {}^*, \ldots,$$
$$1 + c_k - c_q; (-1)^{p-m-n} z],$$

where $z \neq 0$, $1 \le m \le q$, $0 \le n \le p \le q - 1$, and no two of the c_k's ($k = 1, 2, \ldots, m$) differ by zero or an integer. Also, $a_j - c_k \neq 1, 2, 3, \ldots$ for

$j = 1, 2, ..., n$ and $k = 1, 2, ..., m$. If $p = q$, we restrict $|z| < 1$. The prime in the product symbol $\prod'$ denotes omission of the term when $j = k$. An empty product is interpreted as unity and the parameter corresponding to $1 + c_k - c_k$, $k = 1, 2, ..., m$, indicated by * in the ${}_pF_{q-1}$ function, is to be omitted.

$$(MG2):\ G_{p,q}^{m,n}\left(\frac{1}{z}\middle|\begin{matrix}a_p\\c_q\end{matrix}\right) = G_{q,p}^{n,m}\left(z\middle|\begin{matrix}1-c_q\\1-a_p\end{matrix}\right)$$

$$(MG3):\ G_{p,q}^{m,n}\left(z\middle|\begin{matrix}a_p\\c_q\end{matrix}\right) = G_{p-1,q-1}^{m,n-1}\left(z\middle|\begin{matrix}a_1,\ ...,\ a_{j-1},\ a_{j+1},\ ...,\ a_p\\c_1,\ ...,\ c_{k-1},\ c_{k+1},\ ...,\ c_q\end{matrix}\right)$$

where $a_j = c_k$ for some $j = 1, ..., n$ and some $k = m+1, ..., q$

$$(MG4):\ G_{p,q}^{m,n}\left(z\middle|\begin{matrix}a_p\\c_q\end{matrix}\right) = G_{p-1,q-1}^{m-1,n}\left(z\middle|\begin{matrix}a_1,\ ...,\ a_{j-1},\ a_{j+1},\ ...,\ a_p\\c_1,\ ...,\ c_{k-1},\ c_{k+1},\ ...,\ c_q\end{matrix}\right)$$

where $a_j = c_k$ for some , $j = n+1, ..., p,$ and some $k = 1, ..., m$

$$(MG5):\ z^r G_{p,q}^{m,n}\left(z\middle|\begin{matrix}a_p\\c_q\end{matrix}\right) = G_{p,q}^{m,n}\left(z\middle|\begin{matrix}a_p+r\\c_q+r\end{matrix}\right)$$

$$(MG6):\ \frac{d}{dz}\left[z^{-c_1}G_{p,q}^{m,n}\left(z\middle|\begin{matrix}a_p\\c_q\end{matrix}\right)\right] = -z^{-1-c_1}G_{p,q}^{m,n}\left(z\middle|\begin{matrix}a_1,...,a_p\\c_1+1,c_2,...,c_q\end{matrix}\right),\quad m\geq 1$$

$$(MG7):\ z\frac{d}{dz}\left[G_{p,q}^{m,n}\left(z\middle|\begin{matrix}a_p\\c_q\end{matrix}\right)\right] = G_{p,q}^{m,n}\left(z\middle|\begin{matrix}a_1-1,\ a_2,\ ...,\ a_p\\c_1,\ ...,\ c_q\end{matrix}\right) + (a_1-1)G_{p,q}^{m,n}\left(z\middle|\begin{matrix}a_p\\c_q\end{matrix}\right),\quad n\geq 1$$

$$(MG8):\ (1-a_1+c_1)G_{p,q}^{m,n}\left(z\middle|\begin{matrix}a_1,a_2,...,a_p\\c_1,...,c_q\end{matrix}\right) = G_{p,q}^{m,n}\left(z\middle|\begin{matrix}a_1-1,a_2,...,a_p\\c_1,...,c_q\end{matrix}\right) + G_{p,q}^{m,n}\left(z\middle|\begin{matrix}a_1,a_2,...,a_p\\c_1+1,c_2,,...,c_q\end{matrix}\right)$$

$$(MG9):\ (a_p-a_1)G_{p,q}^{m,n}\left(z\middle|\begin{matrix}a_1,\ a_2,\ ...,\ a_p\\c_1,\ ...,\ c_q\end{matrix}\right) = G_{p,q}^{m,n}\left(z\middle|\begin{matrix}a_1-1,\ a_2,\ ...,\ a_p\\c_1,\ ...,\ c_q\end{matrix}\right) + G_{p,q}^{m,n}\left(z\middle|\begin{matrix}a_1,\ a_2,\ ...,\ a_p-1\\c_1,\ c_2,\ ,\ ...,\ c_q\end{matrix}\right)$$

$$(MG10): G_{p,q}^{m,n}\left(z\middle|\begin{matrix}a_p\\c_q\end{matrix}\right)=\frac{\exp(i\pi a_j)}{2\pi i}G_{p-1,q-1}^{m,n}\left(ze^{-i\pi}\middle|\begin{matrix}a_1,\ldots,a_{j-1},a_{j+1},\ldots,a_p\\c_1,\ldots,c_{k-1},c_{k+1},\ldots,c_q\end{matrix}\right)$$
$$-\frac{\exp(-i\pi a_j)}{2\pi i}G_{p-1,q-1}^{m,n}\left(ze^{i\pi}\middle|\begin{matrix}a_1,\ldots,a_{j-1},a_{j+1},\ldots,a_p\\c_1,\ldots,c_{k-1},c_{k+1},\ldots,c_q\end{matrix}\right)$$

Appendix B
Formulas for the Generalized Hypergeometric Functions

Many of the generalized hypergeometric functions can be expressed in terms of elementary functions from calculus and other special functions. We list some of those here.

$$(B1): \ {}_0F_0(-;\ -;\ z) = e^z$$

$$(B2): \ {}_0F_1\left(-;\ \frac{1}{2};\ -\frac{z^2}{4}\right) = \cos z$$

$$(B3): \ {}_0F_1\left(-;\ 1;\ -\frac{z^2}{4}\right) = J_0(z)$$

$$(B4): \ {}_0F_1\left(-;\ 1;\ \frac{z^2}{4}\right) = I_0(z)$$

$$(B5): \ {}_0F_1\left(-;\ \frac{3}{2};\ -\frac{z^2}{4}\right) = \frac{\sin z}{z}$$

$$(B6): \ {}_0F_3\left(-;\ \frac{1}{2},\ \frac{1}{2},\ 1;\ -\frac{z^2}{4}\right) = \mathrm{ber}(z)$$

$$(B7): \ {}_0F_3\left(-;\ \frac{3}{2},\ \frac{3}{2},\ 1;\ -\frac{z^2}{4}\right) = \left(\frac{2}{z}\right)^2 \mathrm{bei}(z)$$

$$(B8): \ {}_1F_0(a;\ -;\ -z) = (1+z)^{-a}$$

$$(B9): \ {}_1F_1\left(\frac{1}{2};\ 1;\ -z\right) = e^{-z} I_0\left(\frac{z}{2}\right)$$

$$(B10): \ {}_1F_1\left(\frac{1}{2};\ \frac{3}{2};\ -z^2\right) = \frac{\sqrt{\pi}}{2z}\mathrm{erf}(z)$$

$$(B11):\quad {}_1F_1\left(\frac{1}{2};\ 2;\ -z\right) = e^{-z/2}\left[I_0\left(\frac{z}{2}\right) + I_1\left(\frac{z}{2}\right)\right]$$

$$(B12):\quad {}_1F_1\left(-\frac{1}{2};\ 1;\ -z\right) = e^{-z/2}\left[(1+z)I_0\left(\frac{z}{2}\right) + zI_1\left(\frac{z}{2}\right)\right]$$

$$(B13):\quad {}_1F_1(1;\ 2;\ -z) = \frac{1-e^{-z}}{z}$$

$$(B14):\quad {}_1F_1(1;\ 3;\ -z) = \frac{2}{z^2}(z-1+e^{-z})$$

$$(B15):\quad {}_1F_1(2;\ 1;\ -z) = (1-z)e^{-z}$$

$$(B16):\quad {}_1F_1(3;\ 2;\ -z) = \left(1-\frac{z}{2}\right)e^{-z}$$

$$(B17):\quad {}_1F_1(a;\ a+1;\ -z) = a\,z^{-a}\gamma(a,\ z)$$

$$(B18):\quad {}_1F_1\left(p+\frac{1}{2};\ 2p+1;\ 2iz\right) = \frac{\Gamma(p+1)}{(z/2)^p}e^{iz}J_p(z)$$

$$(B19):\quad {}_1F_1\left(p+\frac{1}{2};\ 2p+1;\ 2z\right) = \frac{\Gamma(p+1)}{(z/2)^p}e^{z}I_p(z)$$

$$(B20):\quad {}_1F_1\left(-n;\ \frac{1}{2};\ z^2\right) = (-1)^n\frac{n!}{(2n)!}H_{2n}(z), \qquad n=0,\ 1,\ 2,\ \ldots$$

$$(B21):\quad {}_1F_1\left(-n;\ \frac{3}{2};\ z^2\right) = (-1)^n\frac{n!}{(2n+1)!}H_{2n+1}(z), \qquad n=0,\ 1,\ 2,\ \ldots$$

$$(B22):\quad {}_1F_1(-n;\ 1+m;\ z) = \frac{n!\Gamma(m+1)}{\Gamma(n+m+1)}L_n^{(m)}(z), \qquad n=0,\ 1,\ 2,\ \ldots$$

$$(B23):\quad {}_1F_2\left(\frac{1}{4};\ \frac{1}{2},\ \frac{5}{4};\ -\frac{\pi^2z^4}{16}\right) = \frac{1}{z}C(z)$$

$$(B24):\quad {}_1F_2\left(\frac{3}{4};\ \frac{3}{2},\ \frac{7}{4};\ -\frac{\pi^2z^4}{16}\right) = \frac{6}{\pi z^3}S(z)$$

$$(B25):\quad {}_1F_2\left(\frac{1}{2};\ \frac{3}{2},\ \frac{3}{2};\ -\frac{z^2}{4}\right) = \frac{1}{z}\mathrm{Si}(z)$$

$$(B26):\ {}_1F_2\left(1;\ \frac{3}{2},\ p+\frac{3}{2};\ -\frac{z^2}{4}\right) = \frac{\sqrt{\pi}}{2}\left(\frac{2}{z}\right)^{p+1}\Gamma\left(p+\frac{3}{2}\right)\mathbf{H}_p(z)$$

$$(B27):\ {}_1F_2\left(1;\ \frac{3}{2},\ p+\frac{3}{2};\ \frac{z^2}{4}\right) = \frac{\sqrt{\pi}}{2}\left(\frac{2}{z}\right)^{p+1}\Gamma\left(p+\frac{3}{2}\right)\mathbf{L}_p(z)$$

$$(B28):\ {}_2F_1(1-a,\ 1;\ 2;\ -z) = \frac{(1+z)^a-1}{az}$$

$$(B29):\ {}_2F_1\left(a,\ a+\frac{1}{2};\ 2a+1;\ -z\right) = \left(\frac{1+\sqrt{1+z}}{2}\right)^{-2a}$$

$$(B30):\ {}_2F_1\left(a,\ a+\frac{1}{2};\ 2a;\ z\right) = \frac{1}{\sqrt{1-z}}\left(\frac{1+\sqrt{1-z}}{2}\right)^{1-2a}$$

$$(B31):\ {}_2F_1\left(\frac{1}{2},\ 1;\ \frac{3}{2};\ x^2\right) = \frac{1}{2x}\ \ln\frac{1+x}{1-x}$$

$$(B32):\ {}_2F_1\left(\frac{1}{2},\ 1;\ \frac{3}{2};\ -x^2\right) = \frac{1}{x}\tan^{-1}x$$

$$(B33):\ {}_2F_1\left(\frac{1}{2},\ \frac{1}{2};\ \frac{3}{2};\ x^2\right) = \frac{1}{x}\sin^{-1}x$$

$$(B34):\ {}_2F_1\left(\frac{1}{2},\ \frac{1}{2};\ 1;\ z^2\right) = \frac{2}{\pi}K(z)$$

$$(B35):\ {}_2F_1\left(-\frac{1}{2},\ \frac{1}{2};\ 1;\ z^2\right) = \frac{2}{\pi}E(z)$$

$$(B36):\ {}_2F_1\left(-n,\ n+1;\ 1;\ \frac{1-z}{2}\right) = P_n(z), \quad n = 0,\ 1,\ 2,\ \ldots$$

$$(B37):\ {}_2F_1\left(-n,\ n;\ \frac{1}{2};\ \frac{1-z}{2}\right) = T_n(z), \quad n = 0,\ 1,\ 2,\ \ldots$$

$$(B38):\ {}_2F_1\left(-n,\ n+2;\ \frac{3}{2};\ \frac{1-z}{2}\right) = \frac{1}{n+1}U_n(z),\ n = 0,\ 1,\ 2,\ \ldots$$

$$(B39):\ {}_2F_3\left(1,\ 1;\ \frac{3}{2},\ 2,\ 2;\ -\frac{z^2}{4}\right) = \frac{4}{z^2}[\gamma + \ln z - \mathrm{Ci}(z)]$$

$$(B40):\ {}_3F_2(a,\ b,\ c;\ 1+a-b,\ 1+a-c;\ 1) = \frac{\Gamma(1+a/2)\Gamma(1+a-b)\Gamma(1+a-c)\Gamma(1+a/2-b-c)}{\Gamma(1+a)\Gamma(1+a/2-b)\Gamma(1+a/2-c)\Gamma(1+a-b-c)}$$

$$(B41):\ U(1-a;\ 1-a;\ z) = e^z\Gamma(a,\ z)$$

$$(B42):\ U\left(\frac{1}{2};\ \frac{1}{2};\ z^2\right) = \sqrt{\pi}e^{z^2}\mathrm{erfc(z)}$$

$$(B43):\ U(1;\ 1;\ -z) = -e^{-z}\mathrm{Ei}(z)$$

$$(B44):\ U(1;\ 1;\ z) = e^z E_1(z)$$

$$(B45):\ U(1;\ 1;\ -\ln z) = -\frac{1}{z}\mathrm{li}(z)$$

$$(B46):\ U\left(p+\frac{1}{2};\ 2p+1;\ 2z\right) = \frac{e^z}{\sqrt{\pi}(2z)^p}K_p(z)$$

$$(B47):\ M_{-\frac{1}{4},\frac{1}{4}}(z^2) = \frac{1}{2}\sqrt{\pi z}\,e^{z^2/2}\,\mathrm{erf}(z)$$

$$(B48):\ M_{0,m}(2z) = \Gamma(m+1)2^{2m+1/2}\sqrt{z}\,I_m(z)$$

$$(B49):\ M_{k,m}(z) = e^{-z/2}z^{m+1/2}\,{}_1F_1\left(\frac{1}{2}+m-k;\ 2m+1;\ z\right)$$

$$(B50):\ M_{k,m}(z) = e^{z/2}z^{1-k}{}_2F_0\left(\frac{1}{2}-k+m,\ \frac{1}{2}-k-m;\ -;\ \frac{1}{z}\right)$$

$$(B51):\ W_{-\frac{1}{4},\frac{1}{4}}(z^2) = \sqrt{\pi z}\,e^{z^2/2}\,\mathrm{erfc}(z)$$

$$(B52):\ W_{0,m}(2z) = \sqrt{\frac{2z}{\pi}}K_m(z)$$

$$(B53):\ W_{\frac{a-1}{2},\frac{a}{2}}(z) = \frac{1}{z^{(a-1)/2}}e^{z/2}\Gamma(a,\ z)$$

$$(B54):\ W_{k,m}(z) = e^{-z/2}z^{m+1/2}U\left(\frac{1}{2}+m-k;\ 2m+1;\ z\right)$$

$$(B55):\ W_{k,m}(z) = e^{-z/2}z^k{}_2F_0\left(\frac{1}{2}-k+m,\ \frac{1}{2}-k-m;\ -;\ -\frac{1}{z}\right)$$

Appendix C
Formulas for the Meijer *G*-Function

The Meijer *G*-function is of such generality that it incorporates many other functions as special cases. Some of those are listed below.

$$(C1):\; G_{01}^{10}(z|a) = z^a e^{-z}$$

$$(C2):\; G_{11}^{10}\left(z\middle|\begin{matrix} a+b+1 \\ a\end{matrix}\right) = \frac{z^a(1-z)^b}{\Gamma(b+1)}, \qquad |z|<1$$

$$(C3):\; G_{02}^{10}\left(\frac{z^2}{4}\middle|\frac{1}{2},\; 0\right) = \frac{1}{\sqrt{\pi}}\sin z$$

$$(C4):\; G_{02}^{10}\left(\frac{z^2}{4}\middle|0,\; \frac{1}{2}\right) = \frac{1}{\sqrt{\pi}}\cos z$$

$$(C5):\; G_{02}^{10}\left(z^2\middle|0,\; \frac{1}{2}\right) = \frac{2}{\sqrt{\pi}}\cos^2 z - \frac{1}{\sqrt{\pi}}$$

$$(C6):\; G_{02}^{10}\left(-\frac{z^2}{4}\middle|\frac{1}{2},\; 0\right) = \frac{i}{\sqrt{\pi}}\sinh z$$

$$(C7):\; G_{02}^{10}\left(-\frac{z^2}{4}\middle|0,\; \frac{1}{2}\right) = \frac{1}{\sqrt{\pi}}\cosh z$$

$$(C8):\; G_{02}^{10}(z|a,\, b) = z^{(a+b)/2} J_{a-b}(2\sqrt{z})$$

$$(C9):\; G_{02}^{10}\left(\frac{z^2}{4}\middle|\frac{p}{2},\; -\frac{p}{2}\right) = J_p(z)$$

$$(C10):\ G_{22}^{10}\left(z\,\middle|\,\begin{matrix}1,\ 1\\1,\ 0\end{matrix}\right) = \ln(1+z)$$

$$(C11):\ G_{11}^{11}\left(z\,\middle|\,\begin{matrix}1-a\\0\end{matrix}\right) = \Gamma(a)(1+z)^{-a}$$

$$(C12):\ G_{12}^{11}\left(z\,\middle|\,\begin{matrix}\tfrac{1}{2}\\0,\ -\tfrac{1}{2}\end{matrix}\right) = \sqrt{\frac{\pi}{z}}\,\mathrm{erf}(\sqrt{z})$$

$$(C13):\ G_{12}^{11}\left(z\,\middle|\,\begin{matrix}1-a\\0,\ 1-c\end{matrix}\right) = \frac{\Gamma(a)}{\Gamma(c)}\ {}_1F_1(a;\ c;\ -z)$$

$$(C14):\ G_{12}^{11}\left(z\,\middle|\,\begin{matrix}a\\b,\ c\end{matrix}\right) = \frac{\Gamma(1-a+b)}{\Gamma(1-c+b)}\ z^b\ {}_1F_1(1-a+b;\ 1-c+b;\ -z)$$

$$(C15):\ G_{12}^{11}\left(z\,\middle|\,\begin{matrix}\tfrac{1}{2}\\p,\ -p\end{matrix}\right) = \sqrt{\pi}\,e^{-z/2}I_p\left(\frac{z}{2}\right)$$

$$(C16):\ G_{13}^{11}\left(z^2\,\middle|\,\begin{matrix}\tfrac{1}{2}\\p,\ 0,\ -p\end{matrix}\right) = \sqrt{\pi}\,J_p^2(z)$$

$$G_{13}^{11}\left(z^2\,\middle|\,\begin{matrix}\tfrac{1}{2}\\0,\ p,\ -p\end{matrix}\right) = \sqrt{\pi}\,J_p(z)J_{-p}(z)$$

$$(C17):\ G_{13}^{11}\left(\frac{z^2}{4}\,\middle|\,\begin{matrix}\tfrac{p+1}{2}\\\tfrac{p+1}{2},\ -\tfrac{p}{2},\ \tfrac{p}{2}\end{matrix}\right) = \mathbf{H}_p(z)$$

$$(C18):\ G_{22}^{11}\left(z\,\middle|\,\begin{matrix}a,\ b\\c,\ d\end{matrix}\right) = \frac{\Gamma(1+c-a)}{\Gamma(1+c-d)\Gamma(b-c)}z^c$$
$$\times\ {}_2F_1(1+c-a,\ 1+c-b;\ 1+c-d;\ z)$$

$$(C19):\ G_{22}^{12}\left(z\,\middle|\,\begin{matrix}1-a,\ 1-b\\0,\ 1-c\end{matrix}\right) = \frac{\Gamma(a)\Gamma(b)}{\Gamma(c)}\ {}_2F_1(a,\ b;\ c;\ -z)$$

$$(C20):\ G_{22}^{12}\left(-z^2\,\middle|\,\begin{matrix}\tfrac{1}{2},\ \tfrac{1}{2}\\0,\ -\tfrac{1}{2}\end{matrix}\right) = \frac{2\sqrt{\pi}}{z}\sin^{-1}z$$

$$(C21):\ G_{22}^{12}\left(z^2\,\middle|\,\begin{matrix}\tfrac{1}{2},\ 0\\0,\ -\tfrac{1}{2}\end{matrix}\right) = \frac{2}{z}\tan^{-1}z$$

$$(C22):\ G_{24}^{12}\left(z^2\,\middle|\,\begin{matrix} \frac{1-p-q}{2},\ -\frac{p+q}{2} \\ 0,\ -p,\ -q,\ -(p+q)\end{matrix}\right) = \frac{\Gamma\left(\frac{p+q+1}{2}\right)\Gamma\left(\frac{p+q+2}{2}\right)}{\Gamma(p+q+1)}\left(\frac{2}{z}\right)^{p+q} J_p(z)J_q(z)$$

$$(C23):\ G_{02}^{20}(z|p,\ \mu) = 2z^{(p+\mu)/2}K_{p-\mu}(2\sqrt{z})$$

$$G_{02}^{20}\left(\frac{z^2}{4}\,\middle|\,\frac{p}{2},\ -\frac{p}{2}\right) = 2K_p(z)$$

$$(C24):\ G_{12}^{20}\left(2z\,\middle|\,\begin{matrix} \frac{1}{2} \\ p,\ -p\end{matrix}\right) = \frac{1}{\sqrt{\pi}}e^{-z}K_p(z)$$

$$(C25):\ G_{12}^{20}\left(z\,\middle|\,\begin{matrix} a \\ a-1,\ a-\frac{1}{2}\end{matrix}\right) = \sqrt{\pi}\,z^{a-1}\mathrm{erfc}(\sqrt{z})$$

$$G_{12}^{20}\left(z\,\middle|\,\begin{matrix} a \\ b,\ c\end{matrix}\right) = z^{(b+c-1)/2}e^{-z/2}W_{k,m}(z),$$

$$k = \frac{1}{2}(b+c+1) - a,\ m = \frac{1}{2}(b-c)$$

$$(C26):\ G_{22}^{20}\left(z\,\middle|\,\begin{matrix} a+\frac{1}{3},\ a+\frac{2}{3} \\ a,\ a\end{matrix}\right) = z^a\,{}_2F_1\left(\frac{2}{3},\ \frac{1}{3};\ 1;\ 1-z\right), \quad |z|<1$$

$$(C27):\ G_{12}^{21}\left(z\,\middle|\,\begin{matrix} a \\ b,\ c\end{matrix}\right) = \Gamma(b-a+1)\Gamma(c-a+1)z^{(b+c-1)/2}e^{z/2}W_{k,m}(z),$$

$$k = a - \frac{1}{2}(b+c+1),\ m = \frac{1}{2}(b-c)$$

$$(C28):\ G_{12}^{21}\left(z^2\,\middle|\,\begin{matrix} 1 \\ 1,\ \frac{1}{2}\end{matrix}\right) = \pi z e^{z^2}\mathrm{erfc}(z)$$

$$G_{12}^{21}\left(z^2\,\middle|\,\begin{matrix} \frac{1}{2} \\ \frac{1}{2},\ 0\end{matrix}\right) = \pi e^{z^2}\mathrm{erfc}(z)$$

$$(C29):\ G_{13}^{21}\left(z^2\,\middle|\,\begin{matrix} \frac{1}{2} \\ p,\ 0,\ -p\end{matrix}\right) = 2\sqrt{\pi}\,I_p(z)K_p(z)$$

$$(C30):\ G_{22}^{22}\left(z\,\middle|\,\begin{matrix}1-a,\ 1-b\\ 0,\ c-a-b\end{matrix}\right) = \frac{\Gamma(a)\Gamma(b)\Gamma(c-a)\Gamma(c-b)}{\Gamma(c)}\ {}_2F_1(a,\ b;\ c;\ 1-z)$$

$$(C31):\ G_{13}^{30}\left(z^2\,\middle|\,\begin{matrix}a+\frac{1}{2}\\ a+p,\ a-p,\ a\end{matrix}\right) = \frac{2z^{2a}}{\sqrt{\pi}}\,K_p^2(z)$$

$$(C32):\ G_{p,q+1}^{1,p}\left(z\,\middle|\,\begin{matrix}1-a_1,\ldots,\ 1-a_p\\ 0,\ 1-c_1,\ldots,\ 1-c_q\end{matrix}\right) = \frac{\prod_{j=1}^{p}\Gamma(a_j)}{\prod_{j=1}^{q}\Gamma(c_j)}\ {}_pF_q(a_1,\ldots,\ a_p;\ c_1,\ldots,\ c_q;\ -z)$$

$$(C33):\ G_{q+1,p}^{p,1}\left(\frac{1}{z}\,\middle|\,\begin{matrix}1,\ c_1,\ldots,\ c_q\\ a_1,\ldots,\ a_p\end{matrix}\right) = \frac{\prod_{j=1}^{p}\Gamma(a_j)}{\prod_{j=1}^{q}\Gamma(c_j)}\ {}_pF_q(a_1,\ldots,\ a_p;\ c_1,\ldots,\ c_q;\ -z)$$

$$(C34):\ G_{24}^{40}\left(z^2\,\middle|\,\begin{matrix}0,\ \frac{1}{2}\\ \frac{(\mu+\nu)}{2},\ \frac{(\nu-\mu)}{2},\ \frac{(\mu-\nu)}{2},\ -\frac{(\mu+\nu)}{2},\end{matrix}\right) = \frac{2}{\sqrt{\pi}}K_\mu(z)K_\nu(z)$$

Appendix D
Integral Table

Listed below is a short table of integrals that is referenced throughout the chapters in Part II: Applications.

$$(D1):\ \int_0^\infty x^{s-1}e^{-\alpha x^2}J_p(\beta x)\,dx = \frac{\beta^p\Gamma\left(\frac{p+s}{2}\right)}{2^{p+1}\alpha^{(p+s)/2}\Gamma(1+p)}\,{}_1F_1\left(\frac{p+s}{2};\,1+p;\,-\frac{\beta^2}{4\alpha}\right)$$

$$(D2):\ \int_0^\infty \kappa^{\mu}\exp\left(-\frac{\kappa^2}{a^2}\right)d\kappa = \frac{1}{2}a^{(1+\mu)}\,\Gamma\left(\frac{1+\mu}{2}\right)$$

$$(D3):\ \int_0^\infty t^{\beta-1}J_\mu^2(t)\,dt = \frac{1}{2\sqrt{\pi}}\frac{\Gamma\left(\mu+\frac{\beta}{2}\right)\Gamma\left(\frac{1-\beta}{2}\right)}{\Gamma\left(1-\frac{\beta}{2}\right)\Gamma\left(1+\mu-\frac{\beta}{2}\right)}$$

$$(D4):\ \int_0^\infty t^{\beta-1}J_\mu^2(t)J_p(ct)\,dt = \frac{1}{c\sqrt{\pi}}\,G_{33}^{12}\left(\frac{4}{c^2}\middle|\begin{matrix}\frac{1-p}{2},\ \frac{\beta}{2},\ \frac{1+p}{2}\\ \mu+\frac{\beta-1}{2},\ \frac{\beta-1}{2},\ \frac{\beta-1}{2}-\mu\end{matrix}\right)$$

$$(D5):\ \int_0^\infty \kappa^{\mu-1}\frac{\exp(-\kappa^2/\kappa_m^2)}{(\kappa^2+\kappa_0^2)^{11/6}}\,d\kappa = \frac{1}{2}\kappa_0^{\mu-11/3}\Gamma\left(\frac{\mu}{2}\right)U\left(\frac{\mu}{2};\,\frac{\mu}{2}-\frac{5}{6};\,\frac{\kappa_0^2}{\kappa_m^2}\right)$$

$$(D6):\ \int_0^\infty e^{-\beta x}x^{-\lambda}G_{p,q}^{m,n}\left(\alpha x\middle|\begin{matrix}a_1,\ldots,a_p\\ c_1,\ldots,c_q\end{matrix}\right)dx = \beta^{\lambda-1}G_{p+1,q}^{m,n+1}\left(\frac{\alpha}{\beta}\middle|\begin{matrix}\lambda,a_1,\ldots,a_p\\ c_1,\ldots,c_q\end{matrix}\right),$$

$$\beta>0,\ p+q<2(m+n),\ c_j-\lambda>-1,\ j=1,\ldots,m$$

$$(D7):\ \int_0^\infty e^{-st}G_{p,q}^{m,n}\left(bt^2\middle|\begin{matrix}a_1,\ldots,a_p\\ c_1,\ldots,c_q\end{matrix}\right)dt = \frac{1}{\sqrt{\pi}s}G_{p+2,q}^{m,n+2}\left(\frac{4b}{s^2}\middle|\begin{matrix}0,\frac{1}{2},a_1,\ldots,a_p\\ c_1,\ldots,c_q\end{matrix}\right),$$

$$1+2\min(c_j)>0,\ j=1,\ldots,m;\ \ \mathrm{Re}(s)>0,\ \ p+q<2(m+n)$$

$$(D8):\quad \int_0^\infty x^{\lambda-1} G_{p,q}^{m,n}\left(\alpha x\left|\begin{matrix} a_1,\,\ldots,\,a_p \\ c_1,\,\ldots,\,c_q\end{matrix}\right.\right)dx$$

$$= \frac{1}{\alpha^\lambda}\frac{\prod_{j=1}^m \Gamma(c_j+\lambda)\prod_{j=1}^n \Gamma(1-a_j-\lambda)}{\prod_{j=m+1}^q \Gamma(1-c_j-\lambda)\prod_{j=n+1}^p \Gamma(a_j+\lambda)},$$

$$p+q<2(m+n),\quad |\arg(\alpha)|<\left(m+n-\frac{p}{2}-\frac{q}{2}\right)\pi,$$

$$-\min \mathrm{Re}(c_j)<\mathrm{Re}(\lambda)<1-\max\,\mathrm{Re}(a_j),\ j=1,\,2,\,\ldots,\,n$$

$$(D9):\quad \int_0^1 x^{\mu-1}(1-x)^{\nu-1}\, G_{p,q}^{m,n}\left(\alpha x\left|\begin{matrix} a_1,\,\ldots,\,a_p \\ c_1,\,\ldots,\,c_q\end{matrix}\right.\right)dx$$

$$=\Gamma(\nu)G_{p+1,q+1}^{m,n+1}\left(\alpha\left|\begin{matrix} 1-\mu,\,a_1,\,\ldots,\,a_p \\ c_1,\,\ldots,\,c_q,\,1-\mu-\nu\end{matrix}\right.\right),$$

$$p+q<2(m+n),\quad |\arg(\alpha)|<\left(m+n-\frac{p}{2}-\frac{q}{2}\right)\pi,$$

$$\mathrm{Re}(\mu+c_j)>0,\ j=1,\,2,\,\ldots,\,m;\quad \mathrm{Re}(\nu)>0$$

$$(D10):\quad \int_1^\infty \frac{x^{-\mu}}{(x-1)^{1-\nu}}\, G_{p,q}^{m,n}\left(\alpha x\left|\begin{matrix} a_1,\,\ldots,\,a_p \\ c_1,\,\ldots,\,c_q\end{matrix}\right.\right)dx$$

$$=\Gamma(\nu)G_{p+1,q+1}^{m+1,n}\left(\alpha\left|\begin{matrix} a_1,\,\ldots,\,a_p,\,\mu \\ \mu-\nu,\,c_1,\,\ldots,\,c_q\end{matrix}\right.\right),$$

$$p+q<2(m+n),\ |\arg(\alpha)|<\left(m+n-\frac{1}{2}p-\frac{1}{2}q\right)\pi,$$

$$\mathrm{Re}(\mu-\nu-a_j)>-1,\quad j=1,\,\ldots,\,n;\quad \mathrm{Re}(\nu)>0$$

$$(D11):\quad \int_0^\infty x^{\mu-1}(x+\beta)^{-\nu}\, G_{p,q}^{m,n}\left(\alpha x\left|\begin{matrix} a_1,\,\ldots,\,a_p \\ c_1,\,\ldots,\,c_q\end{matrix}\right.\right)dx$$

$$=\frac{\beta^{\mu-\nu}}{\Gamma(\nu)}G_{p+1,q+1}^{m+1,n+1}\left(\alpha\beta\left|\begin{matrix} 1-\mu,\,a_1,\,\ldots,\,a_p \\ \nu-\mu,\,c_1,\,\ldots,\,c_q\end{matrix}\right.\right),$$

$$(p+q)<2(m+n),\quad |\arg(\alpha)|<\left(m+n-\frac{p}{2}-\frac{q}{2}\right)\pi,\ |\arg(\beta)|<\pi,$$

$$\mathrm{Re}(\mu+c_j)>0,\ j=1,\,2,\,\ldots,\,m;\quad \mathrm{Re}(\mu-\nu+a_j)<1,\ j=1,\,2,\,\ldots,\,n$$

$$(D12):\ \int_0^\infty x^{2\lambda} J_\nu(\beta x) G_{p,q}^{m,n}\left(\alpha x^2 \middle| \begin{matrix} a_1, \ldots, a_p \\ c_1, \ldots, c_q \end{matrix}\right) dx$$

$$= \frac{2^{2\lambda}}{\beta^{2\lambda+1}} G_{p+2,q}^{m,n+1}\left(\frac{4\alpha}{\beta^2} \middle| \begin{matrix} \mu, a_1, \ldots, a_p, \sigma \\ c_1, \ldots, c_q \end{matrix}\right),$$

$$\beta > 0, \quad \mu = \frac{1-\nu}{2} - \lambda, \quad \sigma = \frac{1+\nu}{2} - \lambda, p+q < 2(m+n),$$

$$c_j + \lambda + \frac{\nu}{2} > -\frac{1}{2}, \quad j = 1, \ldots, m, \quad a_j + \lambda < \frac{3}{4}, \quad j = 1, \ldots, n$$

$$(D13):\ \int_0^\infty x^{-\lambda} K_\nu(2\sqrt{x}) G_{p,q}^{m,n}\left(\alpha x \middle| \begin{matrix} a_1, \ldots, a_p \\ c_1, \ldots, c_q \end{matrix}\right) dx$$

$$= \frac{1}{2} G_{p+2,q}^{m,n+2}\left(\alpha \middle| \begin{matrix} \lambda - \frac{1}{2}\nu,\ \lambda + \frac{1}{2}\nu,\ a_1, \ldots, a_p \\ c_1, \ldots, c_q \end{matrix}\right),$$

$$p + q < 2(m+n), \quad |\arg(\alpha)| < \left(m + n - \frac{p}{2} - \frac{q}{2}\right)\pi,$$

$$\mathrm{Re}(-\gamma \pm \nu + c_j), \quad j = 1, \ldots, m$$

$$(D14):\ \int_0^\infty x^{s-1} J_\mu(ax) J_\nu\left(\frac{b}{x}\right) dx = \frac{2^{s-1}}{a^s} G_{04}^{20}\left(\frac{a^2b^2}{16} \middle| \frac{(s+\mu)}{2}, \frac{\nu}{2}, -\frac{\nu}{2}, \frac{(s-\mu)}{2}\right),$$

$$a>0, b>0. -\mathrm{Re}\left(\mu+\frac{3}{2}\right) < \mathrm{Re}(s) < \mathrm{Re}\left(\nu+\frac{3}{2}\right)$$

$$(D15):\ \int_0^\infty G_{p,q}^{m,n}\left(\alpha x \middle| \begin{matrix} a_1, \ldots, a_p \\ c_1, \ldots, c_q \end{matrix}\right) G_{\sigma,\tau}^{\mu,\nu}\left(\beta x \middle| \begin{matrix} b_1, \ldots, b_\sigma \\ d_1, \ldots, d_\tau \end{matrix}\right) dx$$

$$= \frac{1}{\beta} G_{p+\tau,q+\sigma}^{m+\nu,n+\mu}\left(\frac{\alpha}{\beta} \middle| \begin{matrix} a_1, \ldots, a_n, -d_1, \ldots, -d_\tau, a_{n+1}, \ldots, a_p \\ c_1, \ldots, c_m, -b_1, \ldots, -b_\sigma, c_{m+1}, \ldots, c_q \end{matrix}\right),$$

$1 \le m \le q,\ 1 \le n \le p,\ p < q$; or $1 \le m \le q,\ 1 \le n \le p,$

$q = p + 1,\ p \ge 1,$

$1 \le \mu \le \tau,\ 1 \le \nu \le \sigma,\ \sigma < \tau$; or $1 \le \mu \le \tau,\ 1 \le \nu \le \sigma,$

$\tau = \sigma + 1,\ \sigma \ge 1,$

$$(D16):\ \frac{1}{2\pi i}\int_{c-i\infty}^{c+i\infty} e^{st}s^{\alpha-1}G_{p,q}^{m,n}\left(\frac{x}{s}\middle|\begin{matrix}a_1,\ \dots,\ a_p\\ c_1,\ \dots,\ c_q\end{matrix}\right)ds$$

$$= t^{-\alpha}G_{p,q+1}^{m,n}\left(xt\middle|\begin{matrix}a_1,\ \dots,\ a_p\\ c_1,\ \dots,\ c_q,\ \alpha\end{matrix}\right),\qquad c>0,\ t>0$$

$$(D17):\ \int_0^\infty \cos(st)G_{p,q}^{m,n}\left(\alpha s^2\middle|\begin{matrix}a_1,\ \dots,\ a_p\\ c_1,\ \dots,\ c_q\end{matrix}\right)ds$$

$$= \sqrt{\pi}\,t^{-1}G_{p+2,q}^{m,n+1}\left(\frac{4\alpha}{t^2}\middle|\begin{matrix}\frac{1}{2},\ a_1,\ \dots,\ a_p,\ 0\\ c_1,\ \dots,\ c_q\end{matrix}\right),\qquad t>0$$

$$(D18):\ \int_0^\infty \sin(st)G_{p,q}^{m,n}\left(\alpha s^2\middle|\begin{matrix}a_1,\ \dots,\ a_p\\ c_1,\ \dots,\ c_q\end{matrix}\right)ds$$

$$= \sqrt{\pi}\,t^{-1}G_{p+2,q}^{m,n+1}\left(\frac{4\alpha}{t^2}\middle|\begin{matrix}0,\ a_1,\ \dots,\ a_p,\ \frac{1}{2}\\ c_1,\ \dots,\ c_q\end{matrix}\right),\qquad t>0$$

NOTE 1: For notational simplicity in the table, restrictions on various parameters (in order for the integrals to converge) have not been given for all cases. Also, when restrictions are listed for the above integrals, these integrals represent only one possible case. Many integrals are valid for several additional cases that we have chosen not to list.

NOTE 2: The Fourier cosine and sine integrals in ($D17$) and ($D18$), when interpreted as transforms or inverse transforms, are usually defined with a multiplicative factor of $\sqrt{2/\pi}$ on both sides.

Appendix E
Asymptotic Expressions

For easy reference, we summarize below the small-argument and large-argument asymptotic relations for the functions discussed in this text.

Generalized Hypergeometric Functions

$$ {}_1F_1(a;\ c;\ -z) \sim 1 - \frac{a}{c} z, \qquad z \to 0 $$

$$ {}_1F_1(a;\ c;\ -z) \sim \frac{\Gamma(c)}{\Gamma(c-a)} z^{-a}, \qquad z \to \infty $$

$$ {}_1F_1(a;\ c;\ z) \sim \frac{\Gamma(c)}{\Gamma(a)} z^{a-c} e^{z}, \qquad z \to \infty $$

$$ {}_1F_k(a;\ c_1, \ldots, c_k;\ -z) \sim 1 - \left(\frac{a}{c_1 \cdots c_k}\right) z, \qquad z \to 0 $$

$$ {}_1F_k(a;\ c_1, \ldots, c_k;\ -z) \sim \frac{\prod_{j=1}^{k} \Gamma(c_j)}{\prod_{j=1}^{k} \Gamma(c_j - a)} z^{-a}, \qquad z \to \infty $$

$$ {}_2F_1(a, b;\ c;\ -z) \sim 1 - \frac{ab}{c} z, \qquad z \to 0 $$

$$ {}_2F_1(a, b;\ c;\ -z) \sim \frac{\Gamma(c)\Gamma(b-a) z^{-a}}{\Gamma(b)\Gamma(c-a)} + \frac{\Gamma(c)\Gamma(a-b) z^{-b}}{\Gamma(a)\Gamma(c-b)}, \qquad z \to \infty $$

$$ {}_2F_2(a,\ b;\ c,\ d;\ -z) \sim 1 - \frac{ab}{cd}z, \qquad z \to 0 $$

$$ {}_2F_2(a,\ b;\ c,\ d;\ -z) \sim \frac{\Gamma(c)\Gamma(d)\Gamma(b-a)}{\Gamma(b)\Gamma(c-a)\Gamma(d-a)} z^{-a} + \frac{\Gamma(c)\Gamma(d)\Gamma(a-b)}{\Gamma(a)\Gamma(c-b)\Gamma(d-b)} z^{-b}, \qquad z \to \infty $$

$$ {}_3F_3(a_1,\ a_2,\ a_3;\ c_1,\ c_2,\ c_3;\ -z) \sim 1 - \frac{a_1 a_2 a_3}{c_1 c_2 c_3} z, \qquad z \to 0 $$

$$ {}_3F_3(a_1,\ a_2,\ a_3;\ c_1,\ c_2,\ c_3;\ -z) \sim \frac{\prod_{j=1}^{3}\Gamma(c_j)}{\prod_{j=1}^{3}\Gamma(a_j)} \left[\frac{\Gamma(a_2-a_1)\Gamma(a_3-a_1)\Gamma(a_1)}{\Gamma(c_1-a_1)\Gamma(c_2-a_1)\Gamma(c_3-a_1)} z^{-a_1} + \frac{\Gamma(a_1-a_2)\Gamma(a_3-a_2)\Gamma(a_2)}{\Gamma(c_1-a_2)\Gamma(c_2-a_2)\Gamma(c_3-a_2)} z^{-a_2} + \frac{\Gamma(a_1-a_3)\Gamma(a_2-a_3)\Gamma(a_3)}{\Gamma(c_1-a_3)\Gamma(c_2-a_3)\Gamma(c_3-a_3)} z^{-a_3} \right], \quad z \to \infty $$

$$ U(a;\ c;\ z) \sim \frac{\Gamma(1-c)}{\Gamma(1+a-c)} + \frac{\Gamma(c-1)}{\Gamma(a)} z^{1-c}, \qquad \mathrm{Re}(z) > 0, \qquad z \to 0 $$

$$ U(a;\ c;\ z) \sim z^{-a}, \qquad z \to \infty $$

$$ M_{k,m}(z) \sim \begin{cases} z^{m+1/2}, & |z| \to 0 \\ \dfrac{\Gamma(2m+1)}{\Gamma\left(\dfrac{1}{2}+m-k\right)} z^{-k} e^{z/2}, & \mathrm{Re}(z) > 0, \quad |z| \to \infty \end{cases} $$

$$ W_{k,m}(z) \sim \begin{cases} \dfrac{\Gamma(2m)}{\Gamma\left(\dfrac{1}{2}+m-k\right)} z^{1/2-m} + \dfrac{\Gamma(-2m)}{\Gamma\left(\dfrac{1}{2}-m-k\right)} z^{1/2+m}, & z \to 0 \\ z^k e^{-z/2}, \quad \mathrm{Re}(z) > 0, & z \to \infty \end{cases} $$

Error Functions

$$\operatorname{erf}(z) \sim \begin{cases} \dfrac{2z}{\sqrt{\pi}}\left(1 - \dfrac{z^2}{3} + \cdots\right), & z \to 0 \\ 1, \quad z \to \infty, \end{cases}$$

$$\operatorname{erfc}(z) \sim \begin{cases} 1, & z \to 0 \\ \dfrac{1}{z\sqrt{\pi}} e^{-z^2}, & z \to \infty \end{cases}$$

Fresnel Integrals

$$C(z) \sim \begin{cases} z, & z \to 0 \\ \dfrac{1}{2}, & z \to \infty \end{cases}$$

$$S(z) \sim \begin{cases} \dfrac{\pi z^3}{6}, & z \to 0 \\ \dfrac{1}{2}, & z \to \infty \end{cases}$$

Incomplete Gamma Functions

$$\gamma(a, z) \sim \begin{cases} \dfrac{z^a}{a}, & z \to 0 \\ \Gamma(a), & z \to \infty \end{cases}$$

$$\Gamma(a, z) \sim \begin{cases} \Gamma(a), & z \to 0 \\ z^{a-1} e^{-z}, & z \to \infty \end{cases}$$

Exponential Integral

$$E_1(z) \sim \begin{cases} -\ln z, \quad \operatorname{Re}(z) > 0, & z \to 0 \\ \dfrac{e^{-z}}{z}, & z \to \infty \end{cases}$$

Sine and Cosine Integrals

$$\mathrm{Si}(z) \sim \begin{cases} z, & z \to 0 \\ \dfrac{\pi}{2}, & z \to \infty \end{cases}$$

$$\mathrm{Ci}(z) \sim \begin{cases} \ln z, & z \to 0 \\ 0, & z \to \infty \end{cases}$$

Complete Elliptic Integrals

$$K(m) \sim \begin{cases} \dfrac{\pi}{2}, & m \to 0 \\ \infty, & m = 1 \end{cases}$$

$$E(m) \sim \begin{cases} \dfrac{\pi}{2}, & m \to 0 \\ 1, & m = 1 \end{cases}$$

Standard Bessel Functions

$$J_p(z) \sim \frac{(z/2)^p}{\Gamma(1+p)}, \quad p \neq -1, -2, -3, \ldots, \quad \mathrm{Re}(z) > 0, \quad z \to 0$$

$$J_p(z) \sim \sqrt{\frac{2}{\pi z}} \cos\left[z - \frac{(2p+1)\pi}{4}\right], \quad p \geq 0, \quad z \to \infty$$

$$Y_0(z) \sim \frac{2}{\pi} \ln z, \quad z \to 0.$$

$$Y_p(z) \sim -\frac{\Gamma(p)}{\pi}\left(\frac{2}{z}\right)^p, \quad p > 0, \quad z \to 0$$

$$Y_p(z) \sim \sqrt{\frac{2}{\pi z}} \sin\left[z - \frac{(2p+1)\pi}{4}\right], \quad p \geq 0, \quad z \to \infty$$

Spherical Bessel Functions

$$j_n(z) \sim \begin{cases} \dfrac{n!(2z)^n}{(2n+1)!}, & n = 0, 1, 2, ..., \quad z \to 0 \\ \dfrac{1}{z}\sin\left(z - \dfrac{n\pi}{2}\right), & n = 0, 1, 2, ..., \quad z \to \infty \end{cases}$$

$$y_n(z) \sim \begin{cases} -\dfrac{2(2n)!}{n!(2z)^{n+1}}, & n = 0, 1, 2, ..., \quad z \to 0 \\ -\dfrac{1}{z}\cos\left(z - \dfrac{n\pi}{2}\right), & n = 0, 1, 2, ..., \quad z \to \infty \end{cases}$$

Modified Bessel Functions

$$I_p(z) = \frac{(z/2)^p}{\Gamma(1+p)}, \; p \neq -1, \; -2, \; -3, \; ..., \quad \text{Re}(z) > 0, \quad z \to 0$$

$$I_p(z) \sim \frac{e^z}{\sqrt{2\pi z}}, \quad \text{Re}(z) > 0, \quad z \to \infty$$

$$K_0(z) \sim -\ln z, \quad \text{Re}(z) > 0, \quad z \to 0$$

$$K_p(z) \sim \frac{\Gamma(p)}{2}\left(\frac{2}{z}\right)^p, \quad p > 0, \quad \text{Re}(z) > 0, \quad z \to 0$$

$$K_p(z) \sim \sqrt{\frac{\pi}{2z}}e^{-z}, \quad \text{Re}(z) > 0, \quad z \to \infty$$

Modified Spherical Bessel Functions

$$i_n(z) \sim \begin{cases} \dfrac{n!(2z)^n}{(2n+1)!}, & n = 0, 1, 2, ..., \quad z \to 0 \\ \dfrac{e^z}{2z}, & n = 0, 1, 2, ..., \quad z \to \infty \end{cases}$$

$$k_n(z) \sim \begin{cases} -\dfrac{2(2n)!}{n!(2z)^{n+1}}, & n = 0, 1, 2, ..., \quad z \to 0 \\ \dfrac{e^{-z}}{z}, & n = 0, 1, 2, ..., \quad z \to \infty \end{cases}$$

Struve Functions

$$\mathbf{H}_p(z) \sim \frac{z^{p+1}}{2^p \sqrt{\pi}\,\Gamma\left(p+\frac{3}{2}\right)}, \quad \text{Re}(z) > 0, \quad z \to 0$$

$$\mathbf{H}_p(z) - Y_p(z) \sim \frac{(z/2)^{p-1}}{\sqrt{\pi}\,\Gamma\left(p+\frac{1}{2}\right)}, \quad \text{Re}(z) > 0, \quad z \to \infty$$

$$\mathbf{L}_p(z) \sim \frac{z^{p+1}}{2^p \sqrt{\pi}\,\Gamma\left(p+\frac{3}{2}\right)}, \quad \text{Re}(z) > 0, \quad z \to 0,$$

$$I_p(z) - \mathbf{L}_p(z) \sim \frac{(z/2)^{p-1}}{\sqrt{\pi}\,\Gamma\left(p+\frac{1}{2}\right)}, \quad \text{Re}(z) > 0, \quad z \to \infty$$

Kelvin's Functions

$$\text{ber}(z) \sim 1 - \frac{z^4}{64}, \quad z \to 0$$

$$\text{bei}(z) \sim \frac{z^2}{4}, \quad z \to 0$$

$$\text{ber}(z) \sim \text{Re}\left(\frac{e^{z\sqrt{i}}}{\sqrt{2\pi z\sqrt{i}}}\right) = \frac{e^{z/\sqrt{2}}}{\sqrt{2\pi z}}\cos\left(\frac{z}{\sqrt{2}} - \frac{\pi}{8}\right), \quad z \to \infty$$

$$\text{bei}(z) \sim \text{Im}\left(\frac{e^{z\sqrt{i}}}{\sqrt{2\pi z\sqrt{i}}}\right) = \frac{e^{z/\sqrt{2}}}{\sqrt{2\pi z}}\sin\left(\frac{z}{\sqrt{2}} - \frac{\pi}{8}\right), \quad z \to \infty$$

$$\mathrm{ker}(z) \sim \gamma - \ln\left(\frac{z}{2}\right) + \frac{\pi z^2}{16}, \qquad z \to 0$$

$$\mathrm{kei}(z) \sim -\ln\left(\frac{z}{2}\right)\left(\frac{z^2}{4}\right) - \frac{\pi}{4}, \qquad z \to 0$$

Airy Functions

$$\mathrm{Ai}(z) \sim \frac{1}{3^{2/3}\Gamma(2/3)} - \frac{z}{3^{1/3}\Gamma(1/3)}, \qquad z \to 0$$

$$\mathrm{Bi}(z) \sim \frac{1}{3^{1/6}\Gamma(2/3)} + \frac{3^{1/6}z}{\Gamma(1/3)}, \qquad z \to 0$$

$$\mathrm{Ai}(-z) \sim \frac{1}{3^{2/3}\Gamma(2/3)} + \frac{z}{3^{4/3}\Gamma(4/3)}, \qquad z \to 0$$

$$\mathrm{Bi}(-z) \sim \frac{z}{3^{5/6}\Gamma(4/3)} + \frac{1}{3^{1/6}\Gamma(2/3)}, \qquad z \to 0$$

$$\mathrm{Ai}(z) \sim \frac{1}{2\sqrt{\pi}\, z^{1/4}} \exp\left(-\frac{2}{3} z^{2/3}\right), \qquad z \to \infty$$

$$\mathrm{Bi}(z) \sim \frac{1}{\sqrt{\pi}\, z^{1/4}} \exp\left(\frac{2}{3} z^{2/3}\right), \qquad z \to \infty$$

$$\mathrm{Ai}(-z) \sim \frac{1}{\sqrt{\pi}\, z^{1/4}} \left[\sin\left(\frac{2}{3} z^{3/2} + \frac{\pi}{4}\right) - \frac{15}{144\, z^{3/2}} \cos\left(\frac{2}{3} z^{3/2} + \frac{\pi}{4}\right)\right], \qquad z \to \infty$$

$$\mathrm{Bi}(-z) \sim \frac{1}{\sqrt{\pi}\, z^{1/4}} \left[\cos\left(\frac{2}{3} z^{3/2} + \frac{\pi}{4}\right) + \frac{15}{144\, z^{3/2}} \sin\left(\frac{2}{3} z^{3/2} + \frac{\pi}{4}\right)\right], \qquad z \to \infty$$

Bessel-Integral Function

$$\mathrm{Ji}_p(z) \sim \begin{cases} -\dfrac{1}{p} + \dfrac{(z/2)^p}{p\Gamma(1+p)}, & p > 0, \quad z \to 0 \\ 0, \quad \mathrm{Re}(z) > 0, & z \to \infty \end{cases}$$

Anger and Weber Functions

$$\mathbf{J}_p(z) = \cos\left(\frac{p\pi}{2}\right) S_1(p, z) + \sin\left(\frac{p\pi}{2}\right) S_2(p, z)$$

$$\mathbf{E}_p(z) = \sin\left(\frac{p\pi}{2}\right) S_1(p, z) - \cos\left(\frac{p\pi}{2}\right) S_2(p, z)$$

$$S_1(p, z) \sim \frac{1}{\Gamma\left(1+\frac{p}{2}\right)\Gamma\left(1-\frac{p}{2}\right)}, \quad z \to 0$$

$$S_2(p, z) \sim \frac{z}{2\Gamma\left(\frac{3+p}{2}\right)\Gamma\left(\frac{3-p}{2}\right)}, \quad z \to 0$$

$$S_1(p, z) \sim \frac{4}{\Gamma\left(\frac{p}{2}\right)\Gamma\left(-\frac{p}{2}\right)z^2}, \quad z \to \infty$$

$$S_2(p, z) \sim \frac{2}{\Gamma\left(\frac{1+p}{2}\right)\Gamma\left(\frac{1-p}{2}\right)z}, \quad z \to \infty$$

Lommel Functions

$$s_{\mu,\nu}(z) \sim \begin{cases} \dfrac{z^{\mu+1}}{(\mu-\nu+1)(\mu+\nu+1)}, & z \to 0 \\ z^{\mu-1}, \quad z \to \infty \end{cases}$$

$$S_{\mu,\nu}(z) \sim z^{\mu-1}, \quad \mu > \frac{1}{2}, \quad z \to \infty$$

Index

C_n^2 profile model, 336
νth-law device, 391

ABCD ray matrices, 291
adaptive optics, 290
Airy disk, 203, 294
Airy functions, 105
Airy, George Biddell, 117
analytic continuation, 8, 22, 86
analytic function, 3
Anger function, 109
angle-of-arrival fluctuations, 276
angular anisoplanatism, 313
anisoplanatic effect, 313
anisoplanatic tilt angle variance, 314
anisoplanatism (G-tilt), 320
anisotropic spectral model, 340
anisotropy coefficients, 341
aperture filter function, 305
argument, 2
astigmatism, 305
atmospheric coherence width, 290

Barnes, Ernest William, 42, 81
Bateman, Harry, 81, 133
Bernoulli, Daniel, 117
Bessel functions, 144, 146, 157, 169–173, 176, 186, 190
Bessel, Friedrich W., 117
Bessel-integral function, 108
beta function, 25
binomial series, 48
bit error rate, 412
branch point, 11
branches, 4

Cauchy distribution, 203
Cauchy product, 28
Cauchy, Augustin-Louis, 16
Cauchy–Riemann equations, 3
characteristic function, 202, 404
Chernov, Lev Aleksandrovich, 286
coherent imaging systems, 291
coherent transfer function, 293
complementary error function, 65
complementary incomplete gamma function, 69
complete elliptic integrals, 74
complex conjugate, 2
complex degree of coherence, 226
complex number, 1
cone effect, 314
confluent hypergeometric function of the first kind, 46, 130, 138
confluent hypergeometric function of the second kind, 61
contour, 5
convolution integral, 33, 39, 194
convolution theorem, 193
correlation coefficient, 398
covariance function of the refractive index, 216
cross correlator, 400
cylinder functions, 83

direct detection system, 407
Dixon's theorem, 156, 166

effective LGS anisoplanatic error, 326
entire function, 3
Erdélyi, Arthur D., 81, 133
error function, 65, 143, 174
essential singularity, 9
Euler, Leonhard, 16, 42, 80
exponential integral, 70
exponential spectrum model, 216, 300, 340
extended Rytov theory, 260, 352

fade, 408
false alarm, 408
false alarm rate, 409
far field of the Rx aperture, 276
fluid mechanics, 414
flux variance, 249, 361
focal anisoplanatism, 313, 326, 333
focus correlation function, 333, 383
Fourier integral theorem, 28, 173, 201
Fourier transform, 201
Fourier, Joseph, 207
free-space optical communication system, 407
Fresnel integrals, 67
Fried parameters, 276, 290, 410

gamma distribution, 78, 195
gamma function, 23
gamma-gamma distribution, 195, 412
Gauss, Carl Friedrich, 16, 80
Gaussian distribution, 40, 78, 389
Gaussian-beam wave, 198, 220, 234, 244, 258, 347, 351
generalized Huygens–Fresnel integral, 292
generalized hypergeometric function, 46, 178
generalized Rytov variance, 343
geometrical optics approximation, 274
gradient-tilt variance, 308, 375
Green's function, 221

half-wave linear detector, 400
Hankel functions, 96
Hankel transform, 189
Hankel, Hermann, 208
HAP model, 336
hard aperture, 292
hard limiter, 404
Heaviside, Oliver, 208
Hufnagel–Valley model, 336
Huygens–Fresnel integral, 220
hybrid imaging methods, 289
hypergeometric function, 45, 131, 135, 161

ideal bandpass limiter, 395
ideal fluid, 414
image jitter, 277
impulse response function, 292, 293
incoherent imaging systems, 291
incomplete gamma function, 69, 182
index-of-refraction parameter, 340
index-of-refraction structure function, 217, 341
inertial range of turbulence, 213
inner scale of turbulence, 213
inverse cosine transform, 206
inverse Fourier transform, 202
inverse Laplace transform, 206
inverse Mellin transform, 28, 183
irradiance correlation width, 249, 251, 256
irradiance covariance function, 249
isolated singularity, 9
isoplanatic angle, 304

isotropic non-Kolmogorov turbulence, 340
isotropic phase structure function, 370

joint moments, 399

Kelvin's functions, 102
Kolmogorov spectrum, 215
Kolmogorov, Andrey Nikolaevich, 285
Kummer functions, 45
Kummer, Ernst Eduard, 80

Laplace inversion formula, 173
Laplace transform, 173
Laplace, Pierre-Simon, 208
laser guide star, 313
Laurent series, 7
Laurent, Pierre Alphonse, 16
Legendre, Adrien Marie, 42, 81
limiter, 395, 404
linear rectifier, 393
linear shift-invariant system, 290
log-amplitude structure function, 274
Lommel function of the first kind, 111
Lommel function of the second kind, 112
long-exposure resolution, 303
longitudinal component, 244
long-term beam radius, 239, 372

Maclaurin, Colin, 42
MacRobert's *E*-function, 132
MacRobert, Thomas Murray, 81, 133
Maiman, Theodore, 286
Meijer, Cornelis Simon, 81, 133
Mellin transform, 28, 183
Mellin transform scaling property, 196
Mellin transform translation property, 196
Mellin, Hjalmar, 42, 208
modified atmospheric spectrum, 215
modified Bessel function, 146
modified Struve function, 100, 148
modified von Kármán spectrum, 215
modulation transfer function, 291
modulus, 2
modulus of the complex degree of coherence, 226, 346
mutual coherence function, 225

narrowband noise, 389
natural guide star, 313
near field of the Rx aperture, 276
Newton, Isaac, 285
nondimensional Gaussian-beam parameters, 221

Obukhov, Alexander Mikhailovich, 285
on-off keying, 408
optical transfer function, 203, 291
optical turbulence, 213
outer scale of turbulence, 213

paraxial wave equation, 198
phase covariance function, 278, 283
phase front radius of curvature, 220, 221
phase structure function, 274, 281, 370
phase temporal covariance function, 279
phase transfer function, 291
phase variance, 277, 371
piston, 305
plane wave, 226, 241, 249, 253, 261, 265, 268, 274, 344, 349, 353, 356, 362, 366, 370
plane wave irradiance covariance function, 363
Pochhammer, Leo August, 42, 81
point spread function, 203, 291, 372

Poisson, Siméon Denis, 208
polarity coincidence correlator, 405
pole of order *m*, 9
postprocessing techniques, 289
power in the bucket (PIB), 410
power in the fiber (PIF), 411
power spectral density, 216, 264, 311, 374
power spectral density of Zernike tilt, 375
probability of detection, 409
probability of false alarm, 409
product of hypergeometric functions, 167
product of three Bessel functions, 172
product of two Bessel functions, 165, 169
products of generalized hypergeometric functions, 163

radial component, 244
Ramanujan theorem, 165
random field, 214
random medium, 212
random process, 214
reduction formula, 123
refractive-index structure parameter, 214, 342
removable singularity, 9
residue, 10
Rice–Nakagami distribution, 390
Riemann zeta function, 197
Riemann, Geor Friedrich Bernhard, 16, 81
round-Earth model, 337
Rytov approximation, 222
Rytov number, 214
Rytov variance, 214
Rytov, Sergei Mikhailovich, 285

Saalschütz's theorem, 168
scintillation index, 241, 349, 352
short exposure, 289
short-exposure image, 296
short-exposure resolution, 303
short-term spot radius, 297
signal-to-noise ratio, 392, 411
simple pole, 10, 29
simply connected domain, 5
sine and cosine integrals, 72
singular point, 3
soft aperture, 292
spatial coherence radius
  Gaussian-beam wave, 238
  plane wave, 227
  spherical wave, 371
spatial frequency, 215, 293
spherical wave, 231, 242, 254, 271, 281, 345, 350, 358, 362, 364, 368
spherical wave phase variance, 284
spot radius, 221, 238, 296, 372
square-law detector, 393
statistically homogeneous random field, 214
statistically isotropic random field, 214
Stirling, James Frazer, 42
Stirling's formula, 27
Strehl ratio, 298, 330, 372, 410
strong irradiance fluctuations, 214
Struve function, 98, 148
Struve, Hermann, 117

Tatarskii, Valerian Il'ich, 286
Taylor series, 7
Taylor, Brook, 42
Taylor's frozen turbulence hypothesis, 264, 366
temperature structure constant, 220
temporal covariance function
  plane wave, 265, 277, 366
  spherical wave, 271, 368
temporal Zernike-tilt covariance, 311

Thompson, William, 117
threshold, 408
threshold-to-noise ratio, 409
tilt, 305
tilt angular anisoplanatism, 326
tip-tilt determination problem, 326
total focus variance, 333, 382
Tricomi, Francesco G., 81

van der Pol, Balthasar, 117
velocity potential function, 416
viscous fluid, 213, 414
von Kármán spectrum, 215, 274, 370
von Lommel, Eugen, 117

Wallis, John, 80
Watson, George Neville, 81, 117
wave structure function, 226, 231, 344, 348
weak irradiance fluctuations, 214
Weber function, 109
Weierstrass, Karl Theodor Wilhelm, 16, 42
Whittaker functions, 64
Whittaker, Edmund Taylor, 81, 133

Zernike polynomials, 36, 305
Zernike-tilt variance, 308, 374
Zuev, Vladimir Yevseyevich, 285

Larry C. Andrews is Professor Emeritus of Mathematics at the University of Central Florida and an associate member of the Townes Laser Institute in the College of Optics/CREOL. He previously was also an associate member of the Florida Space Institute (FSI) and had a joint faculty appointment with the Department of Electrical and Computer Engineering. Before that, he held a faculty position at Tri-State University and was a staff mathematician with the Magnavox Company, antisubmarine warfare (ASW) operation. He received a doctoral degree in theoretical mechanics in 1970 from Michigan State University. Dr. Andrews has been an active researcher in optical wave propagation through random media for more than 35 years and is the author or co-author of 13 textbooks on topics of differential equations, boundary value problems, special functions, integral transforms, wave propagation through random media, and mathematical techniques for engineers. He is a Fellow of SPIE and the author of three Field Guides on Atmospheric Optics, Special Functions, and Random Processes. Along with wave propagation through random media, his research interests include special functions, random variables, atmospheric turbulence, and signal processing.

Melissa K. Beason has a diverse research background spanning more than 20 years that includes laser sources, acoustics, and atmospheric propagation. She is currently an Assistant Research Professor with the Air Force Institute of Technology. She received her Ph.D. in Electrical Engineering from the University of Central Florida (UCF) in 2018. Prior degrees include a Master of Science Degree in Mathematical Sciences from UCF and a Bachelor of Science Degree in Physics from the University of Florida. Dr. Beason also held Research Physicist positions with the Air Force Research Laboratory, Directed Energy Directorate and the Naval Research Laboratory, Underwater Sound Reference Detachment. She was a Preeminent Postdoctoral Scholar with UCF and European Research Consortium in Informatics and Mathematics Fellow during which she collaborated with the Adaptive Optics Group of the Fraunhofer Institute of Optronics, System Technologies and Image Exploitation IOSB. Dr. Beason's research interests include atmospheric propagation through random media and turbulence measurement.